# THE
## INTERNATIONAL SERIES
### OF
## MONOGRAPHS ON PHYSICS

### SERIES EDITORS

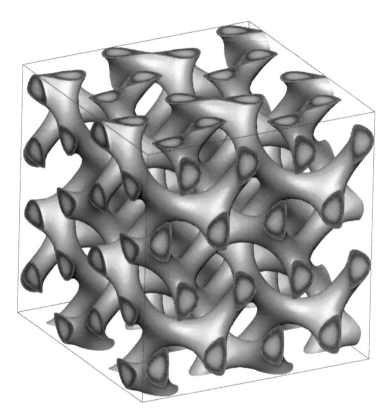

Gyroid mesophase for an AB diblock copolymer melt computed using the self-consistent field theory of Chapter 5. (Figure courtesy of E. Cochran.)

# The Equilibrium Theory of Inhomogeneous Polymers

GLENN H. FREDRICKSON

*Departments of Chemical Engineering and Materials*
*University of California, Santa Barbara, USA*

CLARENDON PRESS · OXFORD
2006

# OXFORD
## UNIVERSITY PRESS

Great Clarendon Street, Oxford OX2 6DP

Oxford University Press is a department of the University of Oxford.
It furthers the University's objective of excellence in research, scholarship,
and education by publishing worldwide in

Oxford New York

Auckland Cape Town Dar es Salaam Hong Kong Karachi
Kuala Lumpur Madrid Melbourne Mexico City Nairobi
New Delhi Shanghai Taipei Toronto

With offices in

Argentina Austria Brazil Chile Czech Republic France Greece
Guatemala Hungary Italy Japan Poland Portugal Singapore
South Korea Switzerland Thailand Turkey Ukraine Vietnam

Oxford is a registered trade mark of Oxford University Press
in the UK and in certain other countries

Published in the United States
by Oxford University Press Inc., New York

© Oxford University Press 2006

The moral rights of the author have been asserted
Database right Oxford University Press (maker)

First published 2006

British Library Cataloguing in Publication Data

Data available

Library of Congress Cataloging in Publication Data

Data available

Typeset by the author
Printed on acid-free paper by
Biddles Ltd., King's Lynn

ISBN 0–19–856729–4   978–0–19–856729–5

1 3 5 7 9 10 8 6 4 2

# FOREWORD

Although polymers are universally present, they became fully brought into scientific consciousness with the discoveries of the 1930s and 1940s, which then led to a major expansion of the field. Once it was recognized that the basic polymer is just a chain of repeated monomers, theoretical descriptions appeared, for example by Yamakawa and in the massive monograph by Flory. The brilliant paper by Rouse introduced dynamics, and papers by Zimm and by Stockmayer and many others produced a basis for a systematic development. This stemmed out of conventional physical chemistry and did not use the mathematical structure of path integrals notably developed by Wiener and the formulation of quantum mechanics, by Feynman. Of course the problems are not the same, for an electron has position $\mathbf{r}(t)$ and gives rise to a (complex) wave function $\psi(\mathbf{r}, t)$, whereas a polymer needs a label $s$ along its length, $\mathbf{r}(s, t)$, and all the analysis is real so that a probability $P([\mathbf{r}], t)$ can be assigned. However, much of the technique is the same and the polymer $\mathbf{r}(s, t)$ has a resemblance to a field variable in quantum field theory $\phi(\mathbf{r}, t)$ ($\mathbf{r}$ is like $\phi$ and $s$ is like $\mathbf{r}$; it is really quite simple!).

Now a whole range of knowledge of what can and cannot be done becomes available, with one glorious addition for polymers. Everything must work. With electrons, photons, pions, etc. things sometimes work and sometimes are plagued by divergences; but there cannot be divergences in polymer science, and the stringency of relativistic invariance is never imposed.

A simple example comes in interactions between monomers. It was well established by Fermi that neutron interactions could be adequately described by a $\delta$ function potential (i.e. a potential zero except at one point, but having a finite integral $\int d\mathbf{r} \, \delta(\mathbf{r}) = 1$). The interaction between monomer $a$ and monomer $b$ is adequately described by such a potential $\delta(\mathbf{r}_a - \mathbf{r}_b)$, which enormously simplifies calculations but which, if one is suspicious, can be worked out in full detail – this detail having essentially no effect on the final answer.

Thus in the period of about 1965–85 polymer theory was brought into the standard structure of theoretical physics, and gained several bonuses such as reptation and polymer screeening, static and dynamic.

At this point polymer theory started to leave conventional theory and take on a life of its own – hence the present book of Prof. Fredrickson. When such a situation arises in physics there are usually two directions in which to go. It has been found that some hitherto complex situations can have a simple explanation, but it is not clear that one really has the correct or complete solution, e.g. is the motion of polymers in a melt solely via reptation? Once one tries a more comprehensive explanation of any complex phenomenon, one finds that new lengths and new times crowd in, so that a vast amount of work is required to produce a convincing improvement on a simple theory such as reptation, theories that lead to reasonably good agreement with experiment and are effective in prediction.

The other direction is not to hone the first theoretical structure into a superior form, but to consider more phenomena at the same level of study. Thus polymers have special surface behavior, have phase changes, and can be designed to do special things. I will confess to finding this latter area of research more in tune with my temperament than the former, but one must not argue that just because one likes an area of research, it is more significant than others which do not fit so well with one's temperament.

Professor Fredrickson also follows this second path, for having given an explanation of how simple ideas in polymer science can solve simple problems, he then proceeds to elucidate a host of new and important aspects of the behavior of polymeric materials. However, this now takes us into new territory of theoretical physics techniques, for the great classical structures of partial differential equation crumbles when one is faced with, in the current jargon, nanotechnology or mesophysics, wherein the "lumpiness" of the material does not easily succumb to the calculus. The way forward must be via new numerical techniques, but these must now handle a field-theoretic environment.

This new world is explored in Prof. Fredrickson's book and offers the challenge of new techniques for new phenomena, as a new phase in polymer materials science is ushered in.

*Cambridge*                                                          S. F. Edwards
25 *August* 2005

# PREFACE

Theory has played an important role in the development of polymer science and technology since its origins in the mid-twentieth-century. The first generation of polymer theorists, such as Flory, Guth, and Stockmayer, brought the background and perspective of physical chemistry to tackling fundamental problems of polymer chain conformational statistics, colligative properties, phase transformations, and phase behavior. In the 1960s and 1970s, a second generation of theorists, notably Edwards and de Gennes, introduced the tools of theoretical physics to the field of polymer science, including field theory techniques, scaling methods, and the renormalization group. The connection between polymer statistical mechanics and other branches of statistical physics and condensed matter theory was elucidated by these researchers and used to analyze a number of important problems, including the excluded volume effect.

The development of a systematic theory for the equilibrium properties of *inhomogeneous* polymers began in earnest in the 1970s, with Edwards, de Gennes, and Helfand playing a key role. By 1975, the functional integral methods, saddle point approximations, and interaction models that form the basis of modern self-consistent field theory (SCFT) were already in place. The 1980s was an exciting period in the further development of the theory, as researchers introduced a variety of analytical techniques for approximately solving the SCFT equations and applied these methods to broad classes of systems, including polymer alloys, block and graft copolymers, and tethered polymer layers. Particularly notable was the application by Leibler of the random phase approximation to treat weakly segregated block copolymers, and the development of methods by Semenov, Milner, Witten, and Cates for the asymptotic analysis of strongly segregated copolymers and polymer brushes.

With the tremendous advance in the availability and power of digital computers, numerical studies of the SCFT equations gradually increased in number over the past two decades. Early work by Helfand and Noolandi and Whitmore utilized finite difference methods and spherical approximations to the unit cells of block copolymer mesophases to render calculations one dimensional. Scheutjens and Fleer developed similar methods for polymers at interfaces, utilizing an underlying lattice. A milestone in 1994 was the introduction by Matsen and Schick of a powerful spectral method for solving the SCFT equations in unit cells of complex block copolymer mesophases. At roughly the same time, Fraaije and collaborators began developing numerical finite difference algorithms for solving the SCFT equations in three dimensions with periodic boundary conditions imposed on a large cubic cell. My own interest in the numerical solution of mean-field equations for inhomogeneous polymers was stimulated in the late 1990s by this work, because it offered for the first time a generic tool that could be used to explore the phase behavior of arbitrarily complicated polymer formulations.

In parallel to these numerical investigations of SCFT, significant strides were made in the 1980s and 1990s by Binder, Kremer, Grest, and others in advancing traditional "particle-based" computer simulation methods to treat inhomogeneous polymers. Methods such as molecular dynamics, Monte Carlo, and Brownian dynamics were fruitfully applied to coarse-grained chain models of polymer brushes, copolymers, and blends. A more recent development was the realization that *field theory models* can also be effectively used as the basis for computer simulations of polymers. Simulation of such models in the mean-field approximation essentially amounts to a numerical implementation of SCFT. However, techniques have been developed for relaxing the mean-field approximation and implementing so-called "field-theoretic simulations" without any simplifying approximations. This innovation has brought together analytical techniques and computer simulation methods in an exciting way for the scientific investigation and practical design of inhomogeneous polymer systems.

The field-based simulation methods have great potential for exploring the equilibrium behavior of a large variety of polymer and soft material systems of contemporary interest, including those relevant to nanotechnology. While these techniques will undoubtedly not replace traditional particle-based simulations, they do represent a powerful new tool with which to examine complex, multi-phase/multi-component systems such as polymer-based nanocomposites, block and graft copolymers, polymer–surfactant and polymer–colloid complexes, polymer alloys, and polyelectrolyte formulations.

The present monograph is an attempt to describe these recent developments in field-based simulations of inhomogeneous polymers and to unify the theoretical framework underlying such computer simulation techniques. This is not a text on the *physics* of inhomogeneous polymers *per se*, but rather on the theoretical and computational methods that can be used to explore the physical behavior of multi-component polymer systems at equilibrium. Nevertheless, along the way, we shall have occasion to discuss a fair amount of inhomogeneous polymer physics in the context of application examples.

I hope that this monograph will appeal to anyone with a serious interest in polymer (and soft condensed matter) theory, and especially those motivated to learn the techniques of field-based computer simulations. I have tried to make the subject matter accessible to first-year graduate students in physics or engineering. Prerequisites would include an introductory course in equilibrium statistical mechanics and a basic knowledge and familiarity with applied math and probability theory at the advanced undergraduate or first-year graduate level. A general course in polymer science would also be helpful and, ideally, exposure to theoretical polymer physics concepts at the level of the classic monographs by de Gennes and Doi & Edwards (de Gennes, 1979; Doi and Edwards, 1986). I assume no prior experience with classical or quantum field theory, path integrals, and the calculus of functionals.

# ACKNOWLEDGEMENTS

A writing project of this scope is not undertaken alone. There are many people to whom I am indebted, including family and professional colleagues and collaborators, as well as students and post-docs.

On the professional side, I am grateful to my thesis advisors Hans Anderson and Curt Frank at Stanford University, who first introduced me to the fascinating subjects of polymers and statistical mechanics and taught me the art of critical thinking and the conduct of research. Upon leaving Stanford, I was extremely fortunate to secure a job at AT&T Bell Laboratories, where young scientists were provided unlimited opportunity to develop knowledge and their careers. For that opportunity I will be eternally grateful, both to the institution and to the managers such as Shiro Matsuoka who supported me during that wonderful six-year period. At Bell Labs, my interest in inhomogeneous polymers emerged from stimulating discussions with Eugene Helfand, Frank Bates, and Ronald Larson, among others, and those interactions led to collaborations that continue to this day. Other collaborations that began at the same time with Ludwik Leibler and Scott Milner have also proved to be lasting and highly beneficial to my education in polymer theory. Since moving to my current position at UCSB, I have enjoyed interactions with an exceptional group of science and engineering faculty. I am particularly grateful to Edward Kramer, Philip (Fyl) Pincus, Hector Ceniceros, and Carlos Garcia-Cervera for enriching my understanding of the subject matter of this book.

I am indebted to the students and post-docs who have worked with me over the past twenty years, some of them very influential in formulating the concepts and approaches that are described herein. I would especially like to acknowledge former group members Jean-Louis Barrat, David Morse, Andrea Liu, David Wu, Francois Drolet, Venkat Ganesan, David (Qiang) Wang, Andre Moreira, Alfredo Alexander-Katz, and Scott Sides. Collaborations with some of these individuals continue to this day. In my current group, I am particularly grateful to Kirill Katsov, Edward Feng, Stephan Baeurle, and Eric Cochran, who have all made significant contributions to the book.

In addition to these individuals, I would like to thank a number of professional colleagues and collaborators at other institutions, including Henri Orland, Hans Fraaije, Mark Matsen, Michael Schick, Friederike Schmid, Andrei Gusev, Takashi Taniguchi, Kurt Binder, and Zhen-Gang Wang, who have greatly expanded my understanding of the subject of inhomogeneous polymers. I am particularly indebted to Sir Sam Edwards for contributing the foreword.

Others at UCSB have made equally important nontechnical contributions to this project. Maureen Evans, Heidi Schow, and Sara Bard, by providing expert and always cheerful administrative support, made it possible for me to find the time to write in spite of numerous competing responsibilities. I am also very grateful to Dottie McLaren for her superb artwork.

I am pleased to acknowledge the financial support of my research in inhomogeneous polymers by the Division of Materials Research of the National Science Foundation and the Petroleum Research Fund administered by the American Chemical Society. Without their sustained commitment to funding basic research in polymer theory, my research activities and those of many others whose work is described in this book would have been seriously curtailed.

Last but not least, I would like to thank my family for their love and support over the years and their patience with me during the writing of this monograph. To Lesley-Anne and Jill I owe much – not only for your love, support, and patience, but also for your help in making me a better person.

# CONTENTS

# 1

# INTRODUCTION

## 1.1 What is a polymer?

A synthetic polymer molecule, or macromolecule, is a species of high molecular weight that is created by chemically coupling large numbers of reactive small molecules – so-called "monomers". For example, the familiar plastic materials HDPE (high-density polyethylene) and LDPE (low-density polyethylene) are produced by reacting ("polymerizing") ethylene gas to produce chain-like macromolecules consisting of 1000 or more ethylene residues. Polymers are typically linear chains created by linking monomers at two sites; however, regularly or statistically branched polymers can be produced by a variety of synthetic methods. In the case of polyethylene, HDPE is manufactured by a process that yields primarily linear chains, while the LDPE process produces a statistical distribution of macromolecules with branched, tree-like chemical connectivity. Linear and branched are two broad terms that are used to characterize the bonding topology, or *architecture*, of a polymer. The classic text of Flory (1953) is a good place to get started in learning about synthetic polymers.

Non-biological methods for synthesizing polymers are not precisely controlled. Polymers produced by the commonly exploited free radical, ionic, or acid–base reactions of monomers are usually imperfect both in architecture and chain length. In the case of LDPE, for example, branched polyethylenes of many topologies are produced by the polymerization process. Even in a process that generates primarily linear polymer molecules, e.g. HDPE, the individual macromolecules can differ widely in molecular weight. Polymer scientists often refer to the *molecular weight distribution* to characterize statistically the molecular weight disorder of a polymer sample (Odian, 1981).

Most commercial synthetic polymers such as the various grades of polyethylene are considered to be *flexible*, because the natural configuration of such a polymer either in the molten state, or dissolved in a solvent, is not a rigid rod but rather a *random coil*. The flexibility necessary for polymer coiling derives in many polymers from relatively unhindered rotations around carbon–carbon single bonds in the polymer backbone (Flory, 1969). The "random" character of a coiled polymer reflects the fact that extremely large numbers of conformational states of the backbone bonds are available, and this produces large fluctuations in the overall size of a polymer coil. The random coil state of a flexible synthetic polymer is similar to the denatured state of a protein. Proteins and other naturally occurring polymers, however, also exhibit *folded* configurations, which are more compact than the random coil and involve significantly smaller numbers of

1

bond conformational states.

The present monograph deals almost exclusively with polymer systems whose physics is dominated by the statistical mechanics of many conformational states. We shall see that an important theme in understanding the equilibrium behavior of inhomogeneous polymers is that flexible polymers resist situations, such as strong extension or compression, where the number of conformational states is reduced from that in the random coil (de Gennes, 1979). Such a reduction in available states is described macroscopically as a decrease in "conformational entropy."

Polymer molecules are also characterized according to their internal composition. Polymers formed by linking together $N$ monomers of one chemical type are referred to as *homopolymers*. $N$ is the so-called *polymerization index* or *degree of polymerization*. A polymer that is produced by linking two or more chemically distinct monomers is referred to as a *copolymer*. For example, SBR (styrene-butadiene rubber) is an important synthetic rubber material that is produced by polymerizing a mixture of styrene monomers with 1,3-butadiene monomers. Such a polymer is classified as a *random* or *statistical copolymer* because there is a statistical distribution associated with the sequence of styrene and butadiene monomers incorporated into the polymer chain (Odian, 1981). SBR is used in applications ranging from automotive tires to adhesives. A second important type of copolymer is a *block copolymer* in which monomers of a given type are grouped into polymerized sequences, or "blocks", along a polymer chain (Hamley, 1998). For example, SBS (styrene-butadiene-styrene) triblock copolymer is a familiar block copolymer material in which (nearly) every molecule has a linear three-block architecture. Each molecule begins and ends with a polymerized sequence (block) of polystyrene, separated by a center block of polybutadiene. Interestingly, random copolymers can have very different properties than block copolymers, even when composed of the same two monomers as in the case of SBR and SBS. For example, SBS has far superior elastic recovery characteristics than SBR.

With modern synthetic methods, it is possible to create macromolecules with a broad variety of monomers and architectures. Homopolymers constructed from a monomer A can be produced with linear, branched, or star architectures. Block copolymers composed of two chemically distinct monomers, A and B, can be synthesized with AB diblock, ABA triblock, or even ABABA pentablock architectures. Closely related to block copolymers are *graft copolymers*, where polymerized sequences of one type of monomer are attached ("grafted") along the backbone of a polymer composed of a second type of monomer. More complex random, block, and graft copolymers involving *three* or more chemically distinct monomers can also be produced. The properties of materials constructed from such exotic macromolecules are just beginning to be explored (Bates and Fredrickson, 1999; Goldacker *et al.*, 1999; Brinkmann-Rengel *et al.*, 1999). Figure 1.1 depicts a number of common homopolymer and copolymer architectures.

homopolymer

AB diblock copolymer

AB statistical copolymer

ABC triblock copolymer

branched polymer

star polymer

AB graft
copolymer

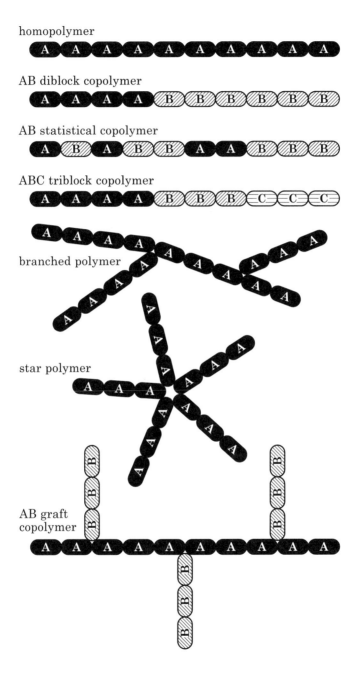

FIG. 1.1. Examples of common homopolymer and copolymer architectures.

## 1.2 Polymeric systems of practical interest

Polymers are ubiquitous in everyday life. Consumer products are rarely without one or more synthetic polymer components, if only in the packaging. Two broad classes of systems are the primary focus of this monograph: *plastic materials* and *solution formulations*.

### 1.2.1 *Plastic materials*

Plastic materials are typically complex mixtures of synthetic macromolecules with colorants, stabilizers, fillers, and other additives (Brydson, 1995). Familiar examples are the plastic cases that enclose our desktop computers, the vinyl materials with which we clad our houses and surface our floors, and the plastic containers that we use to store household items. The compositions of such polymer mixtures can be varied almost continuously in order to achieve an optimal combination of physical properties and price. Plastic materials are attractive not only for this flexibility of formulation and relatively low cost, but also for their ease of processing. The largest class of plastic materials by sales and volume, the so-called thermoplastics, can be melted and formed into useful shapes with widely available processing equipment such as injection molders, extruders, and film blowers.

Many plastic materials in both the molten and solid states are *inhomogeneous* in that their composition, molecular structure, or molecular-level packing varies throughout the material. One example of such inhomogeneity is realized in filled polymer systems, e.g. the familiar case of polypropylene that has been loaded with talc particles or glass fiber to reduce cost and/or modify mechanical properties. In such materials, there are statistical fluctuations in the concentration and orientation of filler particles at a local level, which lead to a *microscopic inhomogeneity*. Moreover, filled polymer systems often exhibit larger-scale *macroscopic inhomogeneities* associated with the history of processing. For example, the surface of a molded polymer part will frequently show a "skin layer" that is depleted of filler particles.

Another common source of inhomogeneity in plastic materials is a liquid–liquid phase separation between different polymeric components that are blended together in the molten state. Due to the small entropy of mixing that characterizes polymeric fluids of high molecular weight, the slightest of chemical differences can cause a phase separation. An important example is the immiscibility of high-density polyethylene (HDPE) with isotactic polypropylene in the melt state. A more extreme example is the remarkable phase separation observed in mixtures of polymers with analogs derived by the substitution of hydrogen with deuterium (Bates and Wignall, 1986). When such molten polymer *blends* or *alloys* are cooled rapidly, the resulting plastic material manifests inhomogeneities that reflect the compositional state of the melt just prior to solidification.

The prevalence of multi-component, multi-phase polymer alloys throughout the spectrum of commodity, engineering, and specialty plastics reflects the versatility of such materials (Richardson and Lokensgard, 1989). By adjusting compositions, molecular weights, and the chemical details of the various polymeric components, it is possible to achieve widely varying sets of physical properties, while retaining attractive economics and ease of processing. Only with an inhomogeneous, multi-phase system can certain combinations of properties be achieved; for example, the combination of high modulus and high toughness is often difficult to obtain in a single-phase plastic material.

### 1.2.2  *Solution formulations*

A second broad class of inhomogeneous polymer systems are solution formulations, which involve one or more polymers, possibly along with particles, surfactants, colorings, and fragrances, suspended or dissolved in a solvent. Some of the most familiar solution formulations are personal care products such as shampoos, conditioners, hairsprays, and cosmetics, as well as liquid detergents, paints, and coatings. Also in this category are certain processed foods, such as mayonnaise, cheese spreads, and ice cream, which are complex mixtures of edible natural and synthetic polymers with food stuffs, oils, and water. Gels, both in food applications and the cremes and lotions of personal care, can also be viewed as solution formulations. The polymers may play one of several roles in these formulations including emulsification and *rheology* control, i.e. control over the viscosity and elasticity of the fluid (Larson, 1999).

Solution formulations can possess inhomogeneities on a variety of length scales. Formulations with suspended particles exhibit heterogeneities associated with non-uniform particle placement, similar to the case of filled plastics. Products such as mayonnaise and hair conditioners are emulsions with hydrophobic and hydrophilic phases, stabilized by surfactants and polymers. These emulsions have droplets ranging in size from $10^{-8}$ m to $10^{-4}$ m and exhibit complex morphologies that reflect the history of preparation and processing.

One feature that distinguishes solution formulations from multi-component plastics is the prevalence of *polyelectrolytes* as ingredients in the former. A powerful way to confer solubility on a polymer is to incorporate acid or base functionality that can dissociate in a high-dielectric solvent such as water, leaving the polymer with a net electrical charge. The solvation energy and entropy of the counter-ions provides sufficient driving force for mixing polymer with solvent that even polymers with extremely hydrophobic backbones can be made to dissolve in water (Barrat and Joanny, 1996). While polyelectrolytes are very useful in formulations including absorbant materials and hair fixatives, they do confer an additional complexity in the design of already complex mixtures. Electrostatic interactions, which can be tuned by adjusting pH and adding salts, must be balanced against dispersive and hydrophobic interactions to control the type and extent of phase separation in the fluid, leading to inhomogeneous structures on multiple length scales.

## 1.3    Macrophase and microphase separation

An important concept in inhomogeneous polymer fluids is the distinction between macroscopic phase separation and *microphase separation* (also referred to
as mesophase or nanophase separation). When two homopolymers, say A and
B, are blended in the melt state or cast from a common solvent, their chemical differences are normally sufficient to induce a liquid–liquid phase separation
(de Gennes, 1979). This is a *macrophase separation* because, given sufficient
time, the emerging A-rich and B-rich phases will coarsen to macroscopic length
scales. The driving force for the phase coarsening is the interfacial free energy
associated with the interfaces separating the two phases. At equilibrium, the
area of the internal interfaces will be minimized subject to mass conservation of
the two species and boundary conditions that reflect the container in which the
fluid is held. This equilibrium condition leads to A-rich and B-rich phase domain
sizes that eventually coarsen to "macroscopic" length scales comparable to the
size of the system.

Melt-compounded polymer blends do not usually have morphologies that
reflect true equilibrium, but rather metastable structures that are created by the
strong flow conditions within polymer processing equipment. A typical situation
is for the minority phase to be dispersed in the form of discrete droplets as
shown in Fig. 1.2. The balance between the shear and interfacial forces present
in a compounder establishes a broad droplet size distribution with mean drop
size in the 0.1–10 $\mu$m range.

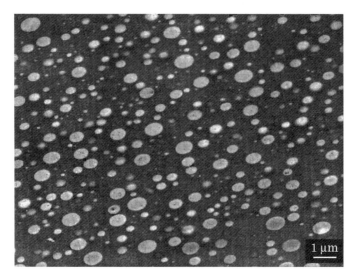

FIG. 1.2. Transmission electron micrograph of an 80/20 composition polyether-
sulfone/polycarbonate blend. Reproduced from Weber (1999).

In contrast to a blend of two homopolymers, a pure molten *block copolymer* fluid such as an AB diblock copolymer or an ABA triblock copolymer cannot exhibit macrophase separation. Such a fluid is a one-component system in which the chemical bonds linking the dissimilar blocks prevent A and B segments of the same copolymer from being separated over macroscopic distances. However, in such fluids an ordering phenomenon known as *microphase separation* is possible (Leibler, 1980; Bates and Fredrickson, 1990; Bates and Fredrickson, 1999). At a local ~1 nm level, the dissimilar blocks in a copolymer melt are driven to phase separate by the same microscopic interactions that produce macrophase separation in an A–B homopolymer alloy. However, unlike the homopolymer blend, the A-rich and B-rich domains created by local phase separation of a copolymer melt cannot coarsen to length scales exceeding the extended chain length of a copolymer, ~100–1000 nm, since this would imply cleavage of bonds connecting the blocks. Moreover, an increase in phase domain size beyond the random coil size of the copolymer, typically of order 10 nm, would imply a conformational entropy penalty. This is a consequence of each copolymer having to deform from its randomly coiled state in order to fill the phase domains with blocks of the appropriate type. *Microphase separation* of a block copolymer melt thus corresponds to a phase separation that is limited in the phase domain size to mesoscopic length scales, typically 5–100 nm. At equilibrium, the competing effects of the interfacial energy, which favors a larger phase domain size, and conformational entropy, which favors a smaller domain size, conspire to establish an optimal (lowest free energy) length scale for the phase separation.

Block copolymers that are produced by well-controlled polymerization methods, such as anionic polymerization (Hsieh and Quirk, 1996), will readily undergo microphase separation. In these systems, the amount of chemical disorder is relatively low in that most of the polymer molecules have the same architecture and the distributions of molecular weight for each block are relatively narrow. Such systems tend to produce spatially periodic microphase-separated composition patterns, so-called block copolymer *mesophases*, which are a type of super-lattice. Indeed, microphase separation of a pure block copolymer melt is more properly viewed as an ordering phenomenon similar to the crystallization of a one-component molecular fluid, rather than a spatially limited type of phase separation. Thermodynamic equilibrium corresponds to mesophases consisting of defect-free, periodic patterns with compositional order in one, two, or three dimensions[1]. Defect states such as lattice dislocations and grain boundaries, however, are relatively low in energy and are easily populated, so micrographs of microphase separated block copolymers usually show only short-range order. Nevertheless, under certain circumstances it it is possible to achieve mesophases that are nearly defect-free and have a high degree of long-range order.

Figure 1.3 shows a transmission electron micrograph (TEM) of a thin film of a styrene-butadiene-styrene (SBS) triblock copolymer exhibiting a highly ordered,

---

[1]True long-range order in dimensions less than or equal to two is destroyed at finite temperatures by thermal fluctuations (Nelson, 2002).

one-dimensional *lamellar* mesophase (Cohen *et al.*, 2000). It is of interest to contrast the scale and features of this example of microphase separation against the macrophase separation pattern shown previously in Fig. 1.2.

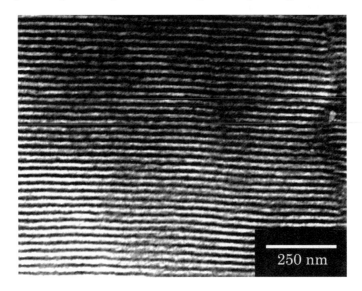

FIG. 1.3. An example of microphase separation. A thin film of a styrene-butadi-ene-styrene (SBS) triblock copolymer as imaged by TEM. Reproduced from Cohen *et al.* (2000).

Block copolymer materials produced by less-controlled synthetic methods, such as the condensation reactions used to create segmented polyurethanes, tend to have broader molecular weight and composition distributions. Such materials rarely exhibit mesophases with long-range order. Indeed, segmented block copolymers and many graft copolymers (which are also prepared by less controlled coupling chemistries) often show microphase separation on a ~10 nm scale with more liquid-like than solid-like short-range order (Holden *et al.*, 1996). An active area of current theoretical polymer research is to understand the type and extent of chemical disorder that can be introduced in a copolymer system and still produce well-ordered mesophases (Dobrynin and Leibler, 1997; Subbotin and Semenov, 2002). The advent of controlled free radical polymerization methods, discussed in Section 1.4.2, has made this line of research particularly relevant to modern polymer technologies.

Some commercial polymer alloys, such as high-impact polystyrene (HIPS), are manufactured by reactive blending processes that yield complex mixtures of homopolymers with block and graft copolymers of various architecture (Baker *et al.*, 2001). These systems can exhibit both macrophase and microphase separation and are conceptually similar to fluid emulsions formed by ternary mixtures of oil, water, and a non-ionic surfactant. In this analogy, oil and water correspond to the A and B homopolymer components of the plastic alloy, and surfactant corresponds to the block or graft copolymer component(s). An idealized version of such a "plastic emulsion" can be prepared by blending two homopolymers with a diblock copolymer as shown in Fig. 1.4 (Mezzenga *et al.*, 2003).

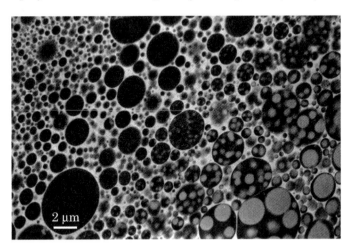

FIG. 1.4. A polymeric emulsion produced by selective solvent casting of a solution of polystyrene (PS) with poly(2-vinylpyridine) (PVP) and PS-PVP diblock copolymer. Reproduced from Mezzenga *et al.* (2003).

*Solutions* of block and graft copolymers can exhibit macrophases, mesophases, or combinations of the two, depending on the composition of the mixture. Phase coexistence between a homogeneous macrophase and an inhomogeneous mesophase is often encountered, as is the possibility of coexistence of two mesophases with different symmetry, lattice constant, or composition (Lodge *et al.*, 2002). Phase behavior in such systems can be exceedingly complex and is dependent on a large number of experimental parameters, including molecular weight, polymer concentration, polymer architecture, and solvent quality.

## 1.4 Modern themes

The science of inhomogeneous polymers is relatively mature, having developed in parallel with technologies for alloying plastics and formulating polymer solutions over the past half-century. Nevertheless, there are several contemporary trends that have significantly elevated the visibility and level of research activity in the field.

### 1.4.1 *Plastics industry trends*

The plastics and rubbers industry has enjoyed remarkable growth since its origins in the early 20th century. Over $1.5 \times 10^8$ metric tons of these materials were manufactured world wide in 2001 (VSPA, 2003), and in the US alone, the industry contributes 1.5 million jobs and 320 billion dollars annually in shipments (SPI, 2003). Twenty-five years ago, the plastics industry was producing approximately $2.5 \times 10^7$ metric tons annually, of which roughly 85% was low-priced "commodity" plastics and the remainder higher-priced "engineering" and "high-performance" plastics. At that time, many companies invested heavily in R&D for engineering and high-performance polymers, anticipating that this specialty sector would grow faster than the market for commodity plastics. The research included significant programs involving new monomer development, polymers and copolymers from these new monomers, and the development of new polymer alloys.

An interesting observation is that the present-day plastics market, in spite of having grown nearly six-fold in size over the past twenty-five years, reflects nearly the same breakdown between commodities and specialties: roughly 85% of all sales represents commodity plastics and rubbers. Evidently the plastics industry has matured to a highly price-sensitive state. Higher-priced plastic materials are tolerated only in applications where commodity plastics cannot deliver the necessary combination of properties. Such high-end applications represent, curiously, a nearly constant 15% of all applications, irrespective of total market size. Responding to the disappointing relative growth of the high-performance polymer sector, many companies dramatically reduced their R&D commitment to developing new monomers and new alloys over the past decade. The focus has instead turned to growth of existing markets through globalization and extending current product lines to new applications as they arise.

One bright spot for polymer R&D, however, comes from the realization that polymers built from readily available, low-priced monomers will continue to rule the marketplace. Rather than seeking advanced plastic materials based on new monomers, polymer scientists and technologists have risen to the challenge of creating new plastic materials from commodity monomer feedstocks. This can be accomplished by combining modern polymer synthesis strategies with theoretical and empirical design tools to construct polymers with new or better-defined molecular architectures. The polymers are then used to formulate plastics with unique combinations of properties that are attractive to the marketplace. Many of these plastics will be multicomponent/multiphase materials created by exercising simultaneous control over macrophase and microphase separation. Current research activity in such "specialty-commodity" plastics, i.e. specialty polymers at commodity prices, is largely focused on commodity monomers derived from petrochemical sources. It is likely, however, that major opportunities for inhomogeneous polymer research will arise from the increasing availability of bioengineered low cost monomers from renewable resources.

### 1.4.2 *New synthetic methods*

An important driver for innovation in inhomogeneous polymers is the rapid pace of development of new catalyst systems and synthetic methods for performing *controlled* polymerizations. Block copolymers are typically prepared by using so-called *living* polymerization systems (Odian, 1981). With such systems, molecular weight advances linearly with reaction time, and polymers with relatively narrow molecular weight distributions can be obtained. Chain transfer and other chain termination mechanisms are suppressed. By exhausting the supply of one monomer and sequentially feeding a second monomer, block copolymers can be readily synthesized. A rather new development is the increasingly widespread appreciation that a polymerization system does not have to maintain perfect control over chain termination processes, i.e. be a strictly living catalyst system, to be effective in creating useful block copolymer materials. This appreciation has arisen from the emergence of several powerful new methods for synthesizing polymers.

Until very recently, virtually all commercial block copolymers were prepared using anionic or cationic polymerization methods, or by condensation of telechelic (bi-functional) pre-polymers. Anionic polymerization is a living polymerization technique that offers excellent control over molecular weight distributions, ready access to a wide variety of polymer and block copolymer architectures, and attractive economics (Hsieh and Quirk, 1996). Nevertheless, the set of monomers that can be polymerized by this method is rather small. Cationic polymerization in principle offers similar molecular weight and architectural control, but is also limited to a relatively small set of monomers. Segmented block copolymers created by reactions of telechelic pre-polymers have been used to produce segmented block copolymers, such as polyurethanes and polyether-polyamide copolymers (Holden *et al.*, 1996). The method, however, cannot produce polymers with narrow molecular weight distributions, and the materials obtained are relatively expensive because of the high cost of the pre-polymers.

*Controlled free radical* polymerization methods, which include atom transfer radical polymerization (ATRP), nitroxide-mediated radical polymerization, and reversible addition-fragmentation chain transfer (RAFT) polymerization (Kamigaito *et al.*, 2001; Hawker *et al.*, 2001; Matyjaszewski and Xia, 2001; Chong *et al.*, 1999), offer the ability to carry out "controlled" polymerizations of a very wide variety of low-cost commodity monomers, including those with useful chemical functionality, such as vinyl acetate and acrylic acid. While in most cases these are not strictly living polymerization techniques, they do provide enough control over termination processes to permit the synthesis of entirely new types of block and graft copolymers. The controlled free radical methods will undoubtedly prove enabling for advancing the application of inhomogeneous polymers in fields ranging from personal care and cosmetics, to medical devices, to molecular electronics.

A second very significant recent advance in polymer synthesis is the advent of *single-site* catalyst systems for carrying out controlled polymerizations. This large class of systems can be further divided into categories according to whether they offer control over polymer molecular weight (i.e. carry out living polymerizations), comonomer incorporation, polymer stereochemistry, or combinations of the three.

Perhaps most exciting are new single-site catalysts that exercise simultaneous control over polymer stereochemistry, comonomer incorporation, and molecular weight (Coates *et al.*, 2002). Heterogeneous, multi-site Ziegler–Natta catalysts are noted for their excellent control over polymer stereochemistry and have long been employed for the preparation of commercial isotactic polypropylene materials. However, these commercial catalyst systems provide limited control over molecular-weight distribution and comonomer incorporation. Single-site catalyst systems have recently been discovered that are both living and provide stereo control. This opens up the possibility of exciting new classes of stereo-regular inhomogeneous polymer materials, such as thermoplastic elastomers based on polypropylene-polyethylene block copolymers.

Another promising and active area in single-site catalysis is in mid- and late-transition metal polymerizations. There has long been commercial interest in identifying catalyst systems that would allow for the controlled incorporation of polar co-monomers, such as acrylates or vinyl acetate, into polyethylene and polypropylene chains. Industrial catalyst systems for synthesizing polyolefins are easily poisoned when such co-monomers are introduced. Recently, a number of catalysts based on group VIII–X metals (including Fe, Co, Ni, and Pd) have been identified that will copolymerize ethylene with polar co-monomers under mild reaction conditions (Ittel *et al.*, 2000; Boffa and Novak, 2000). These catalyst systems open the door to interesting new functional polyethylene random copolymers that will very likely find their way into a variety of polymer formulations. Particularly interesting are recent reports of group VIII living polymerization systems that enable the synthesis of polyethylene block copolymers with functional, polar monomers incorporated into one or more blocks (Gottfried and Brookhart, 2003; Diamanti *et al.*, 2003).

### 1.4.3  *Combinatorial methodologies*

Combinatorial chemistry or "combi-chem" methods are now firmly established as high-throughput experimental techniques for drug discovery and related applications. In the materials science domain, experimental methods for massively parallel synthesis, library design, and analytical screening tools have recently become available (Danielson *et al.*, 1997). These techniques of combinatorial materials discovery are being actively applied to polymeric systems, particularly aimed at the design of polymer formulations and polymerization catalysts. In these areas the experimental parameter space is exceedingly large, so design problems are well suited to massively parallel synthesis and characterization methods. In the case of catalyst discovery, rapid screening tools have been developed to identify catalyst activity, yield, and selectivity (Desrosiers *et al.*, 2003). Polymer formulations design is more problematic since the properties to be optimized are often difficult to measure with small quantities of material and the property sets of interest vary widely among different types of formulations. Nevertheless, scalable analytical tools are beginning to emerge that have broad applicability across many types of polymer systems.

### 1.4.4 *Soft material nanotechnology*

Nanotechnology, a contemporary research and technology craze, is a term used to describe materials and devices that operate by controlling chemistry, physics, and materials structure on nanometer lengthscales. It is interesting to note that polymer technologists have been practicing nanotechnology for decades – long before the explosion in nanotech R&D investment. Indeed, microphase separated styrenic block copolymers, sold by Shell Chemical, Phillips Petroleum, and other companies since the 1970s, are exquisite nanostructures whose properties can be tuned by adjusting polymer architecture. Similarly, the decades-old processes for manufacturing HIPS (high-impact polystyrene) and ABS (acrylonitrile-butadiene-styrene copolymer) plastic materials achieve highly optimized combinations of physical properties by exercising control over phase separation to create unique polymer nanostructures.

Several genuinely new trends, however, are associated with this resurgence of interest in nanotechnology. The first involves the marriage of inorganic nanometer-sized particles with organic polymers. Initiated by ground-breaking research at the Toyota company, the concept of reinforcing polymer resins with natural and synthetic clays has developed into a major R&D endeavor (Vaia and Giannelis, 2001). Exfoliation of natural clays leads to plate-like inorganic nanoparticles of thickness 1 nm and lateral dimensions on the scale of 100 nm. When such particles are dispersed uniformly in a polymer matrix, a variety of interesting properties can potentially result. Among these are improved heat and ignition resistance, higher modulus and stiffness, and higher barrier properties, relative to the pure polymer. Some of these properties, such as barrier, rely on the high aspect ratio and plate-like shape of the nanofiller, and can be obtained at relatively low loadings of the clay, i.e. $< 5\%$ by weight. The challenge in this area is to devise economical schemes for clay platelet exfoliation, stabilization, and dispersion within a broad range of low-cost polymer resins.

A related research theme involves the combination of inorganic nanoparticles that are electronically, optically, or magnetically active with host polymer matrices. In most applications envisioned for such systems, the active particles provide a unique function to the material, while the polymer matrix serves as an inert protective layer to provide mechanical support and isolate the particles (Cheng *et al.*, 2001). Nanocomposites of this type that employ magnetic nanoparticles are under consideration for high-density magnetic storage devices, while similar composites utilizing emissive quantum dot particles could emerge as viable materials for advanced display technologies.

Such nanocomposites present a broad range of challenges that cross traditional organic–inorganic boundaries, but where scientists skilled in the theory and practice of inhomogeneous polymers can contribute greatly. A number of challenges surround the stabilization and passivation of inorganic nanoparticles. Established solution synthesis routes typically employ small organic molecules, such as trioctylphosphine oxide, to stabilize and control the growth of the inorganic particles. These organic species confer solubility of the particles in organic solvents and are sometimes left attached to the particles after synthesis. In some cases, these organics serve also to "passivate" the nanoparticle surfaces, removing low-energy traps and other surface-induced defects that would be deleterious to electronic or optical properties. One active line of research is whether polymers, rather than low molecular weight organics, can be used for stabilization during synthesis, as well as subsequent passivation (Platonova *et al.*, 1997; Diana *et al.*, 2003). Such a strategy leaves the nanoparticles already coated with polymer and can eliminate or aid subsequent steps of dispersing the particles in a polymer matrix. Of key interest is how polymer-stabilized nanoparticles self-assemble into regular arrays and how this assembly depends on polymer molecular weight and architecture, and on long-ranged, e.g. magnetic, inter-particle interactions.

Many other "nanotechnology" applications of inhomogeneous polymers have been envisioned. For example, block copolymers are under active investigation as agents for nanometer-scale patterning of semiconductor surfaces (sub-micron lithography) (Park *et al.*, 1997), and as templates for growing inorganic nanoparticles and nanowires (Thurn-Albrecht *et al.*, 2000). Block and graft copolymers can also be used as substrates for creating nanoporous materials, both organic and inorganic, which can be used in applications ranging from low-$k$ dielectric films to nanoporous molecular sieves, filters, and substrates for catalysis.

Evidently, the rapidly developing field of nanotechnology, and specifically those aspects dealing with soft material and hybrid organic/inorganic nanostructures, offers many exciting opportunities and challenges for researchers of inhomogeneous polymers.

## 1.5   Why theory?

This monograph provides an overview of the theoretical and computational tools that are available for studying the behavior of inhomogeneous polymer systems at equilibrium. From the previous discussion it is evident that many of the most exciting and contemporary systems, such as nanostructured films, composites, and solution formulations, possess a bewildering parameter space for carrying out materials design. While in some sense enabling, the rapid development of new synthetic methods for creating wider varieties of polymer architectures and new types of block copolymers even further exacerbates the problem of a large design space.

The experimental methods of combinatorial materials discovery will undoubtedly prove to be a powerful tool for tackling the design of modern inhomogeneous polymers. However, the equipment required for combinatorial synthesis and screening is expensive and requires a great deal of skill to operate effectively. Indeed, library design demands that the operator use chemical and physical insights to identify the most important experimental parameters to be varied. A person trained in the theory of inhomogeneous polymers as well as in modern experimental techniques would be uniquely qualified as a researcher and technologist in this new era of combinatorial materials design.

A new perspective is that theory and simulation can be effectively applied *in tandem* with high-throughput experimentation to accelerate the materials design process. With a mathematical model in hand that is faithful to the chemistry and physics of a particular polymer system, it is possible to carry out numerical simulations to explore self-assembly behavior and physical properties. One can exploit the availability of inexpensive, yet powerful, distributed computing platforms to tackle large model parameter spaces in a parallel computing environment. Model parameters including molecular weight and composition can be spread across the available computational nodes, while one also utilizes spatial domain decomposition. By such means, it is possible to carry out a virtual *computational-combinatorial* materials design on a high-performance computing cluster. Such a design is valuable only to the extent that model provides an accurate description of the physical system and that the optimal model parameters can be reconciled with experimentally realizable parameters. To ensure that this is the case, it is important that any computational design program be closely coupled with an experimental program – at a minimum to establish the validity of the model assumptions and, ideally, to provide promising leads for a subsequent high-throughput experimental optimization.

An important objective of the present monograph is to describe a theoretical framework and a collection of numerical simulation techniques that are suitable for the computational-combinatorial design of a broad range of inhomogeneous polymer systems.

## 1.6  Modelling perspective and scales

When one approaches a complex material system to be modelled, one of the first decisions to be made is the *perspective* that is to be adopted in building a physical model and translating that model into mathematics. There are two perspectives of primary interest in this monograph: *atomistic* and *mesoscopic*.

### 1.6.1  *Atomistic perspective*

At a reasonably fundamental level, one could describe a polymeric fluid using *ab initio* quantum mechanics and retain the nuclear coordinates and electronic degrees of freedom of the atoms composing the polymer and solvent molecules (Levine, 2000). Such an approach would require essentially no empirical or experimental knowledge to characterize the interactions among the fundamental particles composing the fluid. Unfortunately, this scheme is computationally intractable for systems with more than approximately 1000 atoms, so it is unsuitable for modelling polymeric fluids, where each polymer alone will typically contain more than this number of atoms.

The next higher level of description, which we shall refer to as an *atomistic* perspective, is to eliminate the electronic degrees of freedom and treat the nuclear coordinates classically, rather than quantum mechanically. The elimination of the electronic degrees of freedom leads to classical potentials expressing the potential energy as a function of the nuclear coordinates (Chandler, 1987). These can in principle be obtained by carrying out *ab initio* quantum chemical calculations on small sets of atoms, or more typically by applying empirical potentials that contain parameters characterizing the range and strength of the inter-nuclear interactions. In this classical perspective, each nucleus becomes an "atom" and carries an effective mass that is approximately the nuclear mass. The Hamiltonian of the system is the sum of the kinetic and potential energies of the atoms,

$$H(\mathbf{r}^n, \mathbf{p}^n) = \sum_{i=1}^{n} \frac{\mathbf{p}_i^2}{2m_i} + U(\mathbf{r}^n) \tag{1.1}$$

where $n$ is the total number of atoms in the system, $m_i$ is the effective mass of the $i$th atom, and $\mathbf{r}^n = (\mathbf{r}_1, ..., \mathbf{r}_n)$ and $\mathbf{p}^n = (\mathbf{p}_1, ..., \mathbf{p}_n)$ denote the sets of atomic positions and momenta. The function $U(\mathbf{r}^n)$ is the potential energy, which is often approximated by a sum of spherically symmetric pair potentials $u_{ij}(\mathbf{r}_i, \mathbf{r}_j) = u_{ij}(|\mathbf{r}_i - \mathbf{r}_j|)$ according to

$$U(\mathbf{r}^n) \approx \frac{1}{2} \sum_{i=1}^{n} \sum_{j=1(\neq i)}^{n} u_{ij}(|\mathbf{r}_i - \mathbf{r}_j|) \tag{1.2}$$

This sum over atomic pairs includes effective interactions between bonded and non-bonded atoms. Besides the effective masses of each atom, this atomistic model depends on parameters necessary to specify the various pair potentials. For example, a commonly used two-parameter empirical form for describing non-bonded interactions between a pair of neutral atoms is the Lennard-Jones 6–12 potential

$$u(r) = 4\epsilon \left[ (r/\sigma)^{-12} - (r/\sigma)^{-6} \right] \tag{1.3}$$

where $\epsilon$ is the potential well depth and $\sigma$ is an interaction length scale.

Once the masses and potential parameters that define an atomistic model of a fluid have been specified, a broad range of computer simulation techniques are available (Allen and Tildesley, 1987; Rapaport, 1995; Frenkel and Smit, 1996). For example, a molecular dynamics (MD) simulation can be conducted by numerically integrating Hamilton's classical equations of motion

$$\frac{d\mathbf{p}_i}{dt} = -\frac{\partial H(\mathbf{r}^n, \mathbf{p}^n)}{\partial \mathbf{r}_i} \tag{1.4}$$

$$\frac{d\mathbf{r}_i}{dt} = \frac{\partial H(\mathbf{r}^n, \mathbf{p}^n)}{\partial \mathbf{p}_i} \tag{1.5}$$

forward in time for each atom $i = 1, 2, ..., n$. In three dimensions, this represents a set of $6n$ first-order differential equations that are solved subject to $6n$ initial conditions representing the initial positions and momenta of the atoms. The atoms are normally confined to a cubic computational cell with periodic boundary conditions applied at all faces. "Equilibration" in such a MD simulation requires integration for a time sufficient to allow the phase space $(\mathbf{r}^n, \mathbf{p}^n)$ to be well explored.

It is useful to review the limitations of an atomistic computer simulation. There is an inertial timescale in eqns (1.4)–(1.5) that depends on the form of the potential energy function. For a fluid of identical atoms interacting pairwise via the Lennard-Jones 6–12 potential, this timescale is $\tau = (m\sigma^2/\epsilon)^{1/2}$. In the case of argon, e.g., $\tau \sim 10^{-12}$ s. Accurate numerical integration of eqns (1.4)–(1.5) in a condensed liquid state requires a timestep $\Delta t$ that is approximately two orders of magnitude smaller than $\tau$. Hence, for liquid argon, $\Delta t \sim 10^{-14}$ s. Currently, a state-of-the-art MD simulation of such a fluid would involve tracking the trajectories of approximately $n = 10^6$ atoms for up to $10^6$ timesteps, or 10 ns of real time. Larger-scale MD simulations can be carried out, but they generally require access to supercomputer resources that are not widely available.

With this background, let us assess the computational resources that would be required to carry out a fully atomistic MD simulation of a molten block copolymer film of dimensions comparable to that shown in Fig. 1.3. To simplify the task, we attempt only a two-dimensional simulation and (generously) assign one atom per $0.2 \, \mathrm{nm}^2$. From the field of view shown in Fig. 1.3, it is straightforward to estimate that of order $n \sim 10^8$ atoms would be required. This alone is likely prohibitive. Next, we address the issue of timescale. Adopting an argon-like timestep of $10^{-14}$ s implies that a MD simulation with a duration of up to $10^{-8}$ s would be feasible. Unfortunately, therein lies the difficulty. Local equilibration of a small coherent patch of lamellae in a molten film, say 100 nm on a side, might take place in a time as short as $10^{-3}$ s, but defect migration and larger scale evolution of the composition patterns occur on scales of seconds, hours, or days. The relaxation timescales can be even longer in the vicinity of a glass transition and at high molecular weight, since relaxation times in entangled polymer melts scale grow faster than the third power of the molecular weight (Doi and Edwards, 1986). In short, one would require, at the very least, $10^5$ more computer power than is expended in a state-of-the-art simulation to perform a fully atomistic, two-dimensional MD simulation of a film such as Fig. 1.3. Equilibration of a three-dimensional atomistic model of such an inhomogeneous polymeric fluid would be virtually impossible now and in the foreseeable future.

## 1.6.2 *Mesoscopic perspective*

One strategy for addressing the obvious difficulties associated with fully atomistic simulations of inhomogeneous polymers is to "coarse-grain" the fluid model so that groups of atoms are lumped into larger entities referred to as *particles*. These particles may or may not correspond to molecular species, e.g. solvent molecules, or to the monomer residues within a polymer chain. The particles interact by new effective interaction potentials that must be re-parameterized. Unfortunately, the coarse-graining procedure necessary to carry out this re-parameterization is more an art than a science. At the lowest level, one can group adjacent atoms to form a particle, for example lumping each $CH_2$ unit into a particle along a polyethylene chain, and then using empirical knowledge or quantum chemical calculations to fit parameters in potential functions describing bonded and non-bonded interactions among particles. Such a "united atom" approach has been used quite successfully to simulate oligomeric fluids and single-phase polymeric fluids of low molecular weight, but does not go very far in alleviating the serious spatial and temporal limitations of fully atomistic simulations of inhomogeneous polymers.

Higher levels of coarse-graining are problematic. For example, if we decide to lump ten adjacent monomer repeat units within a polymer backbone into a single particle, then the task of parameterizing the effective interactions between that particle and other such particles is formidable. This is because each particle consists of a sub-chain that can adopt many different conformations. Some success has been achieved by carrying out fully atomistic simulations of small systems of short polymers and then using the simulation data to compute spatial correlation functions among the center-of-mass positions of sub-chains. These correlation functions can in principle be used to build models for effective pair, three-body, and higher inter-particle potential functions that could subsequently provide a basis for particle-based simulations of larger systems. In practice, however, this procedure is fraught with a number of difficulties and is far from routine (Baschnagel *et al.*, 2000).

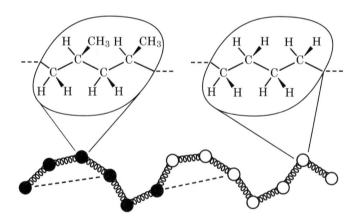

FIG. 1.5. Schematic coarse-graining of a polypropylene-polyethylene block copolymer into a bead-spring particle model. Dark beads denote coarse–grained polypropylene particles; light beads correspond to polyethylene particles. Springs denoted bonded pair potentials between adjacent particles along a polymer. Non-bonded interactions (dashed) are also present in such coarse–grained models.

A more typical coarse-graining procedure for a polymeric fluid is to essentially ignore the atomic details of the polymeric fluid below a threshold of about 1 nm, while preserving larger-scale features of the individual polymers, such as their connectivity, space-filling characteristics, and architecture (Binder, 1995; Baschnagel *et al.*, 2000). For example, Fig. 1.5 depicts the coarse-graining of a polypropylene-polyethylene (PP-PE) diblock copolymer into a "bead-spring" particle model. The positions and momenta of the particles are centered on the beads, and the springs represent effective pair potentials between adjacent (bonded) particles along a chain. Harmonic or anharmonic spring models can be employed. Potentials describing interactions between non-bonded particles on the same or different polymers are taken to be simple empirical forms, usually in the pair approximation. A common choice is the Lennard-Jones potential, or potential functions that are softer at short distances, which allows for larger timesteps (and faster equilibration) in MD simulations. Parameters in these potentials are not determined from first principles, but rather by fitting simulation results to available experimental data. A disadvantage of this approach is that experimental input is needed, so simulation results are predictive only to the extent that a model system has been parameterized. For example, a model parameterized by fitting interaction parameters to thermodynamic or structural data on a PP-PE diblock copolymer melt could be used to make predictions regarding the self-assembly behavior of a PP-PE-PP triblock copolymer melt without any additional experimental input. Another drawback to a coarse-grained description is that phenomena which depend on atomic-scale packing effects in the fluid cannot be captured with this approach. For example, polymer crystallization and glass transition phenomena cannot be quantitatively modelled in a system-specific way by using such a coarse-grained approach. Nevertheless, computer simulations of mesoscopic particle-based models of polymeric fluids have been quite useful in studying a variety of important systems and phenomena such as polymer brushes, block copolymers, and cohesive failure of polymer adhesives and glasses (Binder, 1995; Murat *et al.*, 1999; Müller and Binder, 1998; Rottler *et al.*, 2002).

### 1.6.3    *Particles and fields*

In the mesoscopic approach to modelling polymer fluids described above, the fundamental degrees of freedom are particle positions and momenta. It is the phase space of these coordinates that is explored in a *particle-based* computer simulation. There is an alternative strategy, however, that is the primary focus of the present monograph: *field-based* computer simulations.

To carry out a field-based computer simulation, we require a *statistical field theory model* of a fluid. A statistical field theory is a description of a system in which the fundamental degrees of freedom are not particle coordinates, but rather one or more continuous fields $w(\mathbf{r})$ that vary with position $\mathbf{r}$ (Parisi, 1988). In Chapter 4 we shall see that the relevant $w$ fields for an uncharged polymeric fluid can be interpreted as spatially varying chemical potentials. An additional electrostatic potential field $\Phi(\mathbf{r})$ is present in field theories of charged fluids.

From the perspective of a particle-based model, the fundamental problem of equilibrium statistical mechanics (Chandler, 1987) is to evaluate a configurational partition function of the form

$$\mathcal{Z} = \int d\mathbf{r}^n \; \exp[-\beta U(\mathbf{r}^n)] \tag{1.6}$$

where $\beta \equiv 1/(k_B T)$ is the inverse of the thermal energy and $U(\mathbf{r}^n)$ is the potential energy of an $n$-particle system. The corresponding equilibrium problem for a field theory model is to evaluate an analogous expression

$$\mathcal{Z} = \int \mathcal{D}w \; \exp(-\beta \bar{U}[w]) \tag{1.7}$$

where $\bar{U}[w]$ is an energy functional that depends on the value of a field $w(\mathbf{r})$ at all positions $\mathbf{r}$ within some spatial domain. The expression $\int \mathcal{D}w$ denotes a so-called *functional integral* over all possible $w$ field configurations (see Appendix C). The concept of functional integration will be discussed in detail in subsequent chapters, as will procedures for constructing expressions for $\bar{U}[w]$. Important for the present discussion is the following:

- If one starts from an *atomistic or mesoscopic* particle-based model defined by an expression for $U(\mathbf{r}^n)$, *formally exact* methods are available for transforming eqn (1.6) into a field theory of the form of eqn (1.7).
- Each field $w(\mathbf{r})$, to within a prescribed numerical precision, can be approximately represented by a finite number of degrees of freedom $M$, which could represent the values of $w$ at a set of $M$ spatial grid points, or the coefficients of a Fourier series expansion of $w$ truncated after $M$ terms.

If we denote the $M$ degrees of freedom used to represent a field $w$ by $w^M = (w_1, w_2, ..., w_M)$, eqn (1.7) can be approximated by a conventional $M$-dimensional integral

$$\mathcal{Z} \approx \int dw^M \; \exp[-\beta \bar{U}(w^M)] \tag{1.8}$$

where $\bar{U}$ is now a function of the discrete $w^M$ variables. The limiting behavior of eqn (1.8) for $M \to \infty$ and the extent to which eqn (1.7) is reproduced as the discrete representation of the field is continuously refined are subtle mathematical issues that are deferred to Chapter 6.

The transformations leading from eqn (1.6) to eqn (1.8) provide a model that is suitable for field-based computer simulations of polymeric fluids. In particular, a *field-theoretic computer simulation* is a numerical procedure for evaluating eqn (1.8) and related average quantities by stochastically sampling the phase space of the $w^M$ variables (Fredrickson *et al.*, 2002).

The advantages of field-based simulation methods for studying inhomogeneous polymers can be easily understood. First of all, there is the flexibility

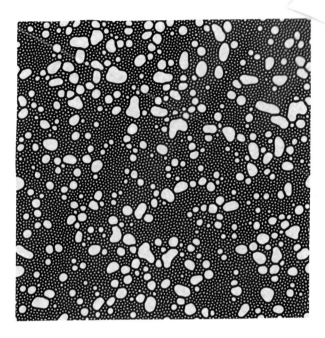

FIG. 1.6. Snapshot of the total A segment density from a field-based simulation of an incompressible melt blend of A homopolymer with ABA triblock copolymer. Higher A density is shown as light. The larger light droplets represent a homogeneous macrophase that is rich in A homopolymer, while the more finely structured background shows a coexisting cylindrical mesophase rich in the triblock copolymer. The dimensions of the simulation cell are $256\,R_g \times 256\,R_g$, where $R_g$ is the radius of gyration of a triblock copolymer. Reproduced from Sides and Fredrickson (2003).

of working with a field theory that originated from either an atomistic *or* a mesoscopic perspective. In addition, the spatial resolution of the field theory can be easily adjusted by giving the relevant fields a finer or coarser representation. There is also the added flexibility in how the fields are represented and discretized (e.g. finite differences, finite elements, or spectral representations), which leverages the large body of knowledge surrounding the numerical solution of partial differential equations. With such an approach, for example, it is possible to simulate a block copolymer mesophase with domain features on the 10 nm scale by using a field-theoretic simulation with a 1 nm spatial resolution, rather than the $\sim 0.01$ nm resolution of a fully atomistic simulation. The possibility of fewer field than particle degrees of freedom, $M \ll n$, evidently provides a significant computational advantage for the field-theoretic technique. This advantage translates into the possibility of simulating mesoscopically ordered polymers with a broad range of structural length scales. For example, Fig. 1.6 shows a snapshot from a high-resolution simulation (using $M \sim 2 \times 10^6$ field variables) of a homopolymer–

...iblock copolymer alloy exhibiting both macrophase and microphase separation (Sides and Fredrickson, 2003). Such a morphology would be very difficult to obtain with an atomistic, or even a coarse-grained, particle-based simulation.

Another important advantage of field-based simulations is that field theory models serve as the basis for most analytical theories of inhomogeneous polymers (de Gennes, 1979; Leibler, 1980; Bates and Fredrickson, 1990). A large body of experimental data has been interpreted using this theoretical framework, and a common language involving parameters in mesoscopic field theories, such as Flory "chi" parameters and statistical segment lengths, has emerged. By providing access to the behavior of these same field theory models across a broader range of parameter space, field-theoretic simulations can couple very effectively with existing theory and experiment.

In the present monograph, we shall adopt the mesoscopic perspective with few exceptions. Chapter 2 describes the statistical properties of coarse-grained, particle-based models of single polymers. These models of single-chain statistical mechanics are extended to include the presence of an external field in Chapter 3. An important component of this chapter is the description of efficient numerical methods for computing the statistical properties of a single polymer in a prescribed potential $w(\mathbf{r})$. Chapter 4 completes the particle-to-field transformation by describing how field theory models can be constructed for a variety of many-chain, inhomogeneous polymer systems. The subsequent two chapters discuss analytical and numerical techniques for analyzing these field theory models and provide application examples. Chapter 5 treats the important case of self-consistent field theory (SCFT), which is obtained by imposing a *mean-field approximation* (Parisi, 1988). This approximation is obtained by assuming that the functional integral in eqn (1.7) is dominated by one extremum field configuration, i.e. a configuration $w^*(\mathbf{r})$ satisfying[2]

$$\left.\frac{\delta \bar{U}[w]}{\delta w(\mathbf{r})}\right|_{w=w^*} = 0$$

so that

$$\mathcal{Z} = \int \mathcal{D}w \, \exp(-\beta\bar{U}[w]) \approx \exp(-\beta\bar{U}[w^*]) \tag{1.9}$$

We shall see that $w^*(\mathbf{r})$ corresponds to a saddle point in the complex plane of the energy functional $\bar{U}[w]$. Finally, Chapter 6 describes analytical and numerical techniques for going beyond the mean-field approximation and examining the effects of field fluctuations on equilibrium thermodynamic and structural quantities relevant to inhomogeneous polymers. When implemented numerically, such techniques amount to field-theoretic computer simulations of discretized field theories such as eqn (1.8).

---

[2]Readers not familiar with functional derivatives should consult Appendix C.

# 2

# IDEAL CHAIN MODELS

## 2.1   Real and ideal chains

The first ingredient in a coarse-grained theoretical description of a polymeric fluid is a mesoscopic model describing the statistical mechanics associated with the conformational states of a single polymer. An excellent introduction to this subject can be found in the recent text by Rubinstein and Colby (2003), and readers interested in more details are encouraged to consult the monographs by Flory (1969) and Volkenstein (1963). The present discussion of the topic will be brief.

Many synthetic polymers, such as the polyethylenes, polypropylenes, and other vinyl polymers such as polystyrene and polyvinylchloride, are considered to be *flexible* by virtue of relatively unhindered rotations about carbon–carbon single bonds along their backbones. This flexibility means that the orientation of a particular segment along a chain is nearly uncorrelated with the orientations of segments that are 10 or more monomer residues removed along the chain. Some specialty synthetic polymers and many biopolymers, such as double-stranded DNA, are much more rigid and bend over large distances by accumulated bond strains. These are termed semiflexible or rigid-rod polymers, depending on their length and degree of flexibility. In the present monograph, we shall be primarily concerned with flexible macromolecules.

In Chapter 1 we discussed the notion of coarse-graining real polymer chains into "bead-spring" chain models with bonded and non-bonded potentials acting between particles (beads). The flexibility of a polymer backbone on mesoscopic scales should be manifest in such a bead-spring description through the expressions for the bonded potential and the non-bonded potentials among particles that are *nearby* along the chain, e.g. second- and third-nearest neighbors. For example, the values of the angle $\theta_3$ in Fig. 2.1 will be largely determined by the interaction potentials acting among particles 2, 3, and 4. At a more atomistic level, such interactions reflect the bonding constraints, associated, say, with the tetrahedral bonding of carbon along a vinyl polymer backbone, and steric interactions among side groups attached to the polymer. The bonding constraints of real polymer chains and the associated close-neighbor potentials that describe them in coarse-grained chain models are referred to as *short-ranged interferences*.

*Long-ranged interferences* are associated with interactions between polymer segments that are separated by large distances along a polymer backbone, but close distances through space. For example, the first segment and the 1000th segment along a polyethylene chain will, to an excellent approximation, be un-

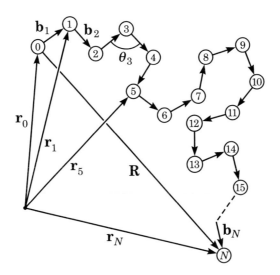

FIG. 2.1. A coarse-grained chain model with $N+1$ particles and $N$ springs (indicated by connecting arrows). The particle positions are denoted by $\mathbf{r}_0, ..., \mathbf{r}_N$, bond vectors by $\mathbf{b}_1, ..., \mathbf{b}_N$, and the chain end-to-end vector by $\mathbf{R}$. An example of a short-ranged interference is the restriction of the angle $\theta_3$ caused by the mutual interactions among particles 2, 3, and 4. A long-ranged interference is exemplified by the through-space interaction between particles 6 and 12.

correlated in orientation since the short-ranged interferences have died off between two segments separated by such a large distance along a chain. Through the random coiling of the polymer, however, those same two segments could find themselves separated by a small distance in space, as with particles 6 and 12 in the coarse-grained chain of Fig. 2.1. The typically strong interactions between two such monomers lead to positional correlations that are referred to as long-ranged interferences. One manifestation of long-ranged interferences is the so-called *excluded volume effect*, which is the constraint that any two segments in a coiled polymer cannot occupy the same location in space (Flory, 1953; de Gennes, 1979).

So-called *ideal chain* models take into account only short-ranged interferences in the statistical mechanics of a polymer chain. An important observation is that regardless of the particular form of the short-ranged interferences, all ideal chain models exhibit *universal* scaling properties at sufficiently large length scales. To clarify this statement, let us consider a measure of the average size of a polymer coil, $R = \sqrt{\langle \mathbf{R} \cdot \mathbf{R} \rangle}$, where $\mathbf{R}$ is the end-to-end vector of the polymer (see Fig. 2.1) and $\langle ... \rangle$ denotes an average over all conformational states of the polymer. This coil size, the "root-mean-squared end-to-end vector," scales with the degree of polymerization $N$ as[3]

$$R \sim bN^{\nu}, \ \nu = 1/2, \ N \to \infty \qquad (2.1)$$

This asymptotic scaling relationship between the average size of an ideal polymer coil, $R$, and the number of segments $N$ is *universal* in the sense that the *scaling exponent* $\nu$ assumes the value $1/2$, regardless of the chemical details of the polymer and the level of coarse-graining employed in modelling the polymer. It is quite remarkable that increasing the molecular weight of a polymer chain by a factor of four doubles the average size of the polymer coil, regardless of whether the polymer is polystyrene or polyethylene! In contrast, the length scale $b$ that appears as a prefactor in eqn (2.1), typically of order $\sim 1$ nm, is sensitive to both the chemical structure of the polymer and the level of coarse-graining, but not the overall molecular weight. It should be emphasized that eqn (2.1) is an asymptotic formula that is valid at large $N$. The approach to this asymptotic limit is not universal and depends on the range of the short-ranged interferences.

A natural question is under what conditions do real polymer chains satisfy the ideal chain scaling law of eqn (2.1)? Since all polymers with only short-ranged interferences exhibit asymptotic ideal chain scaling, this question can be alternatively expressed: under what conditions can long-ranged interferences be neglected for real chains? In principle, long-ranged interferences are never completely absent. However, two situations are commonly encountered in which long-ranged interferences are essentially negligible (de Gennes, 1979; Rubinstein and Colby, 2003):

- A dilute solution of a homopolymer of high molecular weight at the *theta* temperature.
- A homopolymer chain embedded in a melt of chemically identical homopolymers.

It is curious that the apparently simplest case of a linear homopolymer at infinite dilution in a good solvent does not obey ideal chain scaling. For such a system, excluded volume interactions among polymer segments close in space, but remote along the chain, produce a new "universality class" of scaling behavior – that of the self-avoiding random walk (Le Guillou and Zinn-Justin, 1977):

$$R \sim b_s N^{\nu}, \ \nu = 0.588..., \ N \to \infty \qquad (2.2)$$

Again the exponent $\nu$ is universal and depends only on the dimensionality of space,[4] while the length scale $b_s$ depends on the chemical details of both polymer and solvent, but is again independent of the polymer molecular weight. As

---

[3]In this scaling relation and subsequently throughout the monograph, the symbol $\sim$ will be used to indicate a proportionality without regard to the precise numerical coefficient. The symbol $\approx$ will be substituted when an attempt has been made to estimate the value of a numerical prefactor.

[4]$\nu = 0.588...$ in three dimensions, $\nu = 3/4$ in two dimensions.

the solvent quality for the polymer is lowered, usually by decreasing the temperature, attractive interactions among polymer segments are transmitted by the
solvent medium. At the *theta* temperature, these attractive interactions exactly
compensate for the excluded volume repulsion and the ideal chain scaling of
eqn (2.1) is recovered.

A similar phenomenon occurs in a homopolymer melt (de Gennes, 1979). Two
polymer segments that are far removed along the same chain, but separated by
a short distance (of order 0.5 nm) in space, repel each other strongly. However,
this net repulsion is cancelled ("screened") in the asymptotic limit of $N \to \infty$ by
the many other chains (of number $\sim N^{1/2}$) that protrude into the random coil of
the chain of interest. The ideality of a polymer in a melt of identical polymers
is sometimes referred to as the *Flory theorem*.

In physical situations other than these two special cases, long-ranged interferences normally play an important role in the statistical mechanics of real polymer
chains and should be included in atomistic and mesoscopic models through the
specification of appropriate inter-segment potentials. We shall see in Chapter
4 that it is convenient to separate the interactions responsible for short-ranged
and long-ranged interferences when building coarse-grained models of polymers.
Furthermore, this separation will be seen to produce polymer field theories in
which the statistical mechanics of single-chain models plays a central role.

The reader may have noticed that we have been rather vague about the physical interpretation of the non-universal lengths $b$ and $b_s$ that appear as prefactors
in eqns (2.1) and (2.2). Crudely, one can think of $b$ as the length scale at which
the local rigidity of the chain is lost and flexibility of a polymer strand is first
manifest. Its connection to the bonding constraints of a polymer will be explored
below in the context of specific ideal chain models.

## 2.2   Freely jointed chain model

For a general coarse-grained polymer model with $N + 1$ particles, such as that
depicted in Fig. 2.1, the configurational partition function of a single chain can
be expressed by a formula similar to eqn (1.6):

$$Z_0 = \int d\mathbf{r}^{N+1} \ \exp[-\beta U_0(\mathbf{r}^{N+1})] \qquad (2.3)$$

where $\mathbf{r}^{N+1} = (\mathbf{r}_0, \mathbf{r}_1, ..., \mathbf{r}_N)$ denotes the set of $N + 1$ particle positions and
$U_0(\mathbf{r}^{N+1})$ is the potential energy associated with a particular configuration of
the polymer.[5] The notation $\int d\mathbf{r}^{N+1}$ is shorthand for a $3(N + 1)$-dimensional
integral over the $N + 1$ particle positions within a three-dimensional domain of
volume $V$. For an ideal chain model, $U_0$ contains only interaction potential terms
reflecting short-ranged interferences.

---

[5] The subscript 0 is used to indicate that we are discussing the properties of a single, ideal
chain.

The joint probability density of observing a point $\mathbf{r}^{N+1}$ in configuration space is the Boltzmann distribution

$$P_0(\mathbf{r}^{N+1}) = Z_0^{-1} \exp[-\beta U_0(\mathbf{r}^{N+1})] \tag{2.4}$$

This probability weight (or density)[6] is normalized so that $\int d\mathbf{r}^{N+1} P_0(\mathbf{r}^{N+1}) = 1$. The ensemble average of an arbitrary function $f(\mathbf{r}^{N+1})$ over all configurations of the chain[7] can be written

$$\langle f(\mathbf{r}^{N+1}) \rangle_0 = \int d\mathbf{r}^{N+1} P_0(\mathbf{r}^{N+1}) f(\mathbf{r}^{N+1}) \tag{2.5}$$

An alternative representation of the configurational degrees of freedom of an $(N+1)$-particle chain is to retain one "external" coordinate describing the overall location of the polymer and $N$ "internal" coordinates. A particularly convenient choice is the position of a chain end, e.g. $\mathbf{r}_0$, and the set of $N$ *bond vectors* shown in Fig. 2.1, $\mathbf{b}^N = (\mathbf{b}_1, \mathbf{b}_2, ..., \mathbf{b}_N)$, where $\mathbf{b}_i \equiv \mathbf{r}_i - \mathbf{r}_{i-1}$. In the absence of an external potential acting on the polymer, $U_0$ depends only on the internal coordinates $\mathbf{b}^N$. Combining this feature with the fact that the Jacobian of the transformation from coordinates $\mathbf{r}^{N+1}$ to $(\mathbf{r}_0, \mathbf{b}^N)$ is unity allows eqns (2.3)–(2.4) to be reexpressed as

$$Z_0 = V \int d\mathbf{b}^N \exp[-\beta U_0(\mathbf{b}^N)] \tag{2.6}$$

$$P_0(\mathbf{r}_0, \mathbf{b}^N) = Z_0^{-1} \exp[-\beta U_0(\mathbf{b}^N)] \tag{2.7}$$

The distribution is thus uniform in the chain end position $\mathbf{r}_0$.

The *freely jointed chain* model is a very simple ideal chain model in which the bond vectors connecting successive particles are constrained to have a fixed length, $|\mathbf{b}_i| = b$, but the orientations of the $N$ bond vectors are distributed isotropically and independently. The constraint of fixed bond length can in principle be achieved by a spring model for $U_0(\mathbf{b}^N)$ in the limiting case of infinitely stiff springs. A simpler approach is to adopt a representation of the $\mathbf{b}^N$ such that the constraint is automatically satisfied. In particular, we write $\mathbf{b}_i = b\,\mathbf{n}_i$, so that $\mathbf{n}^N = (\mathbf{n}_1, ..., \mathbf{n}_N)$ is a set of $N$ unit vectors distributed independently and uniformly on the unit sphere. It follows that for the freely jointed chain model,

$$P_0(\mathbf{r}_0, \mathbf{n}^N) = \frac{1}{V} \left( \frac{1}{4\pi} \right)^N \tag{2.8}$$

which is normalized so that $\int d\mathbf{r}_0 \int d\mathbf{n}^N P_0(\mathbf{r}_0, \mathbf{n}^N) = 1$, where $\int d\mathbf{n}^N$ denotes $N$ integrals over the unit sphere.

---

[6]The concepts of probability distributions and their moments are discussed in Appendix B.

[7]We shall use the terms 'configurations' and 'conformations' interchangeably in describing the internal degrees of freedom of a coarse-grained polymer model.

Upon application of eqn (2.8) we can examine a variety of statistical properties of the freely jointed chain. Of particular interest are moments of the end-to-end vector $\mathbf{R} = \mathbf{r}_N - \mathbf{r}_0$. This object can be conveniently expressed as $\mathbf{R} = \sum_{i=1}^{N} \mathbf{b}_i = b\sum_{i=1}^{N} \mathbf{n}_i$. The isotropic distribution of the $\mathbf{n}_i$ implies that $\langle \mathbf{n}_i \rangle_0 = 0$ and, hence, a vanishing first moment for $\mathbf{R}$

$$\langle \mathbf{R} \rangle_0 = 0 \tag{2.9}$$

The second moment of the end-to-end vector can be written

$$\langle \mathbf{R}_\alpha \mathbf{R}_\beta \rangle_0 = b^2 \sum_{i=1}^{N} \sum_{j=1}^{N} \langle n_{i\alpha} n_{j\beta} \rangle_0 \tag{2.10}$$

where we use Greek subscripts to denote Cartesian components of vectors and tensors. The terms in the double sum with $i \neq j$ evidently vanish because of the independence of the respective unit vectors. The diagonal terms are evaluated by using $\langle n_{i\alpha} n_{i\beta} \rangle_0 = (1/3)\delta_{\alpha\beta}$, where $\delta_{\alpha\beta}$ is the Kronecker delta.[8] Hence,

$$\langle \mathbf{R}_\alpha \mathbf{R}_\beta \rangle_0 = \frac{b^2 N}{3} \delta_{\alpha\beta} \tag{2.11}$$

The root-mean-square end-to-end vector of the freely jointed chain model is thus fully consistent with the ideal chain scaling law, eqn (2.1):

$$R \equiv \sqrt{\langle \mathbf{R} \cdot \mathbf{R} \rangle_0} = bN^{1/2} \tag{2.12}$$

It is important to note that eqn (2.12) was derived without imposing the asymptotic limit of $N \to \infty$ and the scaling coefficient omitted from eqn (2.1) is exactly unity for the freely jointed chain.

Higher moments of the end-to-end vector are similarly calculated. For example,

$$\langle (\mathbf{R} \cdot \mathbf{R})^2 \rangle_0 = \frac{5}{3} b^4 N \left( N - \frac{2}{5} \right) \tag{2.13}$$

From this result, it is apparent that the probability distribution of the end-to-end vector $\mathbf{R}$ for the freely jointed chain is not a simple Gaussian distribution, since for such a distribution (see Appendix B) $\langle (\mathbf{R} \cdot \mathbf{R})^2 \rangle = (\langle \mathbf{R} \cdot \mathbf{R} \rangle)^2 + 2\langle \mathbf{R}\mathbf{R} \rangle : \langle \mathbf{R}\mathbf{R} \rangle$, which leads to $\langle (\mathbf{R} \cdot \mathbf{R})^2 \rangle_0 = (5/3)b^4 N^2$. However, asymptotically for $N \to \infty$, the fourth moment given in eqn (2.13) is consistent with a Gaussian distribution.

Another important measure of the average size of a polymer chain is the *radius of gyration* $R_g$. This quantity is the root-mean-squared distance between

---

[8]The Kronecker delta function is defined by $\delta_{\alpha\beta} = 1$ for $\alpha = \beta$; $\delta_{\alpha\beta} = 0$ for $\alpha \neq \beta$.

individual segments (particles) and the polymer's center of mass, $\mathbf{R}_c = (N + 1)^{-1} \sum_{i=0}^{N} \mathbf{r}_i$, i.e.

$$R_g^2 \equiv \frac{1}{N+1} \sum_{i=0}^{N} \langle (\mathbf{r}_i - \mathbf{R}_c)^2 \rangle_0 \tag{2.14}$$

$R_g^2$ thus represents the second moment of the mass distribution in a polymer coil and has the advantage that it can be defined for all polymer architectures, e.g. even for ring polymers, which have no chain ends. The squared radius of gyration can be expressed alternatively in the form (Rubinstein and Colby, 2003)

$$R_g^2 = \frac{1}{(N+1)^2} \sum_{i=0}^{N} \sum_{j>i}^{N} \langle (\mathbf{r}_i - \mathbf{r}_j)^2 \rangle_0 \tag{2.15}$$

which is particularly convenient for computation. This expression can be evaluated for the freely jointed chain by expressing $\mathbf{r}_i - \mathbf{r}_j$ as a sum of bond vectors, which leads to $\langle (\mathbf{r}_i - \mathbf{r}_j)^2 \rangle_0 = b^2 |i - j|$. The double sum in eqn (2.15) is then asymptotically evaluated for $N \to \infty$ to yield

$$R_g \approx b(N/6)^{1/2} \tag{2.16}$$

The radius of gyration $R_g$ of the freely jointed chain is thus smaller than the root-mean-square end-to-end vector $R$ by a factor of $1/\sqrt{6}$. This proves to hold for any ideal chain model of a linear polymer in the large $N$ limit.

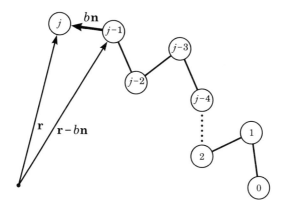

FIG. 2.2. Illustration of the "stochastic process" approach to building up the statistical weight for the end position of a freely jointed chain with $j + 1$ particles from the statistical weight of a chain with $j$ particles.

Rather than calculating moments, another useful way to explore the statistical properties of ideal chain models is to directly examine probability distribution functions for quantities such as the end-to-end vector. Of particular interest is a *reduced distribution function*, $p_0(\mathbf{r}, j)$, which represents the probability density that a polymer chain with $j + 1$ particles has its end (the particle labelled $j$) at position $\mathbf{r}$. The function is normalized so that $\int d\mathbf{r}\, p_0(\mathbf{r}, j) = 1$. In constructing this object, we shall have an opportunity to make an important connection between polymer statistical mechanics and the theory of *stochastic processes* (van Kampen, 1981). Specifically, the random displacements of successive particles along a coarse-grained polymer chain are analogous to random events that occur at regular time intervals in a discrete-time stochastic process.

To exploit the stochastic process analogy, we imagine that the probability density for the end position of a chain with *one particle fewer*, $p_0(\mathbf{r}, j - 1)$, is known. As shown in Fig. 2.2, a $(j + 1)$-particle chain can be built up from a $j$-particle chain by adding one particle and a connecting bond. In the freely jointed chain model, the additional bond is of fixed length, but its orientation is uncorrelated with those of the $j - 1$ bonds that have already been added to the chain. Thus, we can construct the probability density $p_0(\mathbf{r}, j)$ from the known function $p_0(\mathbf{r}, j - 1)$ by multiplying by a transition probability (consistent with the fixed length and uniform orientational distribution of the added bond) and integrating over all possible bond orientations. In particular,

$$p_0(\mathbf{r}, j) = \frac{1}{4\pi} \int d\mathbf{n}\, p_0(\mathbf{r} - b\mathbf{n}, j - 1) \qquad (2.17)$$

The factor $1/(4\pi)$ in this equation represents the uniform transition probability associated with the orientation of the added bond, and the integral is again over the unit sphere. Such an equation is referred to as a *Chapman–Kolmogorov* equation in the theory of stochastic processes (van Kampen, 1981). In this language, the freely jointed chain is an example of a *one-step Markov process*, because the transition probability connects only the positions of adjacent particles along the chain.

To solve an equation like (2.17), we require an "initial condition" $p_0(\mathbf{r}, 0)$, specifying the distribution of position of a 1-particle chain. Equation (2.17) can then be recursively applied for $j = 1, 2, ..., N$ to construct the probability density $p_0(\mathbf{r}, N)$ for the end position of an $(N + 1)$-particle chain. Such recursive schemes are ideally suited for numerical computations and will be heavily exploited throughout this monograph. For the present purposes, eqn (2.17) will be used to derive the full distribution function for the end-to-end vector of a freely jointed chain.

It is convenient to introduce the three-dimensional Fourier transform[9] of a function $f(\mathbf{r})$ by

$$\hat{f}(\mathbf{k}) \equiv \int d\mathbf{r}\, e^{-i\mathbf{k}\cdot\mathbf{r}} f(\mathbf{r}) \qquad (2.18)$$

The inverse transform is given by

---

[9]Definitions and formulas related to Fourier analysis are summarized in Appendix A.

$$f(\mathbf{r}) = \frac{1}{(2\pi)^3} \int d\mathbf{k} \, e^{i\mathbf{k}\cdot\mathbf{r}} \hat{f}(\mathbf{k}) \tag{2.19}$$

Applying the Fourier transform to both sides of eqn (2.17) leads to the following expression:

$$\begin{aligned} \hat{p}_0(\mathbf{k}, j) &= \frac{1}{4\pi} \int d\mathbf{n} \, e^{-ib\mathbf{k}\cdot\mathbf{n}} \, \hat{p}_0(\mathbf{k}, j-1) \\ &= j_0(b|\mathbf{k}|) \, \hat{p}_0(\mathbf{k}, j-1) \end{aligned} \tag{2.20}$$

where $j_0(x) \equiv (\sin x)/x$ is the familiar spherical Bessel function. Recursively applying this equation for $j = 1, 2, ..., N$ leads to

$$\hat{p}_0(\mathbf{k}, N) = [j_0(b|\mathbf{k}|)]^N \hat{p}_0(\mathbf{k}, 0) \tag{2.21}$$

A situation of particular interest corresponds to the initial condition $p_0(\mathbf{r}, 0) = \delta(\mathbf{r})$, where $\delta(\mathbf{r})$ is the three-dimensional Dirac delta function.[10] This implies that the initiating end of the chain (particle 0) is constrained to the origin. With this choice, $\hat{p}_0(\mathbf{k}, 0) = 1$, and $p_0(\mathbf{R}, N)$ can be interpreted as the probability density of a freely jointed chain with $N$ bonds and end-to-end vector $\mathbf{R} = \mathbf{r}_N - \mathbf{r}_0$. It follows that

$$p_0(\mathbf{R}, N) = \frac{1}{(2\pi)^3} \int d\mathbf{k} \, e^{i\mathbf{k}\cdot\mathbf{R}} [j_0(b|\mathbf{k}|)]^N \tag{2.22}$$

is an exact closed-form expression for the probability density of the end-to-end vector. Asymptotic analysis of this integral for $N \gg 1$ and $|\mathbf{R}| \ll Nb$ (Doi and Edwards, 1986) confirms our previous observation that the second and fourth moments of $\mathbf{R}$ are consistent with a Gaussian distribution

$$p_0(\mathbf{R}, N) \approx [3/(2\pi Nb^2)]^{3/2} \exp[-3|\mathbf{R}|^2/(2Nb^2)] \tag{2.23}$$

The important notion of *conformational entropy* is manifest in eqn (2.23). In making the transition between the full phase space distribution function of the freely jointed chain $P_0(\mathbf{r}_0, \mathbf{b}^N)$ and the reduced probability distribution function of the end-to-end vector, $p_0(\mathbf{R}, N)$, we have integrated over all sets of fixed-length bond vectors $\mathbf{b}^N$ consistent with the constraint of a fixed end-to-end vector $\mathbf{R} = \sum_{i=1}^{N} \mathbf{b}_i$. This integration amounts to an enumeration of conformational states. Because all states occur with uniform probability in the freely jointed model, the result is a *purely entropic* contribution to the free energy of a chain with constrained end-to-end vector $\mathbf{R}$:

$$F_0(\mathbf{R}) \equiv -k_B T \ln p_0(\mathbf{R}, N) \approx \frac{3k_B T}{2Nb^2} |\mathbf{R}|^2 \tag{2.24}$$

[10]The Dirac delta function is a generalized function defined by the property $f(\mathbf{r}) = \int d\mathbf{r}' \, \delta(\mathbf{r} - \mathbf{r}')f(\mathbf{r}')$ for any function $f(\mathbf{r})$. See also Appendix A.

The quadratic dependence on chain extension $|\mathbf{R}|$ in this expression can be viewed as an "entropic spring" potential. Fewer conformational states are available for chains with large extensions, so the free energy increases with $|\mathbf{R}|$. Moreover, the "spring constant" $3k_BT/(Nb^2) = k_BT/(2R_g^2)$ softens in inverse proportion to the square of the polymer coil size.

A final topic relates to the applicability of the freely jointed chain model to real chains under ideal conditions. On short length scales, i.e. $\sim 1\,\text{nm}$, the bonding and steric constraints of real polymer chains are clearly oversimplified by the freely jointed chain, which incorporates only a local bond rigidity over a distance $b$. However, on larger scales, i.e. $\sim 5$–$10\,\text{nm}$, real flexible polymers in the ideal state exhibit scaling behavior consistent with eqn (2.12) for the freely jointed chain. We can thus describe the mesoscopic statistical properties of such polymers by defining an *equivalent freely jointed chain*. Namely, one demands that the real polymer in the ideal state has the same mean-squared end-to-end vector, $R^2 = Nb^2$, and maximum end-to-end distance, $R_{\max} = Nb$, as the equivalent freely jointed chain. The two parameters of the freely jointed chain can then be expressed in terms of experimental measurements or estimates of $R^2 = \langle \mathbf{R} \cdot \mathbf{R} \rangle_0$ and $R_{\max}$

$$b = R^2/R_{\max}, \quad N = R_{\max}^2/R^2 \tag{2.25}$$

By following such a procedure, the bond length $b$ of the equivalent chain is referred to as the *Kuhn segment length* and is approximately $1\,\text{nm}$ for most vinyl polymers (Flory, 1969; Rubinstein and Colby, 2003). The number of effective "Kuhn segments" $N$ of the equivalent chain is typically a factor of 50 smaller than the number of skeletal bonds of the real chain.

Obviously the conditions given in eqn (2.25) for mapping a real chain in the ideal state onto the freely jointed model are not unique. Highly extended chain conformations are extremely rare, so the justification for matching $R_{\max}$ between real and equivalent chains is weak. Other methods for defining an equivalent freely jointed chain usually involve a *definition* of $N$ and then a calculation of $b$ by matching $R^2$ (or $R_g^2$) between the real and equivalent chains. Common definitions for $N$ include the average number of monomeric repeat units per chain, or the average number of segments per chain with a prescribed molecular weight (which may or may not correspond to an actual chemical repeat unit). The $b$ values obtained from such procedures are often referred to as *statistical segment lengths*. It is important to note that the values of $b$ and $N$ will vary somewhat depending on the criteria used to define the equivalent chain. Nevertheless, the resulting chain models will generally well describe the mesoscopic ($\gg 1\,\text{nm}$) statistical properties of real polymer chains in the ideal state, irrespective of the criteria applied.

In the present monograph we will adopt a rather casual language when referring to the parameters in coarse-grained, equivalent chain models. $N$ will be interchangeably described as the degree of polymerization, the number of monomers per chain, or the number of statistical segments. Likewise, $b$ will be referred to as the Kuhn length, the statistical segment length, or the monomer length.

## 2.3 Bead-spring models

Another important class of ideal chain models are the so-called bead-spring models. In these models, successive particles along a coarse-grained chain are tethered by "spring potentials" that can be chosen to have a variety of forms. The potential energy $U_0$ defining these models is most conveniently expressed in terms of bond vectors, so the partition function and phase space distribution function of an $(N + 1)$-particle chain (see Fig. 2.1) are commonly written in the form of eqns (2.6)–(2.7). If all of the $N$ bonds of such a chain are equivalent and no external field is present, the potential energy can be expressed as

$$U_0(\mathbf{b}^N) = \sum_{i=1}^{N} h(|\mathbf{b}_i|) \qquad (2.26)$$

where $h(x)$ is the spring potential between adjacent particles along the polymer backbone. It follows that

$$Z_0 = V \left( \int d\mathbf{b} \, \exp[-\beta h(|\mathbf{b}|)] \right)^N \qquad (2.27)$$

$$P_0(\mathbf{r}_0, \mathbf{b}^N) = V^{-1} \prod_{i=1}^{N} \frac{\exp[-\beta h(|\mathbf{b}_i|)]}{\int d\mathbf{b}_i \, \exp[-\beta h(|\mathbf{b}_i|)]} \qquad (2.28)$$

Thus, each bond vector $\mathbf{b}_i$ is distributed independently with a statistical weight proportional to $\exp[-\beta h(|\mathbf{b}_i|)]$.

It remains to specify the functional form of the spring potential $h$ for a bond. The most common choice, which defines the so-called *discrete Gaussian chain model*, is a harmonic bond potential

$$h(x) = \frac{3k_B T}{2b^2} x^2 \qquad (2.29)$$

The parameter $b$ in this potential can be interpreted as the root-mean-square length of a bond, since for any bond $i$

$$\langle \mathbf{b}_i \cdot \mathbf{b}_i \rangle_0 = \frac{\int d\mathbf{b}_i \, (\mathbf{b}_i \cdot \mathbf{b}_i) \exp[-\beta h(|\mathbf{b}_i|)]}{\int d\mathbf{b}_i \, \exp[-\beta h(|\mathbf{b}_i|)]} = b^2 \qquad (2.30)$$

where the required Gaussian integrals have been performed using formulas given in Appendix B.

The mean-squared end-to-end vector of the discrete Gaussian chain can be similarly calculated. In particular,

$$\langle \mathbf{R} \cdot \mathbf{R} \rangle_0 = \sum_{i=1}^{N} \sum_{j=1}^{N} \langle \mathbf{b}_i \cdot \mathbf{b}_j \rangle_0 \qquad (2.31)$$

The terms with $i \neq j$ in this expression vanish because of the independent distribution of the bond vectors, which implies that $\langle \mathbf{b}_i \cdot \mathbf{b}_j \rangle_0 = \langle \mathbf{b}_i \rangle_0 \cdot \langle \mathbf{b}_j \rangle_0 = 0$ for $i \neq j$. Thus,

$$\langle \mathbf{R} \cdot \mathbf{R} \rangle_0 = N \langle \mathbf{b}_i \cdot \mathbf{b}_i \rangle_0 = N b^2 \qquad (2.32)$$

and the ideal chain scaling of eqn (2.12) is exactly recovered, $R = bN^{1/2}$. Apart from the interpretation of $b$ as the fixed bond length in the freely jointed model versus the root-mean-square bond length in the discrete Gaussian model, we see that the expressions for the mean-squared end-to-end vector in the two ideal chain models are identical. It is easy to show that the same equivalence holds for the radius of gyration $R_g$, so that eqn (2.16) also applies to the discrete Gaussian chain.

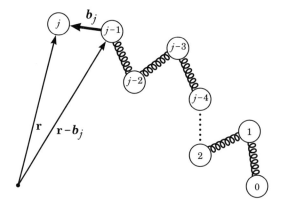

FIG. 2.3. The "stochastic process" approach to constructing the statistical weight for the end position of a discrete Gaussian chain with $j + 1$ particles from the statistical weight of a chain with $j$ particles.

In order to more fully explore the relationship between the freely jointed and discrete Gaussian chain models, it is again useful to exploit the connection with stochastic processes. As for the freely jointed chain, we consider the reduced distribution function $p_0(\mathbf{r}, j)$, expressing the probability density of observing the end particle $j$ of a $(j + 1)$-particle chain at position $\mathbf{r}$. Assuming knowledge of the probability density for a chain with one particle fewer, $p_0(\mathbf{r}, j - 1)$, the object of interest is built up as shown schematically in Fig. 2.3 by means of a Chapman–Kolmogorov equation

$$p_0(\mathbf{r}, j) = \int d\mathbf{b}_j \; \Phi(\mathbf{b}_j; \mathbf{r} - \mathbf{b}_j) \, p_0(\mathbf{r} - \mathbf{b}_j, j - 1) \tag{2.33}$$

In this expression, $\Phi(\mathbf{b}_j; \mathbf{r} - \mathbf{b}_j)$ is the *conditional* probability density that the bond vector connecting particles $j$ and $j - 1$ assumes a value of $\mathbf{b}_j$, given that particle $j - 1$ was located at position $\mathbf{r} - \mathbf{b}_j$. It is normalized so that $\int d\mathbf{b}_j \; \Phi(\mathbf{b}_j; \mathbf{r} - \mathbf{b}_j) = 1$. For a discrete Gaussian chain in the absence of any external potentials, $\Phi$ is independent of both the particle index $j$ along the chain (i.e. the stochastic process is "stationary") and the starting position $\mathbf{r} - \mathbf{b}_j$. The conditional "transition" probability density thus reflects only the Gaussian distribution of bond displacements

$$\Phi(\mathbf{b}_j; \mathbf{r} - \mathbf{b}_j) = \Phi(\mathbf{b}_j) = \frac{\exp[-\beta h(|\mathbf{b}_j|)]}{\int d\mathbf{b}_j \; \exp[-\beta h(|\mathbf{b}_j|)]}$$

$$= \left(\frac{3}{2\pi b^2}\right)^{3/2} \exp[-3|\mathbf{b}_j|^2/(2b^2)] \tag{2.34}$$

Equation (2.33) is easily solved by Fourier transformation, since the form of the transition probability $\Phi$ for the discrete Gaussian chain implies that the right-hand side of the equation is a three-dimensional convolution integral (see Appendix A). The Fourier transform of eqn (2.33) thus yields a product of Fourier transforms

$$\hat{p}_0(\mathbf{k}, j) = \hat{\Phi}(\mathbf{k}) \, \hat{p}_0(\mathbf{k}, j - 1) \tag{2.35}$$

which we solve recursively for $j = 1, 2, ..., N$ to obtain

$$\hat{p}_0(\mathbf{k}, N) = [\hat{\Phi}(\mathbf{k})]^N \hat{p}_0(\mathbf{k}, 0) \tag{2.36}$$

The Fourier transform of the Gaussian transition probability in eqn (2.34) is also Gaussian,[11] $\hat{\Phi}(\mathbf{k}) = \exp(-b^2|\mathbf{k}|^2/6)$, so eqn (2.36) reduces for the discrete Gaussian chain to

$$\hat{p}_0(\mathbf{k}, N) = \exp(-R_g^2|\mathbf{k}|^2) \, \hat{p}_0(\mathbf{k}, 0) \tag{2.37}$$

Here the shorthand $R_g = \sqrt{Nb^2/6}$ has been employed for the radius of gyration of the chain. Finally, we specialize to the case of particle 0 constrained to the origin, $p_0(\mathbf{r}, 0) = \delta(\mathbf{r})$, for which $\hat{p}_0(\mathbf{k}, 0) = 1$ and the inverse Fourier transform of eqn (2.37) is given *exactly* by

$$p_0(\mathbf{R}, N) = [3/(2\pi Nb^2)]^{3/2} \exp[-3|\mathbf{R}|^2/(2Nb^2)] \tag{2.38}$$

It is of interest to compare these results for the discrete Gaussian chain with those obtained previously for the freely jointed chain model. In particular, we see that the Fourier transform of the exact distribution function $\hat{p}_0(\mathbf{k}, N)$ for the freely jointed chain model, eqn (2.21), differs from the corresponding object of

---

[11] The required Gaussian integrals can be found in Appendix B.

the discrete Gaussian chain, eqn (2.37). However, for length scales exceeding $b$, i.e. $b|\mathbf{k}| \ll 1$,

$$[j_0(b|\mathbf{k}|)]^N \approx [1 - b^2|\mathbf{k}|^2/6 + ...]^N \approx \exp(-R_g^2|\mathbf{k}|^2) \qquad (2.39)$$

and the two distribution functions coincide. This observation is consistent with the fact that eqn (2.38) is exact for the discrete Gaussian chain, but is an asymptotic result (requiring $N \gg 1$ and $|\mathbf{R}| \ll Nb$) for the freely jointed chain.

The discrete Gaussian chain model is but one of many bead-spring models that can be devised. It is a particularly convenient model because the distribution of segments is Gaussian at both the level of a coarse-grained bond and the level of the mesoscopic end-to-end vector. This facilitates analytical calculations of a variety of single-chain properties. Nevertheless, a linear (Hookian) spring model utilizing a harmonic potential for the connecting bonds is sometimes inadequate. For example, the discrete Gaussian chain model can exhibit large, unphysical stretching when used to describe polyelectrolytes under low salt conditions (Netz and Orland, 1999). Gaussian chains subjected to strong hydrodynamic flows with extensional character also show unbounded stretching that is not representative of real polymers (Larson, 1988). In such situations a nonlinear spring model can be used to prevent unphysical extended chain conformations. The freely jointed chain can also be applied, but it is often more difficult to implement in polymer dynamics schemes because of the holonomic constraints associated with fixed bond lengths (Doi and Edwards, 1986).

Adopting the perspective that nearly all interesting calculations for inhomogeneous polymers are carried out *numerically*, the use of nonlinear bead-spring models presents no particular difficulties. In particular, eqn (2.36) applies for *any* bead-spring model in which all bonds are identical and are characterized by a spring potential $h(x)$ that depends only on the bond length $x$. It is the form of $\hat{\Phi}(\mathbf{k})$ that distinguishes the statistical properties of different chain models. The Fourier transform of the transition probability density in the case of general $h(x)$ can be written

$$\hat{\Phi}(\mathbf{k}) = \frac{\int_0^\infty dx\, x^2 j_0(|\mathbf{k}|x) \exp[-\beta h(x)]}{\int_0^\infty dx\, x^2 \exp[-\beta h(x)]} \qquad (2.40)$$

For the linear (Gaussian) spring model of eqn (2.29), $\hat{\Phi}_G(\mathbf{k}) = \exp(-b^2|\mathbf{k}|^2/6)$. Comparison of eqns (2.21) and (2.36) indicates that the freely jointed model can be expressed in the same way by means of the choice $\hat{\Phi}_{FJ}(\mathbf{k}) = j_0(b|\mathbf{k}|)$.

As an example of a nonlinear bead-spring model, it is useful to consider a spring potential similar to the Warner spring used in rheological calculations (Bird *et al.*, 1977):

$$h(x) = \frac{3k_BT}{2b^2} \frac{x^2}{(1 - x^2/b_0^2)} \qquad (2.41)$$

This potential contains a parameter $b_0 > 0$, which can be interpreted as the *maximum* length of a bond since $h(x) \to +\infty$ as $x \to b_0$. However, for weak bond extensions, $x/b_0 \ll 1$, eqn (2.41) reduces to the harmonic bond potential given in eqn (2.29). Thus, eqn (2.41) is a two-parameter nonlinear spring law that agrees with the linear spring law of the discrete Gaussian model for small bond displacements, yet saturates the bond length at a finite value $b_0$ for strong applied forces.

When substituted into eqn (2.40), the nonlinear spring model of eqn (2.41) leads to a transformed transition probability $\hat{\Phi}_{NL}(\mathbf{k})$ that can be expressed as a function of the scaled wavevector $b|\mathbf{k}|$ and the dimensionless ratio of lengths $b/b_0$. For $b/b_0 < 0.1$, $\hat{\Phi}_{NL}(b|\mathbf{k}|, b/b_0)$ is nearly indistinguishable from the linear spring function $\hat{\Phi}_G(\mathbf{k})$. At larger values of $b/b_0$, however, the influence of the nonlinear spring appears in the wavevector dependence of $\hat{\Phi}_{NL}$. Figure 2.4 compares the Fourier transforms of the transition probabilities for the discrete Gaussian chain, the freely jointed chain, and the nonlinear bead-spring chain of eqn (2.41) for the case of $b/b_0 = 1.0$. The nonlinear spring law has an oscillatory dependence on $|\mathbf{k}|$ that is qualitatively similar to the freely jointed chain. However, the oscillations are more strongly damped than in the freely jointed model, where bonds are rigidly constrained to have length $b$.

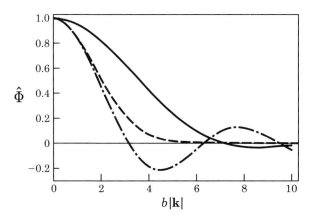

FIG. 2.4. Fourier transforms of the bond transition probabilities, $\Phi_G$, $\Phi_{FJ}$, and $\Phi_{NL}$ for the discrete Gaussian (dashed), freely jointed (dot-dashed), and nonlinear spring models (solid), respectively. The nonlinear spring model is evaluated with $b/b_0 = 1.0$.

## 2.4  Continuous Gaussian chain model

An elegant and particularly convenient ideal chain model for both analytical and numerical calculations is the *continuous Gaussian chain* (Doi and Edwards, 1986). This model can be described as the continuum limit of the discrete Gaussian chain model in which the polymer is viewed as a continuous, linearly elastic filament. As shown in Fig. 2.5, the configuration of the continuous Gaussian chain is specified by a space curve $\mathbf{r}(s)$ in which $s \in [0, N]$ is a contour variable that describes the location of a segment along the backbone of the chain. The position in space of segment $s$ is given by $\mathbf{r}(s)$, and the end-to-end vector $\mathbf{R}$ can be expressed as $\mathbf{R} = \mathbf{r}(N) - \mathbf{r}(0)$.

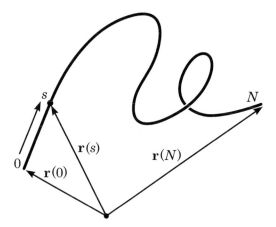

FIG. 2.5. The continuous Gaussian chain model describes the configuration of a polymer as a space curve $\mathbf{r}(s)$, where $s \in [0, N]$ is a contour parameter. The chain end positions correspond to $\mathbf{r}(0)$ and $\mathbf{r}(N)$.

The potential energy of the continuous Gaussian chain can be written

$$U_0[\mathbf{r}] = \frac{3k_B T}{2b^2} \int_0^N ds \left| \frac{d\mathbf{r}(s)}{ds} \right|^2 \tag{2.42}$$

where the square bracket notation in $U_0[\mathbf{r}]$ is used to indicate that $U_0$ is a *functional* of the space curve $\mathbf{r}(s)$ defining the configuration of the polymer.[12] A functional is a mapping between a continuous function and a number (Volterra, 1959) (see Appendix C); in this case the mapping is between $\mathbf{r}(s)$ and the value of $U_0$. The form of the potential energy is closely related to eqns (2.26) and (2.29) for the discrete Gaussian chain. If we view $d\mathbf{r}(s)/ds$ as the local "stretch" in a segment of length $ds$ located at contour position $s$, then eqn (2.42) sums a harmonic potential contribution from each such differential segment over the entire contour of the chain. It is important to note that $s$ does not indicate arc length in the continuous Gaussian chain model, but is simply a parameter indexing the segments along the chain. Thus, the stretch $d\mathbf{r}(s)/ds$ is not constrained to be a unit vector, but is free to fluctuate in magnitude. Equation (2.42) for the potential energy is commonly referred to as the "Edwards Hamiltonian."

The configurational partition function of the continuous Gaussian chain can be written

---

[12]This square bracket notation will be employed throughout the book to indicate a functional dependence. See Appendix C.

$$Z_0 = \int \mathcal{D}\mathbf{r} \, \exp(-\beta U_0[\mathbf{r}])) \tag{2.43}$$

where the notation $\int \mathcal{D}\mathbf{r}$ indicates a *functional integral* over all possible space curves $\mathbf{r}(s)$ describing the configuration of the polymer (see Appendix C). Such functional integrals, also termed *path integrals*, are familiar from the fields of quantum mechanics and probability theory (Feynman and Hibbs, 1965), where $\mathbf{r}(s)$ corresponds to the position of a quantum particle or Brownian particle at time $s$. Indeed, eqn (2.43) is the so-called Wiener measure in the path integral description of classical diffusion (Brownian motion).

Path integrals are complicated mathematical objects that require some so-phistication in definition and manipulation (Simon, 1979). For the purpose of this book, however, we shall treat these objects informally and in physical terms. One approach to defining a path integral is to discretize the path at a set of $N_s + 1$ equally spaced contour points; hence we approximate the continuous function $\mathbf{r}(s)$ by a vector $(\mathbf{r}_0, \mathbf{r}_1, ..., \mathbf{r}_{N_s})$ describing the positions in space of the $N_s + 1$ points. With such a scheme, a path integral can be defined as a $3(N_s + 1)$-dimensional ordinary integral[13]

$$\int \mathcal{D}\mathbf{r} \approx \prod_{i=0}^{N_s} \int d\mathbf{r}_i \tag{2.44}$$

where the quality of the approximation improves as $N_s \to \infty$. At finite $N_s$, such an approximation for the path integral of a continuous Gaussian chain closely resembles the partition function for a discrete Gaussian chain with $(N_s + 1)$ beads. In particular, we have the approximation

$$Z_0 \approx \prod_{j=0}^{N_s} \int d\mathbf{r}_j \, \exp\left(-\frac{3}{2b^2 \Delta s} \sum_{i=1}^{N_s} |\mathbf{r}_{i-1} - \mathbf{r}_i|^2\right) \tag{2.45}$$

where the mean-squared length of one of the $N_s$ bonds is given by $b^2 \Delta s$ and $\Delta s = N/N_s$ is the spacing between contour points.

The subtlety in eqn (2.45) arises from the continuum limit, since $Z_0$ scales as $V N_s^{-(3/2)N_s}$, which vanishes for $N_s \to \infty$. This seemingly disturbing fact turns out to have no consequence in equilibrium statistical mechanics, because we are generally interested in average quantities, which can be expressed as the ratio of two path integrals. For example, the mean-squared end-to-end vector of the continuous Gaussian chain can be expressed as

$$R^2 \equiv \langle \mathbf{R} \cdot \mathbf{R} \rangle_0 = \frac{\int \mathcal{D}\mathbf{r} \, |\mathbf{r}(N) - \mathbf{r}(0)|^2 \exp(-\beta U_0[\mathbf{r}])}{\int \mathcal{D}\mathbf{r} \, \exp(-\beta U_0[\mathbf{r}])} \tag{2.46}$$

where the denominator is simply the partition function $Z_0$. When discretizing the path integrals in both numerator and denominator according to eqn (2.45),

---

[13]It is assumed that the polymer is embedded in a three-dimensional space with volume $V$.

one finds that the singular factors *exactly cancel* so that $R^2$ is well defined in the $N_s \to \infty$ limit. Moreover, in the numerical evaluation of path integrals, one always works with finite $N_s$, so the singularities are naturally avoided. Throughout this book, expressions such as eqn (2.43) will be written and manipulated in a formal way. The reader should be aware that there may indeed be some subtlety in defining and evaluating these objects.

Another useful way to define a path integral is through a *spectral* representation of the path. In particular, we can represent the space curve $\mathbf{r}(s)$ by expanding in a complete set of basis functions $\{\phi_0(s), \phi_1(s), ...\}$ according to

$$\mathbf{r}(s) = \sum_{p=0}^{\infty} \mathbf{a}_p \, \phi_p(s) \tag{2.47}$$

This is a type of generalized Fourier expansion, and the expansion coefficients $\mathbf{a}_p$ can be viewed as generalized Fourier coefficients (see Appendix A). For a polymer that is freely suspended in a fluid medium, a convenient choice of basis functions is the set of cosines consistent with the "no-stretch" boundary conditions $d\mathbf{r}(s)/ds|_{s=0} = d\mathbf{r}(s)/ds|_{s=N} = 0$. This amounts to the cosine Fourier series representation

$$\mathbf{r}(s) = \mathbf{a}_0 + 2\sum_{p=1}^{\infty} \mathbf{a}_p \cos(p\pi s/N) \tag{2.48}$$

The orthogonality of these basis functions allows one to solve for the Fourier coefficients, which are given by

$$\mathbf{a}_p = \frac{1}{N} \int_0^N ds \, \cos(p\pi s/N)\mathbf{r}(s), \quad p = 0, 1, 2, ..., \infty \tag{2.49}$$

These normal modes are known as *Rouse modes* in the polymer literature and play a particularly important role in theories of polymer dynamics (Doi and Edwards, 1986). Physically, $\mathbf{a}_0$ can be interpreted as the position of the center of mass of the polymer, while the $\mathbf{a}_p$ for $p = 1, 2, 3, ...$ provide successively more information about the shape of the polymer on increasingly fine scales.

With the Rouse spectral representation, the integral over all conformations (paths) of the polymer can be interpreted as a product of integrals over all the Rouse modes

$$\int \mathcal{D}\mathbf{r} = \prod_{p=0}^{\infty} \int d\mathbf{a}_p \tag{2.50}$$

The object on the right-hand side of this expression is an infinite-dimensional integral, and so we again encounter problems with the existence of the partition function $Z_0$. Nevertheless, the ratio of two such path integrals proves to be finite, and for numerical purposes we can always *regularize* (remove singularities in) a path integral by retaining only a finite number $P \gg 1$ of normal modes

$$\int \mathcal{D}\mathbf{r} \approx \prod_{p=0}^{P} \int d\mathbf{a}_p \tag{2.51}$$

To illustrate the use of Rouse modes in calculations with the continuous Gaussian chain, it is useful to consider the evaluation of eqn (2.46) for the mean-squared end-to-end vector. Expressed in terms of Rouse modes, the squared end-to-end vector is

$$|\mathbf{r}(N) - \mathbf{r}(0)|^2 = 16 \sum_{p=1,3,\dots}^{\infty} \sum_{q=1,3,\dots}^{\infty} \mathbf{a}_p \cdot \mathbf{a}_q \tag{2.52}$$

where the sums are restricted to odd integers. The average of this quantity can be written as

$$\langle |\mathbf{R}|^2 \rangle_0 = 16 \sum_{p=1,3,\dots}^{\infty} \sum_{q=1,3,\dots}^{\infty} \langle \mathbf{a}_p \cdot \mathbf{a}_q \rangle_0 \tag{2.53}$$

where the average of an object $f(\mathbf{a})$ expressed in Rouse modes is defined as

$$\langle f(\mathbf{a}) \rangle_0 = \frac{\prod_{p=0}^{\infty} \int d\mathbf{a}_p \, f(\mathbf{a}) \exp(-\beta U_0(\mathbf{a}))}{\prod_{p=0}^{\infty} \int d\mathbf{a}_p \, \exp(-\beta U_0(\mathbf{a}))} \tag{2.54}$$

The potential energy of eqn (2.42) is diagonal in the Rouse mode representation

$$\beta U_0(\mathbf{a}) = \frac{1}{2} \sum_{p=1}^{\infty} \alpha(p) \, \mathbf{a}_p \cdot \mathbf{a}_p \tag{2.55}$$

where $\alpha(p) = 6\pi^2 p^2 / (b^2 N)$. Note that the center of mass $p = 0$ mode does not enter the energy and is uniformly distributed. The $p \geq 1$ modes are all statistically independent with a Gaussian distribution, so it follows that (Appendix B)

$$\langle \mathbf{a}_p \cdot \mathbf{a}_q \rangle_0 = \frac{3}{\alpha(p)} \delta_{pq} \tag{2.56}$$

Substitution into eqn (2.53) leads to

$$\langle |\mathbf{R}|^2 \rangle_0 = \frac{8b^2 N}{\pi^2} \sum_{p=1,3,\dots}^{\infty} \frac{1}{p^2} = b^2 N \tag{2.57}$$

We thus conclude that the continuous Gaussian chain shares the property of the discrete Gaussian chain in that its root-mean-square end-to-end vector is given by $R = bN^{1/2}$. Similar arguments can be used to show that the radius

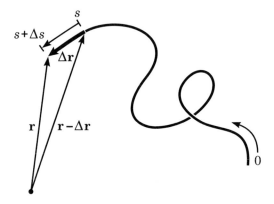

FIG. 2.6. The "stochastic process" approach to constructing the statistical
weight for the end position of a continuous Gaussian chain with $s + \Delta s$
segments from the statistical weight of a chain with $s$ segments.

of gyration of the continuous Gaussian chain is given by the expected formula,
$R_g = b(N/6)^{1/2}$.

In continuing to explore the properties of the continuous Gaussian chain
model, we now return to the stochastic process approach. Specifically, it is useful
to consider a reduced distribution function $p_0(\mathbf{r}, s)$ describing the probability
density that a continuous Gaussian chain of contour length $s$ has its end at
position $\mathbf{r}$. This function is normalized so that $\int d\mathbf{r}\, p_0(\mathbf{r}, s) = 1$. By analogy
with eqn (2.33) for the discrete Gaussian chain, we can build up the distribution
function by means of a Chapman–Kolmogorov equation using information from
a slightly smaller chain

$$p_0(\mathbf{r}, s + \Delta s) = \int d(\Delta \mathbf{r})\, \Phi(\Delta \mathbf{r}; \mathbf{r} - \Delta \mathbf{r})\, p_0(\mathbf{r} - \Delta \mathbf{r}, s) \tag{2.58}$$

Figure 2.6 illustrates the physical content of eqn (2.58), which relies on discretizing the last segment of a continuous chain. The transition probability density
$\Phi(\Delta \mathbf{r}; \mathbf{r} - \Delta \mathbf{r})$ describes the conditional probability of a displacement $\Delta \mathbf{r}$ for
a segment of chain of contour length $\Delta s$, starting from the position $\mathbf{r} - \Delta \mathbf{r}$ at
contour location $s$. The relevant stochastic process for the continuous Gaussian
chain is stationary, so $\Phi = \Phi(\Delta \mathbf{r})$ is again independent of both starting position
and contour location. An expression for $\Phi(\Delta \mathbf{r})$ follows immediately from the
previous discretization of the continuous Gaussian chain, eqn (2.45):

$$\Phi(\Delta \mathbf{r}) = \left(\frac{3}{2\pi b^2 \Delta s}\right)^{3/2} \exp\left(-\frac{3|\Delta \mathbf{r}|^2}{2b^2 \Delta s}\right) \tag{2.59}$$

A useful feature of *continuous* chain models is that Chapman–Kolmogorov
integral equations can be reduced to partial differential equations, which are referred to in probability theory as *Fokker–Planck* equations (van Kampen, 1981;

Wax, 1954) and in quantum theory as *Feynman–Kac* formulas (Feynman and Hibbs, 1965). We illustrate this by deriving the Fokker–Planck equation associated with eqn (2.58). The derivation proceeds by Taylor-expanding both sides of the equation in powers of $\Delta s$ and $\Delta \mathbf{r}$, treating each as small. Noting that $\Phi(\Delta \mathbf{r}; \mathbf{r} - \Delta \mathbf{r})$ is independent of the initial position $\mathbf{r} - \Delta \mathbf{r}$ in this case, we do not expand the transition probability. It follows that

$$p_0(\mathbf{r}, s) + \Delta s \frac{\partial}{\partial s} p_0(\mathbf{r}, s) + O(\Delta s^2) = p_0(\mathbf{r}, s) - \langle \Delta \mathbf{r} \rangle_\Phi \cdot \nabla p_0(\mathbf{r}, s)$$
$$+ \frac{1}{2!} \langle \Delta \mathbf{r} \Delta \mathbf{r} \rangle_\Phi : \nabla \nabla p_0(\mathbf{r}, s)$$
$$+ O(\langle \Delta \mathbf{r} \Delta \mathbf{r} \Delta \mathbf{r} \rangle_\Phi) \qquad (2.60)$$

where the $\Phi$-averages appearing in this equation are defined by

$$\langle f(\Delta \mathbf{r}) \rangle_\Phi = \int d(\Delta \mathbf{r}) \, \Phi(\Delta \mathbf{r}) f(\Delta \mathbf{r}) \qquad (2.61)$$

Making use of the explicit Gaussian form of eqn (2.59) allows the averages on the right-hand side of eqn (2.60) to be evaluated as (Appendix B)

$$\langle \Delta \mathbf{r} \rangle_\Phi = 0 \qquad (2.62)$$

$$\langle \Delta \mathbf{r}_\alpha \Delta \mathbf{r}_\beta \rangle_\Phi = \frac{b^2 \Delta s}{3} \delta_{\alpha\beta} \qquad (2.63)$$

If we insert these expressions into eqn (2.60), it follows that for $\Delta s \to 0$ the reduced distribution function $p_0(\mathbf{r}, s)$ satisfies the Fokker–Planck equation[14]

$$\frac{\partial}{\partial s} p_0(\mathbf{r}, s) = \frac{b^2}{6} \nabla^2 p_0(\mathbf{r}, s) \qquad (2.64)$$

Thus, the Fokker–Planck equation for the continuous Gaussian chain takes the form of a conventional *diffusion equation* with a "diffusion coefficient" given by $b^2/6$. The solution of this equation provides full information about the distribution of end segments, $p_0(\mathbf{r}, s)$.

Fokker–Planck equations are particularly convenient because a wide variety of analytical and numerical techniques are available for solving partial differential equations. In the present case of eqn (2.64), the fundamental (Green function) solution corresponding to the initial condition $p_0(\mathbf{r}, 0) = \delta(\mathbf{r})$ is

$$p_0(\mathbf{r}, s) = [3/(2\pi s b^2)]^{3/2} \exp[-3|\mathbf{r}|^2/(2s b^2)] \qquad (2.65)$$

If we set $s = N$ and $\mathbf{r} = \mathbf{R}$ in this expression, the familiar Gaussian distribution function is recovered for the end-to-end vector [cf. eqn (2.38)]. On scales larger

---

[14]A Fokker–Planck equation that involves only configuration variables and not their conjugate momenta is sometimes referred to as a *Smoluchowski equation*.

than a single bond, the discrete Gaussian chain and the continuous Gaussian chain evidently share the same chain end distribution function. The advantage of working with the continuous chain is that it allows calculations to be performed with partial differential equations such as eqn (2.64). This advantage will become more apparent in Chapter 3, where we consider chains in external potentials.

## 2.5   Wormlike chain model

The continuous Gaussian chain is a particularly convenient ideal chain model for describing the configurational statistics of *flexible* polymer chains. However, many macromolecules encountered in biological systems and certain synthetic polymers, such as liquid crystalline polymers (LCPs) and conjugated polymers, adopt configurational states that more nearly resemble rigid rods than random coils. To describe such systems, which are termed *semiflexible*, a more appropriate continuous chain model is the *Kratky–Porod model* (Kratky and Porod, 1949; Saito *et al.*, 1967). This model is also commonly referred to as the *wormlike chain*.

We recall that the continuous Gaussian chain model incorporates a harmonic energy penalty for local chain stretching, but no energy penalty for chain bending. In contrast, each differential segment of the wormlike chain is constrained to be inextensible, but there is a harmonic energy penalty for local bending. The inextensibility constraint implies that the total contour length of the polymer, $L_c$, is a constant for all chain configurations. The configuration of a wormlike chain is again described by a space curve $\mathbf{r}(s)$ in which $s \in [0, L_c]$ is a parameter denoting arc length along the polymer backbone. The vector $\mathbf{u}(s) = d\mathbf{r}(s)/ds$ is a *tangent vector* to the chain at contour location $s$ and is constrained to be a unit vector, $|\mathbf{u}(s)| = 1$. The magnitude of the vector $d\mathbf{u}(s)/ds = d^2\mathbf{r}(s)/ds^2$ can be similarly interpreted as the local *curvature* of the polymer at contour position $s$. An expression for the bending energy of the wormlike chain is obtained by summing up harmonic curvature contributions along the chain contour:

$$U_0[\mathbf{u}] = \frac{\lambda k_B T}{2} \int_0^{L_c} ds \left| \frac{d\mathbf{u}(s)}{ds} \right|^2 \tag{2.66}$$

The microscopic parameter $\lambda$ has units of length; its physical significance will be discussed shortly. The notation used in eqn (2.66) implies that the bending potential $U_0$ is a functional of $\mathbf{u}(s)$. However, it can also be viewed as a functional of the polymer shape $\mathbf{r}(s)$, since $\mathbf{u}(s) = d\mathbf{r}(s)/ds$. In writing the partition function of the wormlike chain as a path integral, it is convenient to make this constraint and the constraint that $|\mathbf{u}(s)| = 1$ explicit:

$$Z_0 = \int \mathcal{D}\mathbf{r} \, \exp(-\beta U_0[\mathbf{u}]) \prod_s \left[ \delta \left( \mathbf{u}(s) - \frac{d\mathbf{r}(s)}{ds} \right) \delta(|\mathbf{u}(s)| - 1) \right] \tag{2.67}$$

Thus, in computing the wormlike chain partition function, we are to sum over all chain paths $\mathbf{r}(s)$ consistent with $\mathbf{u}(s)$ being a unit tangent vector at all contour locations $s$.

Because of the constraints, eqn (2.67) proves to be difficult to work with directly as a path integral. Instead, we again turn to our stochastic process analogy and consider a reduced distribution function $p_0(\mathbf{r}, \mathbf{u}, s)$. This quantity is defined as the probability density that the *end* of a wormlike chain with contour length $s$ is at position $\mathbf{r}$ *and* the tangent vector of the end segment is $\mathbf{u}$. We assume a normalization such that $\int d\mathbf{r} \int d\mathbf{u}\, p_0(\mathbf{r}, \mathbf{u}, s) = 1$, where the integral over $\mathbf{u}$ denotes an integral on the unit sphere. The motivation for considering a distribution function of both chain end position and orientation is that in building models of semiflexible polymers we will often want to include anisotropic interactions among segments, e.g. nematic interactions in LCPs. As will become apparent in Chapter 4, the description of such interactions requires simultaneous information about segment orientation and position.

The development of a Chapman–Kolmogorov equation for the probability density $p_0(\mathbf{r}, \mathbf{u}, s)$ is closely related to the theory of Brownian motion for a particle with inertia (Chandrasekhar, 1943). In particular, we can invoke the Markov property of eqn (2.67) and write

$$p_0(\mathbf{r}, \mathbf{u}, s + \Delta s) = \int d(\Delta \mathbf{r}) \int d(\Delta \mathbf{u})\ \Psi\left(\Delta \mathbf{r}, \Delta \mathbf{u}; \mathbf{r} - \Delta \mathbf{r}, \mathbf{u} - \Delta \mathbf{u}\right)$$
$$\times\, p_0(\mathbf{r} - \Delta \mathbf{r}, \mathbf{u} - \Delta \mathbf{u}, s) \qquad (2.68)$$

where $\Delta \mathbf{u}$ denotes a differential displacement of the tangent vector and $\Delta \mathbf{r}$ a differential displacement of the end position associated with adding a chain segment of length $\Delta s$. The function $\Psi(\Delta \mathbf{r}, \Delta \mathbf{u}; \mathbf{r}, \mathbf{u})$ describes the conditional transition probability that the added chain segment has positional and orientational displacements $\Delta \mathbf{r}$ and $\Delta \mathbf{u}$, starting from a chain end with position $\mathbf{r}$ and orientation $\mathbf{u}$. This transition probability is normalized so that $\int d(\Delta \mathbf{r}) \int d(\Delta \mathbf{u})\, \Psi = 1$. In the above integrals, it should be noted that $\Delta \mathbf{u}$ is restricted to a rotation, so that $\mathbf{u}$ and $\mathbf{u} - \Delta \mathbf{u}$ reflect nearby points on the unit sphere. From the relation

$$\Delta \mathbf{r} = \int_s^{s+\Delta s} ds\, \mathbf{u}(s) = \mathbf{u}\, \Delta s + O(\Delta s^2) \qquad (2.69)$$

we see that the stochastic nature of the process (2.68) is, to first order in $\Delta s$, restricted to the variable $\Delta \mathbf{u}$. Thus, we can treat $\Delta \mathbf{r}$ as deterministic and reduce the transition probability to the form

$$\Psi(\Delta \mathbf{r}, \Delta \mathbf{u}; \mathbf{r}, \mathbf{u}) = \Phi(\Delta \mathbf{u}; \mathbf{r}, \mathbf{u})\, \delta(\Delta \mathbf{r} - \mathbf{u}\Delta s) \qquad (2.70)$$

where $\Phi(\Delta \mathbf{u}; \mathbf{r}, \mathbf{u})$ is a normalized transition probability for orientational displacements $\Delta \mathbf{u}$ on the unit sphere associated with a contour step of size $\Delta s$. Substitution of eqn (2.70) into eqn (2.68) followed by the shift of $\mathbf{r} \to \mathbf{r} + \mathbf{u}\Delta s$ leads to the simplified Chapman–Komogorov equation

$$p_0(\mathbf{r} + \mathbf{u}\Delta s, \mathbf{u}, s + \Delta s) = \int d(\Delta \mathbf{u})\ \Phi(\Delta \mathbf{u}; \mathbf{r}, \mathbf{u} - \Delta \mathbf{u})\, p_0(\mathbf{r}, \mathbf{u} - \Delta \mathbf{u}, s) \qquad (2.71)$$

We now attempt to derive a corresponding Fokker–Planck equation by expanding both sides of this integral equation for small $\Delta s$ and $\Delta \mathbf{u}$, similar to the treatment of eqn (2.58) for the continuous Gaussian chain. Expanding eqn (2.71) to order $\Delta s$ on the left-hand side and to order $\Delta \mathbf{u}^2$ on the right-hand side, it follows that

$$\Delta s \left[ \frac{\partial}{\partial s} + \mathbf{u} \cdot \nabla_{\mathbf{r}} \right] p_0(\mathbf{r}, \mathbf{u}, s) + O(\Delta s^2) = -\nabla_{\mathbf{u}} \cdot [\langle \Delta \mathbf{u} \rangle_\Phi p_0(\mathbf{r}, \mathbf{u}, s)]$$

$$+ \frac{1}{2!} \nabla_{\mathbf{u}} \nabla_{\mathbf{u}} : [\langle \Delta \mathbf{u} \Delta \mathbf{u} \rangle_\Phi p_0(\mathbf{r}, \mathbf{u}, s)]$$

$$+ O(\langle \Delta \mathbf{u} \Delta \mathbf{u} \Delta \mathbf{u} \rangle_\Phi) \qquad (2.72)$$

where $\nabla_{\mathbf{r}}$ and $\nabla_{\mathbf{u}}$ denote, respectively, gradients with respect to position $\mathbf{r}$ and orientation $\mathbf{u}$. In eqn (2.72), averages over the transition probability density are defined according to

$$\langle f(\Delta \mathbf{u}) \rangle_\Phi \equiv \int d(\Delta \mathbf{u}) \, f(\Delta \mathbf{u}) \Phi(\Delta \mathbf{u}; \mathbf{r}, \mathbf{u}) \qquad (2.73)$$

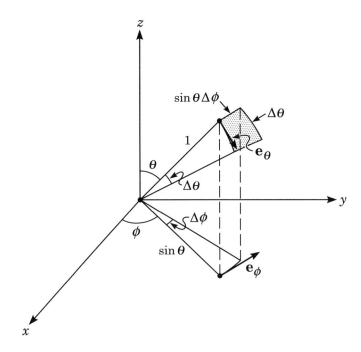

FIG. 2.7. Spherical polar coordinate system used to describe orientational displacements $\Delta \mathbf{u}$ on the unit sphere.

To proceed further, it is necessary to be explicit about the form of the transition probability $\Phi(\Delta \mathbf{u}; \mathbf{r}, \mathbf{u})$ for the wormlike chain. For this purpose, the spherical polar coordinate system shown in Fig. 2.7 is most convenient. Orthogonal unit vectors at a point on the unit sphere in the directions of increasing polar angle $\theta$ and azimuthal angle $\phi$ are denoted by $\mathbf{e}_\theta$ and $\mathbf{e}_\phi$, respectively. Upon discretization of eqn (2.66), the bending energy associated with a small displacement on the unit sphere along $\mathbf{e}_\theta$ and $\mathbf{e}_\phi$, i.e. $\Delta\mathbf{u} = (\Delta\theta, \sin\theta\,\Delta\phi)$, is

$$\beta\Delta U_0 = \frac{\lambda}{2\Delta s}|\Delta\mathbf{u}|^2 = \frac{\lambda}{2\Delta s}[(\Delta\theta)^2 + (\sin\theta\,\Delta\phi)^2] \tag{2.74}$$

The transition probability $\Phi(\Delta\mathbf{u}; \mathbf{r}, \mathbf{u}) \sim \exp(-\beta\Delta U_0)$ is thus a Gaussian distribution with first and second moments given by

$$\langle\Delta\mathbf{u}\rangle_\Phi = 0, \quad \langle\Delta\mathbf{u}\Delta\mathbf{u}\rangle_\Phi = \frac{\Delta s}{\lambda}(\mathbf{e}_\theta\mathbf{e}_\theta + \mathbf{e}_\phi\mathbf{e}_\phi) \tag{2.75}$$

Substituting this result into eqn (2.72) leads to the desired Fokker–Planck equation (Hermans and Ullman, 1952)

$$\frac{\partial}{\partial s}p_0(\mathbf{r}, \mathbf{u}, s) = -\mathbf{u}\cdot\nabla_\mathbf{r}\,p_0(\mathbf{r}, \mathbf{u}, s) + \frac{1}{2\lambda}\nabla_\mathbf{u}^2\,p_0(\mathbf{r}, \mathbf{u}, s) \tag{2.76}$$

where $\nabla_\mathbf{u}^2$ denotes the Laplacian operator on the unit sphere

$$\nabla_\mathbf{u}^2 f \equiv \frac{1}{\sin\theta}\frac{\partial}{\partial\theta}\left(\sin\theta\frac{\partial f}{\partial\theta}\right) + \frac{1}{\sin^2\theta}\frac{\partial^2 f}{\partial\phi^2} \tag{2.77}$$

This operator is commonly referred to as the *rotational diffusion operator* because it generates diffusive motion on the unit sphere (Berne and Pecora, 1976). The parameter $1/(2\lambda)$ multiplying $\nabla_\mathbf{u}^2$ in eqn (2.76) can be viewed as a *rotational diffusion coefficient*.

The Fokker–Planck eqn (2.76) provides full information about the propagation of correlations in segment position and orientation along a wormlike polymer chain. This equation apparently does not have a simple closed-form analytic solution, although its spatial Fourier transform can be developed in a continued fraction expansion (Spakowitz and Wang, 2004). The discussion of numerical methods will be deferred to Section 3.6.

If we restrict attention to orientational correlations, it is convenient to introduce a second reduced distribution function

$$H_0(\mathbf{u}, s) \equiv \int d\mathbf{r}\, p_0(\mathbf{r}, \mathbf{u}, s) \tag{2.78}$$

which can be interpreted as the probability density that the end of a wormlike chain of contour length $s$ has orientation $\mathbf{u}$. $H_0(\mathbf{u}, s)$ is normalized so that $\int d\mathbf{u}\, H_0(\mathbf{u}, s) = 1$. Assuming periodic boundary conditions, eqn (2.76) can be integrated over $\mathbf{r}$ to obtain a Fokker–Planck equation for the reduced orientational distribution function

$$\frac{\partial}{\partial s}H_0(\mathbf{u}, s) = \frac{1}{2\lambda}\nabla_\mathbf{u}^2\, H_0(\mathbf{u}, s) \tag{2.79}$$

This equation is now a simple rotational diffusion equation that is amenable to analytical solution (Berne and Pecora, 1976; Saito *et al.*, 1967). For this purpose, it is convenient to introduce a *spherical harmonic* expansion for $H_0(\mathbf{u}, s)$,

$$H_0(\mathbf{u}, s) = \sum_{l=0}^{\infty} \sum_{m=-l}^{l} h_{lm}(s) Y_{lm}(\mathbf{u}) \qquad (2.80)$$

where the spherical harmonics $Y_{lm}(\mathbf{u})$ are eigenfunctions of the rotational diffusion operator

$$\nabla_{\mathbf{u}}^2 Y_{lm}(\mathbf{u}) = -l(l+1)\, Y_{lm}(\mathbf{u}) \qquad (2.81)$$

These functions are subject to different definitions (Edmonds, 1974). Here we adopt the convention familiar in quantum mechanics (Schiff, 1968) and define $Y_{lm}(\mathbf{u})$ for integer $l, m \geq 0$ according to

$$Y_{lm}(\mathbf{u}) = Y_{lm}(\theta, \phi) = (-1)^m \left[ \frac{2l+1}{4\pi} \frac{(l-m)!}{(l+m)!} \right]^{1/2} P_l^m(\cos\theta)\, \exp(im\phi) \quad (2.82)$$

where $P_l^m(x)$ is the associated Legendre function defined for integer $l, m \geq 0$ by

$$P_l^m(x) = (1 - x^2)^{m/2} \frac{d^m}{dx^m} P_l(x) \qquad (2.83)$$

and $P_l(x)$ is the familiar Legendre polynomial,

$$P_l(x) = \frac{1}{2^l l!} \frac{d^l}{dx^l} (x^2 - 1)^l \quad l \geq 0 \qquad (2.84)$$

The spherical harmonics are defined for negative integer $m$ by the symmetry property

$$Y_{l,-m}(\theta, \phi) = (-1)^m Y_{lm}^*(\theta, \phi) \qquad (2.85)$$

where the asterisk denotes complex conjugation. They are orthonormal in the sense that

$$\int d\mathbf{u}\, [Y_{lm}(\mathbf{u})]^* \, Y_{l'm'}(\mathbf{u}) = \int_0^{2\pi} d\phi \int_0^{\pi} d\theta\, \sin\theta\, [Y_{lm}(\theta, \phi)]^* \, Y_{l'm'}(\theta, \phi)$$
$$= \delta_{ll'} \delta_{mm'} \qquad (2.86)$$

and constitute a complete basis for expanding functions defined on the unit sphere.

Upon substitution of eqn (2.80) into eqn (2.79), the general solution of the rotational diffusion equation can be written

$$H_0(\mathbf{u}, s) = \int d\mathbf{u}'\, G(\mathbf{u}, \mathbf{u}', s) H_0(\mathbf{u}', 0) \qquad (2.87)$$

The kernel of this expression is a Green function with the following spherical harmonic expansion:

$$G(\mathbf{u}, \mathbf{u}', s) = \sum_{l=0}^{\infty} \sum_{m=-l}^{l} [Y_{lm}(\mathbf{u}')]^* \, Y_{lm}(\mathbf{u}) \exp[-sl(l+1)/(2\lambda)] \qquad (2.88)$$

$G(\mathbf{u}, \mathbf{u}', s)$ describes the conditional probability that a chain section of contour length $s$ has its terminal end with orientation $\mathbf{u}$ and its originating end with orientation $\mathbf{u}'$. This Green function is particularly useful for analyzing the statistical properties of ideal wormlike chains.

A quantity of key interest is the *orientational correlation function*

$$\langle \mathbf{u}(s) \cdot \mathbf{u}(s') \rangle = \frac{1}{4\pi} \int d\mathbf{u} \int d\mathbf{u}' \, \mathbf{u} \cdot \mathbf{u}' \, G(\mathbf{u}, \mathbf{u}', |s - s'|) \qquad (2.89)$$

For computing such correlation functions, a convenient formula is the *spherical harmonic addition theorem* (Edmonds, 1974)

$$P_l(\mathbf{u} \cdot \mathbf{u}') = \frac{4\pi}{2l+1} \sum_{m=-l}^{l} [Y_{lm}(\mathbf{u})]^* \, Y_{lm}(\mathbf{u}') \qquad (2.90)$$

Use of this expression and the spherical harmonic expansion of the Green function in eqn (2.89) leads to

$$\langle \mathbf{u}(s) \cdot \mathbf{u}(s') \rangle = \exp(-|s - s'|/\lambda) \qquad (2.91)$$

The physical interpretation of the microscopic length $\lambda$ is now clear; $\lambda$ is the *persistence length*, i.e. the distance along the contour of a wormlike chain over which orientational correlations decay.

The above result can be used to deduce the mean-squared end-to-end vector for the wormlike chain. From the expression

$$R^2 \equiv \langle |\mathbf{r}(L_c) - \mathbf{r}(0)|^2 \rangle = \int_0^{L_c} ds \int_0^{L_c} ds' \, \langle \mathbf{u}(s) \cdot \mathbf{u}(s') \rangle \qquad (2.92)$$

we obtain upon insertion of eqn (2.91)

$$R^2 = 2\lambda\{L_c - \lambda[1 - \exp(-L_c/\lambda)]\} \qquad (2.93)$$

This expression is seen to continuously interpolate between the properties of a flexible, ideal chain and a rigid rod. In the flexible limit, which corresponds to a contour length much greater than the persistence length, $L_c/\lambda \gg 1$, eqn (2.93) reduces to

$$R \approx (2\lambda L_c)^{1/2} \quad (L_c/\lambda \gg 1) \qquad (2.94)$$

This is consistent with the ideal chain scaling formula $R = bN^{1/2}$ for freely jointed chains if we choose to interpret $N = L_c/\lambda$ as the number of statistically independent persistent segments and select $b = \sqrt{2}\lambda$. In the opposite limit of $L_c/\lambda \ll 1$, eqn (2.93) reduces to

$$R \approx L_c \quad (L_c/\lambda \ll 1) \tag{2.95}$$

which is evidently the exact result for a rigid rod polymer. Thus, by varying the two parameters $\lambda$ and $L_c$ in the wormlike chain model it is possible to describe the statistical mechanics of ideal polymers with a wide range of backbone flexibilities.

Before leaving the topic of the wormlike chain, it should be noted that a number of authors have considered variants of the model in which the constraint that $\mathbf{u}(s)$ be a unit vector is locally relaxed (Freed, 1972; Harris and Hearst, 1966). In these variants, a Lagrange multiplier is introduced so that $\mathbf{u}(s)$ has unit magnitude in some global averaged way. A seemingly more natural approach (Tagami, 1969; Saito and Namiki, 1956; Yamakawa, 1997) is to replace the constraint $|\mathbf{u}(s)| = 1$ with a harmonic potential $U(\mathbf{u}) = (3/2)|\mathbf{u}|^2$ and change the rotational diffusion operator on the right-hand side of eqn (2.76) to read

$$\frac{\partial}{\partial s}p_0(\mathbf{r}, \mathbf{u}, s) = -\mathbf{u} \cdot \nabla_\mathbf{r} \, p_0(\mathbf{r}, \mathbf{u}, s) + \frac{1}{2\lambda}\nabla_\mathbf{u} \cdot [(\nabla_\mathbf{u}U + \nabla_\mathbf{u}) \, p_0(\mathbf{r}, \mathbf{u}, s)] \tag{2.96}$$

In this equation, $\mathbf{u}$ is no longer constrained to the unit sphere, so the phase space $(\mathbf{r}, \mathbf{u})$ is six dimensional, in contrast to the five dimensional space for the strict wormlike chain. It is easily verified that eqn (2.96) is consistent with a steady state solution $p_0(\mathbf{r}, \mathbf{u}, \infty) \sim \exp[-U(\mathbf{u})] = \exp[-(3/2)|\mathbf{u}|^2]$. Thus, $\langle|\mathbf{u}|^2\rangle = 1$ in the steady state. This Fokker–Planck equation for an "almost wormlike" chain is closely related to an equation describing the Brownian dynamics of a particle with inertia and has an exact, albeit complicated, closed form solution (Chandrasekhar, 1943). Because this advantage is lost when we consider more realistic cases of chains in external potentials (the subject of Chapter 3), the added computational burden of working with a higher-dimensional phase space makes the model less desirable than the true wormlike chain.

Finally, it is worth noting that the wormlike chain model has been elegantly extended to describe the tendency for helix formation (Yamakawa, 1997). This extension has significance for the construction of statistical mechanical models relevant to proteins and other biomolecules capable of forming secondary structures in solution.

## 2.6  Summary

It is helpful at this stage to review a few important concepts pertaining to ideal chains and models thereof:

- The bonding constraints of real polymer chains lead to *short-ranged interferences*, which we attempt to capture in ideal chain models. Interactions among segments of a polymer that are close in space, but far removed along the chain contour, are referred to as *long-ranged interferences*. These are responsible for the excluded volume swelling of polymers in dilute solution and will be addressed in Chapter 4.

- Chain models incorporating only short-ranged interferences exhibit asymptotic *ideal chain* scaling behavior of the average polymer size, $R \sim R_g \sim bN^{1/2}$, $N \rightarrow \infty$. The scaling exponent of $1/2$ is universal, while the prefactor $b$ depends on the details of the bonding constraints included in the model.

- In studying ideal chain models, we have seen that it is convenient to build up statistical weights using concepts from the theory of stochastic processes. A powerful approach is to write *Chapman–Kolmogorov* integral equations for reduced probability distribution functions like $p_0(\mathbf{r}, s)$ and $p_0(\mathbf{r}, \mathbf{u}, s)$, describing the distributions of position and orientation of the end segment $s$ of a polymer.

- *Continuous* chain models, such as the continuous Gaussian chain and the wormlike chain, are particularly useful because their corresponding Chapman–Kolmogorov equations can be reduced to linear partial differential equations known as *Fokker–Planck* equations. In Chapter 3 we shall see that the availability of asymptotic analytical methods and powerful numerical techniques for solving such equations makes continuous chain models especially well suited for investigations of inhomogeneous polymers.

In the present chapter we have discussed only a small subset of the ideal chain models that have been described in the literature. One important omission relates to chain models produced by inscribing polymers on an underlying *lattice*. Such models of lattice polymers form the basis for the very successful Scheutjens–Fleer (Scheutjens and Fleer, 1979; Fleer *et al.*, 1993) theory of polymer interfaces. These models, however, are not suitable for high-resolution simulations of inhomogeneous polymers in two and three dimensions. Continuum chain models are not only more physically realistic, but they lead to continuum field theories without broken symmetries imposed by an underlying lattice. With a continuum theory, one also has a much wider range of options for representing the fields, including spectral, finite difference, and (structured and unstructured mesh) finite element techniques. We may *choose* to do our numerical calculations on a computational lattice, but the lattice type and spacing is then a matter of numerical convenience and accuracy, rather than a foundation of the model. Indeed, with lattice polymers it is difficult to depart from the notion of each lattice site being occupied by a Kuhn segment, of size $b$. In contrast, using continuum models it is sometimes possible to compute accurately on a computational grid having a spacing comparable to the radius of gyration, $R_g \sim bN^{1/2}$, especially in situations where the inhomogeneities are weak or slowly varying in space. This feature can obviously amount to a very significant computational advantage when attempting high-resolution simulations of a three-dimensional polymeric fluid.

Another important class of ideal chain models are the *rotational isomeric state* (RIS) models (Flory, 1969). These are continuum models that incorporate more realistic local bonding constraints than the freely jointed, bead-spring, or continuous chain models described above. In the RIS scheme, the continuous rotational degrees of freedom about backbone single bonds in the polymer are replaced by a finite number of rotational isomers, i.e. *trans* and *gauche* states. The statistical weights of RIS chains are built up by using transfer matrices that are closely connected to the kernels in Chapman–Komogorov equations. It is indeed possible to carry out field-theoretic computer simulations of polymer models based on RIS chains. However, the short-scale (<1 nm) liquid state physics associated with such local bonding details is not a focus of the present monograph, so it is computationally advantageous to work with the continuous Gaussian and wormlike chain models.

In view of the universal scaling behavior of real and ideal chains on length scales of order 10 nm, one might ask if it matters which chain model we use in modelling inhomogeneous polymers. The answer depends on the characteristic scale of the structures that we hope to describe. In a two-phase polymer melt where the polymers are partially miscible, interfacial widths might be as broad as 5–10 nm. In such a case virtually any of the chain models described in the present chapter could be successfully employed. However, for a strongly immiscible polymer blend where interfacial widths are below 1 nm, a chain model that captures the local rigidity of polymer segments passing through the interface, such as the freely jointed chain or the wormlike chain, would be most appropriate. In such circumstances, we must acknowledge that no *coarse-grained* chain model is going to accurately describe the structure and thermodynamics of individual interfaces. However, if the computational focus is not on the properties of individual interfaces, but rather on the larger-scale self-assembly characteristics of a polymeric fluid, the approximate description of individual polymer–polymer interfaces provided by coarse-grained chain models will generally suffice.

# 3

# SINGLE CHAINS IN EXTERNAL FIELDS

In the previous chapter we described a variety of models for the statistical mechanics of ideal chains. Here the models are generalized to include one or more potential fields that act on individual segments of a polymer chain. In this chapter the added fields will be treated as "external" in that they can be arbitrarily specified. However, we shall see in Chapter 4 that the potential fields of primary importance are those that are generated *self-consistently* by the force fields of surrounding polymer segments.

The exposition in the present chapter is quite detailed because of the importance of the subject matter to the theory of inhomogeneous polymers. Accurate evaluation of the statistical mechanics of a single polymer in a prescribed potential field proves to be the most computationally demanding component of a field-based computer simulation and is an essential ingredient of a successful analytical theory.

## 3.1 Partition and distribution functions

Our first task is to discuss how the partition functions and distribution functions of ideal chain models are modified by the presence of an external field. We begin with the discrete Gaussian chain.

### 3.1.1 *Discrete Gaussian chain*

The external field of primary interest is a spatially varying *chemical potential* field $w(\mathbf{r})$ that acts indiscriminately on the $N + 1$ beads of a discrete Gaussian chain. We again adopt the notation of Fig. 2.1 and Section 2.3. The potential energy can be written[15]

$$U(\mathbf{r}^{N+1}) = U_0(\mathbf{r}^{N+1}) + U_1(\mathbf{r}^{N+1})$$
$$= \sum_{i=1}^{N} h(|\mathbf{r}_i - \mathbf{r}_{i-1}|) + k_B T \sum_{i=0}^{N} w(\mathbf{r}_i) \qquad (3.1)$$

where $h(x) = 3k_B T x^2/(2b^2)$ and $\mathbf{r}^{N+1}$ is shorthand for the $N + 1$ bead coordinates $(\mathbf{r}_0, \mathbf{r}_1, ..., \mathbf{r}_N)$. The first sum over the $N$ bonds is $U_0$, the harmonic stretching energy of the discrete Gaussian chain. The second sum over the $N+1$

---

[15]The subscripts "0" will be omitted in this chapter, because chains subjected to external potentials will in general no longer satisfy ideal chain scaling.

beads accounts for the interaction energy of each bead (segment) with the potential field $k_B T w(\mathbf{r})$. Another way to express the external potential term $U_1(\mathbf{r}^{N+1})$ is

$$\beta U_1(\mathbf{r}^{N+1}) = \int d\mathbf{r} \; w(\mathbf{r})\hat{\rho}(\mathbf{r}) \tag{3.2}$$

where the *microscopic density of segments* is defined by

$$\hat{\rho}(\mathbf{r}) = \sum_{i=0}^{N} \delta(\mathbf{r} - \mathbf{r}_i) \tag{3.3}$$

Such a microscopic density is evidently a very singular function of $\mathbf{r}$ and depends explicitly on the bead coordinates $\mathbf{r}^{N+1}$. Equation (3.2) expresses the fact that the potential $w(\mathbf{r})$ can be viewed as a spatially varying *chemical potential* field that is *thermodynamically conjugate* to the segment density (Chandler, 1987).

Of particular interest is the *ratio* of the partition function for a chain subjected to an external potential $w(\mathbf{r})$, $Z[w]$, to the partition function of an ideal chain, $Z_0$,

$$Q[w] \equiv \frac{Z[w]}{Z_0} = \frac{\int d\mathbf{r}^{N+1} \, \exp[-\beta U(\mathbf{r}^{N+1})]}{V \left( \int d\mathbf{b} \, \exp[-\beta h(|\mathbf{b}|)] \right)^N} \tag{3.4}$$

In the denominator of this expression, use has been made of eqn (2.27), which expresses $Z_0$ as the volume $V$ times $N$ independent (volume) integrals over the bond vectors. As the notation implies, the normalized partition function $Q[w]$ can be viewed as a *functional* of the external potential $w(\mathbf{r})$.

A next step is to associate the $N$ factors of $\int d\mathbf{b} \, \exp[-\beta h(|\mathbf{b}|)]$ in the denominator of eqn (3.4) with the $N$ factors of $\exp[-\beta h(|\mathbf{r}_i - \mathbf{r}_{i-1}|)]$ in the numerator. Recalling the definition of the normalized bond transition probability for the discrete Gaussian chain, eqn (2.34),

$$\Phi(\mathbf{r}) = \frac{\exp[-\beta h(|\mathbf{r}|)]}{\int d\mathbf{r} \exp[-\beta h(|\mathbf{r}|)]} = \left( \frac{3}{2\pi b^2} \right)^{3/2} \exp\left( -\frac{3|\mathbf{r}|^2}{2b^2} \right) \tag{3.5}$$

enables eqn (3.4) to be rewritten as

$$Q[w] = \frac{1}{V} \int d\mathbf{r}^{N+1} \left[ e^{-w(\mathbf{r}_N)} \Phi(\mathbf{r}_N - \mathbf{r}_{N-1}) e^{-w(\mathbf{r}_{N-1})} \Phi(\mathbf{r}_{N-1} - \mathbf{r}_{N-2}) \right.$$
$$\left. \dots \; e^{-w(\mathbf{r}_2)} \Phi(\mathbf{r}_2 - \mathbf{r}_1) e^{-w(\mathbf{r}_1)} \Phi(\mathbf{r}_1 - \mathbf{r}_0) e^{-w(\mathbf{r}_0)} \right] \tag{3.6}$$

This expression can be built up recursively in the following way. We define a functional $q(\mathbf{r}, j; [w])$ for integer $j$ according to

$$q(\mathbf{r}, 0; [w]) = \exp[-w(\mathbf{r})] \tag{3.7}$$

and for $j = 0, 1, 2, ..., N - 1$ by

$$q(\mathbf{r}, j+1; [w]) = \exp[-w(\mathbf{r})] \int d\mathbf{r}' \; \Phi(\mathbf{r} - \mathbf{r}') q(\mathbf{r}', j; [w]) \qquad (3.8)$$

It follows that the normalized partition function can be expressed as

$$Q[w] = \frac{1}{V} \int d\mathbf{r} \; q(\mathbf{r}, N; [w]) \qquad (3.9)$$

In the above, $q(\mathbf{r}, j; [w])$ represents the statistical weight for a chain of $j + 1$ beads to have its end at position $\mathbf{r}$. This object is commonly referred to as a *chain propagator* and is a functional of the external potential field $w(\mathbf{r})$. Equation (3.8) can be viewed as a Chapman–Kolmogorov equation that is strictly analogous to eqn (2.33) for the probability density $p_0(\mathbf{r}, j)$ in the field-free case. Indeed, apart from a different normalization of the functions $p_0(\mathbf{r}, j)$ and $q(\mathbf{r}, j; [w])$, eqn (3.8) is seen to reduce to eqn (2.33) for $w(\mathbf{r}) \to 0$.

It is important to note that eqns (3.7)–(3.9) hold for *any* spring potential $h(x)$, provided that the transition probability density $\Phi(\mathbf{r})$ of the Gaussian model, eqn (3.5), is replaced with a suitable form, cf. eqn (2.40). Thus, eqns (3.7)–(3.9) are applicable for *nonlinear* bead-spring models such as eqn (2.41).

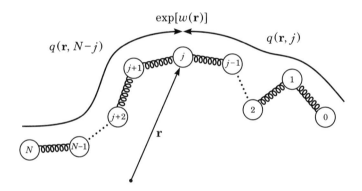

FIG. 3.1. The physical content of the factorization property eqn (3.10). A propagator with $j + 1$ beads can be joined at the position of the $j$th bead, $\mathbf{r} = \mathbf{r}_j$, to a propagator with $N - j + 1$ beads in order to generate the overall statistical weight of a chain. A factor of $\exp[w(\mathbf{r})]$ is required at the junction point.

The normalized partition function $Q[w]$ has an important *factorization property* that is illustrated in Fig. 3.1. If we choose to factor eqn (3.6) at bead $j$, then it is evident that $Q[w]$ can be written

$$Q[w] = \frac{1}{V} \int d\mathbf{r} \; q(\mathbf{r}, N - j; [w]) \; \exp[w(\mathbf{r})] \; q(\mathbf{r}, j; [w]) \qquad (3.10)$$

The physical interpretation of this expression is that the statistical weight of a $(N + 1)$-bead chain can be built up by joining the propagator for the end distribution of a chain with $j+1$ beads and the propagator for a "complementary" chain with $N - j + 1$ beads. The propagators are connected at the position of the $j$th bead, $\mathbf{r} = \mathbf{r}_j$, and a factor of $\exp[w(\mathbf{r})]$ is included to cancel an excess factor of $\exp[-w(\mathbf{r})]$ associated with the two joined ends.

A second type of external potential that is of importance in studying the mechanical response of polymers is a *strain field*. Here we restrict attention to *homogeneous* strains $\epsilon(\mathbf{r}) = \epsilon$. To see how this symmetric tensor field couples to the statistical mechanics of a polymer, it is useful to note that a microscopic expression for the *elastic stress tensor*[16] of a discrete Gaussian chain can be written (Doi and Edwards, 1986)

$$\hat{\sigma}_{\alpha\gamma}(\mathbf{r}) = \frac{3k_BT}{b^2} \sum_{i=1}^{N} \delta(\mathbf{r} - \mathbf{r}_{i-1})(\mathbf{r}_i - \mathbf{r}_{i-1})_\alpha(\mathbf{r}_i - \mathbf{r}_{i-1})_\gamma \qquad (3.11)$$

where Greek subscripts are again used to denote the Cartesian indices of tensors. This expression sums the elastic force per unit area associated with the polymer bonds crossing a prescribed plane.[17] Using the fact that stress and strain are conjugate thermodynamic variables (Landau and Lifshitz, 1986), the elastic energy associated with an imposed strain $\epsilon$ can be written

$$U_{el}(\mathbf{r}^{N+1}) = \int d\mathbf{r} \ \epsilon : \hat{\sigma}(\mathbf{r}) \qquad (3.12)$$

Of particular interest is the partition function, $Z[w, \epsilon]$, of a discrete Gaussian chain subjected to both chemical potential and strain fields. The most convenient normalization in this case is supplied by the partition function of a chain with $w = 0$, but non-vanishing strain:

$$Q[w, \epsilon] \equiv \frac{Z[w, \epsilon]}{Z[0, \epsilon]} = \frac{\int d\mathbf{r}^{N+1} \exp[-\beta U(\mathbf{r}^{N+1}) - \beta U_{el}(\mathbf{r}^{N+1})]}{\int d\mathbf{r}^{N+1} \exp[-\beta U_0(\mathbf{r}^{N+1}) - \beta U_{el}(\mathbf{r}^{N+1})]} \qquad (3.13)$$

The arguments leading to eqns (3.7)–(3.9) still apply, but with a change in the bond transition probability to

$$\Psi(\mathbf{r}; [\epsilon]) = \left(\frac{3}{2\pi b^2}\right)^{3/2} [\det(\mathbf{1} + 2\epsilon)]^{1/2} \exp\left(-\frac{3}{2b^2}(\mathbf{1} + 2\epsilon) : \mathbf{r}\,\mathbf{r}\right) \qquad (3.14)$$

and where $\mathbf{1}$ denotes the unit tensor. Use of eqn (B.11) was made in deriving this expression. As in the strain-free case, $Q[w, \epsilon]$ can be computed according to

$$Q[w, \epsilon] = \frac{1}{V} \int d\mathbf{r} \ q(\mathbf{r}, N; [w, \epsilon]) \qquad (3.15)$$

---

[16]The term "elastic stress" is used with the recognition that the source of the elasticity in a coarse-grained bead-spring model may be conformational entropy.

[17]We have arbitrarily chosen to locate the stress contributed from the $i$th bond on bead $i-1$. Other choices are possible.

with

$$q(\mathbf{r}, 0; [w, \boldsymbol{\epsilon}]) = \exp[-w(\mathbf{r})] \tag{3.16}$$

and

$$q(\mathbf{r}, j+1; [w, \boldsymbol{\epsilon}]) = \exp[-w(\mathbf{r})] \int d\mathbf{r}' \ \Psi(\mathbf{r} - \mathbf{r}'; [\boldsymbol{\epsilon}]) \ q(\mathbf{r}', j; [w, \boldsymbol{\epsilon}]) \tag{3.17}$$

for $j = 0, 1, 2, ..., N-1$. These equations evidently reduce to eqns (3.7)–(3.9) in the limit of vanishing strain, $\boldsymbol{\epsilon} \to 0$, and to eqn (2.33) in the case of vanishing strain and chemical potential, $\boldsymbol{\epsilon}, w \to 0$.

### 3.1.2 Continuous Gaussian chain

The formulas of the previous section can be readily extended to the continuous Gaussian chain model. The microscopic segment density is changed from eqn (3.3) to

$$\hat{\rho}(\mathbf{r}) = \int_0^N ds \ \delta(\mathbf{r} - \mathbf{r}(s)) \tag{3.18}$$

and the potential energy associated with an imposed chemical potential field $w(\mathbf{r})$ becomes a functional of the polymer shape $\mathbf{r}(s)$:

$$\beta U_1[\mathbf{r}, w] = \int d\mathbf{r}' \ w(\mathbf{r}') \hat{\rho}(\mathbf{r}') \tag{3.19}$$

The normalized partition function $Q[w]$ can be expressed as a ratio of path integrals

$$Q[w] \equiv \frac{Z[w]}{Z_0} = \frac{\int \mathcal{D}\mathbf{r} \ \exp(-\beta U_0[\mathbf{r}] - \beta U_1[\mathbf{r}, w])}{\int \mathcal{D}\mathbf{r} \ \exp(-\beta U_0[\mathbf{r}])} \tag{3.20}$$

With this definition, the singularities noted previously for $Z_0$ exactly cancel from numerator and denominator, rendering $Q[w]$ finite in the continuous chain limit. Explicitly, we discretize the two path integrals according to eqn (2.45) by using $N_s + 1$ beads and $N_s$ springs, and retrace the steps for the discrete Gaussian chain leading to eqn (3.6). It follows that eqn (3.20) reduces to

$$Q[w] = \frac{1}{V} \int d\mathbf{r}^{N_s+1} \left[ e^{-\Delta s \, w(\mathbf{r}_{N_s})} \Phi(\mathbf{r}_{N_s} - \mathbf{r}_{N_s-1}) e^{-\Delta s \, w(\mathbf{r}_{N_s-1})} \right.$$
$$\times \quad \Phi(\mathbf{r}_{N_s-1} - \mathbf{r}_{N_s-2}) \ ... \ e^{-\Delta s \, w(\mathbf{r}_2)}$$
$$\times \quad \left. \Phi(\mathbf{r}_2 - \mathbf{r}_1) e^{-\Delta s \, w(\mathbf{r}_1)} \Phi(\mathbf{r}_1 - \mathbf{r}_0) e^{-\Delta s \, w(\mathbf{r}_0)} \right] \tag{3.21}$$

where $\Phi(\mathbf{r} - \mathbf{r}') = \Phi(\Delta \mathbf{r})$ is given by eqn (2.59) and $\Delta s \equiv N/N_s$. As before, the expression for $Q[w]$ can be written as

$$Q[w] = \frac{1}{V} \int d\mathbf{r} \ q(\mathbf{r}, N; [w]) \tag{3.22}$$

where

$$q(\mathbf{r}, 0; [w]) = \exp[-\Delta s\, w(\mathbf{r})] \tag{3.23}$$

and

$$q(\mathbf{r}, s + \Delta s; [w]) = \exp[-\Delta s\, w(\mathbf{r})] \int d\mathbf{r}'\, \Phi(\mathbf{r} - \mathbf{r}') q(\mathbf{r}', s; [w]) \tag{3.24}$$

are used to step forward along the chain from $s = 0$ to $s = N$ in $N_s$ increments of $\Delta s$.

Equation (3.24) is a Chapman–Kolmogorov equation that is closely related to eqn (2.58) in the field-free case. As such, it can be converted in the continuum limit of $\Delta s \to 0$ to a Fokker–Planck equation. Taylor-expanding both sides of eqn (3.24) to first order in $\Delta s$ and the integrand to second order[18] in $\Delta \mathbf{r} = \mathbf{r} - \mathbf{r}'$, one finds that (Edwards, 1965; de Gennes, 1969; Freed, 1972)

$$\frac{\partial}{\partial s} q(\mathbf{r}, s; [w]) = \frac{b^2}{6} \nabla^2 q(\mathbf{r}, s; [w]) - w(\mathbf{r}) q(\mathbf{r}, s; [w]) \tag{3.25}$$

which can be viewed as a generalization of eqn (2.64) to include an external potential. The "initial condition" eqn (3.23) reduces in the continuum limit to the uniform condition

$$q(\mathbf{r}, 0; [w]) = 1 \tag{3.26}$$

The Fokker–Planck equation (3.25) is commonly referred to as a *modified diffusion equation*, and sometimes, by analogy with the path integral formulation of quantum mechanics, as a *Feynman–Kac formula* (Feynman and Hibbs, 1965). Together with eqn (3.22), the modified diffusion equation fully describes the statistical mechanics of the continuous Gaussian chain in an external potential $w(\mathbf{r})$. This is one of the most important results in the theory of inhomogeneous polymers.

The continuous Gaussian chain in a potential $w(\mathbf{r})$ has a *factorization property* analogous to eqn (3.10) for the discrete Gaussian chain. In particular,

$$Q[w] = \frac{1}{V} \int d\mathbf{r}\, q(\mathbf{r}, N - s; [w])\, q(\mathbf{r}, s; [w]) \tag{3.27}$$

where the path integral has been factored at an arbitrary contour position $0 < s < N$. The exponential factor $\exp[\Delta s w(\mathbf{r})]$ necessary to cancel the over-counting by a factor of $\exp[-\Delta s\, w(\mathbf{r})]$ at the joined ends is absent since $\exp[\Delta s\, w(\mathbf{r})] \to 1$ in the continuum limit of $\Delta s \to 0$.

These results are easily extended to the case of a continuous Gaussian chain subjected to a homogeneous *strain field* $\boldsymbol{\epsilon}$, in addition to the chemical potential

---

[18]It should be noted that according to eqn (2.63) $\langle \Delta \mathbf{r} \Delta \mathbf{r} \rangle_\Phi = O(\Delta s)$, so we are effectively expanding both sides to $O(\Delta s)$.

$w(\mathbf{r})$. The microscopic elastic stress expression (3.11) is modified for a continuous Gaussian chain to

$$\hat{\sigma}_{\alpha\gamma}(\mathbf{r}) = \frac{3k_B T}{b^2} \int_0^N ds \ \delta(\mathbf{r} - \mathbf{r}(s)) \frac{d\mathbf{r}_\alpha(s)}{ds} \frac{d\mathbf{r}_\gamma(s)}{ds} \tag{3.28}$$

and the elastic energy can be expressed as a functional of $\mathbf{r}(s)$ and $\boldsymbol{\epsilon}$

$$U_{el}[\mathbf{r}, \boldsymbol{\epsilon}] = \int d\mathbf{r}' \ \boldsymbol{\epsilon} : \hat{\boldsymbol{\sigma}}(\mathbf{r}') \tag{3.29}$$

The normalized partition function for a continuous chain experiencing both strain and chemical potential fields is

$$Q[w, \boldsymbol{\epsilon}] \equiv \frac{Z[w, \boldsymbol{\epsilon}]}{Z[0, \boldsymbol{\epsilon}]} = \frac{\int \mathcal{D}\mathbf{r} \ \exp(-\beta U_0[\mathbf{r}] - \beta U_1[\mathbf{r}, w] - \beta U_{el}[\mathbf{r}, \boldsymbol{\epsilon}])}{\int \mathcal{D}\mathbf{r} \ \exp(-\beta U_0[\mathbf{r}] - \beta U_{el}[\mathbf{r}, \boldsymbol{\epsilon}])} \tag{3.30}$$

Again, the normalization is such that $Q[0, \boldsymbol{\epsilon}] = 1$. Discretizing the path integrals in numerator and denominator according to eqn (2.45) and retracing the steps leading to eqn (3.25) produces the following result (Fredrickson, 2002):

$$Q[w, \boldsymbol{\epsilon}] = \frac{1}{V} \int d\mathbf{r} \ q(\mathbf{r}, N; [w, \boldsymbol{\epsilon}]) \tag{3.31}$$

where the propagator $q(\mathbf{r}, s; [w, \boldsymbol{\epsilon}])$ satisfies the Fokker–Planck equation

$$\frac{\partial}{\partial s} q(\mathbf{r}, s; [w, \boldsymbol{\epsilon}]) = - w(\mathbf{r}) \ q(\mathbf{r}, s; [w, \boldsymbol{\epsilon}])$$
$$+ \frac{b^2}{6} (\mathbf{1} + 2\boldsymbol{\epsilon})^{-1} : \nabla\nabla \ q(\mathbf{r}, s; [w, \boldsymbol{\epsilon}]) \tag{3.32}$$

This equation is subject to the initial condition

$$q(\mathbf{r}, 0; [w, \boldsymbol{\epsilon}]) = 1 \tag{3.33}$$

The Fokker–Planck equation (3.32) is seen to reduce to eqn (3.25) for the case of vanishing strain. At finite $\boldsymbol{\epsilon}$, however, it describes an anisotropic diffusion process consistent with a polymer chain experiencing simultaneous chemical potential and strain fields. These results can be further generalized to the case of an *inhomogeneous* strain field $\boldsymbol{\epsilon}(\mathbf{r})$ (Fredrickson, 2002).

### 3.1.3 Wormlike chain

The wormlike chain model can be similarly extended to include interactions with an external field. Unlike Gaussian chain models, the wormlike chain is often used to describe polymers with local rigidity that are capable of exhibiting liquid crystalline order. Such liquid crystallinity derives at a microscopic level from anisotropic segment–segment interactions. Thus, it is important to allow the external "chemical" potential in the present case to depend on both segment position and orientation, i.e. $w = w(\mathbf{r}, \mathbf{u})$. The interaction energy of a wormlike chain with such a field can be written

$$\beta U_1[\mathbf{r}, w] = \int_0^{L_c} ds\, w(\mathbf{r}(s), \mathbf{u}(s))$$

$$= \int d\mathbf{r}' \int d\mathbf{u}'\, w(\mathbf{r}', \mathbf{u}')\hat{\rho}(\mathbf{r}', \mathbf{u}') \qquad (3.34)$$

where

$$\hat{\rho}(\mathbf{r}, \mathbf{u}) \equiv \int_0^{L_c} ds\, \delta(\mathbf{r} - \mathbf{r}(s))\delta(\mathbf{u} - \mathbf{u}(s)) \qquad (3.35)$$

is a microscopic density of segment position and orientation. The potential $w(\mathbf{r}, \mathbf{u})$ is thermodynamically conjugate to this density.

The partition function of a wormlike chain subjected to a potential $w(\mathbf{r}, \mathbf{u})$, $Z[w]$, normalized by its value in the absence of the potential, $Z_0$, can be written

$$Q[w] \equiv \frac{Z[w]}{Z_0} = \frac{\int^* \mathcal{D}\mathbf{r}\, \exp(-\beta U_0[\mathbf{u}] - \beta U_1[\mathbf{r}, w])}{\int^* \mathcal{D}\mathbf{r}\, \exp(-\beta U_0[\mathbf{u}])} \qquad (3.36)$$

where $U_0[\mathbf{u}]$ is the bending energy given in eqn (2.66). The asterisks on the path integrals in numerator and denominator are a shorthand to indicate that the constraint of eqn (2.67), i.e.

$$\prod_s \delta\left(\mathbf{u}(s) - \frac{d\mathbf{r}(s)}{ds}\right)\delta(|\mathbf{u}(s)| - 1), \qquad (3.37)$$

is imposed on the chain conformations $\mathbf{r}(s)$.

Equation (3.36) can be evaluated by discretizing the path integrals in numerator and denominator and deducing a Chapman–Kolmogorov equation similar to (2.71) for a chain propagator $q(\mathbf{r}, \mathbf{u}, s; [w])$. This quantity describes the statistical weight that a wormlike chain of contour length $s$, and experiencing a potential $w(\mathbf{r}, \mathbf{u})$, has its end segment at position $\mathbf{r}$ and with orientation $\mathbf{u}$. The normalized partition function is obtained from

$$Q[w] = \frac{1}{4\pi V} \int d\mathbf{r} \int d\mathbf{u}\, q(\mathbf{r}, \mathbf{u}, L_c; [w]) \qquad (3.38)$$

In the continuum limit, one finds that the wormlike chain propagator $q(\mathbf{r}, \mathbf{u}, s; [w])$ satisfies a Fokker–Planck equation similar to eqn (2.76) for the field-free case:

$$\frac{\partial}{\partial s} q(\mathbf{r}, \mathbf{u}, s; [w]) = -\, w(\mathbf{r}, \mathbf{u})\, q(\mathbf{r}, \mathbf{u}, s; [w]) - \mathbf{u} \cdot \nabla_{\mathbf{r}}\, q(\mathbf{r}, \mathbf{u}, s; [w])$$

$$+ \frac{1}{2\lambda} \nabla_{\mathbf{u}}^2\, q(\mathbf{r}, \mathbf{u}, s; [w]) \qquad (3.39)$$

This equation is to be solved subject to the initial condition

$$q(\mathbf{r}, \mathbf{u}, 0; [w]) = 1 \qquad (3.40)$$

Equations (3.38)–(3.40) are the fundamental relations defining the statistical mechanics of a wormlike chain experiencing an external potential.

The normalized partition function of the wormlike chain has a *factorization property* analogous to that of the continuous Gaussian chain. From the discretized expression for $Q[w]$ it is straightforward to deduce that

$$Q[w] = \frac{1}{4\pi V} \int d\mathbf{r} \int d\mathbf{u}\, q(\mathbf{r}, -\mathbf{u}, L_c - s; [w])\, q(\mathbf{r}, \mathbf{u}, s; [w]) \qquad (3.41)$$

for any $0 \leq s \leq L_c$. Physically, this formula indicates that the partition function for a polymer of total contour length $L_c$ can be built up by composing a propagator for a chain section of contour length $s$ with a second propagator for a chain of length $L_c - s$. The statistical weight associated with the first propagator $q(\mathbf{r}, \mathbf{u}, s; [w])$ is obtained by integrating eqns (3.39)–(3.40) forward in $s$, starting at the $s = 0$ end of the polymer. The second propagator $q(\mathbf{r}, -\mathbf{u}, L_c - s; [w])$ corresponds to the statistical weight obtained by solving eqns (3.39)–(3.40) for a chain of length $L_c - s$, starting from the *other* end of the polymer. The tangent vector $\mathbf{u}$ in this weight is inverted in sign to account for the difference in direction of propagation along the chain contour.

Although we have referred to $w$ as an external *chemical* potential field, the form $w = w(\mathbf{r}, \mathbf{u})$ adopted is general enough to describe many types of external potentials, including electric and magnetic fields. For example, one case of interest is a chain with permanent electric dipoles along the backbone that are collinear with the tangent vector $\mathbf{u}(s)$, i.e. $\boldsymbol{\mu}(s) = \mu_0 \mathbf{u}(s)$. In this case the electrostatic energy associated with an imposed stationary electric field $\mathbf{E}(\mathbf{r})$ is described by eqn (3.34) with

$$w(\mathbf{r}, \mathbf{u}) = -\beta \mu_0 \mathbf{E}(\mathbf{r}) \cdot \mathbf{u} \qquad (3.42)$$

As another example, a dielectric wormlike chain subjected to a static electric field $\mathbf{E}(\mathbf{r})$ can be described by a potential

$$w(\mathbf{r}, \mathbf{u}) = -\frac{\beta}{2} \mathbf{E}(\mathbf{r}) \cdot \boldsymbol{\alpha}(\mathbf{u}) \cdot \mathbf{E}(\mathbf{r}) \qquad (3.43)$$

where $\boldsymbol{\alpha}(\mathbf{u})$ is the polarizability tensor of a differential chain segment of orientation $\mathbf{u}$. In the special case of cylindrical symmetry of the polarizability tensor about $\mathbf{u}$, $\boldsymbol{\alpha}(\mathbf{u})$ can be expressed as

$$\boldsymbol{\alpha}(\mathbf{u}) = \alpha_{||}\, \mathbf{u}\mathbf{u} + \alpha_\perp \left(1 - \mathbf{u}\mathbf{u}\right) \qquad (3.44)$$

where $\alpha_{||}$ and $\alpha_\perp$ are the polarizabilities of a differential chain segment parallel and perpendicular, respectively, to the tangent vector $\mathbf{u}$.

Similar expressions can be written to describe interactions between wormlike chains and a broad class of static and time-varying external fields. We shall defer the discussion of anisotropic potential fields responsible for liquid crystallinity to Section 4.6.

Unfortunately, the three-body character of bending forces along the wormlike chain complicates the identification of a microscopic stress tensor. Until this difficulty is overcome, it is not possible to explore the equilibrium response of a wormlike chain to spatially varying strain fields.

### 3.1.4 *Rodlike polymer*

The wormlike chain model has the attractive property that it continuously interpolates between a rigid rod and a flexible coil according to the ratio of the persistence and contour lengths, $\lambda/L_c$. It can thus be used to describe macromolecules of arbitrary backbone flexibility. A serious disadvantage of the model, however, is that the evaluation of thermodynamic and structural quantities requires a propagator $q(\mathbf{r}, \mathbf{u}, s)$ that has *two* additional internal coordinates (the angles defining $\mathbf{u}$) beyond those of the propagator $q(\mathbf{r}, s)$ applied in the Gaussian chain model. It will become apparent in Section 3.6 that these extra coordinates add considerably to the cost of numerical computations with the wormlike chain model. Thus for situations of nearly flexible polymers, $\lambda/L_c \ll 1$, it is highly desirable to substitute the continuous Gaussian chain or a bead-spring chain model for the wormlike chain. Similarly, in the case of nearly rigid rod polymers, $\lambda/L_c \gtrsim 1$, a strictly *rodlike polymer model* can be used in place of the wormlike chain. For a rod of length $L_c$ with end at position $\mathbf{r}$ and having orientation $\mathbf{u}$, the normalized partition function is given by

$$Q[w] = \frac{1}{4\pi V} \int d\mathbf{r} \int d\mathbf{u} \, \exp\left[-\int_0^{L_c} ds \, w(\mathbf{r} + s\mathbf{u}, \mathbf{u})\right] \tag{3.45}$$

Given some potential $w(\mathbf{r}, \mathbf{u})$, eqn (3.45) is evidently more straightforward to evaluate than the corresponding eqns (3.38)–(3.40) for the wormlike chain.

## 3.2  Single-chain averages and operators

In the previous sections of this chapter we described how the partition functions for single chains in the presence of external fields can be computed using either Chapman–Kolmogorov or Fokker–Planck equations. While the single-chain partition function and reduced distribution functions (chain propagators) are central objects in the theory of inhomogeneous polymers, it is necessary to be able to express other "observable" quantities as explicit functionals of the external field variables. The quantities of primary interest are related to ensemble averages over the configurational/conformational degrees of freedom of a single polymer experiencing an external potential. For example, in the case of a discrete Gaussian chain subject to a chemical potential $w(\mathbf{r})$, we can define the *single-chain average* of an arbitrary function $f(\mathbf{r}^{N+1})$ of the bead coordinates by

$$\langle f(\mathbf{r}^{N+1}) \rangle_{[w]} = \frac{\int d\mathbf{r}^{N+1} \, f(\mathbf{r}^{N+1}) \exp[-\beta U(\mathbf{r}^{N+1})]}{\int d\mathbf{r}^{N+1} \, \exp[-\beta U(\mathbf{r}^{N+1})]} \tag{3.46}$$

where $U(\mathbf{r}^{N+1})$ is the potential energy given in eqn (3.1). The subscript $[w]$ on the angular brackets indicates that after averaging over the bead coordinates, the averaged quantity can be viewed as a functional of the external field $w(\mathbf{r})$. Similarly, for the *continuous* Gaussian chain in the presence of chemical potential and strain fields $w(\mathbf{r})$ and $\epsilon$, a corresponding single-chain average can be defined as

$$\langle f[\mathbf{r}]\rangle_{[w,\epsilon]} = \frac{\int \mathcal{D}\mathbf{r}\ f[\mathbf{r}] \exp(-\beta U_0[\mathbf{r}] - \beta U_1[\mathbf{r}, w] - \beta U_{el}[\mathbf{r}, \epsilon])}{\int \mathcal{D}\mathbf{r}\ \exp(-\beta U_0[\mathbf{r}] - \beta U_1[\mathbf{r}, w] - \beta U_{el}[\mathbf{r}, \epsilon])} \tag{3.47}$$

where $f[\mathbf{r}]$ is an arbitrary functional of the chain configuration $\mathbf{r}(s)$.

The single-chain average quantities of primary interest are segment densities, density–density correlations, and elastic stresses, all of which can be connected to observables in experiments on inhomogeneous polymers. We shall see that in some cases these can be obtained by taking functional derivatives of the normalized partition functions $Q[w]$ and $Q[w, \epsilon]$ with respect to the field variables $w(\mathbf{r})$ and $\epsilon$. Readers unfamiliar with functional differentiation are advised to consult Appendix C for a brief introduction.

### 3.2.1 Density operators

To begin, we consider the problem of calculating the average segment number density for a single flexible polymer subjected to a chemical potential $w(\mathbf{r})$. This quantity is defined by

$$\rho(\mathbf{r}; [w]) \equiv \langle \hat{\rho}(\mathbf{r}) \rangle_{[w]} \tag{3.48}$$

where the microscopic density $\hat{\rho}(\mathbf{r})$ is given by eqn (3.3) or (3.18), depending on whether the discrete or continuous Gaussian chain model is employed. We shall refer to $\rho(\mathbf{r}; [w])$ as a single-chain density *operator* because it expresses the average segment density as a functional of the imposed field $w(\mathbf{r})$. We shall see in Chapter 4 that the average monomer density in an interacting fluid with many polymers is proportional to an average of the operator $\rho(\mathbf{r}; [w])$ over a suitable distribution of field configurations $w(\mathbf{r})$. The field in this case is not externally imposed but is generated internally by the interactions among non-bonded monomers. Thus, calculation of an "observable" quantity in a many-chain fluid will involve two different kinds of ensemble averages: (i) a single-chain average of the type described in eqns (3.46) and (3.47), and (ii) a field-theoretic average over the distribution of internally generated potential fields. A discussion of the second type of average will be deferred to Chapter 4.

To proceed with the evaluation of eqn (3.48), we again note that $w(\mathbf{r})$ and $\hat{\rho}(\mathbf{r})$ are thermodynamically conjugate variables. Thus, a functional derivative of $-\ln Q[w]$ with respect to $w(\mathbf{r})$ will generate a single-chain average of $\hat{\rho}(\mathbf{r})$. This follows directly, e.g. in the case of the discrete Gaussian chain, from eqn (3.4) for $Q[w]$ and the definition (3.46) of the single-chain average. Specifically,

$$-\frac{\delta \ln Q[w]}{\delta w(\mathbf{r})} = -\frac{1}{Q[w]} \frac{\delta Q[w]}{\delta w(\mathbf{r})} = \langle \hat{\rho}(\mathbf{r}) \rangle_{[w]} \tag{3.49}$$

so that an alternative expression for the segment density operator is

$$\rho(\mathbf{r}; [w]) = -\frac{1}{Q[w]} \frac{\delta Q[w]}{\delta w(\mathbf{r})} \tag{3.50}$$

The right-hand side of this expression can be evaluated (for a discrete Gaussian chain) by direct functional differentiation of eqn (3.6). This leads to

$$\frac{\delta Q[w]}{\delta w(\mathbf{r})} = \frac{1}{V} \sum_{j=0}^{N} \int d\mathbf{r}^{N+1} \left[ e^{-w(\mathbf{r}_N)} \Phi(\mathbf{r}_N - \mathbf{r}_{N-1}) e^{-w(\mathbf{r}_{N-1})} \Phi(\mathbf{r}_{N-1} - \mathbf{r}_{N-2}) \right.$$
$$\times \quad \dots e^{-w(\mathbf{r}_j)}(-1)\delta(\mathbf{r} - \mathbf{r}_j)\Phi(\mathbf{r}_j - \mathbf{r}_{j-1})e^{-w(\mathbf{r}_{j-1})}$$
$$\left. \times \quad \dots e^{-w(\mathbf{r}_2)}\Phi(\mathbf{r}_2 - \mathbf{r}_1)e^{-w(\mathbf{r}_1)}\Phi(\mathbf{r}_1 - \mathbf{r}_0)e^{-w(\mathbf{r}_0)} \right]$$
$$\tag{3.51}$$

The delta function in eqn (3.51) can be used to collapse the integral over $\mathbf{r}_j$, the position of bead $j$. This produces a factorization of the expression, similar to the factorization property of $Q[w]$ described by eqn (3.10) and illustrated in Fig. 3.1. Following the same reasoning, one can rewrite eqn (3.51) by composing a propagator $q(\mathbf{r}, j; [w])$, describing the statistical weight of the chain section commencing with bead 0 and terminating on bead $j$, with a "complementary" propagator $q(\mathbf{r}, N - j; [w])$. The latter describes the statistical weight associated with the remaining chain section *starting* with bead $N$ and *finishing* with bead $j$. It follows that eqn (3.51) can be expressed as

$$\frac{\delta Q[w]}{\delta w(\mathbf{r})} = -\frac{e^{w(\mathbf{r})}}{V} \sum_{j=0}^{N} q(\mathbf{r}, N - j; [w]) \, q(\mathbf{r}, j; [w]) \tag{3.52}$$

and thus

$$\rho(\mathbf{r}; [w]) = \frac{e^{w(\mathbf{r})}}{V Q[w]} \sum_{j=0}^{N} q(\mathbf{r}, N - j; [w]) \, q(\mathbf{r}, j; [w]) \tag{3.53}$$

Equation (3.53) is an important formula in the theory of inhomogeneous polymers because it provides a recipe for calculating the average segment density of a discrete Gaussian chain experiencing an arbitrary potential $w(\mathbf{r})$. Given an expression for $w(\mathbf{r})$, one first solves eqn (3.8) recursively [subject to initial condition (3.7)] to obtain $q(\mathbf{r}, j; [w])$ for $j = 0, 1, 2, ..., N$. Next, $Q[w]$ is calculated by means of eqn (3.9). These results can then be used to evaluate the expression on the right-hand side of eqn (3.53). Thus, knowledge of the chain propagator $q(\mathbf{r}, j; [w])$ for all beads $j$ is sufficient to determine both the normalized partition function $Q[w]$ and the average segment density $\rho(\mathbf{r}; [w])$.

These results are easily extended to other chain models. For the continuous Gaussian chain, discretized according to eqn (3.21), the analog of eqn (3.53) is

$$\rho(\mathbf{r}; [w]) = -\frac{1}{Q[w]} \frac{\delta Q[w]}{\delta w(\mathbf{r})}$$

$$= \frac{\Delta s \exp[\Delta s w(\mathbf{r})]}{V Q[w]} \sum_{j=0}^{N_s} q(\mathbf{r}, (N_s - j)\Delta s; [w]) \, q(\mathbf{r}, j\Delta s; [w]) \quad (3.54)$$

Upon taking the continuum limit, $N_s \to \infty$, $\Delta s = N/N_s \to 0$, we find that this expression reduces to

$$\rho(\mathbf{r}; [w]) = \frac{1}{V Q[w]} \int_0^N ds \, q(\mathbf{r}, N - s; [w]) \, q(\mathbf{r}, s; [w]) \qquad (3.55)$$

which is a well known result, central to the theory of inhomogeneous polymers (Edwards, 1966; de Gennes, 1969; Freed, 1972; Helfand, 1975). Again, we observe a close analogy with the factorization property (3.27) of the continuous Gaussian chain.

The density operator $\rho(\mathbf{r}; [w])$ describes the average density of segments, regardless of their location along a polymer. Another useful density operator, e.g. for the continuous Gaussian chain, is the quantity

$$\rho(\mathbf{r}, s; [w]) \equiv \langle \hat{\rho}(\mathbf{r}, s) \rangle_{[w]} \qquad (3.56)$$

which represents the average density of segments located at contour position $s$. The microscopic density of segments at location $s$ is given in this expression by $\hat{\rho}(\mathbf{r}, s) = \delta(\mathbf{r} - \mathbf{r}(s))$. The average of $\hat{\rho}(\mathbf{r}, s)$ can be calculated by introducing a conjugate field $w_s(\mathbf{r})$ that acts only on monomer $s$ and contributes an additional term to the interaction energy

$$\beta U_1[\mathbf{r}, w, w_s] = \int d\mathbf{r}' \, [\hat{\rho}(\mathbf{r}')w(\mathbf{r}') + \hat{\rho}(\mathbf{r}', s)w_s(\mathbf{r}')] \qquad (3.57)$$

Following arguments similar to those leading to eqn (3.50) gives the result

$$\rho(\mathbf{r}, s; [w]) = \langle \hat{\rho}(\mathbf{r}, s) \rangle_{[w]} = -\left.\frac{\delta \ln Q[w, w_s]}{\delta w_s(\mathbf{r})}\right|_{w_s=0} \qquad (3.58)$$

We now retrace the steps of discretization of the path integral for $Q[w, w_s]$, differentiation with respect to $w_s(\mathbf{r})$, and taking the continuum limit. If we set $w_s(\mathbf{r}) = 0$, it follows that

$$\rho(\mathbf{r}, s; [w]) = \frac{1}{V Q[w]} q(\mathbf{r}, N - s; [w]) \, q(\mathbf{r}, s; [w]) \qquad (3.59)$$

The physical content of eqn (3.59) is described in Fig. 3.2. By comparing this expression with eqn (3.55), we see that the average *total* segment density

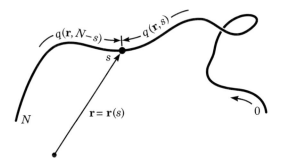

FIG. 3.2. Illustration of the composition formula (3.59) for the average density of segments located at position $s$ along a continuous Gaussian chain. The statistical weight $q(\mathbf{r}, s; [w])$ of a chain section of contour length $s$ is joined at point $\mathbf{r} = \mathbf{r}(s)$ with the statistical weight $q(\mathbf{r}, N - s; [w])$ of the "complementary" chain section having contour length $N - s$.

$\rho(\mathbf{r}; [w])$ is simply the integral of the average segment $s$ density $\rho(\mathbf{r}, s; [w])$ over all contour locations $0 \leq s \leq N$:

$$\rho(\mathbf{r}; [w]) = \int_0^N ds\, \rho(\mathbf{r}, s; [w]) \tag{3.60}$$

This formula is obviously consistent with the relationship between the corresponding microscopic densities $\hat{\rho}(\mathbf{r})$ and $\hat{\rho}(\mathbf{r}, s)$.

A special case of eqn (3.59) is of particular interest. By setting $s = 0$ or $s = N$, one obtains the average density of a chain end segment. In the present case of a homopolymer, the two chain ends are indistinguishable, so a *total chain end density* can be defined by

$$\rho_e(\mathbf{r}; [w]) \equiv \rho(\mathbf{r}, 0; [w]) + \rho(\mathbf{r}, N; [w])$$

$$= \frac{2}{V Q[w]} q(\mathbf{r}, N; [w]) \tag{3.61}$$

where in the second line of this expression we have used $q(\mathbf{r}, 0; [w]) = 1$. Thus, after applying the normalization $2/(V Q[w])$, the propagator $q(\mathbf{r}, N; [w])$ can be interpreted as the average density of chain ends at position $\mathbf{r}$.

The results described in eqns (3.56)–(3.61) can be easily generalized to the discrete Gaussian chain model. We omit this generalization for brevity and turn instead to the *wormlike chain*. For the wormlike chain model, recall that a field $w(\mathbf{r}, \mathbf{u})$ was introduced that is conjugate to the microscopic density of segment position and orientation, $\hat{\rho}(\mathbf{r}, \mathbf{u})$. It follows that an operator corresponding to the single-chain average of the segment position and orientation density can be defined by

$$\rho(\mathbf{r}, \mathbf{u}; [w]) \equiv \langle \hat{\rho}(\mathbf{r}, \mathbf{u}) \rangle_{[w]} = -\frac{\delta \ln Q[w]}{\delta w(\mathbf{r}, \mathbf{u})} \tag{3.62}$$

This expression can be evaluated by discretizing the path integral for $Q[w]$, differentiating with respect to $w(\mathbf{r}, \mathbf{u})$, and restoring the continuum limit. The following result is obtained:

$$\rho(\mathbf{r}, \mathbf{u}; [w]) = \frac{1}{4\pi V Q[w]} \int_0^{L_c} ds\, q(\mathbf{r}, -\mathbf{u}, L_c - s; [w])\, q(\mathbf{r}, \mathbf{u}, s; [w]) \qquad (3.63)$$

This expression is similar to the factorization property (3.41) in that the propagator $q(\mathbf{r}, -\mathbf{u}, L_c - s; [w])$, which describes the statistical weight of the "complementary" chain section, has the sign of $\mathbf{u}$ inverted.

Several related average densities can be obtained from eqn (3.63). Upon integration of both sides over $\mathbf{r}$ or over $\mathbf{u}$, a reduced average density of segment orientation or position is obtained. Thus,

$$\rho(\mathbf{u}; [w]) \equiv \int d\mathbf{r}\, \rho(\mathbf{r}, \mathbf{u}; [w])$$

$$= \frac{1}{4\pi V Q[w]} \int d\mathbf{r} \int_0^{L_c} ds\, q(\mathbf{r}, -\mathbf{u}, L_c - s; [w])\, q(\mathbf{r}, \mathbf{u}, s; [w])$$

$$(3.64)$$

and

$$\rho(\mathbf{r}; [w]) \equiv \int d\mathbf{u}\, \rho(\mathbf{r}, \mathbf{u}; [w])$$

$$= \frac{1}{4\pi V Q[w]} \int d\mathbf{u} \int_0^{L_c} ds\, q(\mathbf{r}, -\mathbf{u}, L_c - s; [w])\, q(\mathbf{r}, \mathbf{u}, s; [w])$$

$$(3.65)$$

provide separate information about the distribution of segment orienton and position. We can also refrain from the chain contour integration in eqn (3.63) and thereby derive an equation for the average density of segment position and orientation at a specified *contour position* $s$:

$$\rho(\mathbf{r}, \mathbf{u}, s; [w]) = \frac{1}{4\pi V Q[w]}\, q(\mathbf{r}, -\mathbf{u}, L_c - s; [w])\, q(\mathbf{r}, \mathbf{u}, s; [w]) \qquad (3.66)$$

Finally, this expression can be used to deduce an average density of *chain end* segment position and orientation

$$\rho_e(\mathbf{r}, \mathbf{u}; [w]) \equiv \rho(\mathbf{r}, \mathbf{u}, 0; [w]) + \rho(\mathbf{r}, \mathbf{u}, L_c; [w])$$

$$= \frac{1}{4\pi V Q[w]}\, [q(\mathbf{r}, -\mathbf{u}, L_c; [w]) + q(\mathbf{r}, \mathbf{u}, L_c; [w])] \qquad (3.67)$$

For the *rodlike polymer model*, the density operator for segment position and orientation is defined exactly as for the wormlike chain, cf. eqn (3.62). Taking a first functional derivative of eqn (3.45) leads to

$$\rho(\mathbf{r}, \mathbf{u}; [w]) = \frac{1}{4\pi V Q[w]} \int_0^{L_c} ds \, \exp\left[-\int_0^{L_c} ds' \, w(\mathbf{r} + (s' - s)\mathbf{u}, \mathbf{u})\right] \quad (3.68)$$

By defining a propagator $q(\mathbf{r}, \mathbf{u}, s; [w])$ for a rodlike polymer according to

$$q(\mathbf{r}, \mathbf{u}, s; [w]) \equiv \exp\left[-\int_0^s ds' \, w(\mathbf{r} - s'\mathbf{u}, \mathbf{u})\right] \quad (3.69)$$

and assuming that $w(\mathbf{r}, \mathbf{u}) = w(\mathbf{r}, -\mathbf{u})$, it follows that the rodlike polymer density operator given in eqn (3.68) can be expressed in exactly the same form as eqn (3.63) for the wormlike chain.

A final topic under the category of single-chain density operators relates to *higher-order moments* of the segment density. The objects $Q[w]$ and $\ln Q[w]$ can be viewed as *generating functionals* (van Kampen, 1981) because functional derivatives of these quantities yield, respectively, moments and cumulant moments of the microscopic single-chain density (see Appendix B). Indeed, for the Gaussian chain models we have seen that the first moment has the alternate representations

$$\langle \hat{\rho}(\mathbf{r}) \rangle_{[w]} = -\frac{\delta \ln Q[w]}{\delta w(\mathbf{r})} = -\frac{1}{Q[w]} \frac{\delta Q[w]}{\delta w(\mathbf{r})} \quad (3.70)$$

Similarly, it follows from the definitions of the normalized partition function and the single-chain average that the second moment of the microscopic density is given by a second functional derivative of $Q$

$$\langle \hat{\rho}(\mathbf{r}) \hat{\rho}(\mathbf{r}') \rangle_{[w]} = \frac{1}{Q[w]} \frac{\delta^2 Q[w]}{\delta w(\mathbf{r}) \delta w(\mathbf{r}')} \quad (3.71)$$

The second cumulant moment is likewise given by a second functional derivative of $\ln Q$

$$\langle \hat{\rho}(\mathbf{r}) \hat{\rho}(\mathbf{r}') \rangle_{[w]} - \langle \hat{\rho}(\mathbf{r}) \rangle_{[w]} \langle \hat{\rho}(\mathbf{r}') \rangle_{[w]} = \frac{\delta^2 \ln Q[w]}{\delta w(\mathbf{r}) \delta w(\mathbf{r}')} \quad (3.72)$$

We recall that eqns (3.53) and (3.55) provide explicit relationships for the Gaussian chain models between a chain propagator $q$ and the first moment of the microscopic segment density. This connection between propagators and moments can be extended to higher-order moments by taking additional functional derivatives and thereby factoring the chain statistical weight at additional points. For example, in the case of the continuous Gaussian chain, the second moment, or *density–density correlation function*, is given by

$$\begin{aligned}
\langle \hat{\rho}(\mathbf{r}) \hat{\rho}(\mathbf{r}') \rangle_{[w]} = {} & \frac{1}{V Q[w]} \int_0^N ds \int_0^s ds' \, q(\mathbf{r}, N - s; [w]) \\
& \times g(\mathbf{r}, \mathbf{r}', s - s'; [w]) q(\mathbf{r}', s'; [w]) \\
& + \frac{1}{V Q[w]} \int_0^N ds' \int_0^{s'} ds \, q(\mathbf{r}', N - s'; [w]) \\
& \times g(\mathbf{r}', \mathbf{r}, s' - s; [w]) q(\mathbf{r}, s; [w])
\end{aligned} \quad (3.73)$$

The function $g(\mathbf{r}, \mathbf{r}', s; [w])$ appearing in this equation is a new chain propagator that satisfies eqn (3.25) but is subject to a delta function initial condition, i.e.

$$\frac{\partial}{\partial s} g(\mathbf{r}, \mathbf{r}', s; [w]) = \frac{b^2}{6} \nabla^2 g(\mathbf{r}, \mathbf{r}', s; [w]) - w(\mathbf{r}) g(\mathbf{r}, \mathbf{r}', s; [w]) \qquad (3.74)$$

$$g(\mathbf{r}, \mathbf{r}', 0; [w]) = \delta(\mathbf{r} - \mathbf{r}') \qquad (3.75)$$

Thus, $g(\mathbf{r}, \mathbf{r}', s; [w])$ is the Green function (or fundamental) solution of eqn (3.25). As illustrated in Fig. 3.3, this propagator generates the statistical weight of a length-$s$ internal section of a polymer that originates at position $\mathbf{r}'$ and terminates at position $\mathbf{r}$.

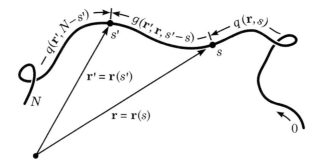

FIG. 3.3. Interpretation of the composition formula (3.73) (last term only) for the segment density–density correlation function of a continuous Gaussian chain. The statistical weight $q(\mathbf{r}, s)$ of a chain end section of contour length $s$ is joined at point $\mathbf{r} = \mathbf{r}(s)$ to a propagator $g(\mathbf{r}', \mathbf{r}, s' - s)$. The propagator describes the statistical weight associated with an internal chain section of length $s' - s$ that starts at $\mathbf{r} = \mathbf{r}(s)$ and ends at $\mathbf{r}' = \mathbf{r}(s')$. The statistical weight of the final "complementary" chain end section is $q(\mathbf{r}', N - s')$.

The physical interpretation of eqn (3.73) is depicted in Fig. 3.3. The equation can be used to compute the second moment of the segment density for a continuous Gaussian chain experiencing a potential $w(\mathbf{r})$. While such formulas for higher-order density correlation functions are convenient for approximate analytical calculations, they should generally be avoided in high-resolution numerical work. This is due to the expense of evaluating a *two-point* propagator such as $g(\mathbf{r}, \mathbf{r}', s; [w])$. If $M$ grid points or spectral components are used to resolve the spatial degrees of freedom, see Section 3.6, the numerical evaluation of $g$ requires at least order $M^2$ operations. Calculations with such $O(M^2)$ scaling are prohibitive in a three-dimensional simulation where $M$ can be as large as $10^6$–$10^7$.

### 3.2.2  Stress operators

Another important quantity is the average *elastic stress* contributed by a polymer chain placed in an inhomogeneous environment. An elastic stress operator can be defined by

$$\boldsymbol{\sigma}(\mathbf{r}; [w, \epsilon]) \equiv \langle \hat{\boldsymbol{\sigma}}(\mathbf{r}) \rangle_{[w,\epsilon]} \tag{3.76}$$

where the single-chain average is calculated according to eqn (3.47). Evaluation of the right-hand side of this expression using the *discrete* Gaussian chain model produces two alternative expressions,

$$\boldsymbol{\sigma}(\mathbf{r}; [w, \epsilon]) = \frac{\int d\mathbf{r}^{N+1} \, \hat{\boldsymbol{\sigma}}(\mathbf{r}) \exp[-\beta U(\mathbf{r}^{N+1}) - \beta U_{el}(\mathbf{r}^{N+1})]}{\int d\mathbf{r}^{N+1} \, \exp[-\beta U(\mathbf{r}^{N+1}) - \beta U_{el}(\mathbf{r}^{N+1})]}$$

$$= \frac{\int d\mathbf{r}^{N+1} \, \hat{\boldsymbol{\sigma}}(\mathbf{r}) \exp[-\beta U(\mathbf{r}^{N+1}) - \beta U_{el}(\mathbf{r}^{N+1})]}{Q[w, \epsilon] \int d\mathbf{r}^{N+1} \, \exp[-\beta U_0(\mathbf{r}^{N+1}) - \beta U_{el}(\mathbf{r}^{N+1})]} \tag{3.77}$$

the second being the most convenient. Inserting the explicit form of the microscopic stress operator, eqn (3.11), into the last expression leads to

$$\sigma_{\alpha\gamma}(\mathbf{r}; [w, \epsilon]) = \frac{3k_B T}{b^2 V Q[w, \epsilon]} \int d\mathbf{r}' \, (\mathbf{r}' - \mathbf{r})_\alpha (\mathbf{r}' - \mathbf{r})_\gamma \Psi(\mathbf{r}' - \mathbf{r}; [\epsilon])$$

$$\times \sum_{j=0}^{N-1} q(\mathbf{r}', N - j - 1; [w, \epsilon]) \, q(\mathbf{r}, j; [w, \epsilon]) \tag{3.78}$$

where the function $\Psi(\mathbf{r}' - \mathbf{r}; [\epsilon])$ was defined in eqn (3.14). This expression can be evaluated for prescribed chemical potential and strain fields by first solving eqns (3.15)–(3.17) to obtain the propagator $q(\mathbf{r}, j; [w, \epsilon])$ and the partition function $Q[w, \epsilon]$. The integrand and summand in eqn (3.78) are then specified, and the remaining operations can be carried out numerically.

A similar expression for the stress operator can be derived on the basis of the *continuous* Gaussian chain model (Fredrickson, 2002). For inhomogeneous strain fields the expression proves to be rather complicated, so it is not reproduced here. In the case of *homogeneous* strains $\epsilon$, one obtains[19]

$$\boldsymbol{\sigma}(\mathbf{r}; [w, \epsilon]) = \frac{k_B T b^2}{3 V Q[w, \epsilon]} \int_0^N ds \, q(\mathbf{r}, s; [w, \epsilon]) \nabla \nabla q(\mathbf{r}, N - s; [w, \epsilon]) \tag{3.79}$$

This expression for the stress operator shows that in the presence of a spatially varying chemical potential $w(\mathbf{r})$, the elastic stress is both anisotropic and inhomogeneous. The stress anisotropy is seen to be quantified by the dyadic quantity $q\nabla\nabla q$, integrated along the chain contour. Closely related formulas have been derived by Tyler and Morse in their studies of the linear elastic properties of block copolymer mesophases (Tyler and Morse, 2003a; Tyler and Morse, 2003b). Equation (3.79) is particularly useful for investigating the non-uniform distribution of chain stretching in meso-structured polymeric fluids.

---

[19] A singular *isotropic* stress term has been omitted from this expression, since it is normally included with the pressure in incompressible fluid models.

## 3.3 Other architectures

The formalism of the previous section assumed the simplest of polymer architectures – a linear homopolymer. Similar results can be obtained for a wide variety of architectures, encompassing those of both practical and academic interest. In this section we derive explicit expressions for branched homopolymers and block and graft copolymers subjected to spatially varying potentials. A prescribed architecture is assumed – the discussion of polymers with chemical disorder, such as randomly branched homopolymers, statistical copolymers, and randomly grafted copolymers, will be deferred to Section 4.7.

### 3.3.1 *Branched homopolymers*

As a first example of a branched homopolymer, we consider a 3-arm star homopolymer as shown in Fig. 3.4. Each arm is described by the continuous Gaussian chain model and has an arbitrary length. The degrees of polymerization of arms 1, 2, and 3 are given, respectively, by $N_1$, $N_2$, and $N_3$. It is assumed that the segments along each arm experience a chemical potential field $w(\mathbf{r})$. Only a single chemical potential is required in this case because the arms are chemically identical.

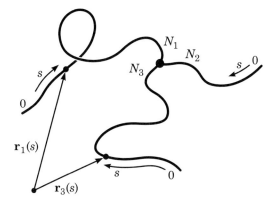

FIG. 3.4. A 3-arm star polymer in the continuous Gaussian chain model. The three arms with degrees of polymerization $N_1$, $N_2$, and $N_3$ are attached at a central branch point. The configuration of the $j$th arm is described by a space curve $\mathbf{r}_j(s)$, $s \in [0, N_j]$.

As illustrated in Fig. 3.4, the configuration of the $j$th arm ($j = 1, 2, 3$) is described by a space curve $\mathbf{r}_j(s)$, where the contour parameter $s$, $0 \le s \le N_j$, is measured from the free end of each arm. The constraint of attachment of the arms at the branch point amounts to the condition $\mathbf{r}_1(N_1) = \mathbf{r}_2(N_2) = \mathbf{r}_3(N_3)$.

The formulas of Section 3.1.2 for a linear homopolymer can be easily extended to the present case. The microscopic segment density can be written as

$$\hat{\rho}(\mathbf{r}) = \sum_{j=1}^{3} \int_0^{N_j} ds\ \delta(\mathbf{r} - \mathbf{r}_j(s)) \tag{3.80}$$

and the potential energy associated with the imposed chemical potential $w(\mathbf{r})$ is a functional of the overall polymer shape $\mathbf{r}(s) \equiv \{\mathbf{r}_1(s), \mathbf{r}_2(s), \mathbf{r}_3(s)\}$:

$$\beta U_1[\mathbf{r}, w] = \int d\mathbf{r}'\ w(\mathbf{r}')\hat{\rho}(\mathbf{r}') \tag{3.81}$$

The normalized partition function $Q[w]$ can again be expressed as a ratio of path integrals

$$Q[w] \equiv \frac{Z[w]}{Z_0} = \frac{\int^{*} \mathcal{D}\mathbf{r}\ \exp(-\beta U_0[\mathbf{r}] - \beta U_1[\mathbf{r}, w])}{\int^{*} \mathcal{D}\mathbf{r}\ \exp(-\beta U_0[\mathbf{r}])} \tag{3.82}$$

where the ideal chain stretching energy is given by

$$\beta U_0[\mathbf{r}] = \frac{3}{2b^2} \sum_{j=1}^{3} \int_0^{N_j} ds\ \left|\frac{d\mathbf{r}_j(s)}{ds}\right|^2 \tag{3.83}$$

and $\int^{*} \mathcal{D}\mathbf{r}$ denotes a path integral over the configurations of the three arms, subject to the branch point constraint

$$\int^{*} \mathcal{D}\mathbf{r} \equiv \int \mathcal{D}\mathbf{r}\ \delta(\mathbf{r}_1(N_1) - \mathbf{r}_2(N_2))\delta(\mathbf{r}_2(N_2) - \mathbf{r}_3(N_3)) \tag{3.84}$$

The path integrals in eqn (3.82) can be evaluated by discretizing the three paths, corresponding to the configurations of the arms of the star, leading to an expression similar to eqn (3.21) for a linear homopolymer. The constraint of eqn (3.84) is imposed by building up the path integrals for the three arms starting at the free ends and then requiring the paths to terminate at the position of a common branch point. It thus follows that the statistical weight of a star polymer with branch point at position $\mathbf{r}$ is given by the product $q(\mathbf{r}, N_1)q(\mathbf{r}, N_2)q(\mathbf{r}, N_3)$, where the propagators each satisfy eqns (3.25)–(3.26). The partition function is given by

$$Q[w] = \frac{1}{V} \int d\mathbf{r}\ q(\mathbf{r}, N_1; [w])q(\mathbf{r}, N_2; [w])q(\mathbf{r}, N_3; [w]) \tag{3.85}$$

It should be noted that application of this formula does not require three independent solutions of the diffusion equation (3.25), but rather *one*, since knowledge of $q(\mathbf{r}, s; [w])$ for $0 \leq s \leq N_{\max}$, where $N_{\max}$ is the largest of the $N_j$, suffices to evaluate all three propagators.

The segment density operator for a 3-arm star homopolymer can be derived by analogy with eqn (3.55) for a linear polymer. In the continuous Gaussian chain model, we have

$$\rho(\mathbf{r}; [w]) = -\frac{1}{Q[w]} \frac{\delta Q[w]}{\delta w(\mathbf{r})} = \rho_1(\mathbf{r}; [w]) + \rho_2(\mathbf{r}; [w]) + \rho_3(\mathbf{r}; [w]) \tag{3.86}$$

where $\rho_j(\mathbf{r}; [w])$ denotes the contribution to the average segment density from the $j$th arm of the star and is given by

$$\rho_j(\mathbf{r}; [w]) = \frac{1}{VQ[w]} \int_0^{N_j} ds \, q_j(\mathbf{r}, N_j - s; [w]) \, q(\mathbf{r}, s; [w]) \qquad (3.87)$$

The "complementary propagator" $q_j(\mathbf{r}, s; [w])$ satisfies the same diffusion equation as $q$, but is subject to a different initial condition, i.e.

$$\frac{\partial}{\partial s} q_j(\mathbf{r}, s; [w]) = \frac{b^2}{6} \nabla^2 q_j(\mathbf{r}, s; [w]) - w(\mathbf{r}) q_j(\mathbf{r}, s; [w]) \qquad (3.88)$$

$$q_j(\mathbf{r}, 0; [w]) = \begin{cases} q(\mathbf{r}, N_2; [w]) q(\mathbf{r}, N_3; [w]) & \text{if } j = 1 \\ q(\mathbf{r}, N_1; [w]) q(\mathbf{r}, N_3; [w]) & \text{if } j = 2 \\ q(\mathbf{r}, N_1; [w]) q(\mathbf{r}, N_2; [w]) & \text{if } j = 3 \end{cases} \qquad (3.89)$$

The physical interpretation of eqn (3.87) for the average density $\rho_3$ contributed by arm $j = 3$ of the star is illustrated in Fig. 3.5. The arm must pass through the point $\mathbf{r} = \mathbf{r}_3(s)$, where $s$ is a contour parameter measured from the free end of the arm. Associated with such an arm configuration is the product of a statistical weight $q(\mathbf{r}, s; [w])$, representing the dangling (free) end of the arm, and a weight $q_3(\mathbf{r}, N_3 - s; [w])$, representing the section of the arm that is attached to the remainder of the star. The complementary propagator $q_3$ is built up by integrating eqn (3.88), starting at the branch point, for a total contour distance of $N_3 - s$. The initial condition for this integration is the statistical weight associated with the remaining two arms at the branch point, which can be expressed by a product of two $q$ propagators as described in eqn (3.89).

The above results can be easily generalized to a star homopolymer with an arbitrary number of arms. In the most general case of a star with $p \geq 3$ arms of different lengths, the evaluation of the total segment density operator requires $p + 1$ solutions of a diffusion equation: one to obtain $q$ for a chain of length $N_{\max}$ (the longest arm length), and one solution for each of the $p$ complementary propagators $q_j(\mathbf{r}, s; [w])$, $s \in [0, N_j]$. In the special case of a $p$-arm star polymer with all arms of equal length, the computational burden is reduced significantly since all the complementary propagators are identical. Hence, only *two* solutions of a diffusion equation are required in this high-symmetry special case.

As a second example of a branched homopolymer, we consider the branched polymer shown in Fig. 3.6. The polymer is composed of five sections, labelled 1–5, which are connected at two branch points A and B. The degree of polymerization of section $j$ is denoted by $N_j$. There is no unique way to build up the statistical weight for such a polymer, but the various alternatives provide equivalent results. To illustrate, we describe how to compute the normalized partition function $Q[w]$ by composing propagators at branch point B. The arrows on the polymer sections of Fig. 3.6 indicate the direction of integration used to compute the propagators, and are also subject to arbitrary assignment. With the directions shown in the

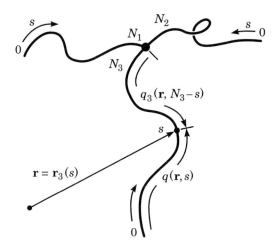

FIG. 3.5. Construction of the density operator $\rho_3(\mathbf{r}; [w])$ for a 3-arm star homopolymer. The density is built up by composing a statistical weight $q(\mathbf{r}, s; [w])$, representing the free end of arm 3, with a complementary propagator $q_3(\mathbf{r}, N_3 - s; [w])$, which represents the statistical weight of the section of the arm connected to the branch point. The latter is obtained by integrating eqn (3.88) away from the branch point, subject to the initial condition (3.89).

figure, and focusing on branch point B, it is apparent that $Q[w]$ can be obtained by combining propagators from sections 3, 4, and 5 according to

$$Q[w] = \frac{1}{V} \int d\mathbf{r}\, q_3(\mathbf{r}, N_3; [w]) q(\mathbf{r}, N_4; [w]) q(\mathbf{r}, N_5; [w]) \qquad (3.90)$$

The $q$ propagators in this expression for sections 4 and 5 initiate at free chain ends, so they satisfy eqns (3.25)–(3.26). In contrast, the $q_3$ propagator for section 3 initiates at branch point A, and thus satisfies diffusion equation (3.88) subject to the initial condition

$$q_3(\mathbf{r}, 0; [w]) = q(\mathbf{r}, N_1; [w]) q(\mathbf{r}, N_2; [w]) \qquad (3.91)$$

This initial condition provides the statistical weight associated with chain sections 1 and 2 at branch point A.

Similar expressions can be written for the various contributions to the average segment density from such a branched polymer. For example, the contribution to the segment density operator from section 3 can be expressed as

$$\rho_3(\mathbf{r}; [w]) = \frac{1}{VQ[w]} \int_0^{N_3} ds\, q_{3c}(\mathbf{r}, N_3 - s; [w])\, q_3(\mathbf{r}, s; [w]) \qquad (3.92)$$

where $q_3(\mathbf{r}, s; [w])$ is the propagator initiating from branch point A just described. The "complementary" propagator $q_{3c}(\mathbf{r}, N_3 - s; [w])$ initiates at branch point B

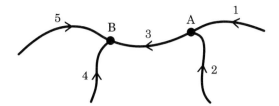

FIG. 3.6. Another example of a branched homopolymer consisting of five sections, 1–5, connected at two branch points, A and B. The arrows indicate one (arbitrary) choice of integration directions for building up the statistical weight of the polymer.

and extends a contour distance $N_3 - s$ to join the propagator $q_3$ at the point $\mathbf{r}$. $q_{3c}$ satisfies eqn (3.88) subject to the initial condition

$$q_{3c}(\mathbf{r}, 0; [w]) = q(\mathbf{r}, N_4; [w]) q(\mathbf{r}, N_5; [w]) \tag{3.93}$$

which provides the statistical weight of branches 4 and 5 at branch point B.

Using related arguments it is possible to construct partition functions, as well as density and stress operators, for branched homopolymers of arbitrary architecture subjected to external potentials. Such expressions can also be readily generalized beyond the continuous Gaussian chain to treat branched polymers within the context of discrete and wormlike chain models.

### 3.3.2 *Block and graft copolymers*

Block and and graft copolymers are materials of particular importance in the polymer industry and of special relevance to nanotechnology. Because such polymers are composed of two or more chemically distinct sections ("blocks" or "grafts"), their statistical mechanics in the presence of potential fields is richer than in the case of homopolymers.

We begin by discussing the case of a continuous Gaussian chain model of an *AB diblock copolymer*, as shown in Fig. 3.7. The copolymer has a total polymerization index of $N$; the section $0 \leq s \leq fN$ (solid) being comprised of type A segments and the section $fN \leq s \leq N$ (dashed) comprised of type B segments. The parameter $f$ can be interpreted as the fraction of the copolymer that is type A. If the A and B segments are further defined to have equal volumes, then $f$ corresponds to the average volume fraction of type A segments on the chain. Associated with the A and B sections of the copolymer are statistical segment lengths $b_A$ and $b_B$, respectively. We also introduce external chemical potential fields $w_A(\mathbf{r})$ and $w_B(\mathbf{r})$ that act separately on the A and B segments of the copolymer.

The stretching potential energy of such a diblock copolymer can be expressed as

$$\beta U_0[\mathbf{r}] = \int_0^N ds \, \frac{3}{2[b(s)]^2} \left| \frac{d\mathbf{r}(s)}{ds} \right|^2 \tag{3.94}$$

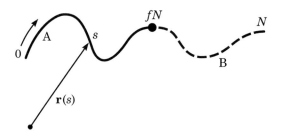

FIG. 3.7. A continuous Gaussian chain model of an AB diblock copolymer. The chain configuration is described by a space curve $\mathbf{r}(s)$, with segments $0 \leq s \leq fN$ corresponding to the A block (solid) and segments $fN < s \leq N$ corresponding to the B block (dashed). $f$ is a parameter indicating the ratio of the A block length to the total chain length.

where

$$b(s) \equiv \begin{cases} b_A, & 0 \leq s \leq fN \\ b_B, & fN < s \leq N \end{cases} \tag{3.95}$$

Similarly, the potential energy associated with the external potentials $w_A$ and $w_B$ can be written as

$$\beta U_1[\mathbf{r}, w_A, w_B] = \int d\mathbf{r}' \, [w_A(\mathbf{r}')\hat{\rho}_A(\mathbf{r}') + w_B(\mathbf{r}')\hat{\rho}_B(\mathbf{r}')] \tag{3.96}$$

where $\hat{\rho}_A$ and $\hat{\rho}_B$ are microscopic segment densities defined by

$$\hat{\rho}_A(\mathbf{r}) = \int_0^{fN} ds \, \delta(\mathbf{r} - \mathbf{r}(s)), \quad \hat{\rho}_B(\mathbf{r}) = \int_{fN}^N ds \, \delta(\mathbf{r} - \mathbf{r}(s)) \tag{3.97}$$

The normalized partition function for such a diblock is given by the expression

$$Q[w_A, w_B] \equiv \frac{Z[w_A, w_B]}{Z_0} = \frac{\int \mathcal{D}\mathbf{r} \, \exp(-\beta U_0[\mathbf{r}] - \beta U_1[\mathbf{r}, w_A, w_B])}{\int \mathcal{D}\mathbf{r} \, \exp(-\beta U_0[\mathbf{r}])} \tag{3.98}$$

The methods described in the previous sections can again be used to relate the partition function and density and stress operators to chain propagators satisfying Fokker–Planck equations. In the present case, a propagator $q(\mathbf{r}, s; [w_A, w_B])$ initiated from the $s = 0$ (A block) end of the copolymer satisfies the diffusion equation

$$\frac{\partial}{\partial s} q(\mathbf{r}, s; [w_A, w_B]) = \frac{[b(s)]^2}{6} \nabla^2 q(\mathbf{r}, s; [w_A, w_B])$$
$$- w(\mathbf{r}, s)q(\mathbf{r}, s; [w_A, w_B]) \tag{3.99}$$

where

$$w(\mathbf{r}, s) \equiv \begin{cases} w_A(\mathbf{r}), & 0 \leq s \leq fN \\ w_B(\mathbf{r}), & fN < s \leq N \end{cases} \tag{3.100}$$

Equation (3.99) is to be solved subject to the initial condition $q(\mathbf{r}, 0; [w_A, w_B]) = 1$.

Another useful quantity is a complementary propagator $q_c(\mathbf{r}, s; [w_A, w_B])$ that is initiated from the $s = N$ end of the copolymer. By building up the copolymer statistical weight starting from the B end, it is straightforward to show that $q_c$ satisfies the analogous diffusion equation

$$\frac{\partial}{\partial s} q_c(\mathbf{r}, s; [w_A, w_B]) = \frac{[b_c(s)]^2}{6} \nabla^2 q_c(\mathbf{r}, s; [w_A, w_B])$$
$$- w_c(\mathbf{r}, s) q_c(\mathbf{r}, s; [w_A, w_B]) \tag{3.101}$$

where

$$b_c(s) \equiv \begin{cases} b_B, & 0 \le s \le (1 - f)N \\ b_A, & (1 - f)N < s \le N \end{cases} \tag{3.102}$$

and

$$w_c(\mathbf{r}, s) \equiv \begin{cases} w_B(\mathbf{r}), & 0 \le s \le (1 - f)N \\ w_A(\mathbf{r}), & (1 - f)N < s \le N \end{cases} \tag{3.103}$$

Equation (3.101) is also integrated forward in $s$ using $q_c(\mathbf{r}, 0; [w_A, w_B]) = 1$.

With the above propagators, the partition function can be computed in two equivalent ways:

$$Q[w_A, w_B] = \frac{1}{V} \int d\mathbf{r} \, q(\mathbf{r}, N; [w_A, w_B])$$
$$= \frac{1}{V} \int d\mathbf{r} \, q_c(\mathbf{r}, N; [w_A, w_B]) \tag{3.104}$$

In other words, the propagator for a diblock copolymer can be built up from either the A end or the B end of the molecule, resulting in equivalent statistical weights for the *overall* chain. As such, it is only necessary to solve *either* eqn (3.99) or eqn (3.101) for the purpose of computing the partition function $Q[w_A, w_B]$.

In contrast, calculation of the density operators for a diblock copolymer requires knowledge of both chain propagators. The average densities of A and B segments are constructed by the now familiar process of composing forward and backward propagators, giving rise to the expressions

$$\rho_A(\mathbf{r}; [w_A, w_B]) = -\frac{1}{Q[w_A, w_B]} \frac{\delta Q[w_A, w_B]}{\delta w_A(\mathbf{r})}$$
$$= \frac{1}{V Q[w_A, w_B]} \int_0^{fN} ds \, q_c(\mathbf{r}, N - s; [w_A, w_B]) \, q(\mathbf{r}, s; [w_A, w_B])$$
$$\tag{3.105}$$

$$\rho_B(\mathbf{r}; [w_A, w_B]) = -\frac{1}{Q[w_A, w_B]} \frac{\delta Q[w_A, w_B]}{\delta w_B(\mathbf{r})}$$
$$= \frac{1}{V Q[w_A, w_B]} \int_{fN}^{N} ds \, q_c(\mathbf{r}, N - s; [w_A, w_B]) \, q(\mathbf{r}, s; [w_A, w_B])$$
$$\tag{3.106}$$

The above results are all well known in block copolymer theory (Helfand and Wasserman, 1976; Hong and Noolandi, 1981; Matsen and Schick, 1994a).

As a final example, we discuss the case of an $A_2B$ *graft copolymer*, with the architecture shown in Fig. 3.8. Such a molecule can be viewed as having an A backbone with a single B block "grafted" to it. Alternatively, the copolymer can be viewed as a type of star copolymer with two A arms, $A_1$ and $A_2$, and a B arm. Taking the latter point of view, the degrees of polymerization of the arms are denoted by $N_{A1}$, $N_{A2}$, and $N_B$, respectively. The continuous Gaussian chain model is adopted for simplicity.

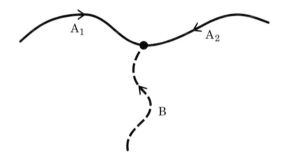

FIG. 3.8. A continuous Gaussian chain model of an $A_2B$ graft copolymer. The copolymer can be viewed as an A homopolymer backbone (solid) with a B homopolymer chain grafted to it (dashed), or as a star copolymer with two A arms and one B arm. The arrows indicate a particular (arbitrary) choice of direction for building up the statistical weight of the molecule.

We require two distinct chain propagators to construct the partition function $Q[w_A, w_B]$ of the graft copolymer shown in Fig. 3.8. These correspond to growing a section of a type A or B homopolymer chain starting from the free ends of the copolymer arms. The propagators, $q_A(\mathbf{r}, s; [w_A])$ and $q_B(\mathbf{r}, s; [w_B])$ satisfy

$$\frac{\partial}{\partial s} q_K(\mathbf{r}, s; [w_K]) = \frac{b_K^2}{6} \nabla^2 q_K(\mathbf{r}, s; [w_K]) - w_K(\mathbf{r}) q_K(\mathbf{r}, s; [w_K]) \qquad (3.107)$$

with K = A or B. Equation (3.107) is to be solved subject to $q_K(\mathbf{r}, 0; [w_K]) = 1$. With these definitions, the overall statistical weight of the graft copolymer is constructed by joining three such propagators at the branch point of the molecule, which implies that the partition function can be written

$$Q[w_A, w_B] = \frac{1}{V} \int d\mathbf{r} \, q_A(\mathbf{r}, N_{A1}; [w_A]) q_A(\mathbf{r}, N_{A2}; [w_A]) q_B(\mathbf{r}, N_B; [w_B]) \qquad (3.108)$$

Density operators can be constructed using methods similar to those employed for branched homopolymers and diblock copolymers. For example, the average segment B density of the graft copolymer can be obtained by composing the (forward) B propagator $q_B$ with a complementary (backward) B propagator $q_{Bc}(\mathbf{r}, s; [w_A, w_B])$ that is initiated from the branch point. The latter propagator satisfies

$$\frac{\partial}{\partial s} q_{Bc}(\mathbf{r}, s; [w_A, w_B]) = \frac{b_B^2}{6} \nabla^2 q_{Bc}(\mathbf{r}, s; [w_A, w_B]) - w_B(\mathbf{r}) q_{Bc}(\mathbf{r}, s; [w_A, w_B])$$
(3.109)

subject to

$$q_{Bc}(\mathbf{r}, 0; [w_A, w_B]) = q_A(\mathbf{r}, N_{A1}; [w_A]) q_A(\mathbf{r}, N_{A2}; [w_A])$$
(3.110)

This initial condition provides the statistical weight of the two A arms that are attached at the branch point. With these definitions, the average density of B segments from the $A_2B$ graft copolymer is given by

$$\rho_B(\mathbf{r}; [w_A, w_B]) = -\frac{1}{Q[w_A, w_B]} \frac{\delta Q[w_A, w_B]}{\delta w_B(\mathbf{r})}$$

$$= \frac{1}{V Q[w_A, w_B]} \int_0^{N_B} ds \, q_{Bc}(\mathbf{r}, N_B - s; [w_A, w_B]) \, q_B(\mathbf{r}, s; [w_B])$$
(3.111)

Similar expressions can be written for the average densities of type A segments contributed by either of the two A arms.

## 3.4 Approximation schemes

In the above Sections 3.1–3.3, the focus was on deriving *exact* expressions for the partition function and average properties for models of single chains in external potentials. With rare exception, these expressions are not amenable to exact analytical evaluation but rather provide a basis for numerical methods that will be taken up in Section 3.6. Although complete numerical solutions are generally desired, in some situations it is helpful to have approximate analytical expressions for objects such as the partition function $Q[w]$ and density operator $\rho(\mathbf{r}; [w])$. These can be used for analytical studies of many-chain field theories of polymeric fluids – the subject of Chapters 4–6. Perhaps more significantly, we shall see that an understanding of the asymptotic behavior of the single-chain averages and operators can assist in the development of effective *numerical methods* for solving many-chain field theories.

The present section will focus on techniques for developing systematic perturbation expansions of the partition function and single-chain averages. Such asymptotic methods are most straightforward to apply to continuous chain models where Fokker–Planck equations are available. Thus for the continuous Gaussian chain and the wormlike chain, one can exploit the extensive literature on regular and singular perturbation methods for partial differential equations (Nahfeh, 1973). The close analogy between the Fokker–Planck equations and the Schrödinger wave equation of quantum mechanics also suggests a number of useful approximation schemes (Schiff, 1968), many of which have been exploited in polymer physics (Edwards, 1966; de Gennes, 1969; de Gennes, 1979). We restrict attention in this section to the continuous chain models and attempt to provide a guide to developing perturbation expansions, rather than an exhaustive list of results.

### 3.4.1   Weak inhomogeneity expansion

A particularly useful perturbation expansion can be derived when the applied potential field $w$ has inhomogeneities that are weak in amplitude. To define such a situation, we introduce the volume average of the potential

$$w_0 \equiv \frac{1}{V} \int d\mathbf{r}\, w(\mathbf{r}) \tag{3.112}$$

and reexpress $w(\mathbf{r})$ according to

$$w(\mathbf{r}) = w_0 + \omega(\mathbf{r}) \tag{3.113}$$

which serves to define the inhomogeneous part of the field, $\omega(\mathbf{r})$. In a situation where the inhomogeneities are weak, a small parameter[20] $\epsilon_a$ ($|\epsilon_a| \ll 1$) describing their characteristic amplitude can be extracted from $\omega(\mathbf{r})$. Thus we replace eqn (3.113) with

$$w(\mathbf{r}) = w_0 + \epsilon_a \omega(\mathbf{r})$$

For the continuous Gaussian chain model of a homopolymer, the Fokker–Planck equation and initial condition (3.25)–(3.26) become

$$\frac{\partial}{\partial s} q(\mathbf{r}, s) = \frac{b^2}{6} \nabla^2 q(\mathbf{r}, s) - w_0 q(\mathbf{r}, s) - \epsilon_a \omega(\mathbf{r}) q(\mathbf{r}, s) \tag{3.114}$$

$$q(\mathbf{r}, 0) = 1 \tag{3.115}$$

where the functional dependence of $q$ on $w = w_0 + \epsilon_a \omega$ has been suppressed in our notation. The term proportional to $w_0$ on the right-hand side of eqn (3.114) can be removed by the substitution

$$q(\mathbf{r}, s) = \exp(-w_0 s)\, p(\mathbf{r}, s) \tag{3.116}$$

which leads to

$$\frac{\partial}{\partial s} p(\mathbf{r}, s) = \frac{b^2}{6} \nabla^2 p(\mathbf{r}, s) - \epsilon_a \omega(\mathbf{r}) p(\mathbf{r}, s) \tag{3.117}$$

$$p(\mathbf{r}, 0) = 1 \tag{3.118}$$

---

[20]The small parameter $\epsilon_a$ employed in this section should not be confused with a strain variable.

A *weak inhomogeneity expansion* can be developed by assuming that $p(\mathbf{r}, s)$ can be expressed as

$$p(\mathbf{r}, s) \sim \sum_{j=0}^{\infty} \epsilon_a{}^j p^{(j)}(\mathbf{r}, s) \tag{3.119}$$

where the $p^{(j)}(\mathbf{r}, s)$ are independent of $\epsilon_a$. In eqn (3.119) we adopt the conventional notation "$\sim$" (Nahfeh, 1973; Bender and Orszag, 1978) to indicate an asymptotic expansion.[21] As such, the infinite series on the right-hand side may be either convergent or divergent. Even when it does not converge, eqn (3.119) can still be useful in truncated form for approximating $p(\mathbf{r}, s)$ at sufficiently small $\epsilon_a$.

The $p^{(j)}$ are calculated by inserting eqn (3.119) into eqns (3.117)–(3.118) and equating terms order by order in $\epsilon_a$. At leading order, $O(\epsilon_a{}^0)$, we have

$$\frac{\partial}{\partial s} p^{(0)}(\mathbf{r}, s) = \frac{b^2}{6} \nabla^2 p^{(0)}(\mathbf{r}, s) \tag{3.120}$$

$$p^{(0)}(\mathbf{r}, 0) = 1 \tag{3.121}$$

which has the trivial solution $p^{(0)}(\mathbf{r}, s) = 1$. At $O(\epsilon_a)$, the corresponding equations are

$$\frac{\partial}{\partial s} p^{(1)}(\mathbf{r}, s) = \frac{b^2}{6} \nabla^2 p^{(1)}(\mathbf{r}, s) - \omega(\mathbf{r}) p^{(0)}(\mathbf{r}, s) \tag{3.122}$$

$$p^{(1)}(\mathbf{r}, 0) = 0 \tag{3.123}$$

Provided the system under consideration is unbounded or subject to periodic boundary conditions, this initial value problem is most easily solved by means of spatial Fourier transforms. Defining Fourier transforms in accordance with eqn (2.18) and assuming that the Fourier transform of $\omega(\mathbf{r})$ exists, denoted by $\hat{\omega}(\mathbf{k})$, one finds that

$$\hat{p}^{(1)}(\mathbf{k}, s) = -\hat{h}_2(\mathbf{k}, s)\hat{\omega}(\mathbf{k}) \tag{3.124}$$

where the carets denote Fourier-transformed quantities and

$$\hat{h}_2(\mathbf{k}, s) \equiv \frac{6}{b^2 k^2}[1 - \exp(-b^2 k^2 s/6)] \tag{3.125}$$

A similar procedure carried out to $O(\epsilon_a{}^2)$ leads to

$$\hat{p}^{(2)}(\mathbf{k}, s) = \frac{1}{V} \sum_{\mathbf{k}'} \hat{h}_3(\mathbf{k}, \mathbf{k}', s)\hat{\omega}(\mathbf{k} - \mathbf{k}')\hat{\omega}(\mathbf{k}') \tag{3.126}$$

where

---

[21] This use of the symbol "$\sim$" differs from the "scales-like" interpretation familiar in polymer physics and used elsewhere in this book. In the present section, coefficients in asymptotic scaling relationships are explicitly calculated.

$$\hat{h}_3(\mathbf{k}, \mathbf{k}', s) \equiv \frac{36}{b^4 k^2 |\mathbf{k} - \mathbf{k}'|^2} \left\{ 1 - \exp(-b^2 k^2 s/6) \right.$$

$$- \frac{k^2}{k^2 - |\mathbf{k} - \mathbf{k}'|^2} \left[ \exp(-b^2 |\mathbf{k} - \mathbf{k}'|^2 s/6) \right.$$

$$\left. - \exp(-b^2 k^2 s/6)] \right\} \tag{3.127}$$

The above expansion, when substituted into eqn (3.22) for the partition function, leads to

$$Q[w] = \frac{1}{V} \int d\mathbf{r} \, q(\mathbf{r}, N)$$

$$\sim e^{-w_0 N} \left[ 1 + \frac{\epsilon_a}{V} \hat{p}^{(1)}(\mathbf{0}, N) + \frac{\epsilon_a{}^2}{V} \hat{p}^{(2)}(\mathbf{0}, N) + \dots \right] \tag{3.128}$$

The $O(\epsilon_a)$ term in this expansion vanishes identically because $\hat{\omega}(\mathbf{0}) = \int d\mathbf{r} \, \omega(\mathbf{r}) = 0$, which follows from the definition (3.113) of $\omega(\mathbf{r})$. One can simplify the $O(\epsilon_a{}^2)$ term by noting that

$$\hat{h}_3(\mathbf{0}, \mathbf{k}', N) = \frac{N^2}{2} \hat{g}_D((k' R_g)^2) \tag{3.129}$$

where $R_g^2 = N b^2/6$ is the unperturbed radius of gyration of the continuous Gaussian chain and $\hat{g}_D(x)$ is a function known as the *Debye function* (de Gennes, 1979; Doi and Edwards, 1986)

$$\hat{g}_D(x) = \frac{2}{x^2} \left[ \exp(-x) + x - 1 \right] \tag{3.130}$$

It follows that the first two terms in the weak inhomogeneity expansion of the partition function can be written

$$Q[w] \sim e^{-w_0 N} \left[ 1 + \frac{\epsilon_a{}^2 N^2}{2V^2} \sum_{\mathbf{k}} \hat{g}_D(k^2 R_g^2) \, \hat{\omega}(\mathbf{k}) \hat{\omega}(-\mathbf{k}) + \dots \right] \tag{3.131}$$

or, by inverting the Fourier transforms,

$$Q[w] \sim e^{-w_0 N} \left[ 1 + \frac{\epsilon_a{}^2 N^2}{2V} \int d\mathbf{r} \int d\mathbf{r}' \, g_D(|\mathbf{r} - \mathbf{r}'|) \, \omega(\mathbf{r}) \omega(\mathbf{r}') + \dots \right] \tag{3.132}$$

where the inverse transform of the Debye function is defined by

$$g_D(|\mathbf{r} - \mathbf{r}'|) = \frac{1}{V} \sum_{\mathbf{k}} e^{i\mathbf{k} \cdot (\mathbf{r} - \mathbf{r}')} \hat{g}_D(k^2 R_g^2)$$

$$= \frac{1}{(2\pi)^3} \int d\mathbf{k} \, e^{i\mathbf{k} \cdot (\mathbf{r} - \mathbf{r}')} \hat{g}_D(k^2 R_g^2) \tag{3.133}$$

The second expression is applicable only in the infinite volume, $V \to \infty$, limit.

A weak inhomogeneity expansion for the segment density operator $\rho(\mathbf{r};[w])$ can be obtained in one of two equivalent ways: substitution into eqn (3.55), or direct functional differentiation of eqn (3.132) according to eqn (3.50). Following the latter approach gives

$$\rho(\mathbf{r};[w]) = -\frac{1}{Q[w]}\frac{\delta Q[w]}{\delta w(\mathbf{r})}$$

$$\sim \rho_0\left[1 - \epsilon_a N\int d\mathbf{r}'\, g_D(|\mathbf{r}-\mathbf{r}'|)\,\omega(\mathbf{r}') + O(\epsilon_a{}^2)\right] \qquad (3.134)$$

where $\rho_0 \equiv N/V$ is the volume-average segment density of a single chain.

If one compares eqns (3.132) and (3.134), an important observation is that the density operator has a non-uniform contribution at $O(\epsilon_a\omega)$, while the first correction to the partition function is $O(\epsilon_a{}^2\omega^2)$. Additionally, $Q[w]$ is proportional to $\exp(-w_0 N)$, while the density operator $\rho(\mathbf{r};[w])$ is *independent* of the average value of the field $w_0$. This proves to be a *general result* that persists to all orders in the weak inhomogeneity expansion. Indeed, under a uniform shift of the potential $w(\mathbf{r}) \to w(\mathbf{r}) + w_u$, $Q$ and $\rho$ have the following exact transformation properties

$$Q[w+w_u] = \exp(-w_u N)Q[w], \quad \rho(\mathbf{r};[w+w_u]) = \rho(\mathbf{r};[w]) \qquad (3.135)$$

By taking one further functional derivative with respect to $w(\mathbf{r})$, it is possible to derive a weak inhomogeneity expansion of the cumulant density–density (or "pair") correlation function. Combining eqns (3.72) and (3.134), we obtain

$$\langle\hat{\rho}(\mathbf{r})\hat{\rho}(\mathbf{r}')\rangle_{[w]} - \langle\hat{\rho}(\mathbf{r})\rangle_{[w]}\langle\hat{\rho}(\mathbf{r}')\rangle_{[w]} = \frac{\delta^2\ln Q[w]}{\delta w(\mathbf{r})\delta w(\mathbf{r}')}$$

$$\sim \rho_0 N\, g_D(|\mathbf{r}-\mathbf{r}'|) + O(\epsilon_a) \qquad (3.136)$$

Thus, the inverse Fourier transform of the Debye function provides the leading $O(\epsilon_a{}^0)$ contribution to the connected density–density correlation function.

The function $g_D(r)$, plotted in Fig. 3.9, is seen to play an important role in the weak inhomogeneity expansion by appearing in leading-order expressions for the partition function, the density operator, and the density–density correlation function. Because $g_D(r)$ arises at $O(\epsilon_a{}^0)$ in the expansion of the density–density correlation function and the latter function is independent of $w_0$, it is apparent that $g_D(r)$ can also be interpreted as the pair correlation function for an *ideal* continuous Gaussian chain.[22] For $r/R_g \ll 1$, the function decays algebraically as $1/r$ according to

$$g_D(r) \sim \frac{3}{\pi b^2 N r} \qquad (3.137)$$

Beyond $r \sim R_g$, it decays monotonically and exponentially to zero as $g_D(r) \sim r^{-1}\exp(-\sqrt{3}r/R_g)$.

---

[22]This is normally how the function $g_D(r)$ is introduced (de Gennes, 1979; Doi and Edwards, 1986).

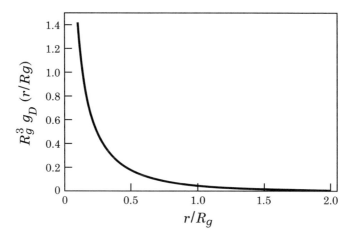

FIG. 3.9. Inverse Fourier transform of the Debye function defined by eqn (3.133) and plotted in dimensionless form as $R_g^3 g_D(r)$ versus $r/R_g$.

It is straightforward to apply the weak inhomogeneity expansion to other polymer architectures, including branched polymers and copolymers. For example, in the case of a 3-arm star homopolymer with equal length arms, $N_j = N$, it follows from eqn (3.85) that

$$Q[w] = \frac{1}{V} \int d\mathbf{r} \, [q(\mathbf{r}, N; [w])]^3 \qquad (3.138)$$

Expanding each factor in the propagator product according to eqns (3.116) and (3.119) leads to

$$Q[w] \sim e^{-3w_0 N} \left[ 1 + \frac{3\epsilon_a^2 N^2}{2V^2} \sum_{\mathbf{k}} \hat{g}_S(k^2 R_g^2) \, \hat{\omega}(\mathbf{k})\hat{\omega}(-\mathbf{k}) + ... \right] \qquad (3.139)$$

where $R_g^2 = Nb^2/6$ is the radius of gyration of *one* arm of the star. The Debye-like function $\hat{g}_S(x)$ is defined for the star by

$$\hat{g}_S(x) = \hat{g}_D(x) + 2[\hat{h}_D(x)]^2 \qquad (3.140)$$

with

$$\hat{h}_D(x) = \frac{1}{x}[1 - \exp(-x)] \qquad (3.141)$$

The leading-order terms in the expansion of the density operator and pair correlation function are thus

$$\rho(\mathbf{r}; [w]) \sim \rho_0 \left[ 1 - \epsilon_a N \int d\mathbf{r}' \, g_S(|\mathbf{r} - \mathbf{r}'|) \, \omega(\mathbf{r}') + O(\epsilon_a^2) \right] \qquad (3.142)$$

and

$$\langle\hat{\rho}(\mathbf{r})\hat{\rho}(\mathbf{r}')\rangle_{[w]} - \langle\hat{\rho}(\mathbf{r})\rangle_{[w]}\langle\hat{\rho}(\mathbf{r}')\rangle_{[w]} \sim \rho_0 N\, g_S(|\mathbf{r} - \mathbf{r}'|) + O(\epsilon_a) \quad (3.143)$$

where $\rho_0 \equiv 3N/V$ is the volume-average segment density and $g_S(r)$ is the inverse Fourier transform of $\hat{g}_S(k^2 R_g^2)$. This expression for the density–density correlation function is in agreement with well known expressions for the scattering function of an ideal star polymer (Benoit and Hadziioannou, 1988).

The weak inhomogeneity expansion can also be extended to the wormlike chain model, defined by the Fokker–Planck equation (3.39) and initial condition (3.40). At $O(\epsilon_a{}^j)$ in the perturbation expansion, it is necessary to solve a Fokker–Planck equation of the form

$$\frac{\partial}{\partial s}p^{(j)}(\mathbf{r}, \mathbf{u}, s; [w]) = -\,\omega(\mathbf{r}, \mathbf{u})\, p^{(j-1)}(\mathbf{r}, \mathbf{u}, s; [w]) - \mathbf{u}\cdot\nabla_{\mathbf{r}}\, p^{(j)}(\mathbf{r}, \mathbf{u}, s; [w])$$

$$+ \frac{1}{2\lambda}\nabla_{\mathbf{u}}^2\, p^{(j)}(\mathbf{r}, \mathbf{u}, s; [w]) \qquad (3.144)$$

This essentially amounts to the ideal chain Fokker–Planck equation (2.76) with an extra inhomogeneous term $\omega(\mathbf{r}, \mathbf{u})\, p^{(j-1)}(\mathbf{r}, \mathbf{u}, s; [w])$ that is determined at the previous order in $\epsilon_a$. The spatial Fourier transform of the above equation can in principle be solved by a continued fraction representation of a spherical harmonic expansion (Spakowitz and Wang, 2004), although such a calculation has not been reported to date.

Finally, we note that the weak inhomogeneity expansion is closely related to an approximation scheme for many-chain systems known as the *random phase approximation (RPA)* (de Gennes, 1969; de Gennes, 1979). This connection will be explored in Section 5.2.1.

### 3.4.2 *Slow gradient expansion*

Another important asymptotic expansion can be derived for external potentials that are *slowly varying in space*. To exploit this property, it is useful to write

$$w(\mathbf{r}) = \psi_w(\mathbf{r}/\xi_w) \qquad (3.145)$$

where $\xi_w$ is a length scale that characterizes the spatial variations of $w(\mathbf{r})$. In situations where $\xi_w$ is much larger than the characteristic size of a polymer, a "slow gradient expansion" can be derived. This asymptotic expansion was first identified (Tang and Freed, 1991) by using a path integral method quite different from that presented below.[23] We begin by considering a homopolymer in the context of the continuous Gaussian chain model.

Substituting eqn (3.145) into eqn (3.25) and scaling the independent variables according to

$$t \equiv s/N, \quad \mathbf{x} \equiv \mathbf{r}/\xi_w \qquad (3.146)$$

---

[23]The approach adopted here is due to V. Ganesan.

produces

$$\frac{\partial}{\partial t}q(\mathbf{x},t) = \epsilon_s\nabla_\mathbf{x}^2 q(\mathbf{x},t) - W(\mathbf{x})q(\mathbf{x},t) \tag{3.147}$$

where $\nabla_\mathbf{x} \equiv \partial/\partial\mathbf{x}$, $W(\mathbf{x}) \equiv N\psi_w(\mathbf{r}/\xi_w)$, and $\epsilon_s$ is defined by

$$\epsilon_s \equiv \frac{Nb^2}{6\xi_w^2} = \left(\frac{R_g}{\xi_w}\right)^2 \tag{3.148}$$

The functional dependence of $q(\mathbf{x},t;[W])$ on $W(\mathbf{x})$ has been suppressed in the scaled diffusion equation (3.147) to simplify the notation. With the above scalings, the initial condition (3.26) transforms to $q(\mathbf{x},0;[W]) = 1$.

Under conditions where $\xi_w$ is large compared with the (ideal) polymer radius of gyration, $R_g$, the parameter $\epsilon_s$ is small and can be used as the basis for a perturbation expansion. This *slow gradient expansion* is defined by

$$q(\mathbf{x},t) \sim \sum_{j=0}^{\infty} \epsilon_s{}^j q^{(j)}(\mathbf{x},t) \tag{3.149}$$

where the $q^{(j)}(\mathbf{x},t)$ coefficient functions are assumed independent of $\epsilon_s$.

Substitution of eqn (3.149) into eqn (3.147) and the initial condition leads to a hierarchy of equations that can be solved for the $q^{(j)}(\mathbf{x},t)$ at each order in $\epsilon_s$. At $O(\epsilon_s{}^0)$, we have

$$\frac{\partial}{\partial t}q^{(0)}(\mathbf{x},t) = -W(\mathbf{x})q^{(0)}(\mathbf{x},t) \tag{3.150}$$

$$q^{(0)}(\mathbf{x},0) = 1 \tag{3.151}$$

which has the trivial solution

$$q^{(0)}(\mathbf{x},t) = \exp[-W(\mathbf{x})t] \tag{3.152}$$

At $O(\epsilon_s)$, the relevant equations are

$$\frac{\partial}{\partial t}q^{(1)}(\mathbf{x},t) = -W(\mathbf{x})q^{(1)}(\mathbf{x},t) + \nabla_\mathbf{x}^2 q^{(0)}(\mathbf{x},t) \tag{3.153}$$

$$q^{(1)}(\mathbf{x},0) = 0 \tag{3.154}$$

These have the solution

$$q^{(1)}(\mathbf{x},t) = \exp[-W(\mathbf{x})t]\left(\frac{t^3}{3}|\nabla_\mathbf{x}W(\mathbf{x})|^2 - \frac{t^2}{2}\nabla_\mathbf{x}^2 W(\mathbf{x})\right) \tag{3.155}$$

Higher-order terms in the gradient expansion of the propagator are straightforward to derive.

Upon substitution into eqn (3.22), a slow gradient expansion of the partition function is obtained:

$$Q[W] = \frac{\xi_w^3}{V} \int d\mathbf{x}\, q(\mathbf{x}, 1)$$

$$\sim \frac{\xi_w^3}{V} \int d\mathbf{x}\, e^{-W(\mathbf{x})} \left\{ 1 + \epsilon_s \left( \frac{1}{3}|\nabla_{\mathbf{x}} W|^2 - \frac{1}{2}\nabla_{\mathbf{x}}^2 W \right) + \ldots \right\}$$

$$\sim \frac{\xi_w^3}{V} \int d\mathbf{x}\, e^{-W(\mathbf{x})} \left\{ 1 - \frac{\epsilon_s}{6}|\nabla_{\mathbf{x}} W|^2 + O(\epsilon_s^2) \right\} \tag{3.156}$$

where the last line follows from an integration by parts.[24] First and second functional derivatives of this expression with respect to $w(\mathbf{r})$ produce similar gradient expansions for the density operator and density–density correlation functions. For example, the expansion of the density operator is given by

$$\rho(\mathbf{x}; [W]) = \frac{\rho_0}{Q[W]} \int_0^1 dt\, q(\mathbf{x}, 1-t)q(\mathbf{x}, t)$$

$$\sim \frac{\rho_0 e^{-W(\mathbf{x})}}{Q[W]} \left\{ 1 + \epsilon_s \left( \frac{1}{6}|\nabla_{\mathbf{x}} W|^2 - \frac{1}{3}\nabla_{\mathbf{x}}^2 W \right) + \ldots \right\} \tag{3.157}$$

where $\rho_0 = N/V$ is the volume-average segment density of the polymer.

Similar slow gradient expansions can be easily derived for essentially any polymer architecture in the continuous Gaussian chain model. Unfortunately, such expansions have rather limited applicability because the potentials relevant to the vast majority of inhomogeneous polymeric fluids have variations on scales well below $R_g$. These rapid potential field variations arise from interfaces of thickness intermediate between the statistical segment length $b$ ($\sim 1$ nm) and the radius of gyration $R_g$ ($\sim 10$ nm). In a few special situations, such as polymer blends in the vicinity of a critical point (de Gennes, 1979), interfaces can be broadened beyond $R_g$, thereby justifying the use of formulas such as (3.156)–(3.157).

The slow gradient expansion is considerably less useful for the wormlike chain model. Unless $w(\mathbf{r}, \mathbf{u})$ is slowly varying in *both* its positional and orientational dependence, so that small parameters can be scaled out in front of both operators $-\mathbf{u} \cdot \nabla_{\mathbf{r}}$ and $\nabla_{\mathbf{u}}^2$ appearing in eqn (3.39), the expansion proves to be intractable. Such a situation of slow $\mathbf{r}$ and $\mathbf{u}$ variations is rarely encountered.

### 3.4.3   *Ground state dominance*

Another important way to analyze the Fokker–Planck equations for the continuous chain models is to utilize *eigenfunction expansions*. Such expansions are familiar methods of analysis in quantum mechanics (Schiff, 1968). To illustrate

---

[24]We assume that the boundary terms arising from the integration by parts can be neglected.

the method for the continuous Gaussian chain, it is helpful to write the diffusion equation (3.25) in operator form:

$$\frac{\partial}{\partial s} q(\mathbf{r}, s) = \mathcal{L}q(\mathbf{r}, s) \tag{3.158}$$

where we have again suppressed the functional dependence of $q$ on $w$. $\mathcal{L}$ is a linear operator defined by

$$\mathcal{L} = \frac{b^2}{6} \nabla^2 - w(\mathbf{r}) \tag{3.159}$$

For real $w(\mathbf{r})$ and suitable boundary conditions, $\mathcal{L}$ is a self-adjoint Sturm–Liouville operator that has real eigenvalues $\Lambda_k$ and eigenfunctions $\psi_k(\mathbf{r})$ that are orthogonal and complete (Riley $et$ $al.$, 1998). These satisfy

$$\mathcal{L}\psi_k(\mathbf{r}) = -\Lambda_k \psi_k(\mathbf{r}), \quad k = 0, 1, 2, ..., \infty \tag{3.160}$$

where the index $k$ is chosen such that $\Lambda_0$ is the smallest eigenvalue. It will be assumed that the $\psi_k(\mathbf{r})$ are normalized. The properties of orthogonality and closure (completeness) are thus

$$\int d\mathbf{r} \, \psi_j(\mathbf{r})\psi_k(\mathbf{r}) = \delta_{jk} \tag{3.161}$$

$$\sum_{k=0}^{\infty} \psi_k(\mathbf{r}')\psi_k(\mathbf{r}) = \delta(\mathbf{r} - \mathbf{r}') \tag{3.162}$$

The solution of eqn (3.158) can be expressed as an expansion in the eigenfunctions of $\mathcal{L}$ according to

$$q(\mathbf{r}, s) = \sum_{k=0}^{\infty} q_k \psi_k(\mathbf{r}) \exp(-s\Lambda_k) \tag{3.163}$$

The expansion coefficients $q_k$ follow from the initial condition (3.26) and the orthonormal properties of the eigenfunctions

$$q_k = \int d\mathbf{r} \, \psi_k(\mathbf{r}) \tag{3.164}$$

Alternatively, the Green function solution to eqn (3.158), i.e. the solution of eqns (3.74)–(3.75), has a closely related expansion

$$g(\mathbf{r}, \mathbf{r}', s) = \sum_{k=0}^{\infty} \psi_k(\mathbf{r}')\psi_k(\mathbf{r}) \exp(-s\Lambda_k) \tag{3.165}$$

The utility of such eigenfunction expansions is limited by our ability to solve the eigenvalue problem (3.160) for appropriate boundary conditions and by the requirement of summing the infinite series. In most realistic situations in polymer physics, we can neither find analytical expressions for the eigenvalues and eigenfunctions nor exactly sum series such as eqn (3.163) and (3.165).

An important approximation scheme known as the *ground state dominance approximation* (Moore, 1977; de Gennes, 1979) applies when an eigenfunction expansion can be truncated after the leading $k = 0$ term. This approximation implies, e.g. for the partition function, that

$$Q[w] = \frac{1}{V} \int d\mathbf{r} \, q(\mathbf{r}, N) = \frac{1}{V} \sum_{k=0}^{\infty} q_k^2 \exp(-N\Lambda_k)$$

$$\sim \frac{q_0^2}{V} \exp(-N\Lambda_0) \tag{3.166}$$

and for the density operator

$$\rho(\mathbf{r}; [w]) = -\frac{\delta \ln Q[w]}{\delta w(\mathbf{r})} \sim \frac{N q_0^2}{V Q[w]} \exp(-N\Lambda_0)[\psi_0(\mathbf{r})]^2$$

$$\sim N[\psi_0(\mathbf{r})]^2 \tag{3.167}$$

The condition (3.161) ensures that the density operator in the ground state approximation has the correct normalization, $\int d\mathbf{r} \, \rho(\mathbf{r}; [w]) = N$.

The ground state approximation is evidently the leading term in an asymptotic expansion for $N \to \infty$. For the approximation to be valid, two conditions are required:

- The eigenvalues must be *discrete*. This implies that the polymer is localized (bound) by the potential to finite region(s) of space.
- The eigenvalue spacing and the chain length should be large enough that the relative contribution of the first neglected mode satisfies

$$\exp[-N(\Lambda_1 - \Lambda_0)] \ll 1.$$

The above conditions are most commonly met in problems involving the absorption of polymers at surfaces or interfaces, or in situations where polymers are confined to a region of characteristic size smaller than their unperturbed (ideal) dimensions, e.g. within nanometer-scale pores (de Gennes, 1979).

From a computational standpoint, the ground state dominance approximation greatly simplifies the evaluation of the partition function and density operator. In place of solving an $s$-dependent partial differential equation (3.158), which is four dimensional (in three spatial dimensions), the ground state dominance approximation requires only the solution of the three-dimensional partial differential equation (3.160) for the lowest eigenvalue and eigenfunction. Moreover, the latter problem has a variational basis because the functional

$$F_1[\psi] = \frac{1}{2} \int d\mathbf{r} \left( \frac{b^2}{6} |\nabla \psi|^2 + w\psi^2 - \Lambda \psi^2 \right) \tag{3.168}$$

has eqn (3.160) as its Euler–Lagrange equation.[25] The eigenvalue $\Lambda$ appearing in this equation plays the role of a Lagrange multiplier, ensuring that the normalization condition $\int d\mathbf{r}\,\psi^2 = 1$ is met. With the use of eqn (3.167), a related functional of $\rho(\mathbf{r};[w])$ can be devised

$$F_2[\rho] = \int d\mathbf{r} \left( \frac{b^2}{24\rho}|\nabla\rho|^2 + w\rho - \Lambda\rho \right) \tag{3.169}$$

The extremum $\rho(\mathbf{r})$ of this functional, with $\Lambda$ adjusted to preserve the normalization $\int d\mathbf{r}\,\rho(\mathbf{r}) = N$, provides the relationship between $\rho(\mathbf{r};[w])$ and $w(\mathbf{r})$ in the ground state approximation. The first term on the right-hand side of eqn (3.169) is sometimes referred to as the *Lifshitz entropy* (Lifshitz et al., 1978). It represents an approximate expression for the conformational entropy (actually, $-S/k_B$) of a continuous Gaussian chain with a prescribed segment density distribution.

To further clarify the conditions under which the ground state dominance approximation applies, it is helpful to discuss an exactly solvable example – a chain subjected to a one-dimensional quadratic potential, which is the analog of the harmonic oscillator in quantum mechanics. Here we choose

$$w(x) = \frac{1}{2}(x/\xi_w)^2 \tag{3.170}$$

where $\xi_w$ is a length scale characterizing the width of the potential basin. The normalized eigenfunctions for this potential and satisfying eqn (3.160) are known as the harmonic oscillator wave functions (Schiff, 1968)

$$\psi_k(x) = \left( \frac{1}{\sqrt{\pi}\,2^k k!\,\xi} \right)^{1/2} e^{-\frac{1}{2}(x/\xi)^2} H_k(x/\xi), \quad k = 0, 1, 2, ..., \infty \tag{3.171}$$

The parameter $\xi \equiv 3^{-1/4}(b\xi_w)^{1/2}$ appearing in this expression is a length scale proportional to the geometric mean of the lengths $b$ and $\xi_w$. The $H_k(x)$ are the familiar $k$th-order Hermite polynomials, defined by

$$H_k(x) = (-1)^k e^{x^2} \frac{\partial^k}{\partial x^k} e^{-x^2} \tag{3.172}$$

Due to the exponential factor in eqn (3.171), the eigenfunctions $\psi_k(x)$ are localized to a region of width $\xi \sim (b\xi_w)^{1/2}$ about the origin. They decay exponentially outside of this region and are normalized in accordance with eqn (3.161), i.e. $\int_{-\infty}^{\infty} dx\,[\psi_k(x)]^2 = 1$. Corresponding to each $\psi_k(x)$ is an eigenvalue

$$\Lambda_k = \frac{b^2}{6\xi^2}(2k+1), \quad k = 0, 1, 2, ..., \infty \tag{3.173}$$

---

[25]Readers unfamiliar with the concepts of Euler–Lagrange equations and functional optimization should consult Appendix C.

The validity of the ground state approximation was seen to depend on the product of the spacing between the zeroth and first eigenvalues and the polymerization index. In the present example we have

$$N(\Lambda_1 - \Lambda_0) = \frac{Nb^2}{3\xi^2} = 2\left(\frac{R_g}{\xi}\right)^2 \tag{3.174}$$

The condition that $\exp[-N(\Lambda_1 - \Lambda_0)] \ll 1$ is thus met when $\xi \ll R_g$. In other words, the ground state dominance approximation is accurate when the length $\xi$ characterizing the width of the region of localization of the polymer is much smaller than the ideal radius of gyration $R_g$. The former length is the geometric mean of the statistical segment length $b$ and the width of the potential $\xi_w$.

The eigenfunction expansion method and ground state dominance approximation also prove useful for the wormlike chain model (Morse and Fredrickson, 1994). Unfortunately the relevant linear operator in this case

$$\mathcal{L} = \frac{1}{2\lambda}\nabla_{\mathbf{u}}^2 - \mathbf{u}\cdot\nabla_{\mathbf{r}} - w(\mathbf{r},\mathbf{u}) \tag{3.175}$$

is not self-adjoint, so the variational method is not immediately applicable.

### 3.4.4   Other approximations

In the previous sections we have discussed the most commonly used analytical methods for approximating the solutions of the Fokker–Planck equations (3.25) and (3.39). However, there are many other possible approximation schemes that can be devised. Here we mention two additional strategies.

#### 3.4.4.1   Fast gradient expansion   

The eigenfunction expansion method, coupled with the ground state dominance approximation, was seen to provide a powerful means for addressing situations of bound states with localization lengths small compared with $R_g$. One might anticipate, however, that it is possible to derive a more general asymptotic expansion, not restricted to bound states, for situations in which the potential is *rapidly varying* on the scale of $R_g$. To investigate this situation, it is again useful to write

$$w(\mathbf{r}) = \psi_w(\mathbf{r}/\xi_w) \tag{3.176}$$

where $\xi_w$ is a length scale characterizing the rapid variations of the potential. Substituting this expression into eqn (3.25) and rescaling the independent variables according to

$$t \equiv b^2 s/(6\xi_w^2), \quad \mathbf{x} \equiv \mathbf{r}/\xi_w \tag{3.177}$$

leads to the dimensionless diffusion equation

$$\frac{\partial}{\partial t}q(\mathbf{x},t) = \nabla_{\mathbf{x}}^2 q(\mathbf{x},t) - \epsilon_f W(\mathbf{x})q(\mathbf{x},t) \tag{3.178}$$

where $\nabla_{\mathbf{x}} \equiv \partial/\partial\mathbf{x}$, $W(\mathbf{x}) \equiv N\psi_w(\mathbf{r}/\xi_w)$, and $\epsilon_f$ is defined by

$$\epsilon_f \equiv \frac{6\xi_w^2}{b^2 N} = \left(\frac{\xi_w}{R_g}\right)^2 = \epsilon_s^{-1} \tag{3.179}$$

The parameter $\epsilon_f$ is clearly small for potentials that are rapidly varying on the scale of $R_g$. The transformed initial condition is $q(\mathbf{x}, 0; [W]) = 1$.

Equation (3.178) has a close resemblance to eqn (3.114), which served as the basis for the weak inhomogeneity expansion. To complete the correspondence, we introduce $W_0 \equiv (\xi_w^3/V) \int d\mathbf{x} W(\mathbf{x})$ and define the non-uniform part of the field $W(\mathbf{x})$ by

$$\Omega(\mathbf{x}) = W(\mathbf{x}) - W_0 \tag{3.180}$$

Equation (3.178) is then transformed by the substitution

$$q(\mathbf{x}, t) = \exp(-\epsilon_f W_0 t)\, p(\mathbf{x}, t) \tag{3.181}$$

where $p(\mathbf{x}, t)$ satisfies

$$\frac{\partial}{\partial t} p(\mathbf{x}, t) = \nabla_{\mathbf{x}}^2 p(\mathbf{x}, t) - \epsilon_f \Omega(\mathbf{x}) p(\mathbf{x}, t) \tag{3.182}$$

and $p(\mathbf{x}, 0) = 1$.

A *fast gradient expansion*, appropriate for situations where the potential is changing rapidly on the scale of $R_g$, can be derived by retracing the steps used in Section 3.4.1. Specifically, one assumes a perturbation expansion of the form

$$p(\mathbf{x}, t) \sim \sum_{j=0}^{\infty} \epsilon_f^j p^{(j)}(\mathbf{x}, t) \tag{3.183}$$

and solves eqn (3.182) order by order in $\epsilon_f$. The following result is obtained for the partition function

$$Q[W] = \frac{\xi_w^3}{V} \int d\mathbf{x}\, q(\mathbf{x}, \epsilon_f^{-1})$$

$$\sim e^{-W_0} \left[ 1 + \frac{\epsilon_f^2 \xi_w^6}{2V^2} \sum_{\mathbf{k}} [\epsilon_f^{-2} \hat{g}_D(k^2 \epsilon_f^{-1})]\, \hat{\Omega}(\mathbf{k})\hat{\Omega}(-\mathbf{k}) + \dots \right] \tag{3.184}$$

where $\hat{\Omega}(\mathbf{k})$ is the Fourier transform of $\Omega(\mathbf{x})$. If one inserts the asymptotic form of the Debye function $\hat{g}_D(k^2 \epsilon_f^{-1}) \sim 2\epsilon_f/k^2$, which is appropriate for $\epsilon_f \to 0$, this expression reduces to

$$Q[W] \sim e^{-W_0} \left[ 1 + \frac{\epsilon_f \xi_w^6}{V^2} \sum_{\mathbf{k}} \frac{1}{k^2}\, \hat{\Omega}(\mathbf{k})\hat{\Omega}(-\mathbf{k}) + \dots \right] \tag{3.185}$$

which, if we invert the Fourier transform, becomes

$$Q[W] \sim e^{-W_0} \left[ 1 + \frac{\epsilon_f \xi_w^3}{V} \int d\mathbf{x} \int d\mathbf{x'} \frac{1}{4\pi|\mathbf{x} - \mathbf{x'}|} \Omega(\mathbf{x})\Omega(\mathbf{x'}) + ... \right]$$

$$(3.186)$$

Finally, restoring the dimensional variables leads to the following fast gradient expansion of the partition function:

$$Q[w] \sim e^{-w_0 N} \left[ 1 + \frac{\epsilon_f N^2}{\xi_w^2 V} \int d\mathbf{r} \int d\mathbf{r'} \frac{1}{4\pi|\mathbf{r} - \mathbf{r'}|} \omega(\mathbf{r})\omega(\mathbf{r'}) + ... \right]$$

$$(3.187)$$

where $\omega(\mathbf{r})$ is the inhomogeneous part of $w(\mathbf{r})$.

It is of interest to compare this result with eqn (3.132) obtained by the weak inhomogeneity expansion. The fast gradient expansion is seen to be a *special case* of eqn (3.132) in that the latter reduces to the former for situations in which $\omega(\mathbf{r})$ varies on scales $\xi_w$ much less than $R_g$. Indeed, if one inserts eqn (3.137) into eqn (3.132) for this situation, the fast gradient expansion (3.187) is recovered. It follows that eqns (3.134) and (3.136) for the density operator and density–density correlation function also reproduce the correct leading-order asymptotic terms in the fast gradient expansion. In Chapters 5 and 6 we shall see that these observations prove useful in developing stable numerical methods for updating the chemical potential fields in field-based computer simulations.

3.4.4.2 *Strong stretching – the classical approximation* Another approximation scheme that is particularly useful in the theory of polymers tethered to a surface (Milner *et al.*, 1988; Skvortsov *et al.*, 1988) and the description of strongly segregated block copolymer mesophases (Semenov, 1985) is the *strong stretching approximation*. Since the approximation is analogous to the classical limit of quantum mechanics, it is also referred to as the *classical approximation*.

The strong stretching approximation is applicable in situations where a polymer chain is strongly extended, so that the available conformational states are greatly reduced. In dense assemblies of polymers that are attached by one of their ends to a surface, so-called *polymer brushes* (Halperin *et al.*, 1992), such a situation arises naturally because the polymers are forced to stretch away from the surface to avoid each other. As illustrated in Fig. 3.10, a similar situation occurs in strongly segregated block copolymer mesophases (Helfand and Wasserman, 1976). Under conditions where the A and B blocks of a diblock copolymer are strongly incompatible, the junction points connecting the two blocks in a melt of similar chains are strongly localized to the interfaces between dissimilar domains. To avoid each other and fill space, the individual A and B blocks must therefore stretch away from the interfaces into the A- and B-rich domains, respectively.

For the purpose of introducing the strong stretching approximation, it is helpful to return to the path integral description of a continuous Gaussian chain subjected to a potential $w(\mathbf{r})$. Here we assume that the stretching of the chain is

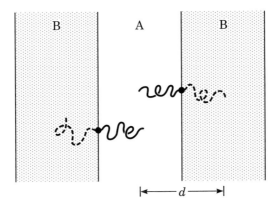

FIG. 3.10. Typical extended ("stretched") chain conformations of individual AB
diblock copolymers in a strongly segregated lamellar mesophase. When the
A and B blocks of the copolymer melt are highly incompatible, the junction
points (solid dots) connecting the blocks are strongly localized at the interfaces between the A and B domains. The chains are also stretched normal to
the interfaces to a characteristic end-to-end distance $d$ that greatly exceeds
the unperturbed (ideal chain) radius of gyration, $R_g$.

due solely to the potential and not to the combined action of the potential and
a tethering constraint, as in the case of polymer brushes.[26] The Green function
$g(\mathbf{r}, \mathbf{r}', N; [w])$, describing the statistical weight of a polymer chain with its ends
at $\mathbf{r}(0) = \mathbf{r}'$ and $\mathbf{r}(N) = \mathbf{r}$, can be expressed as

$$g(\mathbf{r}, \mathbf{r}', N; [w]) \propto \int_{\mathbf{r}(0)=\mathbf{r}'}^{\mathbf{r}(N)=\mathbf{r}} \mathcal{D}\mathbf{r} \, \exp(-\beta U_0[\mathbf{r}] - \beta U_1[\mathbf{r}, w]) \qquad (3.188)$$

where, for simplicity, we have ignored the normalization of $g$. The partition
function is evidently given by

$$Q[w] \propto \int d\mathbf{r} \int d\mathbf{r}' \, g(\mathbf{r}, \mathbf{r}', N; [w]) \qquad (3.189)$$

In a situation where the polymer is strongly stretched with a characteristic
end-to-end separation $|\mathbf{r} - \mathbf{r}'| \sim d$, the contribution to the potential energy from
the stretching term $\beta U_0[\mathbf{r}]$ in eqn (3.188) is of order $\epsilon^{-1} \equiv d^2/R_g^2$, with $|\epsilon| \ll 1$.
Since the potential is responsible for the stretching, it follows that $\beta U_1[\mathbf{r}]$ must
also be of this order of magnitude. We thus introduce a dimensionless contour
variable $t \in [0, 1]$ and a dimensionless polymer shape $\mathbf{x}(t)$,

---

[26]The present situation is a bit contrived because for stretching to be achieved, the potential
must selectively attract different portions of the polymer. While this is natural for a block
copolymer, it is difficult to stretch a homopolymer without an additional tethering constraint.

$$t \equiv s/N, \quad \mathbf{x}(t) \equiv d^{-1}\mathbf{r}(s) \tag{3.190}$$

as well as a scaled potential defined by

$$V(\mathbf{x}(t)) \equiv \frac{R_g^2 N}{d^2} w(\mathbf{r}(s)) = \epsilon N w(\mathbf{r}(s)) \tag{3.191}$$

It follows that the Green function can be expressed as

$$g(\mathbf{x}, \mathbf{x}', 1; [V]) \propto \int_{\mathbf{x}(0)=\mathbf{x}'}^{\mathbf{x}(1)=\mathbf{x}} \mathcal{D}\mathbf{x} \, e^{-\epsilon^{-1} \int_0^1 dt \, [\frac{1}{4}(\frac{d\mathbf{x}(t)}{dt})^2 + V(\mathbf{x}(t))]} \tag{3.192}$$

in which an explicit factor of $\epsilon^{-1} \gg 1$ appears in the exponent of the integrand. For $\epsilon \to 0$, corresponding to a strongly stretched chain, such a path integral over the scaled chain configuration $\mathbf{x}(t)$ can be asymptotically evaluated by Laplace's method (Bender and Orszag, 1978). The leading-order contribution for strong stretching thus arises from the "classical" path $\mathbf{x}^*(t)$ that minimizes the "action"

$$S[\mathbf{x}] = \int_0^1 dt \, \left[ \frac{1}{4} \left( \frac{d\mathbf{x}(t)}{dt} \right)^2 + V(\mathbf{x}(t)) \right]$$

so that

$$g(\mathbf{x}, \mathbf{x}', 1; [V]) \sim \exp\left(-\epsilon^{-1} S[\mathbf{x}^*]\right). \tag{3.193}$$

The classical path satisfies the Euler–Lagrange equation

$$\frac{1}{2} \frac{d^2 \mathbf{x}^*(t)}{dt^2} = \frac{\partial V(\mathbf{x}^*(t))}{\partial \mathbf{x}^*(t)} \tag{3.194}$$

subject to the boundary conditions $\mathbf{x}^*(0) = \mathbf{x}' = \mathbf{r}'/d$ and $\mathbf{x}^*(1) = \mathbf{x} = \mathbf{r}/d$. The term "classical" refers to the close analogy between eqn (3.194) and the Newtonian mechanics of a particle experiencing a potential $U(\mathbf{x}) = -V(\mathbf{x})$.[27] Higher-order asymptotic corrections to eqn (3.193) can be developed by systematic application of Laplace's method to higher order in $\epsilon$.

The evaluation of the classical, or strong stretching, approximation to the Green function proceeds by solving eqn (3.194) with the associated boundary conditions for a prescribed potential $w(\mathbf{r})$. The action $S[\mathbf{x}]$ is then evaluated for the classical path $\mathbf{x}^*(t)$, $t \in [0, 1]$, and used in eqn (3.193) to obtain $g(\mathbf{r}, \mathbf{r}', N; [w])$. Subsequent to evaluating the Green function, eqn (3.189) can be used to obtain an approximation for the partition function, and expressions such as (3.50) and (3.72) can be used to obtain approximations for density operators and density–density correlation functions. Such a procedure is primarily suited for situations in which the potential is simple enough (e.g. harmonic)

---

[27]The analogous approximation in quantum mechanics can be used to derive classical mechanics in the asymptotic limit of $\hbar \to 0$ (Feynman and Hibbs, 1965).

that eqn (3.194) can be integrated analytically (Semenov, 1985; Milner *et al.*, 1988; Netz and Schick, 1998). If numerical methods are required to solve the classical equations of motion, then implementation of the strong stretching approximation is cumbersome and has little advantage over direct numerical attack of eqns (3.22) and (3.25). The latter topic is discussed in Section 3.6.

## 3.5   Boundary conditions

In the preceding sections we have considered models of polymer chains, both in the absence and presence of external potentials, that are free to explore all regions of three-dimensional space (i.e., $\mathbf{r} \in \mathbb{R}^3$). The path integrals, Chapman–Komogorov equations, and Fokker–Planck equations were derived by assuming that space was unbounded. In most situations of interest, however, we would like to impose boundaries on the spatial domain. Sometimes this desire is motivated by computational ease or efficiency; in other cases, a confined geometry is intrinsic to the physical problem of interest. For example, if we are interested in carrying out a field-based numerical simulation of a *bulk* polymeric fluid, there is a practical limitation (controlled by our computational resources) on the number of grid points or spectral elements that can be used to represent the various fields. If we demand a certain spatial resolution to resolve the fine structures within the fluid, then the total size of the computational cell is limited. In particular, if the spatial resolution in a certain direction is $\Delta x$ and we can afford only $N_x$ degrees of freedom, then the length of a side of the cell is limited to $L = \Delta x N_x$. To minimize edge effects and better approximate the properties of a bulk, unbounded fluid, it is thus convenient and appropriate to apply *periodic boundary conditions* on the surface of the computational cell. If the cell is cubic with side length $L$, then we are effectively assuming that all spatially varying fields, e.g. $w(\mathbf{r})$, are triply periodic in the sense that

$$w(\mathbf{r} + L\hat{\mathbf{e}}) = w(\mathbf{r}) \tag{3.195}$$

where $\hat{\mathbf{e}}$ is a unit translation along any of the three edges defining the cube.

In the example just given, the boundary conditions were imposed for reasons of computational necessity, rather than physical circumstance. Examples of the latter include polymeric fluids confined to a rigid container, thin fluid films confined by solid substrates, and films or drops of polymeric liquids confined on one or more sides by a vapor phase, e.g. air. Such bounded systems are evidently of considerable scientific and practical interest, but would seem to fall outside the scope of our formalism because we have consistently assumed that the polymers are free to explore all regions of space. Fortunately, this proves not to be the case because the external potential can be adjusted to confine polymers within a particular domain of interest. A similar situation is encountered in the quantum mechanics of bound systems (Schiff, 1968).

We observed in Section 3.4.3 that a quadratic potential of the form $w(x) = (1/2)(x/\xi_w)^2$ had the effect of localizing a continuous Gaussian chain to within a region of width $\xi \sim (b\xi_w)^{1/2}$ about the origin, $x = 0$. The eigenfunctions $\psi_k(x)$

decayed exponentially as $\sim \exp[-(x/\xi)^2/2]$ outside of this region. A stronger localization can evidently be achieved by selecting a potential that rises faster than $x^2$ for $x \to \pm\infty$. To model a boundary that is impenetrable, such as a solid wall, it thus makes sense to consider the limit of an infinitely steep potential, i.e. a *discontinuous step* in the potential $w(x)$ at the desired boundary location $x = x_0$. Figure 3.11 shows a simple one-dimensional example where we attempt to localize a polymer within a region $0 < x < L$ by imposing a step jump in the potential. Within the region of interest, $x \in (0, L)$, $w(x)$ is arbitrary but finite; outside of the region, $w(x) = W_0 \gg 1$. Because eqn (3.25) is second order in space, we can demand that the solution $q(x, s)$ and its first derivative $\partial_x q(x, s)$ be continuous at the locations of the two steps, $x_0 = 0$ and $x_0 = L$. It is straightforward to show (Schiff, 1968) that in the limit as the step height becomes infinite, i.e. $W_0 \to \infty$, the propagator (for $s > 0$) must *vanish* at each of the step locations, $q(x_0, s) = 0$, while the derivative $\partial_x q(x, s)$ is undetermined at $x = x_0$.

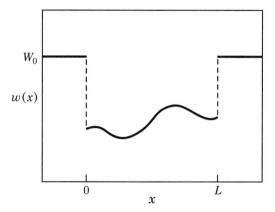

FIG. 3.11. Example of the use of the external potential $w(x)$ to localize a polymer to a finite region $0 < x < L$. Discontinuous steps in $w(x)$ to a large value of $W_0$ are imposed at the boundaries of the region. In the limit that the step height becomes infinite, $W_0 \to \infty$, the polymer is rigorously excluded outside the potential well, and the propagator $q(x, s)$ vanishes at $x = 0$ and $x = L$ for $s > 0$.

It follows more generally that for a polymeric fluid enclosed by a bounding surface $S$ that is *rigid and impenetrable*, the Fokker–Planck equation (3.25) for the continuous Gaussian chain can be solved within the *finite* domain $\mathbf{r} \in \Omega$ bounded by $S$. The solution is subject to the homogeneous Dirichlet boundary condition

$$q(\mathbf{r}, s; [w]) = 0, \quad \text{for } s > 0, \ \mathbf{r} \in S \tag{3.196}$$

Through expressions such as (3.55) and (3.59), it is apparent that this boundary condition on the propagator also implies that the average total, partial, and end segment densities all vanish at a solid boundary.

In the case of a bounding interface between a polymeric fluid and air or another gas phase, the interfacial energies are normally sufficiently high that the width of the vapor–liquid interface is of order 1 nm. Thus, from a mesoscopic perspective, it is reasonable to assume that the gas phase is impenetrable to the polymer and again apply the boundary condition (3.196). However, it should be recognized that liquid–gas interfaces are deformable and, as a consequence, the interfacial shape $\mathcal{S}$ in such a situation must be self-consistently determined by balancing the forces that act upon it.

Similar arguments can be made for justifying the use of Dirichlet boundary conditions in the case of a *wormlike chain*, described by the Fokker–Planck eqn (3.39), confined by rigid and impenetrable boundaries. In some situations, however, it may also be necessary to account for anchoring conditions on the local segment orientation $\mathbf{u}$ at the bounding surfaces (de Gennes and Prost, 1993).

The use of boundary conditions other than eqn (3.196) are sometimes appropriate. For example, consider a chain in the vicinity of an impenetrable wall that weakly attracts the polymer segments (de Gennes, 1979). A sketch of a typical surface potential $w_s(x)$ experienced by the segments is shown in Fig. 3.12. A steep repulsion in the vicinity of the wall at $x = 0$ excludes segments from penetrating the wall, while a potential well of range $b_s$ is responsible for attracting segments near the surface. The range $b_s$ of the attractive well is typically less than 1 nm for neutral (uncharged) polymers and surfaces. One approach to such a problem would be to solve the diffusion equation (3.25) throughout a domain extending into the interior of the wall ($x < 0$), including in the total potential $w(x)$ the surface potential $w_s(x)$, as well as any more slowly varying "bulk" potentials $w_b(x)$ that are present for $x > 0$. Such a numerical solution, however, would require spatial resolution below 1 nm and would severely limit our ability to simulate the structure in the fluid on mesoscopic length scales.

If we are only interested in resolving *mesoscopic* polymer structure away from the wall, a preferable approach would be to replace the surface potential $w_s(x)$ with an *effective boundary condition* at $x = 0$. With this perspective, one solves the diffusion equation (3.25) in the region $x > 0$ outside the wall subject only to the slowly varying bulk potential, $w(x) = w_b(x)$. The net attractive influence of the wall potential is taken into account by imposing a suitable boundary condition on the solution at $x = 0$. For weakly attractive walls, de Gennes (de Gennes, 1979) has argued that the appropriate boundary condition is

$$\left[ \frac{\partial}{\partial x} q(x, s; [w]) + \kappa\, q(x, s; [w]) \right]_{x=0} = 0 \qquad (3.197)$$

where $\kappa^{-1}$ is a positive length scale that characterizes the net strength of the attractive interaction. Specifically, $\kappa b k_B T$ is a characteristic measure of the binding energy per statistical segment at the surface. The derivation of eqn (3.197) invokes the ground state dominance approximation. Augmentation of the "free

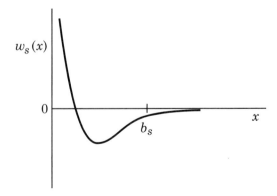

FIG. 3.12. Surface potential $w_s(x)$ leading to the absorption of polymers to a wall located at $x = 0$. The steep repulsion near the wall prevents polymer segments from penetrating into the wall interior $(x < 0)$. The attractive well extending a distance $b_s$ away from the wall is responsible for the net absorption of polymer segments.

energy" functional $F_1[\psi]$ in eqn (3.168) with an attractive delta function surface potential $-(\kappa b^2/12)\delta(x)\psi^2$ leads immediately to the Euler–Lagrange equation (3.160) and the boundary condition (3.197) with $q \to \psi$.

The condition of weak absorption at the segment level, $\kappa b \ll 1$, implies that $\kappa^{-1}$ is a length scale greater than the statistical segment length. However, the ground state dominance approximation requires that the absorbed polymer be confined to a region smaller than its characteristic size $R_g$. Thus, eqn (3.197) is applicable only when the strength of surface attraction produces a mesoscopic length scale $\kappa^{-1}$ in the range $b \ll \kappa^{-1} \ll R_g$.

We shall see in Section 4.9 that similar effective boundary conditions are very convenient for describing polymer melts and concentrated polymer solutions in confined geometries. In such situations, one attempts to replace a rapidly varying surface potential with an effective boundary condition so that the computation of structure in the interior of a fluid is not limited by the necessity of resolving a narrow "boundary layer" at the surface.

## 3.6 Numerical methods

In Section 3.4 we discussed analytical techniques to derive approximate expressions for the chain propagators and partition functions that characterize the statistical mechanics of a polymer in an external potential. Unfortunately, these analytical techniques are quite limited in applicability because they require the potential $w(\mathbf{r})$ to have very particular properties. In most situations of interest, one is therefore resigned to consider *numerical solutions* of the defining Chapman–Kolmogorov and Fokker–Planck equations. Moreover, such a capability turns out to be central to the concept of a "field-theoretic computer simula-tion."

An important factor in assessing the relative merits of particle- and field-based simulations involves an examination of the computationally limiting "inner loop." In a conventional molecular dynamics (MD) simulation, the inner loop amounts to a calculation of the forces acting among the particles of the system. These forces are used to update particle positions and velocities at the next time step. In a field-based simulation of a polymeric fluid, the corresponding "inner loop" involves generating a numerical solution of a Chapman–Kolmogorov or Fokker–Planck equation for a given potential $w(\mathbf{r})$. We shall see in Chapters 5 and 6 that such a solution can be used to define a scheme for updating the potential field at subsequent time steps. Because it is usually the case that a numerical solution of a Fokker–Planck equation such as (3.25) is computationally much more expensive than a potential field update, it is extremely important that efficient numerical procedures be developed for solving such equations.

For the purpose of discussing numerical methods, it proves convenient to focus attention on the simple case of a flexible homopolymer experiencing a potential field $w(x)$ that has variations only along one coordinate $x$. In this case, eqn (3.25) reduces to a one-dimensional diffusion equation

$$\frac{\partial}{\partial s}q(x, s; [w]) = \frac{b^2}{6}\frac{\partial^2}{\partial x^2}q(x, s; [w]) - w(x)q(x, s; [w]) \qquad (3.198)$$

that is subject to the initial condition $q(x, 0; [w]) = 1$ and suitable boundary conditions at the edges of some interval $0 \le x \le L$.

### 3.6.1 Finite difference method

A standard technique for obtaining a numerical solution of an initial value prob-lem such as eqn (3.198) is to apply the *finite difference method* (Press *et al.*, 1992). In this approach one defines a computational grid of discrete points at which the continuous function $q(x, s)$ is sampled. For example, over the interval $x \in [0, L]$ we could use $N_x$ equally spaced points

$$x_j = j\Delta x, \quad j = 0, 1, ..., N_x - 1 \qquad (3.199)$$

where $\Delta x = L/(N_x - 1)$ is the so-called *grid spacing*. A similar $N_s$-point dis-cretization of $s \in [0, N]$ leads to

$$s_n = n\Delta s, \quad n = 0, 1, ..., N_s - 1 \qquad (3.200)$$

where $\Delta s = N/(N_s - 1)$ is the *contour step*. The values of the propagator $q(x, s)$ at these discrete space-contour points will be denoted by

$$q_j^n = q(x_j, s_n) \qquad (3.201)$$

The next step is to introduce approximations for the relevant derivatives. In the case of eqn (3.198), a common choice is a *forward Euler* difference approximation for the $s$ derivative

$$\frac{\partial q}{\partial s}\bigg|_{j,n} = \frac{q_j^{n+1} - q_j^n}{\Delta s} + O(\Delta s) \tag{3.202}$$

and a *centered space* difference approximation for the $x$ derivative

$$\frac{\partial^2 q}{\partial x^2}\bigg|_{j,n} = \frac{q_{j+1}^n - 2q_j^n + q_{j-1}^n}{(\Delta x)^2} + O(\Delta x^2) \tag{3.203}$$

Substitution of these expressions into eqn (3.198) leads to the so-called *forward time-centered space* (FTCS) finite difference scheme

$$\frac{q_j^{n+1} - q_j^n}{\Delta s} = \frac{b^2}{6}\left[\frac{q_{j+1}^n - 2q_j^n + q_{j-1}^n}{(\Delta x)^2}\right] - w_j q_j^n \tag{3.204}$$

The FTCS scheme is a so-called *explicit* algorithm because one can solve directly (explicitly) for the solution of a slightly longer polymer chain, $q_j^{n+1}$, using knowledge of $q_j^n$ for all spatial grid points $j$. It is thus easy to propagate the initial condition, $q_j^0 = 1$, forward to chains of any length. In particular, if one uses the initial data, eqn (3.204) can be solved for $n = 0$ to obtain $q_j^1$ at each $j$. The boundary conditions come into play at this stage, since we are free to impose extra conditions at the "boundary points" $j = 0, N_x - 1$. For example, in the case of the Dirichlet boundary condition (3.196), eqn (3.204) is applied for $n = 0$ at the "interior points" $1 \le j \le N_x - 2$. This nearly complete solution for $q_j^1$ is supplemented with the boundary data

$$q_0^1 = q_{N_x-1}^1 = 0 \tag{3.205}$$

Evidently, this procedure can be repeated for $n = 1, 2, ..., N_s - 2$ to propagate the numerical solution out to polymer chains of length $N$.

While simple to implement, the FTCS scheme has a serious shortcoming. The algorithm is numerically *unstable*, i.e. errors will amplify with increasing $n$, unless the contour step $\Delta s$ is taken to be extremely small. In particular, for the special case of $w = 0$, it is straightforward to show that the so-called Courant-Friedrichs-Lewy (CFL) stability condition is (Press *et al.*, 1992)

$$\Delta s \le 3(\Delta x)^2/b^2 \tag{3.206}$$

Although this stringent stability threshold may be acceptable for certain one-dimensional calculations, the FTCS algorithm is entirely unsuitable for generating high-resolution numerical solutions in two and three dimensions. Besides stability issues, the FTCS scheme suffers from rather low *accuracy*. It is referred to as *first order accurate* in $s$ and *second order accurate* in $x$, since the errors associated with the finite difference approximation to the derivatives are locally $O(\Delta s)$ and $O(\Delta x^2)$, respectively.

The problem of stability can be overcome by switching to a *semi-implicit* finite difference scheme for solving eqn (3.198). A useful algorithm is the *Crank–Nicolson* scheme defined by

$$\frac{q_j^{n+1} - q_j^n}{\Delta s} = \frac{b^2}{12(\Delta x)^2} \left[ (q_{j+1}^{n+1} - 2q_j^{n+1} + q_{j-1}^{n+1}) + (q_{j+1}^n - 2q_j^n + q_{j-1}^n) \right]$$
$$- \frac{1}{2} w_j \left( q_j^{n+1} + q_j^n \right) \tag{3.207}$$

This scheme can be viewed as a variation of the FTCS algorithm in which the finite difference operators on both sides of the diffusion equation are centered at a fictitious contour point $n+1/2$ halfway between contour points $n$ and $n+1$. This contour centering has the advantage of increasing the accuracy of the scheme to *second order* in both $s$ and $x$. More importantly, the Crank–Nicolson algorithm is *unconditionally stable* for any size contour step $\Delta s$. Thus, the choice of contour step is not limited by stability, but is dictated solely by requirements on accuracy.

The solution of eqn (3.207) proceeds by writing the equation in matrix form:

$$\left[ 1 + \frac{\Delta s}{2} \mathbf{w} - \frac{b^2 \Delta s}{12(\Delta x)^2} \mathbf{T} \right] \mathbf{q}^{n+1} = \left[ 1 - \frac{\Delta s}{2} \mathbf{w} + \frac{b^2 \Delta s}{12(\Delta x)^2} \mathbf{T} \right] \mathbf{q}^n \tag{3.208}$$

where $\mathbf{q}^n$ and $\mathbf{q}^{n+1}$ are column vectors containing the values of the propagator at the $N_x$ spatial grid points and at contour steps $n$ and $n+1$, respectively. The $N_x \times N_x$ matrix $\mathbf{T}$ is a linear operator defined by

$$(\mathbf{Tq})_j = q_{j+1} - 2q_j + q_{j-1} \tag{3.209}$$

at the interior spatial grid points $j = 1, 2, ..., N_x - 2$ and supplemented by boundary data at the exterior points $j = 0, N_x - 1$. Two other $N_x \times N_x$ matrices appearing in eqn (3.208) are $\mathbf{w}$, a diagonal matrix with elements $w_j$, and $\mathbf{1}$, the unit matrix. Formally, we can write the solution of eqn (3.208) as

$$\mathbf{q}^{n+1} = \left[ 1 + \frac{\Delta s}{2} \mathbf{w} - \frac{b^2 \Delta s}{12(\Delta x)^2} \mathbf{T} \right]^{-1} \left[ 1 - \frac{\Delta s}{2} \mathbf{w} + \frac{b^2 \Delta s}{12(\Delta x)^2} \mathbf{T} \right] \mathbf{q}^n \tag{3.210}$$

Fortunately the matrix $\mathbf{T}$ is *tridiagonal* for the common cases of Dirichlet ($q = 0$) and Neumann ($\partial_x q = 0$) boundary conditions, so the indicated matrix operations on the right-hand side of eqn (3.210) can be efficiently performed in $O(N_x)$ operations. In the important case of periodic boundary conditions, $q(x + L, s) = q(x, s)$, the linear equations constitute a *cyclic tridiagonal system* that can also be solved efficiently in $O(N_x)$ operations (Press *et al.*, 1992).

The Crank–Nicolson algorithm is attractive for one-dimensional calculations by virtue of its unconditional stability, reasonable (second order) accuracy, and ease of implementation. The computational effort involved in generating a numerical approximation to the solution of eqn (3.198) evidently scales as $N_x N_s$, i.e. $O(N_x)$ operations for each of the $N_s$ iterations of eqn (3.210) required to propagate the initial condition to a chain of contour length $N$.

For numerical solutions of eqn (3.25) in two and three spatial dimensions, the method is less desirable. The analog of the matrix $\mathbf{T}$ arising from finite difference approximations in higher dimensions remains sparse, but is no longer tridiagonal. This can be rectified by an operator-splitting strategy, the alternating-direction implicit method (ADI), which decomposes $\mathbf{T}$ into a sum of $d$ (the number of space dimensions) tridiagonal operators. This leads to a set of $d$ tridiagonal linear systems that can be solved efficiently to generate the overall update from $\mathbf{q}^n$ to $\mathbf{q}^{n+1}$ (Press $et\ al.$, 1992). While this ADI scheme applied in two dimensions is unconditionally stable, in three dimensions it is subject to a severely limiting CFL stability condition of $\Delta s \lesssim (\Delta x/b)^2$. Other semi-implicit finite difference algorithms are available that are unconditionally stable in three dimensions (Ames, 1977), but generally such methods are less attractive than the spectral and pseudo-spectral techniques that will be discussed in the following sections.

It is important to emphasize how the computational cost of generating finite difference solutions varies with the dimensionality of the problem. If we use $N_x$ spatial grid points to resolve the variation of $q(\mathbf{r}, s)$ along each of the $d$ spatial coordinates $\mathbf{r} = (x_1, ..., x_d)$ and $N_s$ contour points to resolve the $s$ dependence, the number of operations required to develop a numerical solution of eqn (3.25) scales as $N_x^d N_s$. For a "high-resolution" three-dimensional ($d = 3$) calculation with $N_x = 256$ and $N_s = 100$, more than $10^9$ operations are required! As this simple example illustrates, the spatial resolution ($\sim N_x^3$) is much more limiting in large three-dimensional calculations than the contour resolution ($\sim N_s$). It is thus advantageous to consider alternatives to finite differencing that provide higher spatial accuracy at a reduced computational cost.

### 3.6.2  $Spectral\ methods$

An alternative to finite difference methods for two- and three-dimensional calculations in regular geometries is the so-called $spectral\ method$ (Gottlieb and Orszag, 1977; Boyd, 2001; Trefethen, 2000). For smooth potential fields $w(\mathbf{r})$, spectral methods tend to converge much more rapidly than finite difference techniques and can provide a significant computational savings. A spectral method for solving modified diffusion equations such as eqns (3.25), (3.99), and (3.101) was first introduced to the polymer physics literature by Matsen and Schick (1994$a$).

### 3.6.2.1  $Choice\ of\ basis$   To introduce the spectral method, we return to the one-dimensional problem of eqn (3.198). The basic idea is to expand the spatial dependence of the unknown function in a complete set of smooth basis functions, $\phi_j(x)$, according to

$$q(x, s) = \sum_{j=0}^{\infty} a_j(s)\, \phi_j(x) \qquad (3.211)$$

where the $a_j(s)$ are $s$-dependent expansion coefficients. Necessary properties of the $\phi_j(x)$ are that they form a complete, linearly independent basis for representing the $x$ dependence of the solution $q(x, s)$, given some smooth $w(x)$ belonging to an appropriate Hilbert space. *Desirable*, but not essential, properties of the $\phi_j(x)$ are

- Basis functions corresponding to different values of $j$ are orthogonal.
- The basis functions individually satisfy the boundary conditions imposed at $x = 0$ and $x = L$.
- The basis functions are eigenfunctions of the operator $\partial^2/\partial x^2$ in the sense that

$$\frac{d^2}{dx^2}\phi_j(x) = -\Lambda_j\phi_j(x)$$

where $\Lambda_j$ denotes the eigenvalue corresponding to $\phi_j(x)$.

We shall proceed by assuming that these desirable properties are met. For the case of *periodic boundary conditions*, the *plane wave basis* satisfies all the above conditions:

$$\phi_j(x) = \exp(i2\pi jx/L), \quad j = 0, \pm 1, \pm 2, ..., \pm\infty \tag{3.212}$$

Correspondingly, for Dirichlet ($q = 0$) boundary conditions, we have

$$\phi_j(x) = \sin(\pi jx/L), \quad j = 1, 2, 3, ..., \infty \tag{3.213}$$

while for Neumann ($\partial_x q = 0$) boundary conditions,

$$\phi_j(x) = \cos(\pi jx/L), \quad j = 0, 1, 2, ..., \infty \tag{3.214}$$

These functions, of course, are the familiar basis functions of trigonometric Fourier series (Riley *et al.*, 1998; Tolstov, 1976).

The power of spectral methods relates to their excellent convergence properties. Consider the following approximation to $q(x)$ for the case of periodic boundary conditions that involves truncating a Fourier series after $M = 2P + 1$ plane waves:[28]

$$q_M(x) = \sum_{j=-P}^{P} a_j \exp(i2\pi jx/L) \tag{3.215}$$

By using the orthogonality of the plane waves, the Fourier coefficients are given by (Appendix A)

$$a_j = \frac{1}{L}\int_0^L dx\, q(x)\exp(-i2\pi jx/L) \tag{3.216}$$

A key theorem of Fourier approximation theory (Gottlieb and Orszag, 1977) is the following:

---

[28]Note that the $s$ dependence of the function has been suppressed for simplicity.

**Theorem 3.1** *If the function $q(x)$ to be approximated is continuous, infinitely differentiable, and periodic with period $L$, then the approximation $q_M(x)$ converges to $q(x)$ more rapidly than any finite power of $1/M$ as $M \to \infty$ for all $x$.*

In practice, the error $|q(x) - q_M(x)|$ can decay as fast as $\sim \exp(-M^2)$ in the number of retained modes for $M \to \infty$. For infinitely smooth functions $q(x)$ this provides a *spectacular* advantage over finite difference methods. Recall that in the case of a second order finite difference scheme, the local error falls off algebraically with the number of grid points as $N_x^{-2}$.

The question arises as to whether the strict condition of infinite smoothness can ever be met in the solution of an equation such as (3.198). Clearly if $w(x)$ is infinitely smooth, the solution of eqn (3.198) subject to the initial condition $q(x,0) = 1$ and periodic boundary conditions will satisfy the conditions of Theorem 3.1. However, we shall see that in most situations of physical interest, we are not free to specify $w(x)$, but it is either self-consistently determined as part of the problem (Chapter 5) or stochastically sampled (Chapter 6). In the former case of polymer self-consistent field theory, when applied in a domain with periodic boundary conditions, the self-consistent solution for $w(x)$ presumably meets the infinite smoothness criterion. In the case of stochastically sampled $w(x)$, the situation is much less clear, although our practical experience to date shows that spectral methods offer far better accuracy than finite difference techniques utilizing comparable number of degrees of freedom, $N_x \sim M$.

In cases of boundary conditions other than periodic, the convergence properties of Fourier series are sensitive not only to the smoothness of $q(x)$ throughout the domain $0 \leq x \leq L$, but also to the values of $q(x)$ and its derivatives at the boundaries. For example, when one applies Neumann boundary conditions, the truncated cosine series

$$q_M(x) = \sum_{j=0}^{M} a_j \, \cos(\pi j x / L) \qquad (3.217)$$

has convergence properties summarized by the following theorem:

**Theorem 3.2** *If $q(x)$ is continuous and infinitely differentiable for $0 \leq x \leq L$ and $q^{(2p+1)}(0) = q^{(2p+1)}(L) = 0$ for $p = 0, 1, 2, ..., \infty$, then the approximation $q_M(x)$ given by eqn (3.217) converges to $q(x)$ more rapidly than any finite power of $1/M$ as $M \to \infty$ for all $x \in [0, L]$.*

Thus, in the case of Neumann conditions, all *odd* derivatives of $q(x)$ must vanish identically at both boundaries in order to achieve "ideal" spectral accuracy.

The case of Dirichlet boundary conditions is similar. Besides the infinite smoothness requirement for $q(x)$, all *even* derivatives of $q(x)$ of second and higher order must vanish at both boundaries in order for a truncated Fourier sine series to exhibit ideal spectral accuracy.

The requirements on higher derivatives of $q(x)$ at the boundaries are usually not met for problems in bounded geometries. Thus, rather than using a sine or cosine basis, it is desirable to use basis functions that do not suffer from these boundary restrictions. Particularly useful are the *Chebyshev polynomials*, defined by $T_n(\cos\theta) = \cos n\theta$. If one rescales $x \in [0, L]$ to a new independent variable $\tilde{x} = 2x/L - 1$ that varies over $-1 \leq \tilde{x} \leq 1$, $q(\tilde{x})$ can be approximated by a truncated Chebyshev series of the form

$$q_M(\tilde{x}) = \sum_{j=0}^{M} a_j T_j(\tilde{x}) \tag{3.218}$$

with expansion coefficients given by

$$a_j = \frac{2}{\pi c_j} \int_{-1}^{1} d\tilde{x}\, q(\tilde{x}) T_j(\tilde{x}) (1 - \tilde{x}^2)^{-1/2} \tag{3.219}$$

where $c_0 = 2$, $c_j = 1$ ($j > 0$). The series (3.218) has a remarkable convergence property analogous to Theorem 3.1 (Gottlieb and Orszag, 1977); namely, if $q(\tilde{x})$ is continuous and infinitely differentiable for $|\tilde{x}| \leq 1$, the error $|q(\tilde{x}) - q_M(\tilde{x})|$ goes to zero at each $\tilde{x} \in [-1, 1]$ more rapidly than any finite power of $1/M$ for $M \to \infty$. Notably absent are any restrictions on the values of $q(\tilde{x})$ or its derivatives at the boundaries $\tilde{x} = \pm 1$. It follows that the convergence properties of Chebyshev series are superior to those of sine and cosine series for problems involving Dirichlet boundary conditions (polymers solutions confined to a slit), Neumann boundary conditions (polymer melt films, see Section 4.9), and mixed boundary conditions such as eqn (3.197) (polymers near attractive surfaces). A final attractive feature of Chebyshev polynomials is that *analysis* (calculation of the $a_j$ expansion coefficients given an approximation to the function at grid points) and *synthesis* (evaluation of the function at grid points given the Chebyshev expansion coefficients) can be performed with efficient fast Fourier transform (FFT) algorithms, as is described in Section 3.6.3 for Fourier bases (Gottlieb and Orszag, 1977; Boyd, 2001; Trefethen, 2000).

Although the convergence properties of Chebyshev series are generally superior to those of sine and cosine series, the Chebyshev basis functions do not possess all of the "desirable" properties noted above. Specifically, the $T_j(\tilde{x})$ are not eigenfunctions of the operator $\partial^2/\partial\tilde{x}^2$, and do not individually satisfy the boundary conditions that are to be imposed at $\tilde{x} = \pm 1$. These difficulties can be overcome, but selection of the Chebyshev basis does lead to other complications in implementation that will be discussed below.

3.6.2.2  *Implementation*  Having settled on a choice of basis, we still have to project the modified diffusion equation (3.198) onto an equation satisfied by a spectral approximation of the form

$$q_M(x, s) = \sum_{j=0}^{M-1} a_j(s)\, \phi_j(x) \tag{3.220}$$

There are several methods, including *Galerkin*, *tau*, and *collocation*, for carrying out such a projection (Gottlieb and Orszag, 1977). Here we introduce the

Galerkin method for the particular case of a plane wave basis and defer discussion of the collocation or "pseudo-spectral" method for more general bases to Section 3.6.3.

The Galerkin method assumes that the basis functions $\phi_j(x)$ are linearly independent and that they *individually satisfy the boundary conditions* to be imposed on $q(x, s)$ at $x = 0, L$. The Galerkin equations for the coefficients $a_j(s)$ of the spectral approximation are obtained by projecting $\phi_j(x)$ onto eqn (3.198) with $q \to q_M$ according to

$$\frac{d}{ds}(\phi_j, q_M) = (\phi_j, \mathcal{L}q_M), \quad j = 0, 1, 2, ..., M - 1, \tag{3.221}$$

where $\mathcal{L} = (b^2/6)\partial^2/\partial x^2 - w(x)$ and $(f, g)$ denotes the $L_2$ inner product of two functions $f(x)$ and $g(x)$ appropriate for the choice of basis.

In the case of *periodic boundary conditions* and a plane wave basis, we have the following orthogonality property:

$$(\phi_j, \phi_k) \equiv \int_0^L dx \ \exp(-i2\pi jx/L)\exp(i2\pi kx/L) = L\delta_{jk} \tag{3.222}$$

This reduces eqns (3.221) for the case of $M = 2P+1$ plane waves to the following set of equations:

$$\frac{d}{ds}a_j(s) = -\frac{2\pi^2 b^2 j^2}{3L^2}a_j(s) - \sum_{l=-P}^{P} d_{j-l}\, a_l(s), \quad j = 0, \pm 1, \pm 2, ..., \pm P \tag{3.223}$$

where $d_j$ is the $j$th Fourier coefficient of $w(x)$ defined by

$$d_j = \frac{1}{L}\int_0^L dx \ \exp(-i2\pi jx/L)w(x) \tag{3.224}$$

The initial condition $q(x, 0) = 1$ corresponds to the condition

$$a_j(0) = \delta_{j0}, \quad j = 0, \pm 1, \pm 2, ..., \pm P \tag{3.225}$$

Together, eqns (3.223) and (3.225) constitute a set of $M$ linear first-order ordinary differential equations and $M$ initial conditions for the Fourier coefficients $a_j(s)$. Once these equations have been solved, the coefficients are substituted into eqn (3.220) with $\phi_j(x) = \exp(i2\pi jx/L)$ to obtain the desired spectral approximation to the solution, $q_M(x, s)$.

In matrix notation, the equation set (3.223) can be written as

$$\frac{d}{ds}\mathbf{a}(s) = \mathbf{L}_M\mathbf{a}(s) \tag{3.226}$$

where the elements of the Hermitian $M \times M$ matrix $\mathbf{L}_M$ are

$$(\mathbf{L}_M)_{jl} = -(2\pi^2 b^2 j^2/3L^2)\delta_{jl} - d_{j-l} \tag{3.227}$$

This operator represents the projection of the original differential operator $\mathcal{L}$ onto the truncated Fourier basis. Because plane waves are eigenfunctions of the operator $\partial^2/\partial x^2$, the first term in eqn (3.227) is *diagonal* in the normal mode representation. Unfortunately, the second term in eqn (3.227), which corresponds to the local operator $w(x)$ in the $x$ representation, has become highly *non-local* in the plane wave representation.

There are several options for solving the above system of equations. The first, the so-called "semi-discrete" method, is the strategy adopted by Matsen and Schick (Matsen and Schick, 1994 a). This approach takes advantage of the linearity of the equations and formally solves eqn (3.226) as

$$\mathbf{a}(s) = \exp(\mathbf{L}_M s)\,\mathbf{a}(0) \qquad (3.228)$$

Evaluation of the time evolution operator $\exp(\mathbf{L}_M s)$ appearing in this expression requires that the matrix $\mathbf{L}_M$ be diagonalized, an expensive $O(M^3)$ operation. In a three-dimensional calculation using $M$ plane waves in each direction, this approach would be hugely expensive since the computational effort would scale as $O(M^9)$. Thus, we do not recommend the semi-discrete strategy for other than small problems.

A preferred method for solving eqn (3.226) is to apply the finite difference technique on the contour variable $s$. As in Section 3.6.1, it is convenient to sample $s \in [0, N]$ at $N_s$ equally spaced points along the chain contour, $s_n = n\Delta s$, $n = 0, 1, ..., N_s - 1$, where the contour step is $\Delta s = N/(N_s - 1)$. The following notation is employed to denote the value of a spectral coefficient evaluated at the $n$th contour grid point:

$$a_j^n = a_j(s_n) \qquad (3.229)$$

Simple *forward Euler* contour-differencing of eqn (3.226) leads to

$$\frac{\mathbf{a}^{n+1} - \mathbf{a}^n}{\Delta s} = \mathbf{L}_M \mathbf{a}^n \qquad (3.230)$$

Starting from the initial vector $\mathbf{a}^0$, one can iterate this equation $N_s - 1$ times to propagate the approximate solution forward to the end of the chain. A solution so generated for an arbitrary potential field $w(x)$ requires order $M^2 N_s$ operations, i.e. one matrix multiplication at a cost of $O(M^2)$ operations[29] for each of the $O(N_s)$ contour steps. This $\sim M^2 N_s$ computational burden would seem to be a modest improvement over the $O(M^3)$ requirement of the semi-discrete approach, since we might expect to set $N_s \sim M \sim 100$ in a typical one-dimensional problem. However, there is a huge advantage to contour-differencing in a large three-dimensional calculation using $M$ plane waves in each direction, since the cost is then $O(M^6 N_s)$ versus $O(M^9)$ for the semi-discrete approach.

---

[29]The potential term in $\mathbf{L}_M \mathbf{a}^n$ naively requires $O(M^2)$ operations to carry out the indicated matrix multiplication. However, this term actually has the form of a convolution sum that can be evaluated in $O(M \log_2 M)$ operations by transforming back into the $x$ representation using fast Fourier transform techniques. We defer discussion of such an approach, which is referred to as *pseudo-spectral*, to Section 3.6.3.

As with the case of the FTCS finite difference scheme (3.204), one might expect that the *explicit* forward Euler algorithm (3.230) would suffer from stability problems. Indeed, it can be shown that eqn (3.230) requires a contour step $\Delta s$ smaller than $O(M^{-2})$ for stability. The resolution of this difficulty is to again adopt an *implicit* or *semi-implicit* contour-stepping scheme.

The *Crank–Nicolson* algorithm, which in the present context can be written

$$\frac{\mathbf{a}^{n+1} - \mathbf{a}^n}{\Delta s} = \frac{1}{2}\mathbf{L}_M(\mathbf{a}^{n+1} + \mathbf{a}^n), \qquad (3.231)$$

is unconditionally stable, but unfortunately is prohibitively expensive to execute because it requires inversion of a non-sparse matrix. This would lead to an overall computational cost of $O(M^3 + M^2 N_s)$.

A more satisfactory semi-implicit scheme is derived by splitting the operator $\mathbf{L}_M = \mathbf{L}_M^D + \mathbf{L}_M^W$ into its diffusional and potential contributions, respectively, defined by the two terms on the right-hand side of eqn (3.227). Because $\mathbf{L}_M^D$ is diagonal in the plane wave representation, the term $\mathbf{L}_M^D \mathbf{a}(s)$ in eqn (3.226) can be treated *implicitly*. The problematic term $\mathbf{L}_M^W \mathbf{a}(s)$ is treated *explicitly*. However, to improve stability, we add and subtract an operator (Gottlieb and Orszag, 1977)

$$\mathbf{L}_M^{\max} = -\frac{1}{2}w_{\max}\mathbf{1}, \qquad (3.232)$$

where $w_{\max} = \max_{x\in[0,L]} w(x)$, at the future $n + 1$ and current $n$ contour locations. This leads to the semi-implicit algorithm

$$\frac{\mathbf{a}^{n+1} - \mathbf{a}^n}{\Delta s} = \mathbf{L}_M^D \mathbf{a}^{n+1} + \mathbf{L}_M^W \mathbf{a}^n + \mathbf{L}_M^{\max}\mathbf{a}^{n+1} - \mathbf{L}_M^{\max}\mathbf{a}^n \qquad (3.233)$$

Because both $\mathbf{L}_M^D$ and $\mathbf{L}_M^{\max}$ are diagonal, eqn (3.233) can be efficiently solved for $\mathbf{a}^{n+1}$ in terms of $\mathbf{a}^n$. A full solution of these equations propagated for $N_s$ contour steps requires $O(M^2 N_s)$ operations in one dimension and $O(M^6 N_s)$ in three dimensions (using $M$ plane waves in each direction). The scheme has the same first order accuracy as the forward Euler algorithm (3.230), but with dramatically improved stability. Higher order semi-implicit contour stepping schemes are straightforward to devise (Ascher *et al.*, 1995), although these are best implemented in a pseudo-spectral framework – the subject of Section 3.6.3.

An example of a numerical solution of eqn (3.198) using the semi-implicit scheme (3.233) with a plane wave basis is shown in Fig. 3.13. A potential $W(x) = Nw(x)$ was employed for a system of size $L/R_g = 10$ that has a deep attractive well centered at $x = 5R_g$. Shown is the propagator $q(x, N)$, which is proportional to the chain end density, cf. eqn (3.61), and exhibits a maximum at the center of the potential well. Note that the maximum of $q(x, N)$ is broader than the potential well and that the end density is significant even outside the well, indicating that the polymer is not completely bound by the potential. The single-chain partition function is obtained from the relation $Q[w] = a_0(N)$ and

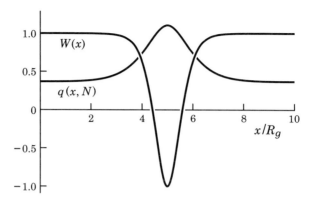

FIG. 3.13. Chain end distribution function $q(x, N)$ obtained by implementing scheme (3.233) with $M = 65$ plane waves and $N_s = 201$ contour steps ($\Delta s = 0.005N$) for the scaled potential function $W(x) = Nw(x) = 1 - 2\left\{\text{sech}[3(x - L/2)/(2R_g)]\right\}^2$. Periodic boundary conditions with period $L = 10R_g$ were assumed.

has a value of 0.539379 for the specified potential. Surprisingly, three-digit accuracy was obtained for $Q[w]$ using a contour step as large as $\Delta s = 0.1N$ for $M = 65$ plane waves.

### 3.6.3  *Pseudo-spectral methods*

In the previous section we applied Galerkin projection and a plane wave basis to illustrate how a spectral approximation can be implemented to solve eqn (3.198) in the case of periodic boundary conditions. Unfortunately, the fact that the linear operator $\mathcal{L}^W = -w(x)$ is highly non-local in the plane wave representation limits the efficiency of the method. In spite of enjoying spectral accuracy in $x$, algorithm (3.233) requires $O(M^2 N_s)$ operations for a solution of the diffusion equation in one dimension. This is far from the computational "ideal" of $O(MN_s)$ operations.

It is possible to devise algorithms that come much closer to this ideal linear scaling with the number of retained modes, while still enjoying spectral accuracy. The strategy is to switch to a different projection technique, the so-called *collocation*, or *pseudo-spectral* method (Lanczos, 1956; Gottlieb and Orszag, 1977; Trefethen, 2000). Recall that the Galerkin technique employs inner products to derive equations satisfied by the expansion coefficients of a spectral approximation $q_M(x, s)$. In the collocation method, one determines the expansion coefficients instead by insisting that $q_M(x, s)$ agree with the exact solution $q(x, s)$ at a set of $M$ "collocation points" $x_0, x_1, ..., x_{M-1}$ that are interior to the domain $[0, L]$. Thus, the spectral coefficients $a_j(s)$ of eqn (3.220) are determined by collocation projection according to

$$\sum_{j=0}^{M-1} a_j(s)\phi_j(x_l) = q(x_l, s), \quad l = 0, 1, ..., M-1 \tag{3.234}$$

where it is assumed that the basis functions $\phi_j(x)$ satisfy the boundary conditions of the problem for each $j$. The results of collocation evidently depend on the choice of the interior points, as well as the choice of basis functions. It is by making an appropriate selection of both that the collocation method can lead to a more efficient algorithm.

We shall begin by discussing the case of periodic boundary conditions (period $L$) and the plane wave basis (3.212). Equation (3.234) becomes

$$\sum_{j=-M/2}^{M/2} a_j(s)\exp(i2\pi jx_l/L) = q(x_l, s), \quad l = 0, 1, ..., M-1 \tag{3.235}$$

A convenient choice of collocation points are the equally spaced set[30] $x_l = lL/M$, $l = 0, 1, ..., M-1$. Substitution into eqn (3.235) leads to

$$\sum_{j=-M/2+1}^{M/2} a_j(s)\exp(i2\pi jl/M) = q_l(s), \quad l = 0, 1, ..., M-1 \tag{3.236}$$

where $q_l(s) \equiv q(x_l, s)$. Note that the term with $j = -M/2$ has been omitted from the sum in the left-hand side of this equation. This is because the collocated basis functions $\exp(i2\pi jl/M)$ are not linearly independent for $j = M/2$ and $j = -M/2$. Thus, we are free to drop the latter term in the sum, so that exactly $M$ modes and $M$ collocation points are retained. A final rewriting of eqn (3.236) invokes the fact that the basis functions $\exp(i2\pi jl/M)$ are periodic in $j$ with period $M$. The sum can therefore be taken over any $M$ successive values of $j$, leading to

$$q_l(s) = \sum_{j=0}^{M-1} a_j(s)\exp(i2\pi jl/M), \quad l = 0, 1, ..., M-1 \tag{3.237}$$

By using the following orthogonality property of the collocated basis functions

$$\sum_{k=0}^{M-1} \exp(-i2\pi jk/M)\exp(i2\pi kl/M) = \delta_{jl}M, \tag{3.238}$$

one can explicitly solve eqn (3.237) for the spectral coefficients

$$a_j(s) = \frac{1}{M}\sum_{l=0}^{M-1} q_l(s)\exp(-i2\pi jl/M), \quad j = 0, 1, ..., M-1 \tag{3.239}$$

---

[30]Notice that one of these points is a boundary point, rather than an interior point, but this proves to be of no consequence for the plane wave basis.

The computational advantage of this particular collocation procedure stems from the fact that eqns (3.239) and (3.237) are *discrete Fourier transform pairs* (Press *et al.*, 1992). Thus, either expression can be efficiently evaluated using $O(M \log_2 M)$ operations by applying a fast Fourier transform (FFT) algorithm. This allows one to spectrally resolve a function that is sampled at discrete points, or go from a specified set of spectral coefficients to values of a function at the collocation points. The ability to switch efficiently between spectral and $x$-space representations of an approximation to a function is responsible for the power and flexibility of the "pseudo-spectral" method.

We now discuss how to apply these collocation formulas for the numerical solution of eqn (3.198). Formally, one can take advantage of the linearity of the equation and "solve" it by integrating along the contour variable from $s$ to some $s + \Delta s$

$$q(x, s + \Delta s) = \exp(\mathcal{L}\Delta s) \, q(x, s) \qquad (3.240)$$

The propagator $\exp(\mathcal{L}\Delta s)$ in this expression is of course difficult to evaluate, because the operator $\mathcal{L} = (b^2/6)\partial^2/\partial x^2 - w(x)$ is not local in either the $x$ space or Fourier representation, cf. eqn (3.227). However, if the diffusive and potential contributions to $\mathcal{L}$ are separated according to $\mathcal{L} = \mathcal{L}^D + \mathcal{L}^W$, then it is straightforward to show by explicit Taylor expansion in $\Delta s$ that

$$\exp(\mathcal{L}\Delta s) = \exp[\mathcal{L}^W \Delta s/2] \exp(\mathcal{L}^D \Delta s) \exp[\mathcal{L}^W \Delta s/2] + O(\Delta s^3) \qquad (3.241)$$

In other words, to *second order accuracy* in $s$, it is possible to use a propagator in which the diffusive and potential contributions can be applied separately. By choosing to sample $s \in [0, N]$ discretely at the contour points $s_n = n\Delta s$, $n = 0, 1, ..., N_s - 1$, where $\Delta s = N/(N_s - 1)$, the above update scheme becomes

$$q^{n+1}(x) = \exp[\mathcal{L}^W \Delta s/2] \exp(\mathcal{L}^D \Delta s) \exp[\mathcal{L}^W \Delta s/2] q^n(x) + O(\Delta s^3) \qquad (3.242)$$

where $q^n(x) \equiv q(x, s_n)$. This update algorithm is well known for solving the time-dependent Schrödinger equation (Fleck *et al.*, 1976; Feit *et al.*, 1982) and was recently introduced (Rasmussen and Kalosakas, 2002) to the polymer physics literature for solving eqn (3.25).

An algorithm for implementing eqn (3.242) takes full advantage of the reciprocal formulas (3.237) and (3.239):

### Pseudo-spectral Algorithm 3.1: Continuous Gaussian chain

1. Starting with a known solution at contour location $s_n$, $q^n(x)$ is evaluated at the $M$ collocation points to define an $M$-element column vector $\mathbf{q}^n$.
2. Because the operator $\exp[\mathcal{L}^W \Delta s/2] = \exp[-w(x)\Delta s/2]$ is diagonal in the $x$ representation, it can be applied to the elements of $\mathbf{q}^n$ by means of $M$ multiplications with $w(x)$ evaluated at each of the collocation points. This results in a column vector

$$\mathbf{q}^{n+1/3} = \exp[\mathcal{L}^W \Delta s/2] \mathbf{q}^n$$

3. An FFT, requiring $O(M \log_2 M)$ operations, is applied to $\mathbf{q}^{n+1/3}$ to obtain its Fourier coefficients $a_j^{n+1/3}$ according to

$$a_j^{n+1/3} = \frac{1}{M} \sum_{l=0}^{M-1} q_l^{n+1/3} \exp(-i2\pi jl/M), \quad j = 0, 1, ..., M-1$$

4. The Fourier representation of the propagator $\exp(\mathcal{L}^D \Delta s)$ is next applied to the vector $\mathbf{a}^{n+1/3}$. This requires only $M$ multiplications because the propagator is diagonal in this representation. If we denote the Fourier coefficients of $\exp(\mathcal{L}^D \Delta s)\mathbf{q}^{n+1/3}$ by the vector $\mathbf{h}$, then

$$h_j = \begin{cases} a_j^{n+1/3} \exp[-2\pi^2 b^2 \Delta s j^2/(3L^2)] & j = 0, 1, ..., M/2 \\ a_j^{n+1/3} \exp[-2\pi^2 b^2 \Delta s(M-j)^2/(3L^2)] & j = M/2+1, ..., M-1 \end{cases}$$

The two expressions for the propagator result from the shift in the $j$ index by $M/2 - 1$ in going from eqn (3.236) to the FFT-standard form of eqn (3.237).

5. An inverse FFT, requiring $O(M \log_2 M)$ operations, is applied to the vector $\mathbf{h}$ to restore the $x$-space representation. This produces a vector $\mathbf{q}^{n+2/3}$ defined on the collocation points by

$$q_l^{n+2/3} = \sum_{j=0}^{M-1} h_j \exp(i2\pi jl/M), \quad l = 0, 1, ..., M-1$$

6. The final step of implementing eqn (3.242) is an application of the operator $\exp[\mathcal{L}^W \Delta s/2] = \exp[-w(x)\Delta s/2]$ identical to that in step 2. Again, this requires $M$ multiplications with $w(x)$ evaluated at each of the collocation points. The result is the desired $M$-element column vector $\mathbf{q}^{n+1}$:

$$\mathbf{q}^{n+1} = \exp[\mathcal{L}^W \Delta s/2]\mathbf{q}^{n+2/3}$$

The overall operation count for executing this algorithm is $O(M \log_2 M)$, since the two FFTs are limiting. By repeating the procedure to step along the contour from the initial condition $(n = 0)$ to the chain end $(n = N_s - 1)$, one finds that a full solution of the diffusion equation (3.25) carries a computational cost that scales like $\sim N_s M \log_2 M$ in one dimension and $\sim N_s M^3 \log_2 M$ in three dimensions with $M$ plane waves per direction. This nearly ideal scaling is clearly far superior to the computational requirements of the "fully spectral" algorithm of eqn (3.233). Moreover, the pseudo-spectral scheme enjoys higher-order accuracy in $s$ and is unconditionally stable. Finally, because the pseudo-spectral algorithm preserves spectral accuracy in the spatial coordinates, allowing one to choose $M \ll N_x$, it is generally superior to finite difference schemes such as eqn (3.210) for problems with periodic boundary conditions. Indeed, Pseudo-spectral Algorithm 3.1 is the method of choice for solving the modified diffusion equation in regular geometries with periodic boundary conditions.[31]

The flexibility of switching between real and reciprocal space in the pseudo-spectral approach has a derivative benefit – auxiliary operators such as the segment density operator $\rho(\mathbf{r}; [w])$ can be efficiently evaluated. For example, in the present one-dimensional case, eqn (3.55) can be evaluated at the collocation points according to

$$\rho(x_l; [w]) = \frac{1}{V Q[w]} \int_0^N ds \, q(x_l, N - s; [w]) \, q(x_l, s; [w]) \qquad (3.243)$$

for $l = 0, ..., M - 1$. The quadrature in this expression can be performed with a Newton–Coates scheme of second order or higher, e.g. Simpson's rule (Press et al., 1992), using the same $N_s$ contour points that were used to integrate the diffusion equation. The partition function is evaluated from the zero-frequency component of the reciprocal space solution as $Q[w] = a_0(N)$. By such means, eqn (3.243) can be evaluated in $O(N_s M)$ operations. It is worth noting that if the full $s$ dependence of $q(\mathbf{r}, s)$ is not required, e.g. if only $Q[w]$ is of interest, then steps 2 and 6 of Algorithm 3.1 can be combined into a single application of the operator $\exp[\mathcal{L}^W \Delta s]$. This will save $O(N_s M)$ multiplications.

Figure 3.14 provides an example of implementing the pseudo-spectral algorithm for the same potential $W(x) = Nw(x)$ that was used in Fig. 3.13. Shown is the chain end distribution function $q(x, N)$ and the dimensionless segment density $\rho(x)V/N$ for a periodic box of length $L = 10R_g$. We see that the total segment density exhibits a more pronounced maximum inside the potential well than the polymer end density.

For problems where either Dirichlet or Neumann boundary conditions are imposed at both $x = 0$ and $x = L$, Algorithm 3.1 can be adapted by substitution of a sine or cosine basis, respectively, for the plane waves. With this modification, fast discrete sine or cosine transforms are substituted for the FFTs. For example, eqn (3.198) can be solved pseudo-spectrally subject to Dirichlet ($q = 0$) boundary conditions at $x = 0, L$ by application of the truncated Fourier sine series

$$q_M(x, s) = \sum_{j=1}^{M} a_j(s) \sin(\pi j x / L) \qquad (3.244)$$

With collocation at the equally spaced points, $x_l = lL/(M + 1)$, $l = 1, 2, ..., M$, which are *interior* to $x \in [0, L]$, we arrive at the following discrete sine transform pair:

---

[31]Garcia-Cervera (2005) has recently developed a fourth order pseudo-spectral algorithm with local contour accuracy that is $O(\Delta s^4)$. This semi-implicit scheme is unconditionally stable and is no more expensive per contour step than Algorithm 3.1. The only disadvantage to the method is a more complex initialization procedure.

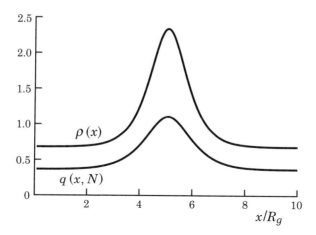

FIG. 3.14. Chain end distribution function $q(x, N)$ and dimensionless segment density $\rho(x)V/N$ obtained by implementing scheme (3.242) pseudo-spectrally with $M = 64$ plane waves and $N_s = 201$ contour steps ($\Delta s = 0.005N$) for the scaled potential function $W(x) = Nw(x) = 1 - 2 \{\text{sech}[3(x - L/2)/(2R_g)]\}^2$ shown in Fig. 3.13. Periodic boundary conditions with period $L = 10R_g$ were assumed.

$$q_l(s) = \sum_{j=1}^{M} a_j(s) \sin(\pi jl/(M + 1)), \quad l = 1, 2, ..., M \qquad (3.245)$$

$$a_j(s) = \frac{2}{M + 1} \sum_{l=1}^{M} q_l(s) \sin(\pi jl/(M + 1)), \quad j = 1, 2, ..., M \qquad (3.246)$$

Fast sine Fourier algorithms (Press *et al.*, 1992) can be used to transform between the collocated values of the function, $\{q_l\}$, and the spectral coefficients, $\{a_j\}$, in $O(M \log_2 M)$ operations.

Algorithm 3.1 for periodic boundary conditions requires only a few modifications to solve the Dirichlet problem with a Fourier sine basis. Equation (3.246) is substituted in step 3 for relating $\mathbf{q}^{n+1/3}$ to $\mathbf{a}^{n+1/3}$. In step 4, the sine Fourier coefficients of $\exp(\mathcal{L}^D \Delta s)\mathbf{q}^{n+1/3}$ are given by

$$h_j = a_j^{n+1/3} \exp[-\pi^2 b^2 \Delta s j^2/(6L^2)] \quad j = 1, 2, ..., M \qquad (3.247)$$

Finally, eqn (3.245) is applied in step 5 to obtain $\mathbf{q}^{n+2/3}$. The overall operation count for a solution of eqn (3.198) using $N_s$ contour points is $O(N_s M \log_2 M)$. The partition function $Q[w]$ can be evaluated by means of the formula

$$Q[w] = \frac{2}{\pi} \sum_{j=1,3,5,...}^{M} j^{-1} a_j(N) \qquad (3.248)$$

Figure 3.15 provides an example of a numerical solution of eqn (3.198) obtained by using the sine pseudo-spectral algorithm with Dirichlet boundary conditions applied at $x = 0, L$. The potential $w(x)$ was the same as that shown in Fig. 3.13, and use of eqn (3.248) leads to a value of $Q[w] = 0.456261$ for the partition function. We see that the quadratic relation between the end distribution function and the segment density implied by eqn (3.243) leads to a segment density profile $\rho(x)$ that vanishes faster near the boundaries than the chain end distribution function $q(x, N)$. Physically this can be understood because segments that are central to a chain experience more *entropic repulsion* near a wall than an end segment. For a middle segment, each of the two polymer "tail" sections emerging from that segment have their conformations restricted by the wall. An end segment in contrast has only one polymer tail that must avoid penetrating the wall.

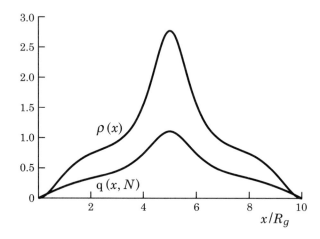

FIG. 3.15. Chain end distribution function $q(x, N)$ and dimensionless segment density $\rho(x)V/N$ obtained by implementing the pseudo-spectral algorithm with $M = 64$ sine basis functions and $N_s = 201$ contour steps ($\Delta s = 0.005N$) for the scaled potential function $W(x) = Nw(x) = 1 - 2\{\mathrm{sech}[3(x - L/2)/(2R_g)]\}^2$ shown in Fig. 3.13. Dirichlet boundary conditions at $x = 0$ and $x = L = 10R_g$ were applied.

Unfortunately, such Fourier sine and Fourier cosine solutions do not achieve the ideal spectral accuracy afforded by a Chebyshev basis, but the contour-stepping can be done more easily. The high accuracy of Chebyshev spectral approximations near the boundaries produces an extreme stiffness of the $a_j(s)$ equations that renders *explicit* contour-stepping virtually useless.[32] Uncondi-

---

[32]The conditional stability condition for forward Euler contour-stepping of the Chebyshev spectral approximation to the Dirichlet problem is $\Delta s < O(M^{-4})$.

tionally stable implicit contour-stepping algorithms can be developed for the Chebyshev basis (Boyd, 2001), but they are more difficult to implement than the sine (and cosine) pseudo-spectral scheme just described. As a result, we recommend using Fourier bases, except in very high-resolution calculations or situations where polymeric fluids are bounded in more than one direction where there is a clear benefit to exploiting the higher spatial accuracy of Chebyshev spectral approximations. Chebyshev pseudo-spectral techniques should also be seriously considered as an alternative to finite difference methods in situations where Fourier bases do not apply, e.g. different boundary conditions on opposing faces of a simulation cell, or Robin's conditions such as eqn (3.197).

### 3.6.4 *Higher dimensions*

The numerical methods described in the above sections can be easily generalized to enable efficient calculations in two or three dimensions. For situations in which the computational domain is geometrically simple, e.g. a square, a circle, a cube, a sphere, or a two- or three-dimensional parallelepiped, pseudo-spectral methods are attractive because they can potentially provide high spatial resolution at reduced computational cost.

As a first example of a higher-dimensional calculation, we consider the problem of solving eqn (3.25) in a three-dimensional cubic cell of side length $L$ that is subject to periodic boundary conditions. The desired solution $q(\mathbf{r}, s)$ can be expanded in a plane wave basis as

$$q(\mathbf{r}, s) = \sum_{\mathbf{k}} a_{\mathbf{k}}(s) \exp(i\mathbf{k} \cdot \mathbf{r}) \tag{3.249}$$

where we adopt standard solid state physics notation (Ashcroft and Mermin, 1976) in which the sum is over all reciprocal lattice vectors ("wavevectors") $\mathbf{k}$ of a simple cubic lattice with lattice constant $L$. The $\mathbf{k}$ are restricted to the values

$$\mathbf{k_j} = \frac{2\pi}{L} (j_1 \hat{\mathbf{x}} + j_2 \hat{\mathbf{y}} + j_3 \hat{\mathbf{z}}) \tag{3.250}$$

where $\hat{\mathbf{x}}$, $\hat{\mathbf{y}}$, and $\hat{\mathbf{z}}$ are orthonormal Cartesian unit vectors defining the edges of the cubic cell and $\mathbf{j} = (j_1, j_2, j_3)$ with $j_i = 0, \pm 1, \pm 2, ..., \pm \infty$ for $i = 1, 2, 3$.

It is useful to consider spectral approximations that involve truncating the sum in eqn (3.249) after $M$ terms in each of the three directions and collocating at points on a uniform grid within the cubic cell defined by

$$\mathbf{r_l} = \frac{L}{M} (l_1 \hat{\mathbf{x}} + l_2 \hat{\mathbf{y}} + l_3 \hat{\mathbf{z}}) \tag{3.251}$$

with $l_i = 0, 1, 2, ..., M - 1$ for $i = 1, 2, 3$. Retracing the steps that led from eqn (3.235) to eqn (3.237) in the one-dimensional case produces

$$q_{\mathbf{l}}(s) = \sum_{j_1=0}^{M-1} \sum_{j_2=0}^{M-1} \sum_{j_3=0}^{M-1} a_{\mathbf{j}}(s) \exp[i2\pi (j_1 l_1 + j_2 l_2 + j_3 l_3)/M] \tag{3.252}$$

where $q_l(s) \equiv q(\mathbf{r_l}, s)$. The inverse relation between the collocated values of the function and its Fourier coefficients is

$$a_{\mathbf{j}}(s) = \frac{1}{M^3} \sum_{l_1=0}^{M-1} \sum_{l_2=0}^{M-1} \sum_{l_3=0}^{M-1} q_l(s) \exp[-i2\pi(j_1 l_1 + j_2 l_2 + j_3 l_3)/M] \qquad (3.253)$$

Equations (3.252) and (3.253) thus correspond to three-dimensional discrete Fourier transform pairs and are in the standard form to allow for efficient calculation in $O(M^3 \log_2 M)$ operations by means of a three-dimensional FFT algorithm (Press $et\ al.$, 1992).

Adaption of the Pseudo-spectral Algorithm 3.1 of Section 3.6.3 for solving eqn (3.25) in a cubic cell with periodic boundary conditions is now straightforward. The real-space multiplications of steps 2 and 6 are carried out on the simple cubic lattice of collocation points $\mathbf{r_l}$ defined above. Three-dimensional FFTs are substituted in steps 3 and 5. In step 4, the Fourier coefficients of $\exp(\mathcal{L}^D \Delta s)\mathbf{q}^{n+1/3}$ are given by

$$h_{\mathbf{j}} = a_{\mathbf{j}}^{n+1/3} g_{j_1} g_{j_2} g_{j_3} \qquad (3.254)$$

where the damping factors, $g_j$, reflect the index shifts necessary to put eqn (3.252) in "standard" FFT form:

$$g_j = \begin{cases} \exp[-2\pi^2 b^2 \Delta s j^2/(3L^2)] & j = 0, 1, ..., M/2 \\ \exp[-2\pi^2 b^2 \Delta s (M-j)^2/(3L^2)] & j = M/2+1, ..., M-1 \end{cases} \qquad (3.255)$$

The above procedure can be easily generalized to retain different numbers of plane waves in the three directions and to orthorhombic cells with varying side lengths, $L_1 \neq L_2 \neq L_3$.

### 3.6.5   Unit-cell calculations

When simulating polymeric fluids such as block copolymers that are capable of forming *spatially periodic mesophases*, it is often convenient to perform the calculations in a computational cell that is a *unit-cell* of the periodic structure. One advantage of such an approach is that it allows for very accurate, high-resolution calculations since all of the computational resources are focused on representing the solution in only one repeated domain of a periodic structure. A second advantage is that defect states, which we shall see are difficult to avoid in "large-cell" numerical simulations that simultaneously resolve the structure in many unit-cells of a mesophase, are much less prevalent in unit-cell calculations.

A disadvantage of unit-cell calculations is that they are difficult to apply when studying fluids for which the candidate mesophases are not known in advance. This is because each calculation assumes a unit-cell shape that is the primitive cell of a particular Bravais lattice. If the lattice is not known in advance, then the calculations must be repeated for many different cell shapes and sizes until the cell that leads to the lowest free energy of the simulated fluid can be identified.

A second disadvantage of unit-cell calculations is that they do not allow for examination of thermal fluctuation effects and the statistical mechanics of defect states. Fluctuation modes with wavelengths greater than the unit cell dimensions are clearly inaccessible with such an approach.

For the case of a simple cubic lattice, the algorithm presented in Section 3.6.4 can be immediately applied for a unit-cell calculation by simply choosing $L$ to be the lattice constant of the periodic mesophase that is to be described. For more general Bravais lattices, where the primitive vectors $\mathbf{e}_1$, $\mathbf{e}_2$, and $\mathbf{e}_3$ are not orthogonal, it is convenient to choose as a unit-cell the parallelepiped that is spanned by these three vectors. This choice of unit-cell, the so-called *primitive cell* (Ashcroft and Mermin, 1976), proves to be much more useful for numerical computations than other common unit-cell definitions, e.g. the Wigner–Seitz cell. The reason is that a parallelepiped cell preserves the ability to apply multi-dimensional FFTs, which we have seen is essential for obtaining efficient pseudo-spectral numerical solutions of the diffusion equation (3.25).

The approach to solving eq (3.25) in a parallelepiped primitive cell with periodic boundary conditions can be nicely illustrated in the two-dimensional oblique cell shown in Fig. 3.16. The primitive lattice vectors $\mathbf{e}_1 = L_1\hat{\mathbf{x}}$ and $\mathbf{e}_2 = L_2(\cos\alpha\,\hat{\mathbf{x}} + \sin\alpha\,\hat{\mathbf{y}})$ define two edges of the parallelogram. The lattice constants are given by $|\mathbf{e}_1| = L_1$ and $|\mathbf{e}_2| = L_2$, and $\alpha$ denotes the angle between $\mathbf{e}_1$ and $\mathbf{e}_2$. The special case of $L_1 = L_2 = L$ and $\alpha = \pi/3$ corresponds to the primitive cell of the *simple hexagonal lattice*.

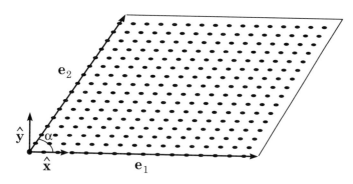

FIG. 3.16. Oblique computational cell to illustrate a unit-cell calculation in a primitive cell of a Bravais lattice. Cartesian unit vectors are denoted by $\hat{\mathbf{x}}$ and $\hat{\mathbf{y}}$, while $\mathbf{e}_1$ and $\mathbf{e}_2$, containing an angle $\alpha$, denote the primitive lattice vectors defining the cell. The lattice constants are $|\mathbf{e}_1| = L_1$ and $|\mathbf{e}_2| = L_2$. The oblique lattice of points are the collocation points defined by eqn (3.257).

As in the previous section, it is useful to adopt a pseudo-spectral approach with a plane wave basis and expand $q(\mathbf{r}, s)$ according to eqn (3.249). The appropriate reciprocal lattice vectors (Ashcroft and Mermin, 1976) correspond to

$$\mathbf{k_j} = \frac{2\pi j_1}{L_1}(\hat{\mathbf{x}} - \cot\alpha\,\hat{\mathbf{y}}) + \frac{2\pi j_2}{L_2}\csc\alpha\,\hat{\mathbf{y}} \qquad (3.256)$$

where $j_i = 0, \pm1, \pm2, ..., \pm\infty$ for $i = 1, 2$. Of interest are spectral approximations to $q(\mathbf{r}, s)$ where $M_1$ plane waves are retained in the direction of the first primitive reciprocal lattice vector[33] and $M_2$ plane waves in the direction of the second. As shown in Fig. 3.16, we then collocate $q(\mathbf{r}, s)$ at points $\mathbf{r_l}$ that correspond to a uniform mesh in the oblique cell:

$$\mathbf{r_l} = (l_1/M_1)\mathbf{e}_1 + (l_2/M_2)\mathbf{e}_2 \qquad (3.257)$$

with $l_i = 0, 1, 2, ..., M_i - 1$ for $i = 1, 2$. This leads to the following two-dimensional discrete Fourier transform, which relates the collocated function to its Fourier coefficients:

$$q_l(s) = \sum_{j_1=0}^{M_1-1}\sum_{j_2=0}^{M_2-1} a_{\mathbf{j}}(s)\exp[i2\pi(j_1 l_1/M_1 + j_2 l_2/M_2)] \qquad (3.258)$$

where $q_l(s) \equiv q(\mathbf{r_l}, s)$. The inverse transform is

$$a_{\mathbf{j}}(s) = \frac{1}{M_1 M_2}\sum_{l_1=0}^{M_1-1}\sum_{l_2=0}^{M_2-1} q_l(s)\exp[-i2\pi(j_1 l_1/M_1 + j_2 l_2/M_2)] \qquad (3.259)$$

These expressions are evidently the two-dimensional analogs of eqns (3.252) and (3.253) and are in "standard" FFT form.

The final ingredient required to adapt the Pseudo-spectral Algorithm 3.1 of Section 3.6.3 is an expression for the Fourier coefficients of the time evolution operator $\exp(\mathcal{L}^D\Delta s)$. It follows from eqn (3.249) that

$$\exp(\mathcal{L}^D\Delta s)q(\mathbf{r}, s)\big|_{\mathbf{r}=\mathbf{r_l}} = \sum_{\mathbf{j}} a_{\mathbf{j}}(s)\exp(-b^2\Delta s|\mathbf{k_j}|^2/6)\exp(i\mathbf{k_j}\cdot\mathbf{r_l}) \qquad (3.260)$$

where $|\mathbf{k_j}|^2 = \mathbf{k_j}\cdot\mathbf{k_j}$ is given explicitly by

$$|\mathbf{k_j}|^2 = \left(\frac{2\pi p_1(j_1)}{L_1}\right)^2 + \left(\frac{2\pi p_2(j_2)}{L_2}\csc\alpha - \frac{2\pi p_1(j_1)}{L_1}\cot\alpha\right)^2 \qquad (3.261)$$

The functions $p_i(j)$ in this expression are defined for $i = 1, 2$ by $p_i(j) = j$ for $0 \le j \le M_i/2$ and $p_i(j) = M_i - j$ for $M_i/2 + 1 \le j \le M_i - 1$.

A pseudo-spectral algorithm for solving eqn (3.25) in the unit-cell shown in Fig. 3.16 thus proceeds by modifying Algorithm 3.1 as follows: Steps 1 and 2 are carried out as simple multiplications by evaluating $q^n(\mathbf{r})$ and $w(\mathbf{r})$ at the

---

[33]The first and second primitive reciprocal lattice vectors correspond to $(2\pi/L_1)(\hat{\mathbf{x}} - \cot\alpha\,\hat{\mathbf{y}})$ and $(2\pi/L_2)\csc\alpha\,\hat{\mathbf{y}}$, respectively.

$M_1 \times M_2$ collocation points $\mathbf{r_l}$ given in eqn (3.257). The FFT in step 3 is performed in two dimensions according to eqn (3.259). The Fourier coefficients of $\exp(\mathcal{L}^D \Delta s)\mathbf{q}^{n+1/3}$ required in step 4 are obtained according to eqn (3.260) as

$$h_{\mathbf{j}} = a_{\mathbf{j}}^{n+1/3} \exp(-b^2 \Delta s |\mathbf{k_j}|^2/6) \tag{3.262}$$

Step 5 is then performed by a two-dimensional FFT using eqn (3.258), while step 6 completes the update with $O(M_1 M_2)$ multiplications at the collocation points. The operation count of $O[M_1 M_2 \log_2(M_1 M_2)]$ for each contour step is thus dominated by the two FFTs. The overall operation count for propagating the solution of eqn (3.25) down the entire chain is $O[N_s M_1 M_2 \log_2(M_1 M_2)]$.

Extension of this algorithm to a parallelepiped cell of arbitrary shape in three dimensions is straightforward.

### 3.6.6   Large-cell calculations

More typical than a unit-cell calculation is a simulation carried out in a *large cell*. In such circumstances, one is often interested in evaluating the structure and thermodynamics of a *bulk* sample of a polymeric fluid that lacks long-range order. Surface effects are minimized by choosing a cell size $L$ that greatly exceeds the structural length scales in the fluid and by applying periodic boundary conditions. A simple cubic simulation cell of length $L$ on each side is the most common choice for this purpose, and the pseudo-spectral algorithm described in Section 3.6.4 is readily applicable.

The optimal choice of $L$ is usually a compromise between the desire to minimize both surface effects and computational requirements. A large value of $L$ relative to structural scales in the fluid is ideal for reducing the influence of the boundaries. However, to accurately simulate the fluid, one must retain plane waves with wavelengths below the finest structures in the fluid. Thus, if the cell dimension $L$ is increased by a factor of 2, double the number of plane waves must be applied in each direction to maintain the same spatial resolution. It follows that the computational cost increases by more than a factor of 8 in a three-dimensional simulation. Large-cell calculations using more than $\sim 64^3$ plane waves can thus easily exceed the computational capability of a single-processor scientific workstation.

Fortunately, distributed memory, parallel computing clusters have become increasing available and affordable in recent years. The key numerical operations involved in implementing the pseudo-spectral algorithm described in Section 3.6.4 are local multiplications at collocation points and multi-dimensional FFTs. The multiplications can be trivially parallelized by domain decomposition as illustrated in Fig. 3.17. Each processor is assigned a "slab" of collocation points and carries out all local multiplications for that set of points. When a "global" calculation is required, e.g. the evaluation of the partition function $Q[w]$, data from different slabs is communicated among processors using the message-passing interface (MPI) protocol (Gropp *et al.*, 1999).

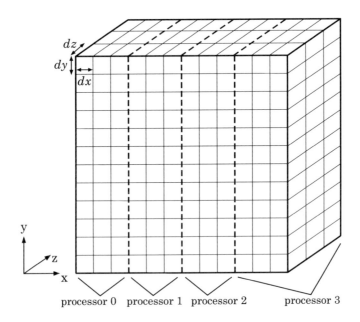

FIG. 3.17. Spatial domain decomposition suitable for implementation of the pseudo-spectral algorithm of Section 3.6.4 on a distributed memory computer cluster. This schematic shows the grid of collocation points $\mathbf{r_l}$ used to sample the fields $w(\mathbf{r})$ and $q(\mathbf{r}, s)$. Each processor is assigned a "slab" of the grid of size $(M_x/n) \times M_y \times M_z$, where $n$ is the number of processors and $M_i$ for $i = x, y, z$ denotes the number of plane waves and collocation points used in each direction. Reproduced from Sides and Fredrickson (2003).

The forward and backward discrete Fourier transforms in the pseudo-spectral algorithm require both local computation and inter-processor communication. A parallel FFT is performed in an analogous manner as a serial FFT on a multi-dimensional array of data. First, a series of one-dimensional FFTs in the $y$ and $z$ directions of Fig. 3.17 is applied to all the data. Since FFT algorithms (Press *et al.*, 1992) require all data in one dimension to be locally available, the slab decomposition is most appropriate. To complete the three-dimensional transform, a matrix transpose operation is performed and then a final one-dimensional FFT is applied along the $x$ direction normal to the slabs. The bulk of the communication cost is incurred through the matrix transpose, which requires sending and receiving data to and from sections of the computational grid that is local to different processors.

This parallelization scheme for solving eqn (3.25) has recently been implemented (Sides and Fredrickson, 2003) using the publicly available FFTW fast Fourier transform libraries (Frigo and Johnson, 1998). Figure 3.18 provides timing results for two- and three-dimensional calculations performed with a Myrinet-

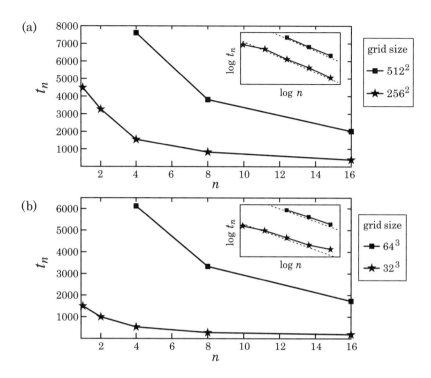

FIG. 3.18. CPU time $t_n$ versus number of processors $n$ for (a) 2D systems and (b) 3D systems. The time shown is the total computation and communication time required to solve the diffusion equation (3.25) for 1000 different realizations of $w(\mathbf{r})$ using the pseudo-spectral algorithm of Section 3.6.4. The algorithm was implemented with the FFTW libraries (Frigo and Johnson, 1998) on a 32-processor Myrinet cluster in the UCSB-MRL Central Computing Facilities (http://www.mrl.ucsb.edu). The insets show the same timing data on a log–log plot. The dotted lines indicate the limit of perfect scaling with number of processors and have slopes of $-1$. Reproduced from Sides and Fredrickson (2003).

based cluster available at UCSB. The timing data represents the sum of the computation and communication time required to solve eqn (3.25) using the pseudo-spectral algorithm for 1000 different realizations of the potential field $w(\mathbf{r})$. The insets in Fig. 3.18 show log–log plots of the same timing data with dotted lines indicating the ideal $\sim n^{-1}$ scaling of run time with number of processors $n$. For large systems, e.g. the case of $64^3$ plane waves, nearly ideal scaling is observed. A decrease in performance for the smaller $32^3$ system is caused by the increased ratio of communication-to-computation time as the number of plane waves (and collocation points) assigned to each processor is decreased. Evidently, the pseudo-spectral algorithm parallelizes well to enable very large-scale simulations on cost-effective distributed memory computer clusters.

### 3.6.7 *Other chain models*

Numerical algorithms can be devised for efficient calculation of the propagators, density operators, and partition functions corresponding to a wide variety of chain models. As we have seen for the continuous Gaussian chain, the pseudo-spectral approach is particularly powerful.

3.6.7.1 *Bead-spring chain* In the case of the discrete bead-spring chain, the relevant equations for computing the propagator $q(\mathbf{r}, j; [w])$ and the partition function $Q[w]$ are eqns (3.7)–(3.9). The Chapman–Kolmogorov equation (3.8) is readily tackled by a pseudo-spectral scheme in which the convolution integral $\int d\mathbf{r}' \Phi(\mathbf{r}-\mathbf{r}')q(\mathbf{r}', j)$ is evaluated in the Fourier domain as a simple multiplication and the factor $\exp[-w(\mathbf{r})]$ is applied in real space at collocation points. For example, if we attempt a calculation in a cubic cell with periodic boundary conditions, the plane wave expansion of eqn (3.249) is applicable:

$$q(\mathbf{r}, j) = \sum_{\mathbf{k}} a_{\mathbf{k}}(j) \exp(i\mathbf{k} \cdot \mathbf{r}) \tag{3.263}$$

Inserting this expansion into the convolution integral leads to

$$\int d\mathbf{r}' \, \Phi(\mathbf{r} - \mathbf{r}')q(\mathbf{r}', j) = \sum_{\mathbf{k}} a_{\mathbf{k}}(j)\phi_{\mathbf{k}} \exp(i\mathbf{k} \cdot \mathbf{r}) \tag{3.264}$$

where

$$\phi_{\mathbf{k}} = \int_{V} d\mathbf{r} \, \Phi(\mathbf{r}) \exp(-i\mathbf{k} \cdot \mathbf{r}) \tag{3.265}$$

and the integral is over the volume $V = L^3$ of the cubic cell. In the typical situation where the box length $L$ greatly exceeds the characteristic bond length $b$, the volume integral in eqn (3.265) may be taken over all space. It follows that

$$\phi_{\mathbf{k}} = \hat{\Phi}(\mathbf{k}) \tag{3.266}$$

which is the characteristic function defined for an arbitrary spring potential $h(x)$ in eqn (2.40). Recall that for the linear (Gaussian) spring model, $\hat{\Phi}(\mathbf{k}) = \exp(-b^2|\mathbf{k}|^2/6)$.

A pseudo-spectral algorithm for solving eqn (3.8) in a cubic cell with periodic boundary conditions thus proceeds as follows:

## Pseudo-spectral Algorithm 3.2: Bead-spring chain

1. Starting with a known solution at bead location $n$, $q(\mathbf{r}, n)$ is evaluated at the $M^3$ collocation points $\mathbf{r_l}$ defined by eqn (3.251) to define a vector $q_\mathbf{l}^n \equiv q(\mathbf{r_l}, n)$.

2. The collocated values of the propagator are Fourier analyzed to obtain a set of coefficients $a_\mathbf{j}^n$ by applying a three-dimensional FFT defined according to eqn (3.253).

3. The Fourier coefficients of the convolution integral in eqn (3.8) are computed by $M^3$ multiplications according to eqns (3.264) and (3.266) as

$$h_\mathbf{j} = a_\mathbf{j}^n \hat{\Phi}(|\mathbf{k_j}|)$$

where

$$|\mathbf{k_j}|^2 = \left(\frac{2\pi}{L}\right)^2 [p(j_1)^2 + p(j_2)^2 + p(j_3)^2]$$

and $p(j) = j$ for $0 \leq j \leq M/2$, $p(j) = M - j$ for $M/2 + 1 \leq j \leq M - 1$.

4. The Fourier representation of the convolution integral is next inverted by use of a FFT algorithm following eqn (3.252):

$$q_\mathbf{l}^{n+1/2} = \sum_{j_1=0}^{M-1} \sum_{j_2=0}^{M-1} \sum_{j_3=0}^{M-1} h_\mathbf{j} \exp[i2\pi(j_1 l_1 + j_2 l_2 + j_3 l_3)/M]$$

5. The final step of the algorithm is to apply the factor $\exp[-w(\mathbf{r})]$ evaluated at the collocation points $\mathbf{r_l}$ with a real-space multiplication. The result is the desired vector $q_\mathbf{l}^{n+1}$ corresponding to the collocated values of the propagator for a chain with $n + 1$ beads:

$$q_\mathbf{l}^{n+1} = \exp[-w(\mathbf{r_l})]q_\mathbf{l}^{n+1/2}$$

This algorithm is applied repeatedly for $n = 0, 1, 2, ..., N - 1$, starting with the initial condition eqn (3.7), to propagate the numerical solution for $q(\mathbf{r}, n)$ to the end of the polymer chain. Each step connecting adjacent beads requires $O(M^3 \log_2 M)$ operations (again dominated by the two three-dimensional FFTs), so the total operation count for a solution of eqns (3.7)–(3.8) is $O(NM^3 \log_2 M)$.

The flexibility of transforming between real and reciprocal space efficiently in the pseudo-spectral approach again makes the evaluation of derivative quantities particularly straightforward. The single-chain partition function $Q[w]$ can be computed in reciprocal space as $Q[w] = a_\mathbf{0}(N)$. The segment density operator for the bead-spring chain is evaluated in real space by applying eqn (3.53) at the collocation points $\mathbf{r_l}$:

$$\rho(\mathbf{r_l}; [w]) = \frac{e^{w(\mathbf{r_l})}}{VQ[w]} \sum_{j=0}^{N} q(\mathbf{r_l}, N - j; [w]) \, q(\mathbf{r_l}, j; [w]) \qquad (3.267)$$

In comparing Algorithms 3.1 and 3.2 for a three-dimensional calculation with a continuous Gaussian chain or a bead-spring chain, one finds that the operation count is $O(N_s M^3 \log_2 M)$ and $O(N M^3 \log_2 M)$, respectively. Thus, the continuous Gaussian chain model only provides a significant computational advantage if we can choose $N_s \ll N$. Indeed, this is often the case if the inhomogeneous structures in a fluid are varying on scales closer to $R_g$ than to $b$. Moreover, there is an aesthetic beauty to the continuous Gaussian chain – we simply smear out chemical details below our spatial and contour resolution and let the potential $w(\mathbf{r})$ act uniformly along the contour. $N_s$ in this case is not a parameter of the model, but simply a numerical parameter that is increased until the results converge to a desired accuracy. In contrast, the parameter $N$ of the bead-spring chain denotes the number of beads, which has physical as well as numerical significance, so it cannot be arbitrarily adjusted.

Nevertheless, the bead-spring chain model can be quite useful for numerical calculations in various contexts. For example, application of a nonlinear spring model such as eqn (2.41) can prevent the non-physical "over-stretching" of the continuous Gaussian chain when used to model polyelectrolytes under low-salt conditions (Netz and Orland, 1999). Moreover, the simple choice of characteristic function $\hat{\Phi}(\mathbf{k}) = j_0(b|\mathbf{k}|)$ in Algorithm 3.2 allows for calculations with a freely jointed chain model in which bonds are rigorously constrained to a length $b$ and the potential is applied at the flexible joints. This freely jointed chain could be useful in a variety of situations, including those where it is desirable to have a model in which chain crossings are prohibited. Such a model would be suitable for investigating the role of topological constraints on the statistical properties of inhomogeneous polymers.

3.6.7.2 *Wormlike chain model*   Finally, we discuss the numerical solution of eqn (3.39) for the wormlike chain model. For this purpose, it is useful to rewrite the equation as

$$\frac{\partial}{\partial s} q(\mathbf{r}, \mathbf{u}, s) = \mathcal{L}\, q(\mathbf{r}, \mathbf{u}, s) \tag{3.268}$$

where the linear operator $\mathcal{L}$ can be decomposed into potential, advective, and diffusive terms, respectively, according to $\mathcal{L} = \mathcal{L}^W + \mathcal{L}^C + \mathcal{L}^D$ with

$$\mathcal{L}^W = -w(\mathbf{r}, \mathbf{u}), \quad \mathcal{L}^C = -\mathbf{u} \cdot \nabla_{\mathbf{r}}, \quad \mathcal{L}^D = \frac{1}{2\lambda} \nabla_{\mathbf{u}}^2 \tag{3.269}$$

This Fokker–Planck equation is considerably more expensive to solve numerically than eqn (3.25) for the continuous Gaussian chain because the propagator $q(\mathbf{r}, \mathbf{u}, s)$ has two additional internal coordinates to describe the $\mathbf{u}$-dependence of the function. Thus, in a three-dimensional simulation it is necessary to solve eqn (3.268) with sufficient resolution in *six* coordinates spanned by $(\mathbf{r}, \mathbf{u}, s)$.

The literature on numerical solutions of eqn (3.268) is quite limited. Matsen and Barrett (1998) adopted a fully spectral approach using a plane wave basis to describe the $\mathbf{r}$ dependence and spherical harmonics to resolve the $\mathbf{u}$ dependence.

With such a choice of basis, a suitable spectral approximation would be a so-called *triangular truncation* (Boyd, 2001)

$$q_M(\mathbf{r}, \mathbf{u}, s) = \sum_{\mathbf{k}} \sum_{l=0}^{M-1} \sum_{m=-l}^{l} a_{\mathbf{k}lm}(s) \exp(i\mathbf{k} \cdot \mathbf{r}) Y_{lm}(\mathbf{u}) \qquad (3.270)$$

where $M^2$ spherical harmonic modes are retained. The spherical harmonics are defined in accordance with eqn (2.82). In eqn (3.270) we assume a cubic cell of side length $L$ with periodic boundary conditions applied. It is further understood that the sum over reciprocal lattice vectors $\mathbf{k}$ is restricted to values given by eqn (3.250) with $j_i = 0, \pm 1, \pm 2, ..., \pm M_x - 1$ for $i = 1, 2, 3$. Thus a total of $M^2 M_x^3$ basis functions are included in the approximation.

Matsen and Barrett (1998) projected eqn (3.268) onto spectral approximations of the form of eqn (3.270) using the Galerkin technique. Since the spherical harmonics are eigenfunctions of $\nabla_{\mathbf{u}}^2$ [see eqn (2.81)], the operator $\mathcal{L}^D$ appears as a diagonal form in the Galerkin equations for the $a_{\mathbf{k}lm}$ coefficients. Unfortunately, the plane wave/spherical harmonic basis does not diagonalize the potential operator $\mathcal{L}^W$. A semi-discrete approach to solving the Galerkin equations, similar to that considered in Section 3.6.2.2, is exceedingly costly as it requires $O(M^6 M_x^9)$ operations in three dimensions. Contour-stepping by applying finite differences in $s$ is somewhat better, but it still amounts to an expensive operation count of $O(N_s M^4 M_x^6)$ for a solution of eqn (3.268) that uses $N_s$ contour steps.

Düchs and Sullivan (2002) adopted a hybrid approach to the solution of eqn (3.268) that involved expanding the $\mathbf{u}$ dependence in spherical harmonics and applying finite differences in both $\mathbf{r}$ and $s$. This scheme has the advantage of rendering the operators band-diagonal in $\mathbf{r}$ and requires $O(N_s M^4 M_x^3)$ operations for a solution, where $M_x$ grid points are applied in each of the three spatial directions. While this scaling is considerably better than the fully spectral approach, it is still far from the computational ideal of linear scaling with the total number of orientational degrees of freedom $M^2$. Moreover, the low spatial accuracy of the finite difference technique is undesirable for three-dimensional calculations.

As with the other chain models, it appears that a properly designed *pseudo-spectral* algorithm is the best strategy for generating high-resolution solutions of the wormlike chain Fokker–Planck equation. An attractive operator splitting scheme with second-order accuracy in $s$ is the following Trotter formula:

$$q(\mathbf{r}, \mathbf{u}, s + \Delta s) = e^{\Delta s \mathcal{L}^W / 2} e^{\Delta s \mathcal{L}^C / 2} e^{\Delta s \mathcal{L}^D} e^{\Delta s \mathcal{L}^C / 2} e^{\Delta s \mathcal{L}^W / 2} q(\mathbf{r}, \mathbf{u}, s)$$
$$+ O(\Delta s^3) \qquad (3.271)$$

Provided that we have fast synthesis and analysis transforms available for switching between grid points in the domains of $\mathbf{r}$ and $\mathbf{u}$ and plane wave and spherical harmonic expansion coefficients, each of the propagator factors in eqn (3.271) can be applied in a representation where it is diagonal and easy to compute.

Evidently the discrete Fourier transforms for going between the $\mathbf{r}$ and $\mathbf{k}$ representations can be done with readily available FFT algorithms. Each transform costs $O(M_x^3 \log_2 M_x)$ operations in three spatial dimensions. In the case of spherical harmonic synthesis and analysis, the situation is not so ideal (Boyd, 2001). As seen in eqn (2.82), the spherical harmonics are products of complex exponentials in the azimuthal (longitudinal) angle, $\exp(im\phi)$, and associated Legendre functions in the polar (latitudinal) angle, $P_l^{|m|}(\cos\theta)$. An FFT can thus be used for the purpose of transforming in the longitudinal direction between values of a function at *uniformly spaced* angles $\phi$ on the unit sphere and coefficients of the azimuthal plane waves $\exp(im\phi)$. If we assume a triangular truncation with $M^2$ spherical harmonics, there are $O(M)$ azimuthal $m$ modes, so this one-dimensional FFT can be performed in $O(M \log_2 M)$ operations. Unfortunately, there is no ideal "fast Legendre transform" for transforming between values of a function at latitudinal grid points and $P_l^{|m|}$ coefficients. The standard approach used in weather forecasting[34] is to apply a Gauss–Legendre quadrature in latitude. This quadrature formula allows one to transform reversibly from $O(M)$ coefficients in the associated Legendre expansion to $O(M)$ values of a function on a non-uniform latitudinal grid (the roots of a Legendre polynomial) in $O(M^2)$ operations – the cost of a matrix multiplication (Boyd, 2001; Swarztrauber, 1979). Fast Legendre transforms with $O(M \log_2 M)$ asymptotic scaling do exist (Orszag, 1986; Driscoll *et al.*, 1997), but they have a very large numerical prefactor. Thus, the straightforward Gauss quadrature approach with $O(M^2)$ scaling proves more efficient up to quite large values of $M$. Overall, we see that the operation count for a single harmonic synthesis or analysis on the unit sphere is $O(M^3 \log_2 M)$ when applying the triangular truncation of eqn (3.270).

A comprehensive software library (SPHEREPACK 3.0) for spherical harmonic analysis and synthesis has been developed by the National Center for Atmospheric Research (Adams and Swarztrauber, 2003). Programs are available for implementing the tandem longitudinal FFT, latitudinal Gauss quadrature algorithm just described. Given the availability of such software and conventional FFT programs (Frigo and Johnson, 1998) for carrying out the spatial transforms, the update scheme given in eqn (3.271) can be implemented pseudo-spectrally. The operation count for each contour step of the scheme will likely be controlled by the two spherical harmonic transforms required to apply the operator $\exp(\Delta s \mathcal{L}^D)$, which cost $O(M_x^3 M^3 \log_2 M)$ operations. The total cost to propagate a solution forward from the initial condition eqn (3.40) to the end of the chain using $N_s$ contour steps is thus $O(N_s M_x^3 M^3 \log_2 M)$.

To our knowledge, a pseudo-spectral implementation of eqn (3.271) for the wormlike chain has not been reported to date. While this operator-splitting strategy does not achieve ideal $O(N_s M_x^3 M^2)$ scaling, it has a more attractive operation count than other numerical schemes that have been proposed in the literature. In addition, the method preserves spectral accuracy in both $\mathbf{r}$ and $\mathbf{u}$, is second-order accurate in $s$, and is expected to be stable. Similar numerical

---

[34] Atmospheric modelling represents one of the most demanding applications for spherical harmonic transforms and has driven the development of this area (Boyd, 2001).

techniques could be developed for related chain models, such as the freely rotating chain and rotational isomeric state (RIS) models (Flory, 1969), which require the propagation of information about segment orientation as well as position.

### 3.6.8  *Other techniques*

As has been described in the previous sections, pseudo-spectral methods, owing to their high accuracy and efficient implementation, are particularly attractive for obtaining numerical solutions of the Fokker–Planck and Chapman–Kolmogorov equations that define the statistical mechanics of coarse-grained chain models. Nevertheless, there are many other numerical techniques that can be used to solve such equations, including finite element, finite volume, and spectral element methods (Chung, 2002; Zienkiewicz and Taylor, 2000). These techniques have the advantage of being able to flexibly locate computational elements at the boundaries of geometrically complex domains. If the fluid domain is a simple box, slab, or parallelepiped, it remains to be seen whether such methods can be competitive with the pseudo-spectral approach. However, in situations where free boundaries are present, e.g. in computing the equilibrium shape of a structured polymer drop, these alternative techniques should be given serious consideration.

## 3.7  Summary

At the end of such a lengthy exposition on the statistical mechanics of polymer chains in external fields it is appropriate to review the key concepts.

- The analysis of the conformational statistics of a polymer chain in an external chemical potential is central to all field-based computer simulations of inhomogeneous polymers.

- One of the most important quantities in this analysis is the normalized *single-chain partition function*, $Q[w]$, which represents the Boltzmann-weighted sum over all chain conformations in the presence of a chemical potential field $w(\mathbf{r})$ that acts on individual segments. The partition function can be evaluated by a simple quadrature from a one-point function, the *chain propagator* $q(\mathbf{r}, s; [w])$, which gives the statistical weight of observing the chain end of a polymer with contour length $s$ at position $\mathbf{r}$.

- The chain propagator for discrete bead-spring chain models satisfies a Chapman–Kolmogorov (CK) integral equation given in eqn (3.8). CK equations for continuous chain models can be further reduced to Fokker–Planck (FP) partial differential equations, such as eqn (3.25) for the continuous Gaussian chain and eqn (3.39) for the wormlike chain. Both types of equations are generalizations of the transfer matrix technique, which is a powerful tool for the statistical mechanical analysis of coupled one-dimensional systems.

- Once a solution to the appropriate CK or FP equation is at hand, the resulting chain propagator can be used to compute various derivative statistical properties, including segment density operators, density–density correlation functions, and elastic stress operators.

- This approach can be generalized to polymer chains of virtually any architecture interacting with one or more potential fields, including branched homopolymers, block copolymers, and graft copolymers.

- When the potential field $w(\mathbf{r})$ has *special* characteristics, it is possible to develop approximate analytical expressions for the chain propagator and related quantities. For example, systematic perturbation expansions can be derived for cases of potential fields with: weak inhomogeneities, slow variation with $\mathbf{r}$, rapid variation with $\mathbf{r}$, deep and narrow minima, and potentials leading to strong chain stretching.

- Under most circumstances, the CK or FP equations must be solved *numerically* to fully represent the statistical properties of polymers in arbitrary potential fields. For this purpose, *pseudo-spectral algorithms* have been devised that combine spectral accuracy in the spatial domain with efficiency. Fourier and Chebyshev bases are preferred for such algorithms because they can be applied in tandem with the fast Fourier transform (FFT), which permits rapid transformations from a set of spectral coefficients to a set of collocated values of a function in the spatial domain. This flexibility of representation allows not only for efficient computation of the chain propagator, but of important derivative quantities such as density operators.

# 4

# MODELS OF MANY-CHAIN SYSTEMS

The previous two chapters dealt exclusively with the statistical properties of a *single* polymer chain in isolation and in the presence of an external potential field. Here, we shall address more realistic situations of many interacting polymers in solution, or under melt conditions. This necessitates a treatment of the interactions among segments belonging to different polymer chains. The intramolecular long-ranged interferences that have been neglected up to this point will also be included in the formalism.

## 4.1 From particles to fields

As discussed in Chapter 1, a fundamental element of the field-based approach to simulating inhomogeneous polymers is the conversion of a particle-based model to a statistical field theory. An important objective of the present chapter is to illustrate how this particle-to-field transformation can be carried out for a broad range of polymer and complex fluid systems. The general approach is to invoke formal techniques related to *Hubbard–Stratonovich transformations*,[35] which have the effect of decoupling interactions among particles (or polymer segments) and replacing them with interactions between the particles and one or more auxiliary fields. We shall see that these fields play the role of the external fields that were considered in the last chapter, although they are internally generated within a fluid. The methods described in the present chapter are well known in the condensed matter (Chaikin and Lubensky, 1995), classical liquid state (Caillol, 2003), and polymer physics communities (Edwards, 1966; Helfand, 1975; Hong and Noolandi, 1981). However, only recently have the resulting field theories been viewed as the basis for computer simulations of soft condensed matter systems.

The particle-to-field transformation can be carried out with either an atomistic or a mesoscopic starting model of a fluid. We begin with the simplest case of a monoatomic fluid such as argon.

### 4.1.1 *Monatomic fluid in the canonical (nVT) ensemble*

In the canonical ensemble, a system is considered to be closed to mass transfer, but it is open to energy transfer through the exchange of heat with a reservoir held at a fixed temperature $T$. The canonical partition function of a classical

---

[35]These auxiliary field techniques are associated with several names, including Kac (1959), Stratonovich (1958), Hubbard (1954), and Edwards (1959).

monatomic fluid with $n$ indistinguishable atoms confined to a volume $V$ can be expressed as (Chandler, 1987)

$$\mathcal{Z}_C(n, V, T) = \frac{1}{n!\lambda_T^{3n}} \int d\mathbf{r}^n \; \exp[-\beta U(\mathbf{r}^n)] \qquad (4.1)$$

where $\lambda_T = h/\sqrt{2\pi m k_B T}$ is the thermal wavelength, $m$ is the mass of an atom, and $h$ is the Planck constant. The potential energy $U(\mathbf{r}^n)$ depends on the relative positions of the $n$ atoms, and by specifying its mathematical form, one defines a particular *atomistic model* of a fluid. It is often the case that the *pair approximation* described by eqn (1.2) is a suitably accurate description of the potential energy. For the purpose of illustration, we shall adopt this perspective and write

$$U(\mathbf{r}^n) = \frac{1}{2} \sum_{j=1}^{n} \sum_{k=1(\neq j)}^{n} u(|\mathbf{r}_j - \mathbf{r}_k|) \qquad (4.2)$$

where $u(r)$ is the familiar pair potential function (Hansen and McDonald, 1986). The factor of $(1/2)$ in this expression corrects for the counting of each pair of particles twice in the double sum. In the case of argon, the Lennard-Jones potential given in eqn (1.3) provides a reasonable two-parameter fit to quantum chemical calculations of $u(r)$. More generally, one can treat a large class of atomic fluids, and even some molecular fluids (such as methane) that are well described by spherically symmetric pair potentials, by keeping the form of $u(r)$ arbitrary.

One restriction on $u(r)$ that is necessary to affect the particle-to-field transformation is that the potential must be *finite on contact*, i.e. $|u(0)| < \infty$. This would seem to rule out many important potential functions, including the Lennard-Jones and the hard sphere potentials, but it is actually straightforward to "regularize" such potentials to render $|u(0)|$ finite with no impact on fluid structure and thermodynamics.[36] For example, the Lennard-Jones potential of eqn (1.3) can be regularized by a simple shift $\delta$ according to

$$u(r) = 4\epsilon \left\{ [(r + \delta)/\sigma]^{-12} - [(r + \delta)/\sigma]^{-6} \right\} \qquad (4.3)$$

The choice $\delta = 0.01\sigma$ produces $u(0) = 4 \times 10^{24}\epsilon$, which is so large that overlapping particle configurations would provide no detectable contribution to $\mathcal{Z}_C$ and thus have no thermodynamic significance. Similarly, a hard sphere potential can be regularized with no impact on the thermodynamic properties of the fluid by simply replacing an infinite potential step at the hard core diameter $r = \sigma$ with a very large, but finite step. In the following, we will assume that $u(r)$ has been regularized by such a procedure.

---

[36]The terms *regularize* and *regularization* will be used throughout the monograph to denote the application of a mathematical procedure that removes a singularity in a quantity of interest.

The next task is to rewrite eqn (4.2) in terms of *microscopic density operators*. By analogy with eqn (3.3), we define the microscopic particle density as a sum of Dirac delta functions centered on the coordinates of each atom

$$\hat{\rho}(\mathbf{r}) = \sum_{j=1}^{n} \delta(\mathbf{r} - \mathbf{r}_j) \tag{4.4}$$

It follows that

$$U(\mathbf{r}^n) = \frac{1}{2} \int d\mathbf{r} \int d\mathbf{r}' \, \hat{\rho}(\mathbf{r}) u(|\mathbf{r} - \mathbf{r}'|) \hat{\rho}(\mathbf{r}') - \frac{1}{2} n \, u(0) \tag{4.5}$$

where the last term subtracts the self-interactions of the $n$ atoms that are included in the first term. Equation (4.1) can thus be written

$$\mathcal{Z}_C = \frac{z_0^n}{n!} \int d\mathbf{r}^n \, \exp\left(-\frac{\beta}{2} \int d\mathbf{r} \int d\mathbf{r}' \, \hat{\rho}(\mathbf{r}) u(|\mathbf{r} - \mathbf{r}'|) \hat{\rho}(\mathbf{r}')\right) \tag{4.6}$$

where $z_0 \equiv \exp(\beta u(0)/2)/\lambda_T^3$. It can be seen from this result that the value of the potential at contact, $u(0)$, only affects the reference chemical potential of the fluid and has no thermodynamic consequences.

The next step in the particle-to-field transformation is to invoke the definition of a *delta functional*, i.e.

$$\int \mathcal{D}\rho \, \delta[\rho - \hat{\rho}] F[\rho] = F[\hat{\rho}] \tag{4.7}$$

for any functional $F[\rho]$.[37] The delta functional $\delta[\rho - \hat{\rho}]$ can be viewed as an infinite-dimensional version of the Dirac delta function that vanishes unless the fields $\rho(\mathbf{r})$ and $\hat{\rho}(\mathbf{r})$ are equal at *all* points $\mathbf{r}$ in the domain of interest. A useful complex exponential representation of the delta functional can be developed by temporarily discretizing space using $M_g$ grid points according to

$$\delta[\rho - \hat{\rho}] \approx \prod_{\mathbf{r}} \delta(\rho(\mathbf{r}) - \hat{\rho}(\mathbf{r}))$$

$$= \frac{1}{(2\pi)^{M_g}} \prod_{\mathbf{r}} \left\{ \int_{-\infty}^{\infty} dw(\mathbf{r}) \, e^{iw(\mathbf{r})[\rho(\mathbf{r}) - \hat{\rho}(\mathbf{r})]} \right\}$$

$$= \int \mathcal{D}w \, e^{i \int d\mathbf{r} \, w(\mathbf{r})[\rho(\mathbf{r}) - \hat{\rho}(\mathbf{r})]} \tag{4.8}$$

The second line of the above expression follows from the application of eqn (A.15) to represent the one-dimensional delta function $\delta(\rho(\mathbf{r}) - \hat{\rho}(\mathbf{r}))$ at grid point $\mathbf{r}$. The prefactor $1/(2\pi)^{M_g}$ contains the normalization factors of $1/(2\pi)$ from all grid points. The final expression in eqn (4.8) results from restoring the continuum

---

[37]See Appendix C for a discussion of functionals and functional integration.

description and can be viewed as a formal definition of the functional integral $\int \mathcal{D}w$ over the auxiliary field $w(\mathbf{r})$. It is important to note that $w(\mathbf{r})$ is a *real* scalar field and that the functional integral in eqn (4.8) is taken along the whole real axis at each $\mathbf{r}$.

The next step in transforming the canonical partition function into a statistical field theory is to insert eqn (4.7) with $F[\rho] = 1$ into the integrand of eqn (4.6). This leads to

$$\mathcal{Z}_C = \frac{z_0^n}{n!} \int \mathcal{D}\rho \int d\mathbf{r}^n \, \delta[\rho - \hat{\rho}] \, \exp\left( -\frac{\beta}{2} \int d\mathbf{r} \int d\mathbf{r}' \, \rho(\mathbf{r})u(|\mathbf{r} - \mathbf{r}'|)\rho(\mathbf{r}') \right) \quad (4.9)$$

where we have used the following property of delta functionals for an arbitrary functional $G[\rho]$: $\delta[\rho - \hat{\rho}]G[\hat{\rho}] = \delta[\rho - \hat{\rho}]G[\rho]$. Next, the representation of eqn (4.8) is inserted for the delta functional, which results in the expression

$$\mathcal{Z}_C = \frac{z_0^n}{n!} \int \mathcal{D}\rho \int \mathcal{D}w \int d\mathbf{r}^n \, e^{i \int d\mathbf{r} \, w(\rho - \hat{\rho}) - (\beta/2) \int d\mathbf{r} \int d\mathbf{r}' \, \rho u \rho} \quad (4.10)$$

It is important to note that as a result of these transformations, the *only* factor in the integrand that depends on the atomic coordinates $\mathbf{r}^n = (\mathbf{r}_1, ..., \mathbf{r}_n)$ is $\exp(-i \int d\mathbf{r} \, w\hat{\rho})$. The integrals over the $n$ particle positions thus factor according to

$$\int d\mathbf{r}^n \, e^{-i \int d\mathbf{r} \, w\hat{\rho}} = \int d\mathbf{r}^n \, e^{-i \sum_{j=1}^{n} w(\mathbf{r}_j)}$$

$$= \prod_{j=1}^{n} \left\{ \int_V d\mathbf{r}_j \, e^{-iw(\mathbf{r}_j)} \right\} = (VQ[iw])^n \quad (4.11)$$

where

$$Q[iw] \equiv \frac{1}{V} \int_V d\mathbf{r} \, e^{-iw(\mathbf{r})} \quad (4.12)$$

The functional $Q[iw]$ can be interpreted as a *single-particle partition function*, i.e. the contribution to the partition function from an atom that interacts not with the other atoms, but only with the purely imaginary field $iw(\mathbf{r})$. This single-atom partition function has the same interpretation and normalization ($Q[0] = 1$) as the single-chain partition function $Q[w]$ discussed in Chapter 3. We thus adopt the same symbol $Q$. Evidently $Q[iw]$ is a *local* functional for an atom, whereas it is a highly *non-local* functional for a polymer molecule. As a result, $Q[iw]$ is much easier to compute for an atom than for a polymer.

Upon combining eqns (4.10) and (4.11), the particle-to-field transformation is complete. The partition function can be expressed as the following statistical field theory:

$$\mathcal{Z}_C(n, V, T) = \mathcal{Z}_0 \int \mathcal{D}\rho \int \mathcal{D}w \, \exp(-H[\rho, w]) \quad (4.13)$$

where the functional

$$H[\rho, w] = - i \int d\mathbf{r}\; w(\mathbf{r})\rho(\mathbf{r}) + \frac{\beta}{2} \int d\mathbf{r} \int d\mathbf{r}'\; \rho(\mathbf{r})u(|\mathbf{r} - \mathbf{r}'|)\rho(\mathbf{r}')$$
$$- n \ln Q[iw] \tag{4.14}$$

is referred to as an "effective Hamiltonian" or "action" (Parisi, 1988; Zee, 2003). The prefactor in eqn (4.13), $\mathcal{Z}_0 \equiv (z_0 V)^n / n!$, is proportional to the partition function of an ideal gas.

Equation (4.13) is the central result of this section, namely that the canonical partition function for a monatomic fluid with pairwise interactions described by an *arbitrary* pair potential $u(r)$ can be expressed as a statistical field theory. The partition function has been shown to be proportional to a functional integral of a Boltzmann-like factor $\exp(-H[\rho, w])$ over two fluctuating fields, $\rho(\mathbf{r})$ and $w(\mathbf{r})$. The first field, $\rho$, can be interpreted as a *fluctuating particle number density*, since it was constrained to the microscopic density $\hat{\rho}$ when introduced into eqn (4.9). The second field $w(\mathbf{r})$ can be viewed as a *fluctuating, inhomogeneous chemical potential field*, because $iw$ appears as the thermodynamically conjugate variable to the field $\rho$ in the first term of the effective Hamiltonian $H$. The correspondence between $w(\mathbf{r})$ and the equilibrium chemical potential of the fluid $\mu$ (a spatially uniform quantity) will be made more precise in Section 4.1.3.

Equation (4.14) shows that the effective Hamiltonian, which has the character of a free energy functional, has three principal contributions. The first term, $-i \int d\mathbf{r}\; w\rho$, can be interpreted as the energy of interaction between the density field $\rho$ and the purely imaginary "chemical potential" field $iw$. The second term proportional to $\int d\mathbf{r} \int d\mathbf{r}'\; \rho u \rho$ represents the energy associated with the particle–particle interactions. Finally, the term $-n \ln Q[iw]$ describes the translational entropy (relative to the ideal gas entropy) of a fluid of $n$ non-interacting atoms experiencing the potential $iw(\mathbf{r})$. While $H[\rho, w]$ has enthalpic and entropic contributions and thus has a "free energy character", it differs from the Helmholtz free energy $A$ of the fluid, because the latter also includes the entropic contributions associated with fluctuations of the fields $\rho$ and $w$. Indeed, evaluation of the Helmholtz free energy involves the use of the familiar thermodynamic connection formula

$$A(n, V, T) = -k_B T \ln \mathcal{Z}_C(n, V, T) \tag{4.15}$$

where $\mathcal{Z}_C$ is obtained by performing functional integrals over the $\rho$ and $w$ fields according to eqn (4.13).

It should now be clear that the chemical potential field $w$ arises naturally in the formalism as a consequence of decoupling interactions among atoms in favor of interactions between individual atoms and the purely imaginary field $iw$. This field is *internally* generated, but from the standpoint of evaluating the single-particle partition function $Q[iw]$, it can be viewed as a prescribed, externally imposed field. Hence the perspective of $w$ as an external field in the previous chapter.

The emergence of a purely imaginary field $iw$ in an otherwise real theory is at first glance surprising. The partition function $\mathcal{Z}_C$ is real, as are the field $\rho$ and the potential function $u$. The effective Hamiltonian $H$, however, is clearly complex and can be decomposed into real and imaginary parts according to $H = H_R + iH_I$. An important consequence is that the Boltzmann-like factor $\exp(-H)$ in the integrand of eqn (4.13) contains a complex phase factor $\exp(-iH_I)$, which can oscillate in sign according to the configurations of the fields $\rho$ and $w$. Thus, the functional integrals defining the partition function contain an *oscillatory integrand*. This feature will be seen in Chapter 6 to produce difficulties (the so-called "sign problem") when attempting to simulate such field theories. Special numerical techniques are required.

One might ask whether it is really necessary to perform complex functional integrals to obtain a real partition function. Indeed, by taking advantage of the fact that $\mathcal{Z}_C$ is real, we can add the complex conjugate of eqn (4.13) to the original equation to obtain

$$\mathcal{Z}_C(n, V, T) = \mathcal{Z}_0 \int \mathcal{D}\rho \int \mathcal{D}w \ \exp(-H_R[\rho, w]) \cos(H_I[\rho, w]) \qquad (4.16)$$

so that the integrand is manifestly real. However, the phase factor $\cos(H_I)$ is not positive definite, and so the integrand is still oscillatory. Thus, there is no escaping the sign problem if we wish to transform from the particle language, where the statistical weight $\exp(-\beta U)$ is positive definite, to the field language, where the corresponding measure $\exp(-H)$ is not. Nevertheless, the benefits of the field-theoretic description generally outweigh the difficulties created by its non-positive definite character, at least for relatively concentrated polymer solutions and polymer melts. In the following, it will prove more convenient to work with the complex field theory given in eqn (4.13), rather than the real field theory of eqn (4.16). They are, of course, equivalent.

Equations (4.13)–(4.14) define a field theory involving two fields that is applicable to monotonic fluids with an arbitrary (but regularized) pair potential $u(r)$. For certain special potential functions, namely those that possess an inverse in the sense that

$$\int d\mathbf{r}' \ u(|\mathbf{r} - \mathbf{r}'|)u^{-1}(|\mathbf{r}' - \mathbf{r}''|) = \delta(\mathbf{r} - \mathbf{r}'') \qquad (4.17)$$

*and* where $u(|\mathbf{r} - \mathbf{r}'|)$ is a positive definite form, the field theory can be considerably simplified. Specifically, if these conditions are met, the $\rho$ integral in eqn (4.13) is a Gaussian integral of the form of eqn (C.28) that can be evaluated analytically. The dual conditions of $u$ possessing an inverse and being positive definite are satisfied if the Fourier transform of $u$

$$\hat{u}(\mathbf{k}) = \int d\mathbf{r} \ u(r) \exp(-i\mathbf{k} \cdot \mathbf{r}) = 4\pi \int_0^\infty dr \ r^2 j_0(kr)u(r) \qquad (4.18)$$

exists and is positive for all $k = |\mathbf{k}|$. For such potentials, eqn (4.13) simplifies to

$$\mathcal{Z}_C(n, V, T) = \mathcal{Z}_0 \int \mathcal{D}w \, \exp(-H[w]) \tag{4.19}$$

where a field-independent normalization factor[38] has been absorbed into the definition of the functional integral over $w$. The new effective Hamiltonian depends only on the single field $w$

$$H[w] = \frac{1}{2\beta} \int d\mathbf{r} \int d\mathbf{r}' \, w(\mathbf{r})u^{-1}(|\mathbf{r} - \mathbf{r}'|)w(\mathbf{r}') - n \ln Q[iw] \tag{4.20}$$

The particle–particle interactions are now contained in the first term of this Hamiltonian, which involves the inverse of the pair potential.

Important examples of pair potentials for which the simplified description of eqns (4.19)–(4.20) is applicable include the one-parameter repulsive delta function potential

$$u(r) = u_0 \, \delta(\mathbf{r}) \tag{4.21}$$

where $u_0 > 0$, and the repulsive Yukawa (or Debye–Hückel) potential

$$u(r) = \frac{u_0}{4\pi r} \exp(-\kappa r) \tag{4.22}$$

with $u_0 > 0$, $\kappa \geq 0$. The latter reduces to a repulsive Coulomb potential for $\kappa = 0$. These potentials are not regularized at $r = 0$, but this is easily accomplished. For example, the repulsive step potential

$$u(r) = \begin{cases} 3u_0/(4\pi\delta^3), & 0 \leq r \leq \delta \\ 0, & r > \delta \end{cases} \tag{4.23}$$

with $u_0, \delta > 0$ has the same Fourier transform for $\delta \to 0+$ as the delta function potential (4.21), but is finite at contact. For very small, but non-vanishing $\delta$, the potential of eqn (4.23) also meets the other criteria of positivity and invertibility [for wavenumbers $k \lesssim O(1/\delta)$].

Unfortunately, most realistic pair potentials with a hard core, e.g. the regularized Lennard-Jones potential of eqn (4.3), do not satisfy the conditions for eqns (4.19)–(4.20) to be applicable. In such circumstances, the full field theory of eqns (4.13)–(4.14) must be used. From the derivation of the full theory, it should also be clear that the method presented for decoupling interactions is not restricted to pair-decomposable potential functions. Indeed, the particle-to-field transformation can be carried out even when *three-body interactions* are present in the model for $U(\mathbf{r}^n)$, although such interactions would preclude the simplification to a field theory involving only a single $w$ field.

### 4.1.2 *Monatomic fluid in the grand canonical ($\mu VT$) ensemble*

The decoupling technique just described can be extended to derive statistical field theories for other important ensembles. For example, the grand canonical

---

[38] This factor is given explicitly by $\exp[-(1/2)\text{Tr}\ln(\beta u/2\pi)]$.

ensemble is convenient for studying the statistical thermodynamics of fluid systems that are open to both mass and energy transfer (Chandler, 1987; McQuarrie, 1976; Hansen and McDonald, 1986). The relevant grand canonical partition function can be written

$$\mathcal{Z}_G(\mu, V, T) = \sum_{n=0}^{\infty} \exp(\beta\mu n)\mathcal{Z}_C(n, V, T) \tag{4.24}$$

where $\mu$ is the chemical potential and the sum is over all numbers of atoms. Insertion of eqns (4.10) and (4.11) into the right-hand side of this expression leads immediately to

$$\mathcal{Z}_G = \int \mathcal{D}\rho \int \mathcal{D}w \; e^{i \int d\mathbf{r} \, w\rho - (\beta/2) \int d\mathbf{r} \int d\mathbf{r}' \, \rho u \rho} \sum_{n=0}^{\infty} \frac{(zVQ[iw])^n}{n!} \tag{4.25}$$

where $z \equiv z_0 \exp(\beta\mu)$ is the *activity*. Evaluation of the sum over the number of particles results in the desired field theory for the grand canonical ensemble:

$$\mathcal{Z}_G(\mu, V, T) = \int \mathcal{D}\rho \int \mathcal{D}w \; \exp(-H_G[\rho, w]) \tag{4.26}$$

with

$$H_G[\rho, w] = - i \int d\mathbf{r} \; w(\mathbf{r})\rho(\mathbf{r}) + \frac{\beta}{2} \int d\mathbf{r} \int d\mathbf{r}' \; \rho(\mathbf{r})u(|\mathbf{r} - \mathbf{r}'|)\rho(\mathbf{r}')$$
$$- zV \; Q[iw] \tag{4.27}$$

The thermodynamic connection in this ensemble is through the equation of state

$$pV = k_B T \; \ln \mathcal{Z}_G(\mu, V, T) \tag{4.28}$$

where $p$ is the pressure, and the average number of particles in the system is controlled by adjusting the chemical potential $\mu$ or activity $z$ according to

$$\langle n \rangle = \left( \frac{\partial \ln \mathcal{Z}_G(\mu, V, T)}{\partial \ln z} \right)_{V,T} \tag{4.29}$$

Comparing eqns (4.14) and (4.27) for the effective Hamiltonians of the canonical and grand canonical ensembles, respectively, we see that they differ only in the form of the last translational entropy term. Thus, it is straightforward to switch ensembles in the field-theoretic framework. Finally, if $u(r)$ is positive definite and invertible as described in the previous section, the grand canonical theory can be similarly reduced to a statistical field theory involving only the $w$ field:

$$\mathcal{Z}_G(\mu, V, T) = \int \mathcal{D}w \; \exp(-H_G[w]) \tag{4.30}$$

with effective Hamiltonian

$$H_G[w] = \frac{1}{2\beta} \int d\mathbf{r} \int d\mathbf{r}' \; w(\mathbf{r})u^{-1}(|\mathbf{r} - \mathbf{r}'|)w(\mathbf{r}') - zV \; Q[iw] \tag{4.31}$$

Again, eqns (4.20) and (4.31) differ only in the translational entropy terms.

### 4.1.3  *Averages and operators for monatomic fluids*

Having demonstrated how to convert an atomistic particle-based model of a simple fluid into a field theory, it is now appropriate to discuss how ensemble averages of quantities of interest are to be calculated. For some arbitrary observable $G[\rho, w]$ that can be expressed as a functional of the fluctuating $\rho$ and $w$ fields, the *ensemble average* of $G$ is defined by

$$\langle G[\rho, w] \rangle \equiv \frac{\int \mathcal{D}\rho \int \mathcal{D}w \; G[\rho, w] \exp(-H[\rho, w])}{\int \mathcal{D}\rho \int \mathcal{D}w \; \exp(-H[\rho, w])} \qquad (4.32)$$

This expression is appropriate for computing averages using the full field theory in the canonical ensemble. The corresponding average for the grand canonical ensemble is defined by the substitution of $H_G[\rho, w]$ for the effective Hamiltonian $H[\rho, w]$. In situations where the pair potential is invertible and the simpler theories of eqns (4.19)–(4.20) or eqns (4.30)–(4.31) can be applied, the ensemble average of an observable $G[w]$ is defined by

$$\langle G[w] \rangle \equiv \frac{\int \mathcal{D}w \; G[w] \exp(-H[w])}{\int \mathcal{D}w \; \exp(-H[w])} \qquad (4.33)$$

where it is understood that $H_G[w]$ is substituted for $H[w]$ in the grand canonical case.

In the canonical ensemble, an important thermodynamic quantity is the chemical potential defined by $\mu = (\partial A/\partial n)_{T,V}$. It follows from eqns (4.13)–(4.15) that

$$\mu = \mu_0 - k_B T \langle \ln Q[iw] \rangle \qquad (4.34)$$

where $\mu_0 = -k_B T(\partial \ln \mathcal{Z}_0/\partial n)_{T,V}$ is the chemical potential of the ideal gas. This expression shows that the excess contribution to the chemical potential is given by the average of the "operator" $-k_B T \ln Q[iw]$ over the field fluctuations. For a typical realization of $w(\mathbf{r})$, this operator has both real and imaginary parts. However, the imaginary part averages to zero, leaving a purely real contribution to the chemical potential.

A closely related thermodynamic quantity in the grand canonical ensemble is the average particle density, defined as $\rho_0 \equiv \langle n \rangle/V$ with $\langle n \rangle$ given by eqn (4.29). By performing the indicated derivative in that equation, it follows that $\langle n \rangle = zV \langle Q[iw] \rangle$, so

$$\rho_0 \equiv \frac{\langle n \rangle}{V} = z \langle Q[iw] \rangle \qquad (4.35)$$

Thus, $zQ[iw]$ can be viewed as an operator, which, when we average over the $w$ field fluctuations in the grand canonical ensemble, leads to the average particle density.

Another important class of ensemble averages are those related to correlation functions of particle density. In the full theories involving both $\rho$ and $w$ fields, density correlation functions can be directly calculated as averages over appropriate factors of $\rho(\mathbf{r})$. For example, in the canonical ensemble, the *density–density correlation function* $\langle \rho(\mathbf{r})\rho(\mathbf{r}')\rangle$ can be evaluated according to

$$\langle \rho(\mathbf{r})\rho(\mathbf{r}')\rangle = \frac{\int \mathcal{D}\rho \int \mathcal{D}w \ \rho(\mathbf{r})\rho(\mathbf{r}') \exp(-H[\rho, w])}{\int \mathcal{D}\rho \int \mathcal{D}w \ \exp(-H[\rho, w])} \qquad (4.36)$$

This object can be related to the scattered radiation intensity in x-ray and neutron scattering experiments on simple liquids (Hansen and McDonald, 1986). Similarly, in the grand canonical ensemble, the *average local density* of atoms is given by

$$\langle \rho(\mathbf{r})\rangle = \frac{\int \mathcal{D}\rho \int \mathcal{D}w \ \rho(\mathbf{r}) \exp(-H_G[\rho, w])}{\int \mathcal{D}\rho \int \mathcal{D}w \ \exp(-H_G[\rho, w])} \qquad (4.37)$$

The derivation of an alternative expression for $\langle \rho(\mathbf{r})\rangle$ in the grand canonical ensemble is instructive.[39] Note that eqn (4.37) can be reexpressed as

$$\langle \rho(\mathbf{r})\rangle = \frac{1}{\mathcal{Z}_G} \int \mathcal{D}\rho \int \mathcal{D}w \ e^{zVQ[iw]-(\beta/2)\int d\mathbf{r} \int d\mathbf{r}' \ \rho u \rho} \frac{\delta}{i\,\delta w(\mathbf{r})} e^{i\int d\mathbf{r} \ w\rho} \qquad (4.38)$$

A functional integration by parts, assuming that $\int \mathcal{D}\rho \exp(-H_G)$ decays to zero as $w(\mathbf{r}) \to \pm\infty$ for all points $\mathbf{r}$ in the domain, leads to

$$\begin{aligned}
\langle \rho(\mathbf{r})\rangle &= -\frac{1}{i\,\mathcal{Z}_G} \int \mathcal{D}\rho \int \mathcal{D}w \ e^{i\int d\mathbf{r} \ w\rho - (\beta/2)\int d\mathbf{r} \int d\mathbf{r}' \ \rho u \rho} \frac{\delta}{\delta w(\mathbf{r})} e^{zVQ[iw]} \\
&= \left\langle izV\frac{\delta Q[iw]}{\delta w(\mathbf{r})} \right\rangle = \langle z\exp[-iw(\mathbf{r})]\rangle \qquad (4.39)
\end{aligned}$$

where in deriving the last expression, we have used the explicit form given in eqn (4.12) for the single-particle partition function. The final expression shows that averaging the operator $\rho(\mathbf{r})$ in the grand canonical ensemble is equivalent to averaging the operator $z\exp[-iw(\mathbf{r})]$. Moreover, the volume average of both sides of eqn (4.39) reproduces eqn (4.35), which is consistent with the expectation that

$$\rho_0 = \frac{1}{V} \int d\mathbf{r} \ \langle \rho(\mathbf{r})\rangle \qquad (4.40)$$

Another important thermodynamic quantity of interest is the *pressure p*. Evaluation of the pressure is immediate in the grand canonical ensemble via eqn (4.28), provided $\mathcal{Z}_G$ can be computed. In field theories formulated in the canonical ensemble, the pressure is a more elusive object. One strategy that

---

[39]I am grateful to H. Orland for teaching me this trick.

is applicable for differentiable pair potentials $u(r)$ is to utilize the *virial theorem* derived from the particle representation of $\mathcal{Z}_C$ (Chandler, 1987; McQuarrie, 1976)

$$\beta p/\rho_0 = 1 - \frac{\beta}{6n} \sum_{j=1}^{n} \sum_{k=1(\neq j)}^{n} \langle v(|\mathbf{r}_j - \mathbf{r}_k|) \rangle \qquad (4.41)$$

In this expression, $\rho_0 = n/V$ is the average density in the canonical ensemble and $v(r) = r\, du(r)/dr$ is the virial function. Equation (4.41) can be rewritten in terms of an average over microscopic particle densities, which in turn can be replaced by an average over $\rho$ fields in the field-theoretic description:

$$\beta p/\rho_0 = 1 - \frac{\beta}{6n} \int d\mathbf{r} \int d\mathbf{r}' \, v(|\mathbf{r} - \mathbf{r}'|)[\langle \rho(\mathbf{r})\rho(\mathbf{r}') \rangle - \delta(\mathbf{r} - \mathbf{r}')\langle \rho(\mathbf{r}) \rangle] \qquad (4.42)$$

### 4.1.4 *Averages and operators for the simplified field theories*

It is important to discuss how average densities and density correlation functions can be calculated using the simplified field theories of eqns (4.19)–(4.20) and (4.30)–(4.31) in which the density field $\rho$ does not explicitly appear.

Our starting point is the canonical ensemble, where it proves convenient to augment eqn (4.1) with a "source term" involving a field $J(\mathbf{r})$ that is conjugate to the microscopic density:

$$\mathcal{Z}_C[J] = \frac{1}{n!\lambda_T^{3n}} \int d\mathbf{r}^n \, \exp\left[ -\beta U(\mathbf{r}^n) - \int d\mathbf{r} \, J(\mathbf{r})\hat{\rho}(\mathbf{r}) \right] \qquad (4.43)$$

The logarithm of $\mathcal{Z}_C[J]$ is a *generating functional* in the sense that functional derivatives with respect to $J(\mathbf{r})$ provide expressions for the connected (cumulant) correlation functions of density.[40] In particular,

$$\langle \hat{\rho}(\mathbf{r}) \rangle = - \left. \frac{\delta \ln \mathcal{Z}_C[J]}{\delta J(\mathbf{r})} \right|_{J=0} \qquad (4.44)$$

where the average on the left-hand side of this equation denotes an ensemble average over the particle coordinates $\mathbf{r}^n$. Similarly, the density pair correlation function is given by

$$\langle \hat{\rho}(\mathbf{r})\hat{\rho}(\mathbf{r}') \rangle - \langle \hat{\rho}(\mathbf{r}) \rangle \langle \hat{\rho}(\mathbf{r}') \rangle = \left. \frac{\delta^2 \ln \mathcal{Z}_C[J]}{\delta J(\mathbf{r})\delta J(\mathbf{r}')} \right|_{J=0} \qquad (4.45)$$

In order to compute the derivatives on the right-hand sides of the above equations, it is helpful to transform to a field-theoretic representation of $\mathcal{Z}_C[J]$

---

[40]Generating functions and cumulant moments are discussed in Appendix B.

by retracing the steps that led from eqn (4.1) to eqn (4.19). Evidently $u(r)$ must again satisfy the necessary conditions. One obtains the following field theory:

$$\mathcal{Z}_C[J] = \mathcal{Z}_0 \int \mathcal{D}w \, \exp(-H[w, J]) \tag{4.46}$$

$$H[w, J] = \frac{1}{2\beta} \int d\mathbf{r} \int d\mathbf{r}' \, w(\mathbf{r})u^{-1}(|\mathbf{r} - \mathbf{r}'|)w(\mathbf{r}') - n \ln Q[iw + J] \tag{4.47}$$

where $J$ enters the effective Hamiltonian only as a shift in the argument of $Q$. It follows that for $J = 0$, $\mathcal{Z}_C[0] = \mathcal{Z}_C$ and $H[w, 0] = H[w]$. The right-hand side of eqn (4.44) can thus be evaluated as

$$\langle \hat{\rho}(\mathbf{r}) \rangle = -n \left\langle \left. \frac{\delta \ln Q[iw + J]}{\delta J(\mathbf{r})} \right|_{J=0} \right\rangle = -n \left\langle \frac{\delta \ln Q[iw]}{\delta iw(\mathbf{r})} \right\rangle \tag{4.48}$$

where the averages in the last two expressions are defined in accordance with eqn (4.33). This result suggests that we can define a *particle density operator*

$$\tilde{\rho}(\mathbf{r}; [iw]) \equiv -n \frac{\delta \ln Q[iw]}{\delta iw(\mathbf{r})} = \frac{\rho_0}{Q[iw]} \exp[-iw(\mathbf{r})] \tag{4.49}$$

so that

$$\langle \hat{\rho}(\mathbf{r}) \rangle = \langle \tilde{\rho}(\mathbf{r}; [iw]) \rangle \tag{4.50}$$

In other words, $\tilde{\rho}$ is an operator whose average over the $w$ field fluctuations reproduces the average local density $\langle \hat{\rho}(\mathbf{r}) \rangle$ of the original particle-based model. The alert reader will notice a striking similarity between the intermediate expression in eqn (4.49) and the segment density operator for a *single polymer* introduced in eqn (3.50). Indeed, for the present case of a monatomic fluid in the canonical ensemble, the *total* particle density operator $\tilde{\rho}$ is simply the *single-particle density operator*

$$\rho(\mathbf{r}; [iw]) \equiv -\frac{\delta \ln Q[iw]}{\delta iw(\mathbf{r})} \tag{4.51}$$

multiplied by the number of particles $n$.

An alternative way of computing the average local density follows from a procedure analogous to that used to derive eqn (4.39). The final expression in eqn (4.48) can be written as

$$\langle \hat{\rho}(\mathbf{r}) \rangle = \frac{i}{\mathcal{Z}_C} \int \mathcal{D}w \, e^{-(1/2\beta) \int d\mathbf{r} \int d\mathbf{r}' \, wu^{-1}w} \frac{\delta}{\delta w(\mathbf{r})} e^{n \ln Q[iw]} \tag{4.52}$$

Functional integration by parts leads immediately to

$$\langle \hat{\rho}(\mathbf{r}) \rangle = \frac{i}{\beta} \int d\mathbf{r}' \, u^{-1}(|\mathbf{r} - \mathbf{r}'|) \langle w(\mathbf{r}') \rangle \tag{4.53}$$

Thus, an "alternative" expression for a particle density operator is

$$\tilde{\rho}_a(\mathbf{r};[iw]) \equiv \frac{i}{\beta} \int d\mathbf{r}'\, u^{-1}(|\mathbf{r} - \mathbf{r}'|)w(\mathbf{r}') \tag{4.54}$$

with the property that $\langle\hat{\rho}(\mathbf{r})\rangle = \langle\tilde{\rho}_a(\mathbf{r};[iw])\rangle$. Note that $\tilde{\rho} \neq \tilde{\rho}_a$, in spite of the fact that they are both acceptable operators for computing the average local density in the field theory described by eqns (4.19)–(4.20).

The "alternative approach" is particularly convenient for the evaluation of higher-order density correlation functions. For example, the two-point function can be written

$$\langle\hat{\rho}(\mathbf{r})\hat{\rho}(\mathbf{r}')\rangle = \frac{1}{\mathcal{Z}_C[J]}\frac{\delta^2 \mathcal{Z}_C[J]}{\delta J(\mathbf{r})\delta J(\mathbf{r}')}\bigg|_{J=0}$$

$$= -\frac{\mathcal{Z}_0}{\mathcal{Z}_C}\int \mathcal{D}w\, e^{-(1/2\beta)\int d\mathbf{r}\int d\mathbf{r}' wu^{-1}w}\frac{\delta^2 \exp(n\ln Q[iw])}{\delta w(\mathbf{r})\delta w(\mathbf{r}')} \tag{4.55}$$

Two functional integrations by parts leads to the expression

$$\langle\hat{\rho}(\mathbf{r})\hat{\rho}(\mathbf{r}')\rangle = -\beta^{-2}\int d\mathbf{r}_1 \int d\mathbf{r}_2\, u^{-1}(|\mathbf{r} - \mathbf{r}_1|)u^{-1}(|\mathbf{r}' - \mathbf{r}_2|)\langle w(\mathbf{r}_1)w(\mathbf{r}_2)\rangle$$
$$+ \beta^{-1}u^{-1}(|\mathbf{r} - \mathbf{r}'|) \tag{4.56}$$

so that the density–density pair correlation function in the simplified field theory can be computed from the pair correlation function of the $w$ field.

The above derivations of density operators and correlation function formulas are easily extended to the *grand canonical* version of the simplified theory, eqns (4.30)–(4.31). The particle density operator is given in the grand canonical ensemble by

$$\tilde{\rho}_G(\mathbf{r};[iw]) \equiv -zV\frac{\delta Q[iw]}{\delta iw(\mathbf{r})} = z\exp[-iw(\mathbf{r})] \tag{4.57}$$

so that the average local particle density is

$$\langle\hat{\rho}(\mathbf{r})\rangle = \langle\tilde{\rho}_G(\mathbf{r};[iw])\rangle \tag{4.58}$$

The average on the right is carried out with the statistical weight $\exp(-H_G[w])$. This expression is evidently consistent with eqn (4.39) for the full theory involving both $\rho$ and $w$ fields.

Interestingly, the alternative particle density operator $\tilde{\rho}_a(\mathbf{r};[iw])$ has the same form of eqn (4.54) in both canonical and grand canonical ensembles. It follows that the average local density and the density–density correlation function can be computed in the grand canonical theory by using the "alternative" eqns (4.53) and (4.56), respectively. It is understood that when these expressions are applied in the grand canonical case, the averages on the right-hand sides are carried out with the appropriate statistical weight, $\exp(-H_G[w])$.

We shall make considerable reference to the above formulas throughout the book. Particularly important are the various forms of the particle density operators for the *simplified* field theories involving only the $w$ field. These operator forms and the relevant statistical weights $P[w]$ are summarized in Table 4.1. Ensemble averages of an observable $G[w]$ are defined by applying $P[w]$ in accordance with eqn (4.33) as

$$\langle G[w] \rangle = \frac{\int \mathcal{D}w \; G[w] P[w]}{\int \mathcal{D}w \; P[w]} \qquad (4.59)$$

**Table 4.1** *Density operators and statistical weights for monatomic fluids*

|  | canonical | grand canonical |
|---|---|---|
| statistical weight | $P[w] = \exp(-H[w])$ | $P[w] = \exp(-H_G[w])$ |
| density operator | $\tilde{\rho} = -n\frac{\delta \ln Q}{\delta i w} = \frac{\rho_0}{Q}e^{-iw}$ | $\tilde{\rho}_G = -zV\frac{\delta Q}{\delta i w} = ze^{-iw}$ |
| alternative form | $\tilde{\rho}_a = (i/\beta)\int d\mathbf{r}' \; u^{-1}w$ | $\tilde{\rho}_a = (i/\beta)\int d\mathbf{r}' \; u^{-1}w$ |

## 4.2  Neutral polymer solutions

The particle-to-field transformation technique introduced in Section 4.1 for the monatomic fluid is immediately applicable to more complex fluids, including polymer solutions and melts. Indeed, we could proceed by building a fully *atomistic* model of a polymeric fluid and deduce a corresponding field theory. Such an approach would have limited utility, however, because the field theory would have to be resolved with atomic resolution in order to make quantitative predictions. The resulting field-based simulations would be prohibitively expensive – probably more expensive than a direct particle-based simulation of the starting model. Thus, from this point on we adopt the *mesoscopic perspective* described in Chapters 1 and 2, and smear out atomic details below about 1 nm.

The definition of a *mesoscopic model* of a polymeric fluid requires two ingredients:

1. A coarse-grained *chain model* of the type described in Chapter 2, including a specification of polymer architecture
2. An *interaction model* describing the potential energy associated with the interactions among the coarse-grained entities in the fluid. These could include polymer segments, solvents, colloidal particles, nanoparticles, ions, etc.

Because there are many well-established chain models that can be employed, several examples of which were given in Chapter 2, and essentially an infinite variety of polymer architectures and interaction potentials, it is not possible to enumerate all possible mesoscopic models of polymer fluids. As a result, in the present chapter we will provide some representative examples of models that serve to illustrate the technique[41] of model-building, as well as lay a foundation for subsequent discussions of how to analyze and simulate field-theoretic models.

In the present section on neutral (non-electrolyte) polymer solutions, two models will be discussed:

- **Model A:** homopolymer in a good solvent, treated implicitly
- **Model B:** homopolymer solution, explicit solvent

Both of these models have precedence in the polymer physics literature. Model A was introduced many years ago by Edwards (Edwards, 1965) and is sometimes referred to as the "Edwards model." The field theory corresponding to Model B is not as well known, but its mean-field approximation is the continuum analog of the Flory–Huggins lattice theory of polymer solutions (Flory, 1953; de Gennes, 1979). We begin our discussion with the simpler Model A.

### 4.2.1  Model A: homopolymer in a good solvent

Model A, or the "Edwards model," is the standard continuum model of flexible polymers dissolved in a good solvent. Since its introduction in the mid-1960s (Edwards, 1965), hundreds of published theoretical studies on its behavior have appeared (Doi and Edwards, 1986; des Cloizeaux and Jannink, 1990; Freed, 1987). The polymers in Model A are assumed to be linear homopolymers of uniform length (monodisperse) with statistical properties described by the *continuous Gaussian chain model*. The polymers are dissolved in a good solvent and the solvent degrees of freedom are not explicitly included in the statistical mechanical description ("implicit solvent"). Interactions among segments are assumed to be characterized by a *pair potential of mean force* $\bar{u}(r)$, following the McMillan–Mayer description of non-electrolyte solutions (Hill, 1960).

The McMillan–Mayer theory shows rigorously that if the solvent degrees of freedom are integrated out of a classical statistical mechanical description of a solution, the remaining interactions among solute molecules are properly described by a potential of mean force (Chandler, 1987). In the particular case of a polymer solution, the relevant solute degrees of freedom are the segment coordinates, and in Model A it is further assumed that the potential of mean force $\bar{U}$ can be decomposed into a sum over all segment pairs of a pair potential of mean force, $\bar{u}(r)$.[42] In the $nVT$ canonical ensemble of $n$ continuous Gaussian chains contained in a solution of volume $V$, the segment coordinates are the set of conformations of $n$ space curves

$$\mathbf{r}^{nN} \equiv \{\mathbf{r}_j(s)\} \quad s \in [0, N], \; j = 1, 2, ..., n \tag{4.60}$$

The pair approximation thus amounts to

---

[41] "Art" may be a more accurate term than technique.

[42] We use overbars to distinguish potentials of mean force from bare (vacuum) potential functions.

$$\bar{U}[\mathbf{r}^{nN}] \approx \frac{1}{2} \sum_{j=1}^{n} \sum_{k=1}^{n} \int_{0}^{N} ds \int_{0}^{N} ds' \ \bar{u}(|\mathbf{r}_{j}(s) - \mathbf{r}_{k}(s')|) \qquad (4.61)$$

where it is understood that when the integrals over $s$ and $s'$ for $j = k$ are performed, the self-interaction of a segment with itself, $s = s'$, is excluded. It is important to note that eqn (4.61) includes both interactions between segments on *different* polymer chains, e.g. the interaction $\bar{u}(|\mathbf{r}_3 - \mathbf{r}_4|)$ between segments at points $\mathbf{r}_3$ and $\mathbf{r}_4$ in Fig. 4.1, as well as interactions between non-adjacent segments on the *same* chain. The latter intramolecular contributions include the long-ranged interferences, e.g. $\bar{u}(|\mathbf{r}_1 - \mathbf{r}_2|)$ in Fig. 4.1, that produce the "excluded volume effect" discussed in the context of eqn (2.2). Finally, it is important to realize that eqn (4.61) (and Model A) assumes that the intramolecular and intermolecular pair potentials of mean force are described by a single function $\bar{u}(r)$.

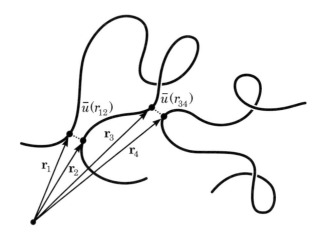

FIG. 4.1. Intramolecular and intermolecular interactions between segments in a polymer solution. The potential of mean force between two segments belonging to the same chain and residing at positions $\mathbf{r}_1$ and $\mathbf{r}_2$ is denoted by $\bar{u}(r_{12})$. This is an example of a "long-ranged interference." An intermolecular interaction between segments located at positions $\mathbf{r}_3$ and $\mathbf{r}_4$ and belonging to different chains is denoted by $\bar{u}(r_{34})$.

Unlike a bare pair potential $u(r)$, the pair potential of mean force $\bar{u}(r)$ depends on the thermodynamic state of the pure solvent separating two segments, namely on its temperature and chemical potential. At close separations $r$, $\bar{u}(r)$ will exhibit complicated oscillations associated with the solvent packing about the two referenced chain segments. From a mesoscopic perspective, we are not interested in this local liquid structure but rather hope to capture the net attractive or repulsive interaction associated with the close approach of two chain

segments mediated by solvent. It is thus common in polymer physics (Zimm
et al., 1953; Doi and Edwards, 1986), and in Model A, to adopt a simple delta
function model for the potential of mean force

$$\bar{u}(r) = k_B T\, u_0 \delta(\mathbf{r}) \tag{4.62}$$

The prefactor, $u_0$, is known as the *excluded volume parameter* and has the dimensions of volume. It assumes positive values for good solvents, expressing the
fact that there is a net repulsion between two segments if they are separated by
a solvent that they both like. Similarly, $u_0 < 0$ for a poor solvent, since there
is then a net attraction between polymer segments. At the compensation point
$u_0 = 0$, the net attractions and repulsions are balanced and the solution is said
to be at the "theta" ($\Theta$) point (Flory, 1953; de Gennes, 1979). Since $\bar{u}(r)$ and
hence $u_0$ depend on temperature, the theta point can often be approached by
adjusting $T$.

Equations (4.61)–(4.62) only constitute a well-defined statistical mechanical
model if $u_0 > 0$.[43] Thus, *Model A is restricted to good solvents*.

The definition of Model A is now complete and we can proceed with converting the coarse-grained particle model just described to a statistical field theory.
The canonical partition function can be expressed in terms of $n$ path integrals
over a Boltzmann factor containing the various energetic contributions:

$$\mathcal{Z}_C(n, V, T) = \frac{1}{n!(\lambda_T^3)^{nN}} \prod_{j=1}^{n} \int \mathcal{D}\mathbf{r}_j \, \exp\left(-\beta U_0[\mathbf{r}^{nN}] - \beta \bar{U}[\mathbf{r}^{nN}]\right) \tag{4.63}$$

As discussed above, the intermolecular interactions among segments and the
long-ranged interferences are included in the potential of mean force $\bar{U}$. The
energetic contribution from intramolecular, short-ranged interferences is contained in $U_0$, which for the continuous Gaussian chain is obtained by extending
eqn (2.42) to an $n$-chain system:

$$U_0[\mathbf{r}^{nN}] = \frac{3k_B T}{2b^2} \sum_{j=1}^{n} \int_0^N ds \left|\frac{d\mathbf{r}_j(s)}{ds}\right|^2 \tag{4.64}$$

If we retrace the steps used to go from eqn (4.1) to eqn (4.13), the particle-based
description of Model A summarized by eqn (4.63) can be converted into a field
theory. We shall keep the form of $\bar{u}(r)$ arbitrary until the end in order to keep
the results as general as possible.

The first step is to introduce a *microscopic segment density* operator that
extends the definition (3.18) to a many-chain system:

$$\hat{\rho}(\mathbf{r}) = \sum_{j=1}^{n} \int_0^N ds \, \delta(\mathbf{r} - \mathbf{r}_j(s)) \tag{4.65}$$

---

[43] A repulsive three-body potential is needed if $u_0 < 0$. Otherwise, the segments will collapse
together in solution with nothing to bound the increase in local segment density.

With this definition, eqn (4.61) can be rewritten as

$$\bar{U}[\mathbf{r}^{nN}] = \frac{1}{2} \int d\mathbf{r} \int d\mathbf{r}' \; \hat{\rho}(\mathbf{r})\bar{u}(|\mathbf{r} - \mathbf{r}'|)\hat{\rho}(\mathbf{r}') - \frac{1}{2}nN \, \bar{u}(0) \qquad (4.66)$$

where the second term subtracts the self-interactions present in the first term and, as in the monatomic fluid case, we assume that $\bar{u}(r)$ has been regularized so that $\bar{u}(0)$ is finite. The next few steps for the monatomic fluid carry over immediately to the polymer case, so that the analog of eqn (4.10) is

$$\mathcal{Z}_C = \frac{z_0^n}{n!} \int \mathcal{D}\rho \int \mathcal{D}w \prod_{j=1}^{n} \int \mathcal{D}\mathbf{r}_j \; e^{-\beta U_0 + i \int d\mathbf{r} \, w(\rho - \hat{\rho}) - (\beta/2) \int d\mathbf{r} \int d\mathbf{r}' \, \rho \bar{u} \rho} \qquad (4.67)$$

where $z_0 \equiv \exp(\beta N\bar{u}(0)/2)/\lambda_T^{3N}$. Just as in the case of the atomic fluid, the introduction of the auxiliary $\rho$ and $w$ fields has served to decouple the interactions between different polymers but also the intramolecular long-ranged interferences. As a result, the path integrals over the conformations of the $n$ polymers factor by chain according to

$$\prod_{j=1}^{n} \int \mathcal{D}\mathbf{r}_j \; e^{-\beta U_0 - i \int d\mathbf{r} \, w\hat{\rho}} = \left\{ \int \mathcal{D}\mathbf{r} \; e^{-3/(2b^2) \int_0^N ds \, |d\mathbf{r}(s)/ds|^2 - i \int_0^N ds \, w(\mathbf{r}(s))} \right\}^n$$

$$= (Z_0 Q[iw])^n \qquad (4.68)$$

where in the last line we have applied the definition of the normalized *single-chain partition function* $Q[w]$ given in eqn (3.20). The factor $Z_0$ is the path integral of a free polymer in zero field, defined by eqn (2.43). It is helpful to recognize that $Z_0$ is extensive, so we reexpress it as $Z_0 \equiv V g_N$, where $g_N$ is independent of $n$, $V$, $w$, and $\rho$.[44] Upon combining eqns (4.67) and (4.68), we have thus succeeded in deriving a statistical field theory for Model A:

$$\mathcal{Z}_C(n, V, T) = \mathcal{Z}_0 \int \mathcal{D}\rho \int \mathcal{D}w \; \exp(-H[\rho, w]) \qquad (4.69)$$

with effective Hamiltonian

$$H[\rho, w] = -i \int d\mathbf{r} \, w(\mathbf{r})\rho(\mathbf{r}) + \frac{\beta}{2} \int d\mathbf{r} \int d\mathbf{r}' \, \rho(\mathbf{r})\bar{u}(|\mathbf{r} - \mathbf{r}'|)\rho(\mathbf{r}')$$

$$- n \ln Q[iw] \qquad (4.70)$$

and where $\mathcal{Z}_0 \equiv (z_0 g_N V)^n / n!$ can be interpreted as the partition function for an ideal gas of non-interacting polymers.

---

[44] $g_N$ strictly does not exist for a continuous Gaussian chain, but this has no thermodynamic consequences. A simple regularization procedure is to replace the chain with a bead-spring chain of $N$ beads, in which case $g_N = g_M^{N-1}$, where $g_M$ is a volume of order the segment volume.

One difference between eqns (4.69)–(4.70) for a polymer solution and the corresponding eqns (4.13)–(4.14) for a monatomic fluid arises from the distinction between the bare potential $u(r)$ of the fluid and the potential of mean force $\bar{u}(r)$ of the solution. The latter of course depends on the characteristics of the solvent. Another important difference between the two field theories is associated with the two forms of $Q[iw]$, namely eqn (4.12) for an atom and eqn (3.20) for a polymer. The *non-locality* of the functional $Q[iw]$ in the polymer case provides for a much richer structure of the field theory.

Up to this point, we have not assumed a specific form for $\bar{u}(r)$, so eqns (4.69)–(4.70) describe a very general model of a polymer solution consisting of continuous Gaussian polymers whose segments interact via an arbitrary potential of mean force $\bar{u}(r)$. The field theory has even broader applicability than this would imply, however, because it can be applied to polymer solutions described by *any chain model*, provided the appropriate expression for $Q[iw]$ is substituted into eqn (4.70). Thus, eqns (4.69)–(4.70) would represent a field-theoretic model for a solution of bead-spring chains with arbitrary spring potential $h(|\mathbf{r}|)$ if eqn (3.4) is applied for $Q[iw]$.

Finally, we specialize to the particular Model A potential given in eqn (4.62). Because this form of $\bar{u}(r)$ for good solvents ($u_0 > 0$) is positive definite and has an inverse, the $\rho$ field in eqn (4.69) can be integrated out of the theory to derive a simplified field theory analogous to eqn (4.19). This theory, which will be referred to as the *canonical Model A field theory*, is given by

$$\mathcal{Z}_C(n, V, T) = \mathcal{Z}_0 \int \mathcal{D}w \, \exp(-H[w]) \tag{4.71}$$

with

$$H[w] = \frac{1}{2u_0} \int d\mathbf{r} \, [w(\mathbf{r})]^2 - n \ln Q[iw] \tag{4.72}$$

The corresponding *grand canonical Model A field theory* follows immediately from eqns (4.30) and (4.31) as

$$\mathcal{Z}_G(\mu, V, T) = \int \mathcal{D}w \, \exp(-H_G[w]) \tag{4.73}$$

with

$$H_G[w] = \frac{1}{2u_0} \int d\mathbf{r} \, [w(\mathbf{r})]^2 - zVQ[iw] \tag{4.74}$$

and where the polymer *activity* $z$ has been defined as $z \equiv z_0 g_N \exp(\beta\mu)$.

Segment density operators and correlation functions for the above theories can be deduced by following the arguments of Section 4.1.4. In the canonical case, the segment density operator is given by

$$\tilde{\rho}(\mathbf{r}; [iw]) = -n\frac{\delta \ln Q[iw]}{\delta iw(\mathbf{r})} = n\rho(\mathbf{r}; [iw])$$

$$= \frac{n}{VQ[iw]} \int_0^N ds \, q(\mathbf{r}, N - s; [iw]) q(\mathbf{r}, s; [iw]) \tag{4.75}$$

where in the last expression of the first line, eqn (3.50) was used to identify the *single-chain* segment density operator $\rho(\mathbf{r}; [iw])$. In the final expression, eqn (3.55) was applied to relate the single-chain operator to a solution $q(\mathbf{r}, s; [iw])$ of the Fokker–Planck equation (3.25) with $w \to iw$. The segment density operator has an important conservation property that its volume integral must equal the number of statistical segments, i.e. $\int d\mathbf{r} \, \tilde{\rho}(\mathbf{r}; [iw]) = nN$. This property can be derived by combining the factorization formula (3.27) with the last expression in eqn (4.75).

Ensemble averages of operators over all realizations of the potential field are evaluated as in Section 4.1.4. The average segment density in the canonical Model A is given by $\langle \tilde{\rho}(\mathbf{r}; [iw]) \rangle$, where the average is computed according to eqn (4.33). Similarly, in the grand canonical version of Model A, the segment density operator is

$$\tilde{\rho}_G(\mathbf{r}; [iw]) = -zV \frac{\delta Q[iw]}{\delta iw(\mathbf{r})} = zV Q[iw] \rho(\mathbf{r}; [iw])$$

$$= z \int_0^N ds \, q(\mathbf{r}, N - s; [iw]) q(\mathbf{r}, s; [iw]) \qquad (4.76)$$

and the average segment density is obtained from $\langle \tilde{\rho}_G(\mathbf{r}; [iw]) \rangle$, where the average is computed according to eqn (4.59) with $P[w] = \exp(-H_G[w])$.

An alternative segment density operator that applies in both ensembles is

$$\tilde{\rho}_a(\mathbf{r}; [iw]) = u_0^{-1} iw(\mathbf{r}) \qquad (4.77)$$

and the density–density correlation function can be computed in either ensemble according to

$$\langle \hat{\rho}(\mathbf{r}) \hat{\rho}(\mathbf{r}') \rangle = u_0^{-1} \delta(\mathbf{r} - \mathbf{r}') - u_0^{-2} \langle w(\mathbf{r}) w(\mathbf{r}') \rangle \qquad (4.78)$$

The canonical Model A field theory has been known since the mid-1960s (Edwards, 1965) and is the most studied of all polymer field theory models. In the case of a bulk *homogeneous* solution, Model A has been tackled with a battery of analytical techniques including the Gaussian approximation (Edwards, 1966; Edwards, 1975; Doi and Edwards, 1986), variational methods (Muthukumar and Edwards, 1982), scaling methods (de Gennes, 1979), and renormalization group theory (Freed, 1987; des Cloizeaux and Jannink, 1990; Oono, 1985; Ohta and Nakanishi, 1983). In spite of this large body of work, reliable analytical results for Model A are restricted to fairly limited regions of its two-parameter space. The grand canonical version of the theory has been considerably less studied. Moreover, beyond scaling arguments and mean-field solutions, much less is known about the properties of either field theory in *inhomogeneous* situations such as bound geometries.

### 4.2.2 Model B: homopolymer solution, explicit solvent

While Model A is the simplest field theory model of a polymer solution, it is unsatisfactory in certain parameter regimes. For example, if the solvent is so poor that $u_0 < 0$, the model is not well defined. Also, if the polymer concentration is sufficiently high, the pair approximation for the potential of mean force is of questionable validity. One approach to remedy these deficiencies in Model A is to add a local three-body term to $\bar{U}$ of the form $\bar{U}_3 = u_1 \int d\mathbf{r}[\hat{\rho}(\mathbf{r})]^3$ with coefficient $u_1 > 0$ (des Cloizeaux and Jannink, 1990). This stabilizes the theory against poor solvent conditions and provides an additional parameter to adjust beyond $u_0$ in order to match experimental solution data.

An alternative one-parameter approach that we adopt here is to more closely follow the Flory–Huggins theory of polymer solutions (Flory, 1953; de Gennes, 1979). In this theory, local contacts between polymer and solvent are parameterized with a Flory "chi" parameter, and the solution is assumed to be locally incompressible. Polymer segments and solvent molecules are placed to fill the sites of a uniform lattice and the mean-field approximation is invoked. Our Model B shares the interaction model and incompressibility assumption of the Flory–Huggins theory, but it represents a continuum field theory (rather than a lattice model) that is formulated beyond the mean-field approximation.

In the canonical ensemble, Model B consists of a mixture of $n_S$ solvent molecules and $n_P$ polymer molecules in a system of fixed volume $V$ and temperature $T$. The conformational properties of the polymers are described by the continuous Gaussian chain model, and each chain is assumed to occupy a volume $v_0 N$, where $v_0$ is the volume of a statistical segment. For simplicity, each solvent molecule is assumed to occupy the same volume $v_0$ in the solution.[45] Furthermore, the solution is taken to be locally incompressible with a fixed total number density (of solvent and segments) of $\rho_0 = 1/v_0$, implying no volume change on mixing. By analogy with the Flory–Huggins lattice theory, the interaction energy between solvent molecules and polymer segments is described by a local, pairwise interaction of the form

$$\beta U_1[\mathbf{r}^{n_S + n_P N}] = v_0 \chi_{PS} \int d\mathbf{r}\, \hat{\rho}_S(\mathbf{r})\hat{\rho}_P(\mathbf{r}) \qquad (4.79)$$

where $\chi_{PS}$ is the Flory "chi" parameter between polymer segments and solvent molecules and the microscopic densities of solvent and polymer segments are defined, respectively, by

$$\hat{\rho}_S(\mathbf{r}) = \sum_{j=1}^{n_S} \delta(\mathbf{r} - \mathbf{r}_j) \qquad (4.80)$$

$$\hat{\rho}_P(\mathbf{r}) = \sum_{j=1}^{n_P} \int_0^N ds\, \delta(\mathbf{r} - \mathbf{r}_j(s))) \qquad (4.81)$$

The interaction energy $U_1$ is treated in the theory as a bare potential, rather than as a potential of mean force, since the solvent degrees of freedom $\mathbf{r}^{n_S} =$

---

[45]Actually, we can choose to *define* statistical segments to have the same volume as a solvent molecule. This of course would impact the values of the statistical segment length $b$ and the number of segments $N$.

$(\mathbf{r}_1, ..., \mathbf{r}_{n_S})$ are explicitly included. The Flory parameter $\chi_{PS}$ is a dimensionless phenomenological parameter that is used to describe the energetic strength (in units of $k_B T$) of local contacts between solvent molecules and polymer segments, relative to solvent–solvent and polymer–polymer contacts (de Gennes, 1979).

The canonical partition function of Model B can be written

$$\mathcal{Z}_C(n_S, n_P, V, T) = \frac{1}{n_S! n_P! (\lambda_T^3)^{n_S + n_P N}} \prod_{j=1}^{n_P} \int \mathcal{D}\mathbf{r}_j \prod_{k=1}^{n_S} \int d\mathbf{r}_k$$
$$\times \exp(-\beta U_0[\mathbf{r}^{n_P N}] - \beta U_1[\mathbf{r}^{n_S + n_P N}])$$
$$\times \delta[\hat{\rho}_S + \hat{\rho}_P - \rho_0] \tag{4.82}$$

where $U_0$ is given by eqn (4.64). The object $\delta[\hat{\rho}_S + \hat{\rho}_P - \rho_0]$ denotes a functional delta function that imposes a local incompressibility constraint. This constraint mandates that the total microscopic density $\hat{\rho}_S(\mathbf{r}) + \hat{\rho}_P(\mathbf{r})$ is equal to $\rho_0 = 1/v_0$ at each point $\mathbf{r}$. We shall adopt a more direct technique than in the previous section in order to convert this coarse-grained particle model into a field theory. The first step is to introduce two new microscopic densities $\hat{\rho}_+$ and $\hat{\rho}_-$ by means of the definition

$$\hat{\rho}_\pm(\mathbf{r}) \equiv \hat{\rho}_P(\mathbf{r}) \pm \hat{\rho}_S(\mathbf{r}) \tag{4.83}$$

The delta functional in eqn (4.82) can thus be rewritten as $\delta[\hat{\rho}_+ - \rho_0]$ and constrains only the total microscopic density $\hat{\rho}_+(\mathbf{r})$. Noting the presence of this delta functional, we can transform the interaction term $U_1$ according to

$$\beta U_1[\mathbf{r}^{n_S + n_P N}] = \frac{1}{4} v_0 \chi_{PS} \int d\mathbf{r} \ ([\hat{\rho}_+(\mathbf{r})]^2 - [\hat{\rho}_-(\mathbf{r})]^2)$$
$$= \frac{1}{4} v_0 \chi_{PS} \int d\mathbf{r} \ (\rho_0^2 - [\hat{\rho}_-(\mathbf{r})]^2) \tag{4.84}$$

With the assumption that $\chi_{PS} > 0$, these quadratic interactions in the $\hat{\rho}_-$ field can be decoupled by means of the Gaussian functional integral given in eqn (C.27):

$$e^{-\beta U_1} = e^{-\chi_{PS}(n_S + n_P N)/4} \int \mathcal{D}w_- \ e^{\int d\mathbf{r} \ [\hat{\rho}_- w_- - (\rho_0/\chi_{PS})w_-^2]} \tag{4.85}$$

In this expression, which is an example of a Hubbard–Stratonovich transformation, we have absorbed the normalizing denominator in eqn (C.27) into the definition of the functional integral over the auxiliary field $w_-(\mathbf{r})$ on the right-hand side. A second auxiliary field $w_+(\mathbf{r})$ is introduced by giving the delta functional an exponential representation

$$\delta[\hat{\rho}_+ - \rho_0] = \int \mathcal{D}w_+ \ e^{-i \int d\mathbf{r} \ w_+(\hat{\rho}_+ - \rho_0)} \tag{4.86}$$

The two *real* auxiliary fields $w_\pm$ can be viewed as fluctuating chemical potential fields because they are conjugate to the microscopic densities $\hat{\rho}_\pm$. The field $w_+$ can be interpreted as a *total chemical potential* or *pressure-like potential*, since it is responsible for enforcing local incompressibility. Similarly, the field $w_-$ represents an *exchange chemical potential*, because it is conjugate to the local density difference between the polymer and solvent species.

Upon substitution of eqns (4.85) and (4.86) into eqn (4.82), the configurational integrals for each solvent and polymer molecule become independent and can be factored. A field-theoretic representation of the *canonical Model B partition function* is immediately obtained:

$$\mathcal{Z}_C(n_S, n_P, V, T) = \mathcal{Z}_0 \int \mathcal{D}w_+ \int \mathcal{D}w_- \, \exp(-H[w_+, w_-]) \qquad (4.87)$$

with the effective Hamiltonian

$$H[w_+, w_-] = \int d\mathbf{r} \, [(\rho_0/\chi_{PS})w_-^2 - i\rho_0 w_+] \\ - n_P \ln Q_P[iw_+ - w_-] - n_S \ln Q_S[iw_+ + w_-] \qquad (4.88)$$

In eqn (4.87), $\mathcal{Z}_0$ represents the partition function of an *ideal gas* of $n_S$ solvent molecules and $n_P$ polymers in a volume $V$. The functional $Q_S[iw_+ + w_-]$ corresponds to the partition function of a *single* solvent molecule subjected to the complex field $iw_+(\mathbf{r}) + w_-(\mathbf{r})$. It is the same functional that was derived for the monatomic fluid and is given by eqn (4.12) as

$$Q_S[iw_+ + w_-] = \frac{1}{V} \int_V d\mathbf{r} \, e^{-[iw_+(\mathbf{r}) + w_-(\mathbf{r})]} \qquad (4.89)$$

Similarly, $Q_P[iw_+ - w_-]$ is the normalized *single-polymer* partition function $Q[w]$ defined in eqn (3.20) with $w(\mathbf{r})$ replaced by the complex field $iw_+(\mathbf{r}) - w_-(\mathbf{r})$.

It is important to discuss the physical content of the canonical Model B theory and its relationship to the analytic structure of the model. Two fluctuating chemical potential fields were required to effect the particle-to-field transformation – one for each species in the system. In contrast, Model A involved only a single $w$ field because the solvent degrees of freedom were not explicitly included in the description. Another important point is that the field $w_+$ enters the effective Hamiltonian of Model B only as the purely imaginary object $iw_+$. This is analogous to the way in which the field $w$ appears in Model A. In both cases, a purely imaginary field arises because it is serving to decouple a *repulsive* interaction in the starting model. In Model A, this is apparent because $u_0 > 0$ corresponds to a net repulsion between polymer segments. In Model B, the $w_+$ field arose from the representation (4.86) of the incompressibility constraint. The role of this constraint is to stabilize against local increases in total (segment plus solvent) density, which again implies repulsive forces. In contrast, the $w_-$ field contributes *real* contributions to the effective Hamiltonian of Model B. This can be traced to the final expression in eqn (4.84), which indicates that the polymer–solvent interaction, when expressed in the $\hat{\rho}_\pm$ variables and for $\chi_{PS} > 0$, is a net *attraction*. In other words, with the assumed form of the interaction, the solvent and polymer components prefer to be surrounded by similar rather than dissimilar molecules. *A general observation is that purely imaginary and purely real auxiliary fields, respectively, arise from decoupling repulsive and attractive interactions.*

Model B contains a single parameter $\chi_{PS}$ that characterizes the strength of the local interactions between polymer segments and solvent molecules and thus dictates the quality of the solvent. Similarly, Model A contains the excluded volume parameter $u_0$, which plays a similar role. It is not possible to exactly connect the two parameters, because Models A and B have different analytic forms, and field fluctuations will renormalize $u_0$ and $\chi_{PS}$ differently. Nevertheless, in the mean-field approximation, which neglects all fluctuations in the $w$ and $w_\pm$ fields, both models simplify considerably (see Chapter 5). By expanding the free energy density of the mean-field approximation to Model B (equivalent to the Flory–Huggins theory) to second order in polymer volume fraction (de Gennes, 1979), the following correspondence with Model A is obtained:

$$u_0 = v_0(1 - 2\chi_{PS})$$

Thus at the mean-field level, $\chi_{PS} > 1/2$ corresponds to poor solvent conditions, $\chi_{PS} = 1/2$ reflects theta solvent conditions, and $0 < \chi_{PS} < 1/2$ is characteristic of a good solvent. The present formulation of Model B is restricted to $\chi_{PS} > 0$, but other representations of the model can be derived for solvents that are better than "athermal," i.e. $\chi_{PS} < 0$.

The *grand canonical Model B* follows from eqns (4.87) and (4.88) by the now familiar modification of the single-particle and single-chain entropy terms:

$$\mathcal{Z}_G(\mu_S, \mu_P, V, T) = \int \mathcal{D}w_+ \int \mathcal{D}w_- \, \exp(-H_G[w_+, w_-]) \qquad (4.90)$$

where

$$H_G[w_+, w_-] = \int d\mathbf{r} \, [(\rho_0/\chi_{PS})w_-^2 - i\rho_0 w_+]$$
$$- z_P V Q_P[iw_+ - w_-] - z_S V Q_S[iw_+ + w_-] \qquad (4.91)$$

and the polymer and solvent activities are denoted, respectively, by $z_P$ and $z_S$. A subtle point is that due to the incompressibility constraint in Model B, which implies that $\langle n_S \rangle + \langle n_P \rangle N = \rho_0 V$, the two activities are not independent variables. It is thus a common practice (Matsen, 1995a) to fix either $z_P$ or $z_S$ at a convenient value and then vary the other parameter to adjust the concentration of polymers in the solution.

Finally, it is appropriate to present the forms of the polymer segment and solvent density operators in both ensembles. In the canonical ensemble, we have

$$\tilde{\rho}_P(\mathbf{r}; [iw_+ - w_-]) = -n_P \frac{\delta \ln Q_P[iw_+ - w_-]}{\delta[iw_+(\mathbf{r}) - w_-(\mathbf{r})]}$$

$$= \frac{n_P}{VQ_P[iw_+ - w_-]} \int_0^N ds\, q(\mathbf{r}, N - s; [iw_+ - w_-])$$
$$\times q(\mathbf{r}, s; [iw_+ - w_-]) \qquad (4.92)$$

$$\tilde{\rho}_S(\mathbf{r}; [iw_+ + w_-]) = -n_S \frac{\delta \ln Q_S[iw_+ + w_-]}{\delta[iw_+(\mathbf{r}) + w_-(\mathbf{r})]}$$

$$= \frac{n_S}{VQ_S[iw_+ + w_-]} \exp\{-[iw_+(\mathbf{r}) + w_-(\mathbf{r})]\} \quad (4.93)$$

Ensemble averages of these operators over the $w_\pm$ fields using the statistical weight $\exp(-H[w_+, w_-])$ lead to the average densities $\langle \hat{\rho}_P(\mathbf{r}) \rangle$ and $\langle \hat{\rho}_S(\mathbf{r}) \rangle$, respectively.

Similarly, in the grand canonical ensemble, the density operators for Model B are

$$\tilde{\rho}_{P,G}(\mathbf{r}; [iw_+ - w_-]) = -z_P V \frac{\delta Q_P[iw_+ - w_-]}{\delta[iw_+(\mathbf{r}) - w_-(\mathbf{r})]}$$

$$= z_P \int_0^N ds\, q(\mathbf{r}, N - s; [iw_+ - w_-])q(\mathbf{r}, s; [iw_+ - w_-])$$

$$(4.94)$$

$$\tilde{\rho}_{S,G}(\mathbf{r}; [iw_+ + w_-]) = -z_S V \frac{\delta Q_S[iw_+ + w_-]}{\delta[iw_+(\mathbf{r}) + w_-(\mathbf{r})]}$$
$$= z_S \exp\{-[iw_+(\mathbf{r}) + w_-(\mathbf{r})]\} \qquad (4.95)$$

and the relevant statistical weight is $\exp(-H_G[w_+, w_-])$.

## 4.3 Polymer blends

Another important class of inhomogeneous polymer fluids are molten mixtures of polymers, commonly referred to as *blends* or *alloys*. Many thermoplastics, for example, are multi-component blends that can be melt-processed into useful plastic parts. In this class, we consider two illustrative models:

- **Model C:** incompressible binary homopolymer blend
- **Model D:** compressible binary homopolymer blend

### 4.3.1 *Model C: incompressible homopolymer blend*

Model C is the most commonly encountered coarse-grained model of a binary blend in the polymer physics literature. Its origins date back to early work of Edwards and de Gennes in the mid- to late 1960s (de Gennes, 1969). Model C is a simple extension of Model B to the situation in which both components are homopolymers. It adopts a Flory-type interaction between dissimilar polymer segments and imposes local incompressibility in the melt. In the mean-field approximation, Model C reduces for homogeneous phases to the familiar Flory–Huggins theory for a binary blend (de Gennes, 1979).

Specifically, in the canonical ensemble, Model C consists of an incompressible mixture of $n_A$ type "A" homopolymers and $n_B$ type "B" homopolymers (see Fig. 4.2), each of which is described by the continuous Gaussian chain model. The A homopolymers have $N_A$ statistical segments of length $b_A$. Correspondingly, the B homopolymers have $N_B$ segments of length $b_B$. The dissimilar segments are defined so that their volumes, denoted $v_0$, are equal. The incompressibility constraint again implies that the sum of the A and B segment densities is everywhere equal to $\rho_0 = 1/v_0$. The interaction energy between dissimilar segments is modelled by eqn (4.79) with the simple relabelling of $P \to A$, $S \to B$. In this expression, the microscopic densities of segments of species $K = A$ or $B$ are defined by analogy with eqn (4.81):

$$\hat{\rho}_K(\mathbf{r}) = \sum_{j=1}^{n_K} \int_0^{N_K} ds \, \delta(\mathbf{r} - \mathbf{r}_j^K(s))) \tag{4.96}$$

FIG. 4.2. Binary blend of homopolymers considered in Models C and D. The type A homopolymers (dark) and type B homopolymers (light) are both modelled as continuous Gaussian chains. The melt is assumed to be locally incompressible or compressible, respectively, in Models C and D.

The *canonical partition function for Model C* can be transformed to a field theory by following the same steps that led from eqn (4.82) to eqn (4.87) in Model B. The result is

$$\mathcal{Z}_C(n_A, n_B, V, T) = \mathcal{Z}_0 \int \mathcal{D}w_+ \int \mathcal{D}w_- \, \exp(-H[w_+, w_-]) \tag{4.97}$$

where $\mathcal{Z}_0$ denotes the ideal gas partition function of $n_A + n_B$ non-interacting continuous Gaussian chains, and the effective Hamiltonian is given by

$$H[w_+, w_-] = \rho_0 \int d\mathbf{r} \, [(1/\chi_{AB})w_-^2 - iw_+]$$
$$- n_A \ln Q_A[w_A] - n_B \ln Q_B[w_B] \tag{4.98}$$

The *complex* fields $w_K$ (for $K = A$ or $B$) in this expression represent the following linear combinations of the $w_\pm$:

$$w_A \equiv iw_+ - w_-, \quad w_B \equiv iw_+ + w_-$$

The single-chain partition functions for species $K$ are given by eqn (3.22):

$$Q_K[w_K] = \frac{1}{V} \int d\mathbf{r} \, q_K(\mathbf{r}, N_K; [w_K]) \tag{4.99}$$

where the propagator for polymer species $K$ satisfies a slightly modified version of eqn (3.25)

$$\frac{\partial}{\partial s} q_K(\mathbf{r}, s; [w_K]) = \left[\frac{b_K^2}{6}\nabla^2 - w_K(\mathbf{r})\right] q_K(\mathbf{r}, s; [w_K]) \tag{4.100}$$

subject to $q_K(\mathbf{r}, 0) = 1$. Because the $w_K$ fields are complex, these diffusion equations are now complex equations that lead in general to a partition function $Q_K$ with both real and imaginary parts.

The *grand canonical Model C field theory* is obtained by making the obvious replacement of $z_K V Q_K$ for $n_K \ln Q_K$ in eqn (4.98):

$$H_G[w_+, w_-] = \rho_0 \int d\mathbf{r} \, [(1/\chi_{AB})w_-^2 - iw_+]$$
$$- z_A V Q_A[w_A] - z_B V Q_B[w_B] \tag{4.101}$$

where $z_K$ denotes the activity of polymer species $K = A$ or $B$. As in the grand canonical Model B, the incompressibility constraint implies that the thermodynamic properties of the grand canonical Model C are dictated by one independent activity parameter. Thus, either $z_A$ or $z_B$ can be fixed to a convenient value and the other varied to adjust the composition of the blend.

Segment density operators in the two ensembles follow immediately from eqns (4.92) and (4.94):

$$\tilde{\rho}_K(\mathbf{r}; [w_K]) = -n_K \frac{\delta \ln Q_K[w_K]}{\delta w_K(\mathbf{r})}$$
$$= \frac{n_K}{V Q_K[w_K]} \int_0^{N_K} ds \, q_K(\mathbf{r}, N_K - s; [w_K]) q_K(\mathbf{r}, s; [w_K])$$
$$\tag{4.102}$$

$$\tilde{\rho}_{K,G}(\mathbf{r};[w_K]) = -z_K V \frac{\delta Q_K[w_K]}{\delta w_K(\mathbf{r})}$$

$$= z_K \int_0^{N_K} ds \, q_K(\mathbf{r}, N_K - s; [w_K]) q_K(\mathbf{r}, s; [w_K]) \quad (4.103)$$

### 4.3.2    Model D: compressible homopolymer blend

Although Model C is the standard and most studied model of binary polymer blends, it is difficult to apply in some circumstances. For example in the case of a confined polymer melt, the boundary condition eqn (3.196) imposed on each species propagator is incompatible with the incompressibility constraint at the bounding surface. One resolution of this incompatibility is to relax the strict constraint of no total segment variations near the surface (Wu *et al.*, 1995). Another situation where Model C is inconvenient relates to non-equilibrium extensions of the formalism, where it is often desirable to formulate the dynamical equations in conventional density variables, rather than chemical potential ($w_+$) variables (Fraaije and Sevink, 2003).[46] For a strictly incompressible model, the inversion from $w_+$ fields to $\rho_+$ fields using formulas analogous to eqn (4.77) is ill defined.

In such circumstances, a compressible binary blend model due to Helfand (Helfand, 1975) provides a useful alternative that we designate as Model D. Model D is identical to Model C, except that in place of the incompressibility constraint, the following energetic term is included:

$$\beta U_2(\mathbf{r}^{n_A N_A + n_B N_B}) = \frac{1}{2}\zeta v_0 \int d\mathbf{r} \, [\hat{\rho}_+(\mathbf{r}) - \rho_0]^2 \quad (4.104)$$

The dimensionless parameter $\zeta > 0$ describes the strength of the harmonic energy penalty for local density fluctuations away from $\rho_0$. $\zeta^{-1}$ is a measure of the compressibility of the model, so that the incompressible limit is approached as $\zeta \to \infty$. Instead of the representation (4.86) for the delta functional used in Model C, the quadratic interactions manifest in eqn (4.104) for Model D are decoupled by means of the Hubbard–Stratonovich transformation

$$e^{-\beta U_2} = \int \mathcal{D}w_+ \, e^{-\int d\mathbf{r} \, [i(\hat{\rho}_+ - \rho_0)w_+ + (\rho_0/2\zeta)w_+^2]} \quad (4.105)$$

It follows that the effective Hamiltonian for the *canonical Model D* is given by

$$H[w_+, w_-] = \rho_0 \int d\mathbf{r} \, [(1/\chi_{AB})w_-^2 + (1/2\zeta)w_+^2 - iw_+]$$
$$- n_A \ln Q_A[w_A] - n_B \ln Q_B[w_B] \quad (4.106)$$

This expression differs from the corresponding Model C eqn (4.98) only by the presence of the term proportional to $\zeta^{-1}$. Model D thus contains two interaction

---

[46]This topic is discussed briefly in Section 5.3.6.

parameters, $\chi_{AB}$ and $\zeta$, rather than the single parameter $\chi_{AB}$ of Model C. By adjusting the value of $\zeta$, we generate a family of models that range from very stiff to very soft against variations in total segment density.

The *grand canonical Model D* has the effective Hamiltonian

$$H_G[w_+, w_-] = \rho_0 \int d\mathbf{r} \, [(1/\chi_{AB})w_-^2 + (1/2\zeta)w_+^2 - iw_+]$$
$$- z_A V Q_A[w_A] - z_B V Q_B[w_B] \qquad (4.107)$$

In this model there is no incompressibility constraint, so $z_A$ and $z_B$ can be varied independently to adjust the average segment densities of the two species. The segment density operators for Model D in the two ensembles are unchanged from eqns (4.102) and (4.103).

## 4.4 Block and graft copolymers

The interaction models just discussed for polymer blends are easily extended to melts of block or graft copolymers. Here we discuss the cases of melts of AB diblock copolymers and $A_2B$ graft copolymers, whose architectures are shown in Figs. 3.7 and 3.8, respectively.

### 4.4.1 *Model E: diblock copolymer melt*

Model C can be straightforwardly adapted to treat an incompressible melt of AB diblock copolymers (see Fig. 4.3). The resulting Model E is the "standard" model of block copolymers that has been adopted in the literature (Leibler, 1980; Hong and Noolandi, 1981; Matsen and Schick, 1994a). A fundamental distinction between a binary homopolymer blend and a diblock copolymer melt is that the latter is a *one-component system*. Thus, there is only one single-chain partition function $Q[w_A, w_B]$ describing a diblock melt. This function was defined in eqn (3.98) and can be expressed in terms of a forward propagator $q$ or a "complementary" propagator $q_c$ by means of eqn (3.104). For the case of $n$ diblock chains, each with $N$ statistical segments, the microscopic segment density operators are obtained by a simple generalization of eqn (3.97) to a multi-chain system:

$$\hat{\rho}_A(\mathbf{r}) = \sum_{j=1}^{n} \int_0^{fN} ds \, \delta(\mathbf{r} - \mathbf{r}_j(s))$$
$$\hat{\rho}_B(\mathbf{r}) = \sum_{j=1}^{n} \int_{fN}^{N} ds \, \delta(\mathbf{r} - \mathbf{r}_j(s)) \qquad (4.108)$$

in which $f$ denotes the fraction of statistical segments belonging to the A block. If segments of each species are again defined to occupy a common volume $v_0$ in the melt, then $f$ can also be interpreted as the average volume fraction occupied by the type A segments.

FIG. 4.3. Block copolymer melt considered in Model E. The type A blocks
(dark) and type B blocks (light) are both modelled as continuous Gaussian
chains. The melt is assumed to be locally incompressible.

By adopting the same interaction model as in Model C, namely an incom-
pressibility constraint on $\hat{\rho}_+ = \hat{\rho}_A + \hat{\rho}_B$ and a local A–B interaction proportional
to $\chi_{AB}\hat{\rho}_A\hat{\rho}_B$, then the steps leading to eqn (4.97) for the blend case are easily retraced for a copolymer melt. The following *canonical Model E field theory*
describes an incompressible melt of AB diblock copolymers:

$$\mathcal{Z}_C(n, V, T) = \mathcal{Z}_0 \int \mathcal{D}w_+ \int \mathcal{D}w_- \, \exp(-H[w_+, w_-]) \qquad (4.109)$$

where $\mathcal{Z}_0$ denotes the ideal gas partition function of $n$ non-interacting continuous
Gaussian chains, and the effective Hamiltonian is given by

$$H[w_+, w_-] = \rho_0 \int d\mathbf{r} \, [(1/\chi_{AB})w_-^2 - iw_+] - n \ln Q[w_A, w_B] \qquad (4.110)$$

The complex fields $w_K$ have the same interpretation as in eqn (4.98), and the
single-chain partition function $Q[w_A, w_B]$ is evaluated according to eqn (3.104),
by solving either of the two *complex* diffusion equations (3.99) or (3.101). The
functional $Q$ for diblock copolymers has a significantly different non-local character than the $Q_K$ functionals that enter Model C. This difference arises from
the A and B block connectivity in the copolymer case.

The *grand canonical Model E field theory* is defined by the simple change of
effective Hamiltonian to

$$H_G[w_+, w_-] = \rho_0 \int d\mathbf{r} \, [(1/\chi_{AB})w_-^2 - iw_+] - zVQ[w_A, w_B] \qquad (4.111)$$

where $z$ is the diblock copolymer activity.

Monomer density operators for Model E are defined as in Model C but are computed differently because of the difference in single-chain partition functions. It follows from eqns (3.105), (3.106), and (4.102) that in the canonical ensemble,

$$\tilde{\rho}_K(\mathbf{r}; [w_A, w_B]) = -n \frac{\delta \ln Q[w_A, w_B]}{\delta w_K(\mathbf{r})}$$

$$= \frac{n}{VQ[w_A, w_B]} \int_{D_K} ds \, q_c(\mathbf{r}, N - s; [w_A, w_B]) q(\mathbf{r}, s; [w_A, w_B])$$

(4.112)

where $\int_{D_K} ds$ denotes an integration over the A block, $s \in [0, fN]$, for $K = A$, and over the B block, $s \in (fN, N]$, for $K = B$. Similarly, in the grand canonical ensemble,

$$\tilde{\rho}_{K,G}(\mathbf{r}; [w_A, w_B]) = -zV \frac{\delta Q[w_A, w_B]}{\delta w_K(\mathbf{r})}$$

$$= z \int_{D_K} ds \, q_c(\mathbf{r}, N - s; [w_A, w_B]) q(\mathbf{r}, s; [w_A, w_B])$$

(4.113)

While Model E is standard for describing flexible AB diblock copolymer melts, many related models can be generated by modifying either the underlying chain or interaction model. In the case of a modification to the chain model, e.g. bead-spring diblock copolymer chains, the expressions for $Q$ and the density operators are easily adapted by means of the formulas presented in Chapter 3. For a change to the interaction model, the terms in $H$ and $H_G$ *not* involving $Q$ are subject to modification. For example, a model of a *compressible* AB diblock copolymer melt is obtained by simply adding the term proportional to $\zeta^{-1}$ in eqn (4.106) of Model D to the Model E expressions for $H$ or $H_G$.

### 4.4.2 $A_2B$ graft copolymer melt

Finally, we discuss the case of an *architectural* change from a diblock copolymer melt to a melt of $A_2B$ graft copolymers depicted in Fig. 3.8. If we adopt the same incompressible interaction model that was used in Models C and E, it follows that the effective Hamiltonians $H$ and $H_G$ are unchanged from eqns (4.110) and (4.111). However, the single-chain partition function $Q[w_A, w_B]$ entering these equations is evaluated for a graft copolymer by means of eqn (3.108). The expressions for the segment density operators must also be appropriately modified. For example, the B segment density operator is given in the canonical ensemble by eqns (3.111) and (4.112) as

$$\tilde{\rho}_B(\mathbf{r}; [w_A, w_B]) = -n \frac{\delta \ln Q[w_A, w_B]}{\delta w_B(\mathbf{r})}$$

$$= \frac{n}{V Q[w_A, w_B]} \int_0^{N_B} ds \ q_{Bc}(\mathbf{r}, N_B - s; [w_A, w_B]) q_B(\mathbf{r}, s; [w_B])$$

$$(4.114)$$

## 4.5 Polyelectrolytes

Another important class of polymeric fluids are solutions of polyelectrolytes. These macromolecules contain chemical groups that are capable of dissociation in polar solvents, producing charged species. Polyelectrolytes are commonly used in personal care formulations, as adsorbent materials, and as components of aqueous coatings (Hara, 1993; Dautzenberg *et al.*, 1994). Many naturally occurring biopolymers are also polyelectrolytes, so the contemporary focus on everything "bio" has led to much interest in the chemistry and physics of charged macromolecules. Some useful reviews dealing primarily with the physics of polyelectrolytes include those by: Förster and Schmidt (1995), Barrat and Joanny (1996), and Netz and Andelman (2003).

Polyelectrolytes acquire a net electrostatic charge because upon dissolution in a polar solvent, usually water, the functional groups along the chain can dissociate to release free *counterions*, leaving bound residues along the chain carrying the opposite charge. For example, in the case of poly(acrylic acid) (PAA), each chemical repeat unit has an acid functionality COOH that can dissociate in water to release a hydrogen ion $H^+$ and leave a negatively charged $COO^-$ residue bound to the chain (see Fig. 4.4). The excellent water solubility of such polymers is a consequence of this dissociation event. Water has a large electric dipole moment, which produces a high dielectric constant and thereby lessens the Coulomb interactions between bound ions and counterions. Moreover, the water dipoles "hydrate" or cluster around the ions, creating stable complexes that further lower the energy of the dissolved ions (Pauling, 1988). This is especially important for polyelectrolytes such as PAA, which have hydrophobic backbones, since the hydration of the bound ions effectively gives the polymer a hydrophilic coating that improves compatibility with water. The entropy gain by the released counterions is of course another strong driving force for the dissolution of the polymer.

Polyelectrolyte formulations tend to be very complicated because salts are often added to control ionic strength, and acids and bases can be added to adjust pH. Moreover, in a personal care product such as a hair spray or cosmetic cream, the polyelectrolyte is often only one of many ingredients, including surfactants, oils, alcohols, etc. The multi-component nature of such formulations and the competing effects of short-ranged and long-ranged interactions can produce complex self-assembly and phase behavior that is often unanticipated. First-principles design of these types of fluid mixtures for a particular product application is obviously very challenging.

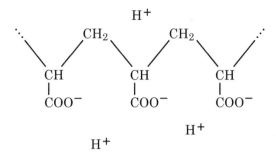

FIG. 4.4. Poly(acrylic acid) dissociates in aqueous solution to release hydrogen ions H$^+$ and leave negatively charged COO$^-$ ions bound to the polymer backbone.

In our discussion of polyelectrolyte models, we shall restrict attention to the simplest case of a linear homo-polyelectrolyte dissolved in a polar solvent, without added salts, acids, or bases. In addition, only "strong" polyelectrolytes, which fully dissociate when dissolved in the solvent, will be considered. Consistent with the previous sections, a mesoscopic perspective will be adopted throughout. The resulting model, Model F, has precedence in the literature (Borukhov *et al.*, 1998; Shi and Noolandi, 1999).

### 4.5.1 *Model F: polyelectrolyte solution, implicit solvent*

Model F is a simple extension of Model A to include groups that can dissociate in the solvent. We treat the solvent, usually water, *implicitly* as a dielectric continuum that separates the free counterions and the polymer segments containing bound charges. The chains are again taken to be continuous Gaussian chains, and the so-called *smeared* charge model is adopted. In this model, the charge associated with the ions bound to the polymers is smeared out uniformly along the chain contours, rather than localized at specific sites. From a mesoscopic viewpoint, this smeared model makes most sense for polymers like PAA that in solution are uniformly and densely charged along their backbones, and is less reliable for chains with sparse, randomly placed charges, such as a polyethylene-poly(acrylic acid) statistical copolymer. The charge of each counterion is denoted by $eZ_C$, where $e$ is the fundamental charge of a proton and $Z_C$ is an integer that denotes the valence of the counterion, e.g. $Z_C = +1$ for H$^+$ ions and $Z_C = +2$ for Ca$^{2+}$ ions. Correspondingly, in the smeared model, it is assumed that each statistical segment of the polymer carries a charge $epZ_P$, where $p$ is the number of charged residues bound to the segment and $Z_P$ is the valence of each residue. For example, in the case of 3 COO$^-$ groups per statistical segment, $Z_P = -1$ and $p = 3$. By *electroneutrality*, in the canonical ensemble with $n$ polymer chains in a volume $V$, it follows that the number of free counterions $n_C$ is given by

$$n_C Z_C + n N p Z_P = 0 \qquad (4.115)$$

In Model F the solvent is treated implicitly, so only the polymer segment and counterion coordinates enter the statistical mechanical description. A pair approximation is adopted for the potential of mean force that acts among these retained degrees of freedom. Two types of interactions are included in the pair potential of mean force: a *short-ranged* interaction of the type assumed in eqn (4.62) of Model A, and a *long-ranged* Coulomb interaction. The short-ranged interaction is parameterized by an excluded volume parameter $u_0 > 0$ that describes the net repulsive interaction between polymer segments, both intra- and intermolecular, mediated by the solvent. Because of the hydration of the charge residues, the net segment–segment interaction can be repulsive even for polymers with hydrophobic backbones. In principle, $u_0$ should be a function of $p$, because as $p$ is lowered, the level of hydration decreases and $u_0$ will be reduced.

The most distinctive feature in polyelectrolyte models is the long-ranged Coulomb interaction. The Coulomb potential acting between an ion with charge $eZ_j$ and a second ion with charge $eZ_k$, separated by a distance $r$ in a uniform dielectric medium, can be written (Eyges, 1972)

$$\bar{u}_e(r) = \frac{e^2 Z_j Z_k}{\epsilon r} \qquad (4.116)$$

where cgs units are employed. The dielectric constant $\epsilon$ in the present case corresponds to the dielectric constant of the pure solvent, e.g. $\epsilon \approx 80$ in the case of water at room temperature. The Coulomb interaction is often rewritten in the form

$$\bar{u}_e(r) = k_B T \frac{l_B Z_j Z_k}{r} \qquad (4.117)$$

where $l_B \equiv e^2/(\epsilon k_B T)$ is the so-called *Bjerrum length*. It defines a length scale at which the electrostatic interaction is comparable to the thermal energy $k_B T$.

For the purpose of expressing the form of the electrostatic contribution to the potential of mean force, $\bar{U}_e$, it is helpful to define a *microscopic charge density*

$$\hat{\rho}_e(\mathbf{r}) = p Z_P \sum_{j=1}^{n} \int_0^N ds\, \delta(\mathbf{r} - \mathbf{r}_j^P(s)) + Z_C \sum_{j=1}^{n_C} \delta(\mathbf{r} - \mathbf{r}_j^C) \qquad (4.118)$$

which is expressed in units of $e$. The two terms in this expression are contributions, respectively, from the bound ions and the counterions. From eqn (4.115), it follows that the volume integral of $\hat{\rho}_e$ vanishes by electroneutrality, i.e. $\int_V d\mathbf{r}\, \hat{\rho}_e = 0$. The electrostatic interaction energy can thus be written

$$\beta \bar{U}_e[\mathbf{r}^{nN+n_C}] = \frac{l_B}{2} \int d\mathbf{r} \int d\mathbf{r}'\, \hat{\rho}_e(\mathbf{r}) \frac{1}{|\mathbf{r} - \mathbf{r}'|} \hat{\rho}_e(\mathbf{r}') \qquad (4.119)$$

This expression evidently sums all the pairwise Coulomb interactions between counterions, between counterions and bound charges, and between charged polymer segments (both intra- and intermolecular). It also includes the singular interactions of each ion with itself, but we have previously seen that these spurious terms can be removed by a regularization of the potential and a subtraction. The net effect is a shift in chemical potential of each species that has no thermodynamic consequence. We thus subsequently ignore the contributions of the self-interactions.

It follows that the total potential of mean force in Model F can be expressed as the sum of short-ranged and long-ranged contributions:

$$\beta \bar{U}[\mathbf{r}^{nN+n_C}] = \frac{u_0}{2} \int d\mathbf{r} \, [\hat{\rho}_P(\mathbf{r})]^2 + \frac{l_B}{2} \int d\mathbf{r} \int d\mathbf{r}' \, \hat{\rho}_e(\mathbf{r}) \frac{1}{|\mathbf{r} - \mathbf{r}'|} \hat{\rho}_e(\mathbf{r}') \quad (4.120)$$

where $\hat{\rho}_P(\mathbf{r})$ is the microscopic polymer segment density defined in eqn (4.65). Notice that no short-ranged interactions have been included between counterions and between counterions and polymer segments, because normally the counterion concentration is sufficiently low that such interactions are negligible compared with the Coulomb interaction. Obviously Model F contains a highly simplified and idealized description of the the the potential of mean force in a strong polyelectrolyte solution. Nevertheless, it is sufficient to capture many of the salient characteristics of such solutions.

The canonical partition function for Model F can be converted to a field theory by following the familiar strategy of decoupling the quadratic interactions present in $\bar{U}$ with a Hubbard–Stratonovich transformation. In the present case, two auxiliary fields are required because of the two types of interactions. The *canonical Model F field theory* is given by

$$\mathcal{Z}_C(n, V, T) = \mathcal{Z}_0 \int \mathcal{D}w \int \mathcal{D}\Phi \, \exp(-H[w, \Phi]) \quad (4.121)$$

where $\mathcal{Z}_0$ denotes the ideal gas partition function of a collection of $n + n_C$ noninteracting polymers and counterions. The effective Hamiltonian assumes the form

$$H[w, \Phi] = \frac{1}{2} \int d\mathbf{r} \left( \frac{1}{u_0}[w(\mathbf{r})]^2 + \frac{1}{4\pi l_B}|\nabla\Phi(\mathbf{r})|^2 \right)$$
$$- n \ln Q_P[iw + ipZ_P\Phi] - n_C \ln Q_C[iZ_C\Phi] \quad (4.122)$$

where it is understood that the number of counterions $n_C$ is related to the number of polymers $n$ by eqn (4.115). The single-polymer and single-counterion partition functions, $Q_P$ and $Q_C$, are defined by eqns (3.20) and (4.12), respectively.

In the above field theory, the auxiliary field $w(\mathbf{r})$, which served to decouple the first interaction term in eqn (4.120), can again be interpreted as a fluctuating chemical potential field for polymer segments. The second field $\Phi(\mathbf{r})$, which was responsible for decoupling the electrostatic interactions in eqn (4.120), can be viewed as a *fluctuating electrostatic potential field*. This interpretation can be understood by introducing a *microscopic electrostatic potential* $\hat{\Phi}$ according to the conventional definition (Eyges, 1972)

$$\hat{\Phi}(\mathbf{r}) = \int d\mathbf{r}' \, \frac{\hat{\rho}_e(\mathbf{r}')}{|\mathbf{r} - \mathbf{r}'|} \quad (4.123)$$

which permits eqn (4.119) to be rewritten as[47]

$$\beta \bar{U}_e [\mathbf{r}^{nN+n_C}] = \frac{l_B}{8\pi} \int d\mathbf{r} \; |\nabla \hat{\Phi}(\mathbf{r})|^2 \tag{4.124}$$

Apart from the factor of $l_B$, this expression has the same form as the "square-gradient" term involving $\Phi$ appearing in the effective Hamiltonian of the field theory. Thus, we can loosely interpret $\Phi(\mathbf{r})$ in the field theory of eqns (4.121) and (4.122) as a fluctuating electrostatic potential. Of course, the interpretation of both fields is complicated by the *complex* nature of the field theory; both $Q_P$ and $Q_C$ are to be evaluated for purely imaginary arguments involving $iw$ and $i\Phi$. It should be noted that this analytic structure is consistent with the fact that both interactions in eqn (4.120) were repulsive (positive definite), as per the discussion in Section 4.2.2.

A final comment about the form of eqn (4.122) is that the particle-to-field transformation has the effect of replacing a long-ranged Coulomb interaction $|\mathbf{r}-\mathbf{r}'|^{-1}$ with a short-ranged interaction of the form $\nabla^2 \delta(\mathbf{r}-\mathbf{r}')$. This is beneficial from the standpoint of numerical simulations, because Ewald sums and other computationally expensive techniques for dealing with long-ranged interactions (Allen and Tildesley, 1987; Frenkel and Smit, 1996) are avoided.

Ensemble averages are defined in the usual way for the canonical Model F field theory, i.e. for an observable $G[w, \Phi]$

$$\langle G[w, \Phi] \rangle \equiv \frac{\int \mathcal{D}w \int \mathcal{D}\Phi \; G[w, \Phi] \exp(-H[w, \Phi])}{\int \mathcal{D}w \int \mathcal{D}\Phi \; \exp(-H[w, \Phi])} \tag{4.125}$$

Density operators can be deduced by using the techniques described in Section 4.1.4. Operators for the densities of polymer segments and counterions, respectively, can be expressed as functionals of $w$ and $\Phi$ according to

$$\tilde{\rho}_P(\mathbf{r}; [w_P]) = -n \frac{\delta \ln Q_P[w_P]}{\delta w_P(\mathbf{r})}$$

$$= \frac{n}{V Q_P[w_P]} \int_0^N ds \; q(\mathbf{r}, N - s; [w_P]) q(\mathbf{r}, s; [w_P]) \tag{4.126}$$

$$\tilde{\rho}_C(\mathbf{r}; [w_C]) = -n_C \frac{\delta \ln Q_C[w_C]}{\delta w_C(\mathbf{r})}$$

$$= \frac{n_C}{V Q_C[w_C]} \exp[-w_C(\mathbf{r})] \tag{4.127}$$

where complex fields conjugate to the respective densities have been introduced according to $w_P \equiv iw + ipZ_P\Phi$ and $w_C \equiv iZ_C\Phi$. Ensemble averages of the

---

[47]To derive eqns (4.122) and (4.124) it is helpful to note that the functional inverse of $1/|\mathbf{r} - \mathbf{r}'|$, defined according to eqn (C.29), is $-(1/4\pi)\nabla^2 \delta(\mathbf{r} - \mathbf{r}')$.

*complex* operators $\tilde{\rho}_P$ and $\tilde{\rho}_C$ following eqn (4.125) produce *real* densities that coincide with the physical average densities $\langle\hat{\rho}_P(\mathbf{r})\rangle$ and $\langle\hat{\rho}_C(\mathbf{r})\rangle$.

An operator for the charge density can be easily constructed from the above two species densities:

$$\tilde{\rho}_e(\mathbf{r};[w,\Phi]) = pZ_P\tilde{\rho}_P(\mathbf{r};[w_P]) + Z_C\tilde{\rho}_C(\mathbf{r};[w_C]) \qquad (4.128)$$

### 4.5.2  *Extensions*

Many extensions of Model F are obviously possible. A grand canonical version is straightforward to derive, as is a version with additional short-ranged interactions acting among counterions and between counterions and polymer segments. Added salt could also be easily incorporated, with the resulting ions treated as point-like entities that interact via Coulomb interactions with the other charged species.

Some changes to the model are required when the charge density on the polymer chains is low and the backbones are hydrophobic. Such systems are capable of forming complex structures, including aggregates, globules, and microphases, across a broad range of concentration and solution conditions (Dobrynin and Rubinstein, 2001; Baigl *et al.*, 2002; Joanny and Leibler, 1990). In such cases of "hydrophobic polyelectrolytes," the parameter $u_0$ can change sign, and Model F becomes ill defined. One simple resolution is analogous to the approach taken in Model B; namely, to treat the solvent explicitly and impose an overall incompressibility constraint. Such variants of Model F are well known (Shi and Noolandi, 1999; Wang *et al.*, 2004) and require at least three fields in comparison to the two fields of the present theory.

Model F assumes strong polyelectrolyte conditions where full dissociation is achieved and the "smeared charge" model is appropriate. However, a number of important polyelectrolyte systems, either by virtue of strong ion binding, low dielectric solvents, or extreme pH conditions, exhibit incomplete dissociation. Under such conditions, an "annealed charge" model[48] is more appropriate in which the dissociated or undissociated state of each ion pair is traced over in the partition function and a Boltzmann factor accounts for the difference in energy between the two states (Borukhov *et al.*, 1998).

Another generalization of Model F that is important for systems capable of phase separation (or microphase separation) into hydrophobic and hydrophilic domains is to account for the nearly two orders of magnitude difference in dielectric constant between the two domain types. A simple modification of the effective Hamiltonian eqn (4.122) that addresses this shortcoming (Netz and Orland, 2000) is to replace the Bjerrum length $l_B$ with a local Bjerrum length $l_B(\mathbf{r}) = e^2/[\epsilon(\mathbf{r})k_BT]$. A model can then be made for how the local dielectric constant $\epsilon(\mathbf{r})$ depends on the local density of each species (Wang *et al.*, 2004).

A final extension of Model F is to incorporate chain models other than the continuous Gaussian chain. Gaussian chain models are inappropriate for very

---

[48]The treatment of such cases of "annealed disorder" is discussed in Section 4.7.

dilute polyelectrolyte solutions at low salt concentrations, where the Coulomb forces can "overstretch" individual polymer chains (Netz and Orland, 1999). The wormlike chain or a nonlinear bead-spring chain model would be more suitable for building a field theory to describe such situations. Moreover, the wormlike chain could also be used to incorporate the effect of inherent chain rigidity into the model.

## 4.6   Liquid crystalline polymers

Liquid crystalline polymers (LCPs) are polymeric materials containing rod-like "mesogenic" units that are capable of producing spontaneous orientational order within a fluid – the so-called liquid crystalline state. The nature of liquid crystallinity can be extremely varied depending on the shapes and chemical details of the mesogen units. Nematic, smectic, and cholesteric phases are the most familiar cases of liquid crystalline order, but more complicated examples can be found, even in liquid crystals of low molecular weight. The monograph by de Gennes and Prost (1993) provides an excellent introduction to the physics of liquid crystalline fluids.

When mesogenic units are incorporated into polymers, the possibilities for interesting orientationally ordered phases are significantly enhanced (Ciferri *et al.*, 1982; Blumstein, 1985). Mesogens can be incorporated into the backbone in a homogeneous fashion, or can be attached to side chains, producing *main-chain LCPs* or *side-chain LCPs*, respectively. A second broad classification refers to whether the liquid crystalline order can be achieved by changing temperature as in so-called *thermotropics*, or by changing polymer concentration in a solution, in which case the fluids are *lyotropics* (Odijk, 1986). Mesogens can also be incorporated into macromolecules, either as main-chain or side-chain moieties, by random copolymerization or block/graft copolymerization. This architectural complexity can lead to a variety of interesting phenomena in which orientational ordering competes with compositional ordering, e.g. macrophase separation or microphase separation.

Because LCPs represent such a diverse set of materials, it is not possible in a monograph of this scope to present a comprehensive set of models that encompass all types of LCP systems. Thus, we focus on a simple representative model of a main-chain homopolymer system that can exhibit *nematic* order. This model, Model G, can be viewed as a model of a thermotropic, main-chain LCP melt, or alternatively, as a model of a lyotropic solution of semiflexible polymers. Our formalism is similar to descriptions that have appeared previously in the literature (Wang and Warner, 1986; Lansac and Maissa, 1992; Gupta and Edwards, 1993; Matsen, 1996; Düchs and Sullivan, 2002)

### 4.6.1   *Model G: polymer nematics*

Model G is an extension of Model A, but here the wormlike chain model is substituted for the continuous Gaussian chain, and additional anisotropic interactions among polymer segments are included. Main-chain nematic polymers tend to be semiflexible due to the mesogens that are incorporated in their backbones. By applying the wormlike chain model, it is tacitly assumed that the resistance to bending is homogeneous along the chain. This assumption is clearly unrealistic for many types of main-chain LCPs, but it provides a simple starting point for understanding the behavior of polymer nematics.

In order to express the anisotropic interactions that lead to nematic order, it is helpful to generalize the definition (3.35) of a *microscopic density of segment position and orientation* to a system containing $n$ wormlike polymers:

$$\hat{\rho}(\mathbf{r}, \mathbf{u}) \equiv \sum_{j=1}^{n} \int_0^{L_c} ds \, \delta(\mathbf{r} - \mathbf{r}_j(s)) \delta(\mathbf{u} - \mathbf{u}_j(s)) \tag{4.129}$$

If we invoke the pair approximation for the interaction potential, the contribution to the potential energy from non-bonded interactions among segments can be described as a quadratic form in $\hat{\rho}(\mathbf{r}, \mathbf{u})$:

$$U_1 = \frac{1}{2} \int d\mathbf{r} \int d\mathbf{r}' \int d\mathbf{u} \int d\mathbf{u}' \, \hat{\rho}(\mathbf{r}, \mathbf{u}) v_2(\mathbf{r}, \mathbf{r}', \mathbf{u}, \mathbf{u}') \hat{\rho}(\mathbf{r}', \mathbf{u}') \tag{4.130}$$

where $v_2$ is a suitable pair potential function that is assumed to be regularized at contact. It is also understood that in the case of a lyotropic system, $v_2$ should be interpreted as a pair potential of mean force and will depend on the properties of the pure solvent.

A typical further approximation in a mesoscopic description of a LCP system is to assume that the interactions are local in space, which amounts to a delta function approximation for $v_2$:

$$v_2(\mathbf{r}, \mathbf{r}', \mathbf{u}, \mathbf{u}') \approx \delta(\mathbf{r} - \mathbf{r}')v(\mathbf{u}, \mathbf{u}') \tag{4.131}$$

where $v(\mathbf{u}, \mathbf{u}')$ is a local orientation-dependent interaction. The form of $v$ is not arbitrary but should be symmetric and invariant to sign inversion of $\mathbf{u}$, $\mathbf{u}'$, or both. Common expressions for $v$ are the *Onsager form*

$$\beta v(\mathbf{u}, \mathbf{u}') = u_1 |\mathbf{u} \times \mathbf{u}'| \tag{4.132}$$

which derives from the Onsager theory of lyotropic LCPs (Onsager, 1949), and the *Maier–Saupe form*

$$\beta v(\mathbf{u}, \mathbf{u}') = u_0 - u_2 P_2(\mathbf{u} \cdot \mathbf{u}') \tag{4.133}$$

which is fundamental to the Maier–Saupe theory of thermotropic liquid crystals (Maier and Saupe, 1958). The Onsager form arises from the calculation of the excluded volume between two slender rods with orientations $\mathbf{u}$ and $\mathbf{u}'$ and leads to a prefactor $u_1 > 0$ that corresponds to a temperature-independent (athermal) microscopic length. The Maier–Saupe form contains an *isotropic* excluded volume term $u_0 > 0$ analogous to eqn (4.62) and an *anisotropic* quadrapolar term proportional to $u_2 > 0$. The latter multiplies $P_2(\mathbf{u} \cdot \mathbf{u}') \equiv (1/2)[3(\mathbf{u} \cdot \mathbf{u}')^2 - 1]$, which is the second-order Legendre polynomial defined in eqn (2.84). The anisotropic coefficient $u_2$ was supposed in the original theory to derive from van der Waals interactions and thus vary as $1/T$. In reality, both the Onsager and Maier–Saupe forms should be considered to be phenomenological, because the values of the parameters $u_0$, $u_1$, and $u_2$ obtained by fitting experimental data rarely conform to these purely entropic or purely enthalpic expectations. Essential in both cases for the stabilization of nematic order is that $v(\mathbf{u}, \mathbf{u}')$ decreases with mutual alignment of two nearby polymer segments.

The canonical partition function for a collection of $n$ interacting wormlike chains can be expressed as a product of $n$ path integrals of the form of eqn (2.67) over a Boltzmann factor $\exp(-\beta U_1)$ that incorporates the non-bonded segmental interactions described above. This expression can be converted to a field theory by repeating the steps that led to eqns (4.69) and (4.70) for Model A. We obtain the following *canonical Model G field theory*:

$$\mathcal{Z}_C(n, V, T) = \mathcal{Z}_0 \int \mathcal{D}\rho \int \mathcal{D}w \, \exp(-H[\rho, w]) \qquad (4.134)$$

with effective Hamiltonian

$$H[\rho, w] = -i \int d\mathbf{r} \int d\mathbf{u} \, w(\mathbf{r}, \mathbf{u})\rho(\mathbf{r}, \mathbf{u}) - n \ln Q[iw]$$
$$+ \frac{\beta}{2} \int d\mathbf{r} \int d\mathbf{u} \int d\mathbf{u}' \, \rho(\mathbf{r}, \mathbf{u})v(\mathbf{u}, \mathbf{u}')\rho(\mathbf{r}, \mathbf{u}') \qquad (4.135)$$

where $Q[iw]$ is the normalized partition function for a single wormlike polymer chain in the purely imaginary potential $iw(\mathbf{r}, \mathbf{u})$ defined by eqn (3.36). In this model, $\mathcal{Z}_0$ is interpreted as the partition function for an ideal gas of $n$ non-interacting wormlike polymer chains.

Provided that $v(\mathbf{u}, \mathbf{u}')$ is positive definite and has a functional inverse, it is possible to carry out the Gaussian integral over the fluctuating $\rho(\mathbf{r}, \mathbf{u})$ field to achieve a simpler field theory in the single field $w(\mathbf{r}, \mathbf{u})$. Unfortunately, $v$ is not positive definite for either the Onsager or Maier–Saupe forms, as can be easily verified by a spherical harmonic expansion. Thus, it appears that the simplified form of the Model G field theory, analogous to eqns (4.71) and (4.72) for Model A, has limited applicability to nematic polymers.

In the full canonical Model G field theory, the segment density $\rho(\mathbf{r}, \mathbf{u})$ appears explicitly, so averages and moments of segment number density and orientation are directly accessible. A second important density operator, which has functional dependence *only* on the $w$ field, can be defined according to

$$\tilde{\rho}(\mathbf{r}, \mathbf{u}; [iw]) \equiv -n\frac{\delta \ln Q[iw]}{\delta iw(\mathbf{r}, \mathbf{u})} \qquad (4.136)$$

which is the *single-chain* density operator $\rho(\mathbf{r}, \mathbf{u}; [iw])$ of eqn (3.62), multiplied by the number of polymers, $n$. It follows that this operator can be evaluated for a particular realization of $w$ from solutions of the Fokker–Planck eqn (3.39) combined with eqn (3.63). Other reduced segment density operators can be defined according to eqns (3.64)–(3.67).

Another operator that is of specific interest in LCPs is a *nematic order parameter tensor* $S_{\alpha\beta}(\mathbf{r}; [\rho])$ (de Gennes and Prost, 1993). This operator, while not fundamental to the theory, can be defined by

$$S_{\alpha\beta}(\mathbf{r}; [\rho]) \equiv \frac{V}{nL_c} \int d\mathbf{u} \, \rho(\mathbf{r}, \mathbf{u}) \left( u_\alpha u_\beta - \frac{1}{3}\delta_{\alpha\beta} \right) \qquad (4.137)$$

Its expectation value over the $\rho$ field fluctuations provides a useful measure of the extent and spatial variation of nematic order in the system. Finally, it should be noted that a *grand canonical* version of the Model G field theory is obtained by the usual replacement of $-n \ln Q$ with $-zVQ$ in the effective Hamiltonian (4.135).

## 4.7   Disorder effects

In the above sections, a number of models were presented for idealized many-chain polymer systems in which the polymer chains were imagined to be of uniform length, composition, and architecture. Most commercial polymers, however, possess a distribution of lengths (so-called "polydispersity"), variable compositions and sequence distributions in the case of copolymers, and "architectural complexity" in the case of branched or grafted systems. This *chemical disorder* arises from the statistical nature of the polymerization and coupling reactions that are used to synthesize polymers (Odian, 1981). Given the ubiquity of chemical disorder in synthetic macromolecules and that such disorder can produce significant changes in the structure and thermodynamics of a polymeric fluid, it is essential that disorder effects be captured in our field-theoretic models.

A useful way to characterize chemical disorder in polymeric fluids is to adopt the terminology of *quenched* and *annealed* disorder that is familiar in magnetic systems (Grinstein, 1985; Mezard *et al.*, 1987). In the present context, a polymer fluid with *quenched disorder* has disorder that is *permanently affixed* to the macromolecules by virtue of the chemical bonding in the system. Usually the quenched disorder is established at the time of synthesis, but it can also be developed by post-polymerization reactions such as crosslinking. A simple example of a polymer system with quenched disorder is a polydisperse linear homopolymer melt that was produced by a step polymerization. Provided that the covalent bonds defining the various polymers in the sample are permanent and no new bonds are created in the melt (the melt is unreactive), the distribution of molecular weight is stationary in time and serves to characterize the extent of quenched chemical disorder in the system. As a second example, a graft copolymer that was synthesized by incorporating a "macromonomer" into a step or addition polymerization of a second monomer has two types of quenched disorder. In such a material, the individual polymers vary not only in the number of macromonomer "grafts" that they contain, but also in the overall length of the polymer backbones. Such a material is said to possess quenched *compositional disorder*, as well as quenched *chain length disorder* (polydispersity). A third type of quenched disorder is commonly encountered when using coupling reactions to produce star polymers. Due to incomplete coupling, a mixture of star polymers with varying numbers of arms is often obtained, e.g. 3-arm stars mixed with 4-arm and 5-arm stars. Such disorder is referred to as *architectural disorder*.

Polymer systems with *annealed* disorder are defined by having chain length, composition, and architecture variations that are capable of changing with time and in response to external stimuli. These systems are deemed "equilibrium" or "living" polymers (Wittmer *et al.*, 1998; Greer, 1996). The transient character associated with annealed disorder arises from the presence of *reversible* chemical bonds. Examples of polymer systems with annealed disorder include liquid sulfur (Scott, 1965; Wheeler *et al.*, 1980), which undergoes an equilibrium polymerization to form a complex mixture of chains and rings at elevated temperatures, some protein filaments (Oozawa and Asakura, 1975), polyesters capable of trans-esterification reactions in the melt (Brydson, 1995), and "supramolecular polymers", which are macromolecular assemblies created through the use of multiple hydrogen bonds and other reversible linkages (Bosman *et al.*, 2004). While quenched disorder is more ubiquitous than annealed disorder in synthetic polymer systems, it is important to understand how to build models for both types of systems. We provide some representative examples in the following sections.

### 4.7.1 *Model H: quenched polydispersity*

A typical case of quenched disorder is a solution or melt of a polydisperse homopolymer. Here we consider a solution in a good solvent and discuss how Model A can be modified to account for polydispersity. The approach follows closely that in a recent publication (Fredrickson and Sides, 2003).

In the $nVT$ canonical ensemble, the only change required to adapt Model A to a polydisperse solution is that each chain in the system potentially has a different length, $N_j$, for $j = 1, ..., n$. By tracing this simple change through the particle-to-field transformation, one recovers a field theory of the same form as eqn (4.71), but where the effective Hamiltonian is given by

$$H[w] = \frac{1}{2u_0} \int d\mathbf{r} \, [w(\mathbf{r})]^2 - \sum_{j=1}^{n} \ln Q(N_j; [iw]) \qquad (4.138)$$

In this expression, $Q(N_j; [iw])$ is the partition function for a *single* polymer chain of length $N_j$ in the purely imaginary field $iw(\mathbf{r})$. It is defined by eqn (3.22) according to

$$Q(N; [iw]) = \frac{1}{V} \int d\mathbf{r} \, q(\mathbf{r}, N; [iw]) \qquad (4.139)$$

For a macroscopic sample with $n \sim 10^{23}$, the sum over chains in eqn (4.138) contains all realizations of the quenched distribution of chain lengths. Thus, with vanishing error in the thermodynamic limit,[49] the effective Hamiltonian of the *canonical Model H field theory* can be written

$$H[w] = \frac{1}{2u_0} \int d\mathbf{r}\, [w(\mathbf{r})]^2 - \int_0^\infty dN\, p_Q(N) \ln Q(N; [iw]) \qquad (4.140)$$

where $p_Q(N)dN$ is the quenched distribution of chain lengths, representing the number of chains with lengths between $N$ and $N + dN$. As such, the zeroth moment of the distribution recovers the total number of chains $n$ in the system:

$$\int_0^\infty dN\, p_Q(N) = n \qquad (4.141)$$

The *number average chain length* is given by the ratio of first to zeroth moments

$$N_n \equiv \frac{\int_0^\infty dN\, N p_Q(N)}{\int_0^\infty dN\, p_Q(N)} \qquad (4.142)$$

and the *weight average chain length* reflects the ratio of second to first moments

$$N_w \equiv \frac{\int_0^\infty dN\, N^2 p_Q(N)}{\int_0^\infty dN\, N p_Q(N)} \qquad (4.143)$$

For a specified form of the distribution function $p_Q(N)$, the Model H field theory defined by eqns (4.71) and (4.140) is completely determined. In addition to the excluded volume parameter $u_0$ and the average polymer density $n/V$, which is dictated by the normalization of $p_Q$, the theory has a functional dependence on the *shape* of the distribution function $p_Q(N)$. In special cases, e.g. a homogeneous bulk solution, a few low-order moments of $p_Q$, such as those determining $N_n$ and $N_w$, may prove sufficient to parameterize the thermodynamic properties of the model (Stockmayer, 1949). In general inhomogeneous situations, however, this is not guaranteed. Modern polymer characterization methods that combine multiple-angle light scattering with size exclusion chromotography are fortunately capable of providing detailed information about the full molecular weight distribution of a sample.

Evaluation of eqn (4.140) requires that $Q(N; [iw])$ be available for arbitrary $N$ so that the indicated quadrature over chain length can be performed. It is rare that such integrals can be done analytically, although progress can be made for Model H under homogeneous bulk conditions (Ohta and Oono, 1982). Numerically, however, the integral can be efficiently performed (Fredrickson and Sides, 2003) with a Gauss quadrature appropriate to the interval $N \in [0, \infty]$. The partition function $Q(N; [iw])$ is evaluated at the set of (unequally spaced)

---

[49] A further assumption in eqn (4.140) is that the discrete chain length distribution function of a real polymer sample can be approximated by a continuous distribution. This is normally an excellent approximation.

abscissas $N_\alpha$, $\alpha = 1, 2, ..., M$ in an $M$-point Gauss quadrature formula by means of a *single* solution of the diffusion eqn (3.25) out to the *largest* abscissa to obtain $q(\mathbf{r}, s; [iw])$ for $0 \leq s \leq N_M$. $Q(N_\alpha; [iw])$ is then determined for $\alpha < M$ by interpolation of this solution. For a typical monomodal distribution function such as the Shulz distribution (Rubinstein and Colby, 2003), it is our experience that less than 10 Gauss points ($M \lesssim 10$) are required for 4 to 6 digit accuracy.

A remaining task is to define operators whose average over the fluctuating $w$ field with the statistical weight $\exp(-H[w])$ produces average densities of interest. An operator for the *local segment density*, including segments from chains of all lengths, is given by

$$
\begin{aligned}
\tilde{\rho}(\mathbf{r}; [iw]) &\equiv - \int_0^\infty dN \, p_Q(N) \frac{\delta \ln Q(N; [iw])}{\delta iw(\mathbf{r})} \\
&= \int_0^\infty dN \, \frac{p_Q(N)}{VQ(N; [iw])} \int_0^N ds \, q(\mathbf{r}, N - s; [iw]) q(\mathbf{r}, s; [iw])
\end{aligned}
$$

$$(4.144)$$

Similarly, an operator for the *local density of segments contributed by chains of length $N$* is obtained by simply holding off on the final integral over $N$:

$$
\tilde{\rho}(\mathbf{r}, N; [iw]) \equiv \frac{p_Q(N)}{VQ(N; [iw])} \int_0^N ds \, q(\mathbf{r}, N - s; [iw]) q(\mathbf{r}, s; [iw]) \qquad (4.145)
$$

By means of these operators, it is possible to examine the equilibrium segment densities in an inhomogeneous system and establish the contributions from chains of different lengths. Calculations of this type have been recently reported for a polydisperse version of Model C for a binary blend (Fredrickson and Sides, 2003) and a polydisperse version of Model E for a diblock copolymer melt (Sides and Fredrickson, 2004).

### 4.7.2 *Model I: annealed polydispersity, equilibrium polymers*

In the quenched polydispersity example just described, the distribution of chain lengths $p_Q(N)$ was assumed to be *specified* and determined at the time of synthesis. Here, we discuss an equilibrium polymer system in which the *annealed distribution of chain lengths*, $p_A(N)$ is determined by the principles of chemical equilibria.

As a simple model of equilibrium polymers, it is useful to consider a collection of difunctional monomers (shown as open circles in Fig. 4.5) that are dispersed in a good solvent. The monomers can link *reversibly* at two sites (shown as black dots in Fig. 4.5) to create linear polymer chains of varying lengths. Each linking event lowers the energy of the system by an amount $2k_B Th$, where $k_B Th$ (with $h > 0$) can be viewed as the energy of an unsatisfied (unlinked) binding site. We ignore linkages between the two ends of a polymer to form cycles (ring polymers); the population of such species is normally much smaller than that of linear chains.

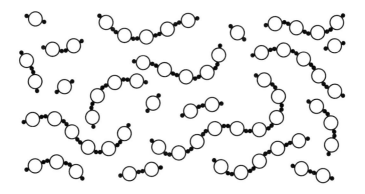

FIG. 4.5. Simple model of an equilibrium polymer system. Each monomer (open circle) is difunctional and has two sites (black dots) at which a reversible bond can be formed with a site from another segment. Ring polymers (cycles) are ignored, so the population of macromolecules consists of unreacted monomers, $N = 1$, and linear homopolymer chains with any degree of polymerization $N$.

The chain length distribution for some *instantaneous* state of association in the system can be characterized by a function $p(N)$, where $p(N)dN$ describes the number of chains with lengths between $N$ and $N + dN$. Here it is assumed that the polymers are sufficiently long on average that a continuous distribution function provides an accurate description of the chain population. In a system of specified volume $V$, the total number of polymers, $n_P$, and the total number of monomers (both reacted and unreacted), $n_M$, can be expressed as functionals of the chain length distribution according to

$$n_P = \int_0^\infty dN\, p(N) \qquad (4.146)$$

$$n_M = \int_0^\infty dN\, N p(N) \qquad (4.147)$$

The potential energy $U_h$ associated with the linkages among segments can also be expressed as a functional of $p(N)$, because each monomer or polymer contains exactly *two* unsatisfied linkage sites (black dots). Thus,

$$\beta U_h = 2h n_P = 2h \int_0^\infty dN\, p(N) \qquad (4.148)$$

For simplicity, the chain and interaction models applied in Model A are used to describe the solution of equilibrium polymers depicted in Fig. 4.5. Each polymer is treated as a continuous Gaussian chain with a specified length $N$ that is taken to be a continuous random variable. The polymers are dissolved in an (implicit) good solvent and the segment–segment interactions are described by

the pair potential of mean force given in eqn (4.62). The most convenient ensemble for such an equilibrium polymer problem is a $\mu_M VT$ grand canonical
ensemble in which $\mu_M$ is the chemical potential conjugate to the number of *segments* $n_M$ that are introduced into the system. It should be emphasized that the
average number of polymers, and indeed, the *equilibrium* chain length distribution $p_A(N)$, are not prescribed but are determined by the chemical equilibrium
associated with the reversible linkages among monomers.

Upon implementing the particle-to-field transformation, we obtain the following result for the grand canonical partition function:

$$\mathcal{Z}_G(\mu_M, V, T) = \int \mathcal{D}p \int \mathcal{D}w \left( \prod_N \frac{1}{p(N)!} \right) \left( \frac{e^{\beta \mu_M}}{\lambda_T^3} \right)^{n_M} e^{-2hn_P}$$

$$\times\, e^{-(1/2u_0) \int d\mathbf{r}\, w^2 + \int_0^\infty dN\, p(N)\, \ln[Z_0(N)Q(N;[iw])]} \qquad (4.149)$$

The functional integral over $p(N)$ in this expression effectively sums over all
numbers of segments and polymers by virtue of eqns (4.146) and (4.147), but
also over all possible forms that the chain length distribution might assume. It
is understood that in performing this functional integral, the physical requirement of $p(N) \geq 0$ must be met for all $N \in [0, \infty]$. The binding energy of the
linked polymers has been included through the factor $\exp(-2hn_P)$, while the
segment–segment interaction energy is accounted for in the term involving the
excluded volume parameter $u_0$. The translational and configurational entropy of
the polymers is captured in the last factor involving $Z_0(N)Q(N; [iw])$, the product of the partition function of a single polymer of length $N$ in zero external
field, eqn (2.43), and the normalized single-chain partition function of a length
$N$ polymer experiencing the purely imaginary potential $iw(\mathbf{r})$. Finally, the factor $\prod_N 1/p(N)!$ accounts for the indistinguishability of identical polymers (and
segments) of the same length $N$. This factor can be rewritten according to

$$\prod_N \frac{1}{p(N)!} \equiv e^{-\int_0^\infty dN\, \ln p(N)!}$$

$$\approx e^{-\int_0^\infty dN\, p(N)\ln p(N)} e^{n_P} \qquad (4.150)$$

where Stirling's approximation has been applied in the last line, as is appropriate
in the thermodynamic limit ($n_M \to \infty$, $V \to \infty$, $n_M/V$ fixed).

Equation (4.149) can be expressed in a more convenient form by means of
some definitions and by making the functional dependence of the integrand on
$p(N)$ explicit. It is useful to write $Z_0(N) = V g_M^{N-1}$, where $g_M$ is a microscopic
volume of order the segment volume[50] that is independent of $N$ and $V$. A dimensionless *activity* $z_M$ for *monomers* can be defined according to

---

[50]This formula follows from a discrete bead-spring chain of $N$ segments, since $Z_0(N)$ does
not strictly exist for a continuous Gaussian chain. With this exception, the continuous chain
description is applied throughout.

$$z_M \equiv \frac{g_M \exp(\beta\mu_M)}{\lambda_T^3} \tag{4.151}$$

This activity should not be confused with the activity of *chains* z introduced in the context of Model A. Upon introduction of these definitions into eqn (4.149), the field theory can be rewritten as

$$\mathcal{Z}_G(\mu_M, V, T) = \int \mathcal{D}p \int \mathcal{D}w \ \exp(-H[p, w]) \tag{4.152}$$

where the effective Hamiltonian $H[p, w]$ is defined by

$$H[p, w] = \int_0^\infty dN \ p(N)\{\ln p(N) + 2h - 1 - \ln[VQ(N; [iw])/g_M]\}$$
$$- \ln z_M \int_0^\infty dN \ Np(N) + \frac{1}{2u_0} \int d\mathbf{r} \ [w(\mathbf{r})]^2 \tag{4.153}$$

If we confine our discussion to situations where the thermodynamic limit applies, this two-field theory can be further simplified by noting that the $p(N)$ functional integral can be evaluated by the maximum term method (Laplace's method) (Bender and Orszag, 1978).[51] The extremum chain length distribution function, $\tilde{p}_A(N; [iw])$, defined by

$$\left.\frac{\delta H[p, w]}{\delta p(N)}\right|_{p=\tilde{p}_A} = 0 \tag{4.154}$$

turns out to be

$$\tilde{p}_A(N; [iw]) = \frac{z_M^N V e^{-2h}}{g_M} Q(N; [iw]) \tag{4.155}$$

This object is to be viewed as an *operator* in the sense that it is a functional of the $w(\mathbf{r})$ field; the physical *annealed chain length distribution* is obtained from

$$p_A(N) \equiv \langle \tilde{p}_A(N; [iw]) \rangle = \frac{z_M^N V e^{-2h}}{g_M} \langle Q(N; [iw]) \rangle \tag{4.156}$$

where the angular brackets denote an average with statistical weight $\exp(-H[w])$ over the $w$ field fluctuations. The effective Hamiltonian for performing such an average is obtained by substitution of eqn (4.155) into eqn (4.153), leading to

$$H[w] = -\int_0^\infty dN \ \tilde{p}_A(N; [iw]) + \frac{1}{2u_0} \int d\mathbf{r} \ w^2$$
$$= -\frac{V e^{-2h}}{g_M} \int_0^\infty dN \ z_M^N Q(N; [iw]) + \frac{1}{2u_0} \int d\mathbf{r} \ w^2 \tag{4.157}$$

[51]The applicability of the maximum term method in the thermodynamic limit can be seen because $p(N)\Delta N$, the number of chains in any small chain length fraction between $N$ and $N + \Delta N$, diverges as $n_M, V \to \infty$ and multiplies all terms in eqn (4.153) that involve $p(N)$.

and the partition function reduces to

$$\mathcal{Z}_G(\mu_M, V, T) = \int \mathcal{D}w \, \exp(-H[w]) \tag{4.158}$$

Equations (4.157)–(4.158) thus define the *grand canonical Model I field theory.*

The average number of monomers in the system (linked and unlinked), $\langle n_M \rangle$, is determined by the thermodynamic condition

$$\langle n_M \rangle = \left( \frac{\partial \ln \mathcal{Z}_G(\mu_M, V, T)}{\partial \ln z_M} \right)_{V,T} = -\left\langle \frac{\partial H[w]}{\partial \ln z_M} \right\rangle \tag{4.159}$$

This is seen to recover the expected interpretation of $\langle n_M \rangle$ as the first moment of the annealed distribution function:

$$\langle n_M \rangle = \int_0^\infty dN \, N \, \langle \tilde{p}_A(N; [iw]) \rangle = \int_0^\infty dN \, N p_A(N) \tag{4.160}$$

Various definitions of an average chain length can be applied. For example, a *number average* chain length can be defined according to

$$N_n \equiv \frac{\int_0^\infty dN \, N p_A(N)}{\int_0^\infty dN \, p_A(N)} = \frac{\int_0^\infty dN \, N \, z_M^N \langle Q(N; [iw]) \rangle}{\int_0^\infty dN \, z_M^N \langle Q(N; [iw]) \rangle} \tag{4.161}$$

It should be noted in the above expressions that the physical range of the dimensionless segment activity is $0 \leq z_M < 1$, so a factor of $z_M^N$ implies an *exponential decay* with increasing $N$. Thus, $p_A(N)$ is normalizable and has finite moments.

Local segment density operators are also of interest if the equilibrium polymer system is inhomogeneous. An operator whose average over the $w$ field fluctuations gives the *total density of segments* is

$$\begin{aligned}
\tilde{\rho}(\mathbf{r}; [iw]) &\equiv -\int_0^\infty dN \, \frac{\delta \tilde{p}_A(N; [iw])}{\delta iw(\mathbf{r})} \\
&= \frac{e^{-2h}}{g_M} \int_0^\infty dN \, z_M^N \int_0^N ds \, q(\mathbf{r}, N - s; [iw]) q(\mathbf{r}, s; [iw])
\end{aligned} \tag{4.162}$$

Similarly, an operator for the *local density of segments* contributed by chains of length $N$ is

$$\tilde{\rho}(\mathbf{r}, N; [iw]) = \frac{e^{-2h} z_M^N}{g_M} \int_0^N ds \, q(\mathbf{r}, N - s; [iw]) q(\mathbf{r}, s; [iw]) \tag{4.163}$$

Theoretical investigations of equilibrium polymer systems have apparently been limited to *homogeneous* systems, so we are not aware of a precedent for the above field theory in the literature. Nevertheless, the theory reproduces well-known results in the homogeneous limit. For example, in the case of a bulk, homogeneous solution with vanishingly small excluded volume, $u_0 \to 0+$, the $w$ field is confined to fluctuate in a narrow harmonic basin of width $\sim \sqrt{u_0}$. Under such conditions, $Q(N; [iw])$ can be replaced by $Q(N; [0]) = 1$ and the equilibrium chain length distribution reduces to $p_A(N) \sim z_M^N$. Thus, the chain length distribution is *exponential* – a well known and distinguishing feature of equilibrium polymer systems (Cates, 1988; Wittmer *et al.*, 1998). Under inhomogeneous conditions, e.g. the above solution model in a confined geometry, or in more general cases of excluded volume correlations (Cates, 1988), the chain length distribution will in general differ from a simple exponential. Such effects are currently under investigation (Feng and Fredrickson, 2005).

### 4.7.3   *Model J: random graft copolymers*

As a final example of a system with chemical disorder, we consider a polymeric fluid with *architectural disorder*, rather than chain length disorder. A common strategy for creating copolymers (so-called "compatibilizers") in industrial practice is to reactively couple an end-functional polymer to a second polymer that has complementary reactive functional sites along its backbone. Such *reactive blending* (Baker *et al.*, 2001) is normally carried out in the melt using an extruder and results in a complex mixture of graft copolymers and, often, unreacted functional polymers. Such a situation is depicted in Fig. 4.6, where end-functional B polymer is coupled to a linear A polymer to produce AB graft copolymer. Normally, the reactive functionality is introduced into the A polymer by polymerization of A monomers with a small amount of functional co-monomer to produce a statistical copolymer with a nearly uniform distribution of functional groups. This leads to graft copolymers with a random, but relatively uniform, placement of B grafts along the A backbones. In most cases, the grafting reaction is irreversible, so the architectural disorder is *quenched* rather than annealed.

To describe the equilibrium structure and thermodynamic properties of such a graft copolymer alloy in the melt state, a simple extension of Model C suffices. It proves convenient to work in a canonical ensemble in which the number of polymers *prior* to reaction are specified; namely, a collection of $n_A$ reactive A polymers and $n_B$ end-functional B polymers in a volume $V$. For simplicity, these polymers are assumed to be monodisperse, continuous Gaussian chains with chain lengths $N_A$ and $N_B$, respectively. The interaction model is the same as Model C – a Flory–Huggins-type interaction between dissimilar segments parameterized by $\chi_{AB}$ and the assumption of local incompressibility. We must also specify the state of chemical disorder created by the grafting reactions that are assumed to have taken place. A simple way to characterize the state of grafting is to associate with each differential segment $ds$ of the $j$th A polymer backbone a discrete random variable $\theta_j(s)$. This variable assumes the value 1 if a B graft is attached at that site, and the value 0 if there is no B graft. The total number of grafted B polymers in the system is thus

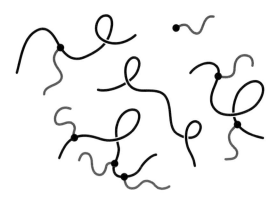

FIG. 4.6. Products of reactive coupling of A homopolymer containing uniformly placed reactive functional groups (dark) with end-functional B homopolymer (light). The resulting alloy contains graft copolymers with varying number and placement of B grafts, along with unreacted A and B homopolymer.

$$n_G = \sum_{j=1}^{n_A} \int_0^{N_A} ds\, \theta_j(s) \tag{4.164}$$

and the number of free (unreacted) B chains is thus $n_B - n_G$.

For the above model, the particle-to-field transformation described in the context of Model C leads to the following expression for the *canonical Model J field theory*:

$$\mathcal{Z}_C(n_A, n_B, V, T) = \mathcal{Z}_0 \int \mathcal{D}w_+ \int \mathcal{D}w_- \exp(-H[w_+, w_-]) \tag{4.165}$$

where $\mathcal{Z}_0$ denotes the ideal gas partition function of a total of $n_A$ non-interacting graft copolymers and free A polymers, and $n_B - n_G$ free B polymers. The effective Hamiltonian is given by

$$H[w_+, w_-] = \rho_0 \int d\mathbf{r} \left[(1/\chi_{AB})w_-^2 - iw_+\right]$$
$$- \sum_{j=1}^{n_A} \ln Q_G[w_A, w_B, \theta_j] - (n_B - n_G) \ln Q_B[w_B] \tag{4.166}$$

where $Q_B[w_B]$ is the normalized partition function for a single free (ungrafted) B polymer and is defined by eqns (4.99) and (4.100). The normalized partition function for the $j$th graft copolymer (or free A polymer), $Q_G[w_A, w_B, \theta_j]$, can be expressed as

$$Q_G[w_A, w_B, \theta] = \frac{1}{V} \int d\mathbf{r}\, q_G(\mathbf{r}, N_A; [w_A, w_B, \theta]) \tag{4.167}$$

where the propagator $q_G$ satisfies

$$\frac{\partial}{\partial s} q_G(\mathbf{r}, s; [w_A, w_B, \theta]) = \left[\frac{b_A^2}{6}\nabla^2 - w_G(\mathbf{r}, \theta(s))\right] q_G(\mathbf{r}, s; [w_A, w_B, \theta]) \quad (4.168)$$

subject to $q_G(\mathbf{r}, 0; [w_A, w_B, \theta]) = 1$. The field $w_G(\mathbf{r}, \theta(s))$ depends explicitly on the state $\theta(s)$ of the $s$th segment:

$$w_G(\mathbf{r}, \theta(s)) \equiv w_A(\mathbf{r}) - \ln[\theta(s)q_B(\mathbf{r}, N_B; [w_B]) + 1 - \theta(s)] \quad (4.169)$$

where $q_B(\mathbf{r}, N_B; [w_B])$ is the propagator defined in eqn (4.100) for a B graft and $w_A = iw_+ - w_-$ and $w_B = iw_+ + w_-$ are the same complex fields introduced in Model C.

According to the above equations, the statistical weight of a graft copolymer is built up by integrating eqn (4.168) along the A backbone of the copolymer. At each contour location $s$, the chemical potential field experienced by a differential segment $ds$, $w_G(\mathbf{r}, \theta(s))$, is either $w_A(\mathbf{r}) - \ln q_B(\mathbf{r}, N_B)$ or $w_A(\mathbf{r})$, respectively, depending on whether a B graft is attached or not attached at that segment.

For a system of finite size, $\mathcal{Z}_C$ and $H$ are both functionals of the particular state of the disorder variables $\{\theta_j(s)\}$, although this dependence is suppressed in our notation. However, in the thermodynamic limit for which $n_A \to \infty$, the sums over chains in eqns (4.164) and (4.166) include *all* possible realizations of the disorder variables. Thus, in the thermodynamic limit it follows that

$$n_G = n_A N_A p \quad (4.170)$$

where $p \equiv \overline{\theta_j(s)}$ is the *average fraction of A segments that contain a graft*. The overbar denotes a quenched ensemble average over the $\theta_j(s)$ random variables. In the thermodynamic limit, the effective Hamiltonian for Model J can thus be rewritten as

$$H[w_+, w_-] = \rho_0 \int d\mathbf{r} \, [(1/\chi_{AB})w_-^2 - iw_+]$$
$$- n_A \overline{\ln Q_G[w_A, w_B, \theta]} - (n_B - n_A N_A p) \ln Q_B[w_B] \quad (4.171)$$

and $H$ and $\mathcal{Z}_C$ no longer depend on the specific configuration of the disorder variables. The net effect is that Model J assumes the form of a disorder-free field theory, but where the *logarithm* of the single-chain partition function is to be averaged over the quenched disorder. Indeed, this is the *same* structure that we observed in Model H for a case of quenched chain length disorder. The integral over $N$ in eqn (4.140) affects an average of $\ln Q$ over the quenched distribution of chain lengths, $p_Q(N)$. In contrast, Model I has a structure that involves the average of $Q$, rather than $\ln Q$, over the *annealed* distribution of chain lengths. This observation is *general*: polymer systems with quenched disorder in macromolecular size or architecture produce effective field theories containing $\overline{\ln Q}$; systems with annealed disorder lead to field theories involving $\overline{Q}$.

Problems with *quenched* architectural disorder prove to be difficult and computationally expensive, because it is normally not possible to analytically perform the average of $\ln Q$ over the disorder variables. In the case of quenched chain length disorder, the situation is tractable because a one-dimensional quadrature over $N$ suffices to evaluate $\overline{\ln Q}$. For the present Model J, however, evaluation of $\overline{\ln Q}$ involves summing over all possible grafting patterns $\theta(s)$ along a chain. Because this effectively infinite-dimensional sum cannot be performed analytically, and its numerical evaluation would add a significant computational burden to an already challenging field theory, we would seem to be at a crossroads. Fortunately, there are a few options available. The first is to apply one of the schemes described in Section 3.4 in order to develop an analytical approximation for $\ln Q$. It might then be possible to explicitly carry out the quenched disorder average on the approximated theory. For example, if the weak inhomogeneity expansion of Section 3.4.1 is applied, the disorder averages relevant to Model J can be performed analytically (Fredrickson *et al.*, 1992; Foster *et al.*, 1995; Qi *et al.*, 1999). A second strategy of general utility is to *approximate the quenched average by an annealed average*, i.e. make the approximation

$$\overline{\ln Q} \approx \ln \overline{Q} \qquad (4.172)$$

since $\overline{Q}$ is usually much easier to evaluate. This scheme was recently shown to produce surprisingly accurate results for a random graft copolymer model similar to Model J (Patel and Fredrickson, 2003). It has the advantage that no assumption is required about the strength or length scales of the inhomogeneities in the fluid.

A third strategy for dealing with quenched disorder is to adopt the so-called "replica trick" (Mezard *et al.*, 1987) and write

$$\overline{\ln Q} = \lim_{m \to 0} (\overline{Q^m} - 1)/m \qquad (4.173)$$

The object $Q^m$ is obtained by introducing $m$ identical "replicas" of the system that are non-interacting but share the same set of disorder variables $\{\theta_j(s)\}$. For this purpose, one pretends that $m$ is an integer ($> 1$). The next step is to perform the quenched disorder average $\overline{Q^m}$, assuming that this is possible, which has the effect of coupling the various replicas. The final step is to analytically continue the resulting formula to the case of $m \to 0$. Unfortunately, this protocol can usually only be carried out when combined with an approximation scheme, such as the weak inhomogeneity expansion (Shakhnovich and Gutin, 1989). The couplings between replicas that are introduced by the disorder average destroys the single-chain nature of the polymer field theory and makes the full theory computationally problematic.

To complete the discussion of Model J, we implement the annealed approximation (4.172). In order to evaluate $\overline{Q}_G$, it is necessary to be more precise about defining the statistical properties of the $\{\theta(s)\}$ variables for a single chain. The simplest model is to assume that these are stationary random variables that are uncorrelated from one differential segment to the next. Thus, $\theta(s)$ is distributed independently of $\theta(s+ds)$, and the probability distribution for the value of $\theta$ at *any* differential segment is given by

$$P(\theta) = p\,\delta_{\theta,1} + (1-p)\,\delta_{\theta,0} \tag{4.174}$$

so that $\bar{\theta} \equiv \sum_{\theta=0}^{1} \theta P(\theta) = p$. This disorder model is consistent with the notion that the grafting sites are distributed with a uniform probability along the A polymer backbones. Models with correlated grafting sites can be similarly constructed (Fredrickson and Milner, 1991).

The disorder average of $Q_G$ is obtained by temporarily discretizing its defining path integral by analogy with eqn (3.21). The exponential factors involving $w_G$ for each differential segment can be averaged independently by means of eqn (4.174). For segment $k$, we have

$$\sum_{\theta=0}^{1} P(\theta) e^{-ds\,w_G(\mathbf{r}_k,\theta)} = e^{-ds\,\overline{w}_G(\mathbf{r}_k)} + O(ds^2) \tag{4.175}$$

where $\overline{w}_G$ is a disorder-averaged chemical potential field defined by

$$\overline{w}_G(\mathbf{r};[w_A,w_B]) \equiv \sum_{\theta} P(\theta)\,w_G(\mathbf{r},\theta)$$
$$= w_A(\mathbf{r}) - p\,\ln q_B(\mathbf{r},N_B;[w_B]) \tag{4.176}$$

When the continuum limit is restored, it follows that the disorder-averaged, single-chain partition function $\overline{Q}_G$ can be obtained from an averaged version of eqns (4.167) and (4.168). In particular,

$$\overline{Q}_G[w_A,w_B] = \frac{1}{V} \int d\mathbf{r}\,\bar{q}_G(\mathbf{r},N_A;[w_A,w_B]) \tag{4.177}$$

where the averaged propagator $\bar{q}_G$ satisfies

$$\frac{\partial}{\partial s}\bar{q}_G(\mathbf{r},s;[w_A,w_B]) = \left[\frac{b_A^2}{6}\nabla^2 - \overline{w}_G(\mathbf{r};[w_A,w_B])\right]\bar{q}_G(\mathbf{r},s;[w_A,w_B]) \tag{4.178}$$

The interpretation of the above equations is that the statistical weight for an "average" AB graft copolymer can be built up by solving diffusion eqn (4.178) along the A backbone, with the average potential $\overline{w}_G$ experienced by a differential A segment given by the weighted average of $w_A - \ln q_B$ (graft present) and $w_A$ (no graft present).

With the above annealed approximation, it is straightforward to evaluate the effective Hamiltonian of Model J for any realization of the fields $w_+$ and hence $w_A$ and $w_B$. The first step is to solve eqn (4.100) to obtain the free B chain propagator $q_B(\mathbf{r},N_B)$ and partition function $Q_B$. The average field $\overline{w}_G$ can then be evaluated by means of eqn (4.176). Finally, eqn (4.178) is solved to obtain

$\bar{q}_G$ and $\overline{Q}_G$. All the terms in eqn (4.171) are thus determined, provided that we approximate $\overline{\ln Q_G}$ by $\ln \overline{Q}_G$.

Density operators for A and B segments can be defined by close analogy with eqn (4.102) for Model C. The local density of type A segments is obtained by averaging the operator

$$
\tilde{\rho}_A(\mathbf{r}; [w_{A/B}]) = -n_A \frac{\delta \ln \overline{Q}_G[w_{A/B}]}{\delta w_A(\mathbf{r})}
$$

$$
= \frac{n_A}{V\overline{Q}_G[w_{A/B}]} \int_0^{N_A} ds\, \bar{q}_G(\mathbf{r}, N_A - s; [w_{A/B}])\bar{q}_G(\mathbf{r}, s; [w_{A/B}])
$$

$$(4.179)$$

over the fluctuating $w_\pm$ fields. The notation $[w_{A/B}]$ implies a functional dependence on both $w_A = iw_+ - w_-$ and $w_B = iw_+ + w_-$. The local density of type B segments has contributions from free B chains, as well as from B chains grafted to A polymers. The segment density operator for the *free* B chains is given by

$$
\tilde{\rho}_{Bf}(\mathbf{r}; [w_B]) = -(n_B - n_A N_A p) \frac{\delta \ln Q_B[w_B]}{\delta w_B(\mathbf{r})}
$$

$$
= \frac{n_B - n_A N_A p}{V Q_B[w_B]} \int_0^{N_B} ds\, q_B(\mathbf{r}, N_B - s; [w_B])q_B(\mathbf{r}, s; [w_B])
$$

$$(4.180)$$

For the *grafted* B chains, the local density operator of type B segments is

$$
\tilde{\rho}_{Bg}(\mathbf{r}; [w_{A/B}]) = -n_A \frac{\delta \ln \overline{Q}_G[w_{A/B}]}{\delta w_B(\mathbf{r})}
$$

$$
= \frac{n_A}{V\overline{Q}_G[w_{A/B}]} \int_0^{N_B} ds\, q_{Bc}(\mathbf{r}, N_B - s; [w_{A/B}])q_B(\mathbf{r}, s; [w_B])
$$

$$(4.181)$$

The "complementary" propagator $q_{Bc}$ appearing in this equation satisfies the same differential equation as $q_B$, eqn (4.100), but is subject to an *inhomogeneous initial condition* that reflects the fact that the graft could be connected at any site along the A backbone of the copolymer

$$
q_{Bc}(\mathbf{r}, 0; [w_{A/B}]) = p \int_0^{N_A} ds\, \bar{q}_G(\mathbf{r}, N_A - s; [w_{A/B}])\bar{q}_G(\mathbf{r}, s; [w_{A/B}]) \quad (4.182)
$$

Finally, the sum of the free and grafted B chain operators, $\tilde{\rho}_B \equiv \tilde{\rho}_{Bf} + \tilde{\rho}_{Bg}$, defines an operator that can be used to examine the spatial distribution of *all* type B segments.

## 4.8 Tethered polymer layers

Another important class of polymeric systems are fluids in which macromolecules are "tethered" to a surface by an end. When multiple chains are tethered at high areal densities, such layers are referred to as *polymer brushes* and are important in a number of physical contexts, including colloidal stabilization, lubrication, and surface modification (Dolan and Edwards, 1975; Alexander, 1977; de Gennes, 1980a; Milner *et al.*, 1988; Milner, 1991; Halperin *et al.*, 1992; Grest and Murat, 1995). Examples of both "solution" and "melt" polymer brushes are pervasive in the modern polymer science literature. Here, we restrict attention to the simplest case of a solution polymer brush consisting of end-tethered polymers of the same length, immersed in a good solvent.

### 4.8.1 *Model K: solution polymer brush*

A standard model of a solution polymer brush is constructed by adopting the same chain and interaction model as in Model A, except that each polymer chain is assumed to be attached by one end to a flat surface. The points of attachment are normally taken to be uniformly distributed in the plane of the surface (see Fig. 4.7), although the validity of this assumption rests on the synthetic techniques used to carry out the tethering of the chain ends to the surface. If the chemical bonds between the surface and chain ends are *permanent*, the polymer brush possesses *quenched disorder* associated with the random attachment points in the plane of the surface. Alternatively, if the bonds are *reversible*, then the set of attachment points are *annealed* random variables. Such a situation is realized, e.g., when end-functionalized polymer chains are reversibly adsorbed to a surface from solution. In defining Model K, we shall adopt the more typical assumption of permanent bonding and hence quenched disorder.

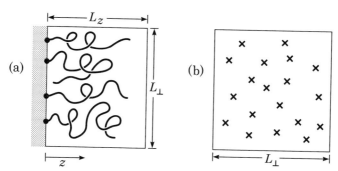

FIG. 4.7. Schematic of a tethered polymer layer or "polymer brush". Polymer chains are attached by one end to a surface as shown in the side view (a). The points of attachment (denoted by crosses) are distributed randomly in the plane, as shown in the plane view (b).

It is convenient to work in a canonical ensemble in which $n$ continuous Gaussian chains of length $N$ are tethered to a flat surface located at $z = 0$. The surface is assumed to be impenetrable by the polymer segments; the latter are surrounded by implicit solvent and are neither attracted nor repelled by the surface apart from the impenetrability constraint. The length of the system in the direction of the surface normal coordinate $z$ is denoted by $L_z$, while $L_\perp$ denotes the lateral system size along each of two orthogonal coordinates in the surface plane. The system volume is thus $V = L_z L_\perp^2$. In order to specify the quenched disorder in the system, it is helpful to introduce a set of two-dimensional vectors $\mathbf{r}_{\perp,j}$, $j = 1, 2, ..., n$, which denote the positions of the grafted chain ends in the plane of the surface. For a particular realization of these random vectors, the particle-to-field transformation for Model K follows closely that of Model A and leads to a field theory of the same form of eqn (4.71), but where the effective Hamiltonian is given by

$$H[w] = \frac{1}{2u_0} \int d\mathbf{r} \, [w(\mathbf{r})]^2 - \sum_{j=1}^{n} \ln Q(\mathbf{r}_{\perp,j}; [iw]) \qquad (4.183)$$

The object $Q(\mathbf{r}_{\perp,j}; [iw])$ in this expression represents a normalized partition function for the $j$th polymer whose grafted chain end is located at position $\mathbf{r}_{\perp,j}$ in the plane $z = 0$, and whose segments experience the purely imaginary potential $iw(\mathbf{r})$. This functional can be computed by means of the formula

$$Q(\mathbf{r}_\perp; [iw]) = \int d\mathbf{r} \, q_{\mathbf{r}_\perp}(\mathbf{r}, N; [iw]) \qquad (4.184)$$

where $q_{\mathbf{r}_\perp}(\mathbf{r}, s; [iw])$ denotes a propagator for a chain of length $s$ that originated at $\mathbf{r}_\perp$ in the grafting plane and has its free end at position $\mathbf{r}$. This propagator satisfies eqn (3.25), but with the initial condition $[\mathbf{r} = (x, y, z), \mathbf{r}_\perp = (x_\perp, y_\perp)]$

$$q_{\mathbf{r}_\perp}(\mathbf{r}, 0; [iw]) = \delta(x - x_\perp)\delta(y - y_\perp)\delta(z - \epsilon) \qquad (4.185)$$

which constrains the initiating chain end to the tethering point. A parameter $\epsilon \to 0+$ is used to displace the tethering point an infinitesimal distance off the surface into the physical domain. The impenetrability of the surface to polymer segments is captured by imposing the Dirichlet boundary condition eqn (3.196) on the propagator at $z = 0$ for all $s > 0$.

In the thermodynamic limit of $(n \to \infty, L_\perp \to \infty, \sigma$ finite), where $\sigma \equiv n/L_\perp^2$ is the surface density of tethered polymers, we can again replace the sum over chains in eqn (4.183) by a quenched average over the locations of the tethering points. It follows that the effective Hamiltonian for the *canonical Model K field theory* is given by

$$H[w] = \frac{1}{2u_0} \int d\mathbf{r} \, [w(\mathbf{r})]^2 - \sigma \int d\mathbf{r}_\perp \, \ln Q(\mathbf{r}_\perp; [iw]) \qquad (4.186)$$

The corresponding segment density operator is obtained by differentiation with respect to $w(\mathbf{r})$:

$$\tilde{\rho}(\mathbf{r}; [iw]) = -\sigma \int d\mathbf{r}_\perp \frac{\delta \ln Q(\mathbf{r}_\perp; [iw])}{\delta iw(\mathbf{r})}$$

$$= \int d\mathbf{r}_\perp \frac{\sigma}{Q(\mathbf{r}_\perp; [iw])} \int_0^N ds \, q_{\mathbf{r}_\perp}(\mathbf{r}, s; [iw]) q(\mathbf{r}, N - s; [iw])$$

$$(4.187)$$

where $q(\mathbf{r}, s; [iw])$ is a complementary propagator that initiates from the free end of a tethered chain. $q$ satisfies eqn (3.25), subject to the familiar initial condition $q(\mathbf{r}, 0; [iw]) = 1$.

Equations (4.186) and (4.187) at first glance appear to be very computationally demanding because each evaluation of $q_{\mathbf{r}_\perp}(\mathbf{r}, s; [iw])$ and $Q(\mathbf{r}_\perp; [iw])$ for a prescribed $\mathbf{r}_\perp$ requires an independent solution of the diffusion equation (3.25). A large number of such solutions would be necessary to accurately perform the two-dimensional quadratures over $\mathbf{r}_\perp$ in order to evaluate $H[w]$ and $\tilde{\rho}(\mathbf{r}; [iw])$. For high-resolution simulations, the cost would be prohibitive. One strategy for addressing this dilemma would be to follow the procedure of the last section and approximate the quenched average over grafting positions $\mathbf{r}_\perp$ by an annealed average. However, a clever alternative that avoids any additional approximations was identified by Müller (2002).

Müller noted that the partition function for a tethered polymer chain has a factorization property analogous to eqn (3.27), namely

$$Q(\mathbf{r}_\perp; [iw]) = \int d\mathbf{r} \, q_{\mathbf{r}_\perp}(\mathbf{r}, s; [iw]) q(\mathbf{r}, N - s; [iw]) \qquad (4.188)$$

for all $s \in [0, N]$. The choice $s = N$ in eqn (4.188) recovers eqn (4.184), while setting $s = 0$ yields a computationally attractive alternative

$$Q(\mathbf{r}_\perp; [iw]) = q((x_\perp, y_\perp, \epsilon), N; [iw]) \qquad (4.189)$$

In this expression, the partition function is evaluated by integrating the diffusion equation starting at the free chain end, rather than at the grafting point. The advantage is that expressions such as eqn (4.186) can be evaluated from a *single* solution of the diffusion eqn (3.25) for $q(\mathbf{r}, N; [iw])$ by applying that solution at a series of quadrature points in the grafting plane.

Müller further showed that a related simplification is possible for the density operator (Müller, 2002). Equation (4.187) can be rewritten as

$$\tilde{\rho}(\mathbf{r}; [iw]) = \int_0^N ds \, q_c(\mathbf{r}, s; [iw]) q(\mathbf{r}, N - s; [iw]) \qquad (4.190)$$

where $q_c$ is a new complementary propagator defined by

$$q_c(\mathbf{r}, s; [iw]) \equiv \int d\mathbf{r}_\perp \frac{\sigma}{Q(\mathbf{r}_\perp; [iw])} q_{\mathbf{r}_\perp}(\mathbf{r}, s; [iw]) \qquad (4.191)$$

Because $q_c$ is linearly related to $q_{\mathbf{r}_\perp}$, $q_c$ also satisfies the diffusion eqn (3.25). However, upon combining eqns (4.191), (4.185), and (4.189), the initial condition for $q_c$ is seen to be

$$q_c(\mathbf{r}, 0; [iw]) = \frac{\sigma \, \delta(z - \epsilon)}{q((x, y, \epsilon), N; [iw])} \qquad (4.192)$$

Thus, evaluation of the density operator according to eqn (4.190) requires just two solutions of the diffusion eqn (3.25). A first solution is used to obtain $q(\mathbf{r}, s; [iw])$, which is needed both in eqn (4.190) and to determine the initial condition for $q_c$. The second solution of the diffusion equation, starting with the initial condition eqn (4.192), results in $q_c(\mathbf{r}, s; [iw])$, which can be used in eqn (4.190) to complete the evaluation of the density operator.

## 4.9   Boundary conditions and confinement effects

In Section 3.5, we introduced boundary conditions that are to be applied to single-chain propagators in confined geometries. Having just completed a discussion of multi-chain models, it is appropriate that we revisit the issue of boundary conditions in this context, i.e. models of molten polymers or polymer solutions confined to a pore or slit.

An important observation is that in all the models described in the present chapter, other than Model F, *the spatial non-locality of the effective Hamiltonian arises from the connectivity of the polymers and is solely contained within the single-chain partition function Q*. Thus the only objects requiring the application of boundary conditions are the *single-chain propagators* that are used to construct $Q$. No explicit boundary conditions need be imposed on the chemical potential "$w$" fields for these models. In the case of Model F, the effective Hamiltonian contains an additional non-local term $\sim |\nabla \Phi|^2$ associated with the fluctuating electrostatic potential field $\Phi$. This term mandates that a boundary condition on $\Phi$ also be applied.

### 4.9.1   *Bulk fluids*

If the objective is to describe a *bulk*, unbound polymeric solution or melt, the models described in the present chapter require little if any modification. Analytical calculations are normally performed in the thermodynamic limit of the relevant model, or in a large cubic cell subject to periodic boundary conditions. Numerical simulations are restricted to finite domains, so a cubic cell with periodic boundary conditions is a convenient choice. For models that do not possess spatially periodic mesophases, which includes all of the above models except E and J, the use of periodic boundary conditions suppresses "finite size effects" associated with confining field variables to a finite domain. Intensive thermodynamic properties computed in a periodic cell of size $L$ for such systems are typically observed to differ from their values in the thermodynamic limit ($L \to \infty$)

by a "finite size correction" that decays *exponentially* with $L$. An exception to this exponential decay is observed in simulations of systems that are in close proximity to a second-order phase transition or critical point, where a divergent correlation length arises. In such situations, which include Models B, C, and D near their critical consolute points, finite size corrections are expected to decay *algebraically* with $1/L$.

Model F deserves special mention because of the long-ranged nature of the Coulomb interactions that are included in the polyelectrolyte model. In conventional particle simulations of electrolyte solutions (Allen and Tildesley, 1987; Frenkel and Smit, 1996), the Coulomb potential is problematic because its slow decay with distance makes it necessary to compute interactions between particles in the simulation cell and their distant periodic images. Specialized techniques such as Ewald summation have been developed for this purpose, but they carry a significant computational burden and complicate otherwise straightforward algorithms. By converting a particle-based model of an electrolyte to a field theory, as in the case of Model F, the Coulomb operator is replaced by its functional inverse – the operator $\nabla^2$, which acts locally rather than at a distance. Thus, in the field-theoretic representation a straightforward application of periodic boundary conditions to a finite cell automatically includes all distant interactions of charges with their periodic images. No special techniques are required.

Bulk simulations of models such as E and J that are capable of exhibiting spatially periodic mesophases require special attention. Mesophases formed by block and graft copolymers are solid-like in at least one direction, implying that they can sustain a stress or strain indefinitely. If the simulation cell is incommensurate in size or shape with the periodic order of the mesophase, then the equilibrated fluid structure will be subject to a strain that is incapable of relaxing. In such cases, the free energy of the phase under investigation is elevated, and incorrect conclusions regarding its stability relative to other structured and unstructured phases might be drawn. There are several strategies for avoiding this situation. The first is to conduct a "large-cell" simulation of the type described in Section 3.6.6. Upon increasing the cell size to a value much larger than the mesophase period, any residual strain or stress will be significantly diminished, although this comes at a significant computational price because the minimum spatial resolution is dictated by the local structure of the mesophase and not by the cell size.

A second strategy, which requires that the symmetry of the mesophase be known in advance, is to choose a computational cell (with periodic boundary conditions) that is a primitive unit-cell of the periodic mesophase structure. For example, in the case of the familiar cylindrical mesophase, consisting of a hexagonally packed array of cylindrical aggregates, the oblique computational cell of Fig. 3.16 would be suitable. With the cell shape selected by symmetry considerations, the cell volume can be adjusted to minimize the free energy and thereby remove any residual stress. Clearly such a "unit-cell" is optimal from a computational standpoint, because it focuses the computational effort on spatially resolving only one primitive element of the periodic mesophase.

A third option is to adopt a *variable cell shape and size* simulation technique in which the geometrical characteristics and size of the computational cell are adjusted in the course of a simulation to minimize the free energy. Such methods are well known in particle simulations of atomic and molecular solids (Parrinello and Rahman, 1981; Ray and Rahman, 1984) and have been recently been adapted to field-based simulations of polymeric fluids (Barrat *et al.*, 2005). In the latter case it is particularly convenient to restrict the cell geometry to a continuous family of parallelepipeds so that multi-dimensional FFTs can be applied at any stage as the cell shape evolves. An important benefit of variable cell techniques is their *flexibility*; they can be applied to unit-cell calculations with or without advance knowledge of the mesophase symmetry, and they can also be used to effectively and automatically lower residual stress in large-cell calculations. The downside is simply the burden of developing a more complex simulation code and a small computational overhead associated with the box size and shape adjustments.

### 4.9.2 *Confined solutions*

Boundary conditions other than periodic are required when simulating coarse-grained models of polymeric fluids in confined geometries. A useful place to start the discussion of such systems is in the context of a polymer solution confined to a slit between two parallel surfaces (Alexander-Katz *et al.*, 2003). For simplicity, we consider a flexible homopolymer dissolved in a good solvent so that Model A applies. The two surfaces defining the slit are assumed to be impenetrable to the solvent and polymer, and neutral to both species. A field-based simulation of such a system would utilize a rectangular parallelepiped cell in which periodic boundary conditions are applied in the two directions transverse to the slit (in the plane of the surfaces) and the homogeneous Dirichlet condition (3.196) is applied at each of the bounding surfaces. Because the non-locality of Model A is confined to the functional $Q[iw]$, these boundary conditions need only be applied to the single-chain propagator $q(\mathbf{r}, s; [iw])$.

A schematic of the average segment density $\langle \hat{\rho}(z) \rangle$ obtained from a simulation of the above model is shown in Fig. 4.8, where $z$ denotes the coordinate normal to the bounding surfaces of the slit and $L$ is the distance between the surfaces. Three length scales are in evidence: $L$, defining the confinement length scale; $\xi$, which is the scale on which the segment density goes from zero at the wall to a nearly uniform value $\rho_b$ in the center of the slit; and $R_g$, the radius of gyration of the polymers in the solution. It is well known from the theory of polymer solutions (de Gennes, 1979; Rubinstein and Colby, 2003) that the length $\xi$, which is comparable to the *correlation length* of a bulk solution at the same concentration, decreases rapidly with increasing segment concentration $\rho_0 = nN/V$ in the important semidilute and concentrated regimes. In the semidilute regime, the thermodynamic properties are dictated by this length scale, with both the bulk free energy density and the osmotic pressure scaling as $k_B T/\xi^3$.

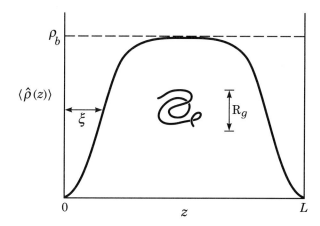

FIG. 4.8. Schematic of the average segment density profile across a polymer solution confined to a slit of width $L$. The density profile obtained with the use of homogeneous Dirichlet boundary conditions is shown as the solid curve. The dashed line is the profile obtained by imposing the homogeneous Neumann conditions (4.193).

Although $\xi \ll R_g$ in this regime, it is necessary to resolve the finest scale $\xi$ for an accurate description of the thermodynamics and structure of the confined solution. Dirichlet boundary conditions applied to the propagator $q$ in this case produce average density profiles (such as the solid curve in Fig. 4.8) that go smoothly to zero at each wall over a distance of order $\xi$.

As the segment concentration $\rho_0$ is further increased and the concentrated regime is entered, the concentration healing length $\xi$ becomes much smaller than either $R_g$ or $L$, and the density profile rises very sharply away from each sur-face. To resolve this sharp rise in a numerical simulation, while still capturing the structure of the solution on scales of $R_g$ and $L$, requires considerably more computational effort and eventually the cost becomes prohibitive. At this stage, use of an *effective boundary condition* is advised.[52] Rather than resolve the nar-row boundary layers near the two surfaces, which serve only to repel polymers and establish an interfacial free energy contribution that can be attributed to the surfaces, one can simply focus on the larger-scale structure and properties of the fluid by imposing the effective Neumann boundary conditions (de Gennes, 1980b; Wu *et al.*, 1995)

$$\frac{\partial}{\partial z} q(\mathbf{r}, s; [iw]) \bigg|_{z=0,L} = 0 \qquad (4.193)$$

[52] Alternatively, an adaptive, non-uniform computational grid can be utilized, although im-plementation of such a scheme in tandem with a pseudo-spectral solution of the diffusion equation has not yet been achieved.

These effective boundary conditions can be viewed as conditions on the "outer" part of the propagator solution, i.e. the solution outside of the two boundary layers (Nahfeh, 1973). For the present problem, replacing the Dirichlet conditions of eqn (3.196) with the Neumann conditions (4.193) produces the flat average density profile shown as a dashed line in Fig. 4.8. Simulation of Model A in this case allows for the investigation of equilibrium phenomena associated with the interplay of the two length scales $R_g$ and $L$ without the need for explicit resolution of the sharp $\xi$ scale.

A similar effective boundary condition strategy could be used to efficiently simulate Model B in a confined geometry at high polymer concentration. For both models, it should be understood that the elimination of fluctuation modes with wavelengths comparable to and smaller than $\xi$ will have the effect of renormalizing parameters such as $u_0$ and $\chi_{PS}$ in the field theory (see Section 6.1.5). This renormalization may have to be accounted for when interpreting results from simulations that employed effective boundary conditions and a coarse field resolution.

Effective boundary conditions other than eqn (4.193) are useful under certain circumstances. For example, the mixed boundary condition (3.197) can be applied to the various solution models of this chapter when the polymers are weakly attracted to a surface. In the case of Model F, it is also necessary to specify a condition on the electrostatic potential field $\Phi$ at all bounding surfaces. For this purpose, standard electrostatic considerations apply, e.g. the normal derivative of $\Phi$ should vanish at a surface carrying no charge (Eyges, 1972).

### 4.9.3 Confined melts

The effective boundary condition approach is particularly useful for the treatment of molten polymers in confined geometries. In a description with full atomic resolution, the average segment density of a polymer melt in the close vicinity of an impenetrable surface exhibits an oscillatory profile that is illustrated in Fig. 4.9. From a *mesoscopic* perspective, the details of this profile are not of intrinsic interest, since the oscillations are confined to scales below $\sim 1\,\mathrm{nm}$. Thus, mesoscopic models of polymer melts, such as Model D for a molten blend, normally do not contain the "hard-core" repulsive interactions that lead to such an oscillatory profile. Instead, a soft repulsion is included (parameterized by $\zeta$ in Model D) that produces a monotonic increase in the total segment density $\langle \hat{\rho}_+(z) \rangle$ shown as the dashed curve in Fig. 4.9. The healing length $\xi_m$ that characterizes this melt density profile is dictated by the strength of the soft repulsion, scaling as $\xi_m \sim \zeta^{-1/2}$ for Model D.

In the simplest case, a study of a coarse-grained blend model, such as Model D, in a confined environment proceeds by imposing the Dirichlet condition eqn (3.196) on the polymer propagators at the confining surfaces. This requires that boundary layers of thickness $\xi_m$ be resolved near the melt surfaces. Alternatively, one could ignore such boundary layers and, in the case of "neutral" surfaces with no preferential affinity for one of the blend components, apply the

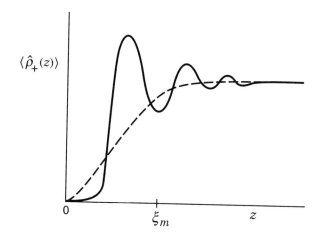

FIG. 4.9. Sketch of the total segment density of an equilibrium polymer melt in the vicinity of a surface. In a fully atomistic description with hard-core repulsions, an oscillatory profile is observed (solid curve). A mesoscopic description of a melt with soft repulsions, e.g. Model D, produces a monotonic profile with a "healing length" $\xi_m$ (dashed curve).

effective Neumann boundary conditions (4.193) to all the species propagators. The latter approach is clearly preferable, because it does not require numerical resolution of a thin boundary layer for the typical case of a weakly compressible melt, $\zeta \gg 1$. Moreover, for a strictly incompressible model, such as Models C, E, and J, the first strategy is *inapplicable* because the assumption of no total density variation is mathematically incompatible with the Dirichlet conditions $q_K = 0$. A neutral surface is thus treated in incompressible models by applying Neumann boundary conditions to all the chain propagators.

A more typical situation is that of a bounding surface (solid or vapor) at which one of the components of a melt blend, or one of the blocks/grafts of a copolymer, is energetically preferred. In a compressible melt model, a tendency for surface enrichment by one of the components can be taken into account by including a short-ranged attractive potential between the surface and the segments of the preferred species. However, to avoid resolving this surface potential and associated boundary layer, a better alternative is to apply mixed boundary conditions similar to eqn (3.197) on each propagator species (Wu *et al.*, 1996). This strategy is also suitable for incompressible melt models, such as C, E, and J. For example, in the case of Model C and a surface at $z = 0$ that preferentially attracts one of the two blend components, an appropriate set of boundary conditions is

$$\left[ \frac{\partial}{\partial z} q_K(\mathbf{r}, s; [w_K]) + \kappa_K q_K(\mathbf{r}, s; [w_K]) \right]_{z=0} = 0 \qquad (4.194)$$

for $s > 0$ and $K = $ A and B. The parameters $\kappa_A$ and $\kappa_B$ have units of inverse length and characterize the strengths of delta function potentials that attract segments of type A and B, respectively, to the surface. The derivation of eqn (4.194) invokes the ground state dominance approximation, which requires that $b_K \ll |\kappa_K|^{-1} \ll R_{g,K}$. An additional condition on $\kappa_A$ and $\kappa_B$ is imposed by the incompressibility constraint of Model C. Using the fact that in the ground state approximation, $\tilde{\rho}_K(\mathbf{r}) \propto q_K(\mathbf{r})^2$, the constraint $\sum_K \langle \tilde{\rho}_K(\mathbf{r}) \rangle = \rho_0$ implies the condition

$$[\kappa_A \langle \tilde{\rho}_A(\mathbf{r}; [w_A]) \rangle + \kappa_B \langle \tilde{\rho}_B(\mathbf{r}; [w_B]) \rangle]_{z=0} = 0 \qquad (4.195)$$

Since the average segment densities are positive definite, it follows that the $\kappa_K$ must be of mixed sign, or both identically zero, to be consistent with incompressibility in the surface layer. The case of $\kappa_K = 0$ recovers a pair of Neumann boundary conditions of the form of eqn (4.193), which are suitable for a melt blend at a neutral surface. In contrast, if component A is enriched at the surface, then $\kappa_A > 0$ and $\kappa_B$ is fixed by eqn (4.195) to a negative value. The magnitude of $\kappa_A$ can be adjusted to vary the strength of the preferential attraction.

Equation (4.195) is a *nonlinear* condition relating $\kappa_A$ to $\kappa_B$ that can be difficult to implement because it requires that the ensemble averages of the density operators over the $w_\pm$ field fluctuations be computed. These in turn depend on the $\kappa_K$ through the boundary conditions (4.194) applied to the chain propagators. However, if the preferential attraction of the surface for one of the components is very *weak*, so that the surface density $\langle \tilde{\rho}_K(\mathbf{r}) \rangle|_{z=0}$ deviates only slightly from the average bulk density $\rho_{K0} = n_K N_K / V$, then eqn (4.195) can be *linearized* to obtain

$$\kappa_A \rho_{A0} + \kappa_B \rho_{B0} = 0 \qquad (4.196)$$

This linearized boundary condition can be applied in tandem with eqn (4.194), with no iterative adjustment of either of the $\kappa_K$ required.

The boundary conditions (4.194)–(4.196) are readily applicable to other incompressible binary blend and copolymer models, including Models E and J.

### 4.9.4   *Polymer brushes*

Model K was introduced in Section 4.8 as a coarse-grained model of solution polymer brushes. It was assumed in this model that the surface is impenetrable and neutral to the solvent and polymer species, so that the Dirichlet boundary condition (3.196) is appropriate. At small values of the surface density of grafted chains, $\sigma = n/L_\perp^2$, the average segment density $\langle \hat{\rho}(z) \rangle$ for this model rises smoothly with distance $z$ from the surface and then decays on a length scale $h$ that is comparable to the radius of gyration $R_g$ of a free polymer in the same solvent. However, as $\sigma$ is increased to values satisfying $\sigma R_g^2 \gg 1$, a dense brush is created that is characterized by a large separation of the length scales describing the rise, $\xi_b$, and the decay, $h$, of the average segment density. This situation is shown schematically in Fig. 4.10. Numerical resolution of both length scales becomes increasingly difficult as $\sigma$ grows, because $\xi_b$ and $h$ move in opposite directions with $\sigma$.

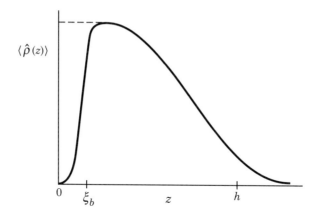

FIG. 4.10. Sketch of the average segment density in a polymer brush with a high surface concentration of grafted chains. The density rises to a maximum value on a length scale $\xi_b$ that is much smaller than the typical size of an extended polymer or the average brush height $h$.

One solution to this computational dilemma is to ignore the $\xi_b$ scale at high grafting densities and replace the Dirichlet condition with a suitable effective boundary condition. The appropriate condition is evidently the same as that applied to concentrated polymer solutions near a neutral surface, namely the Neumann boundary condition given in eqn (4.193). The resulting density profile is shown as the dashed curve in Fig. 4.10.

Other effective boundary conditions can be applied to the tethering surfaces of polymer brushes. The Neumann condition (4.193) is appropriate for solution brushes under the conditions described above, but also for *melt* brushes when the tethered polymer layer is taken to be incompressible. In the case of a solution brush where the polymer is weakly attracted to the surface, eqn (3.197) can be substituted for the Dirichlet condition. Similarly, eqns (4.194) and (4.195) would be appropriate boundary conditions for an incompressible melt brush consisting of a mixture of A and B chains grafted to a surface that prefers one of the species.

### 4.9.5   Irregularly shaped domains

Imposing the boundary conditions described above for polymer solutions or melts can be difficult if the fluid domain has a complicated shape. One approach in this situation is to utilize finite element or finite volume numerical techniques, which are readily adaptable to geometrically complex domains. Unfortunately the flexibility of grid design with these techniques is offset by the loss of spectral accuracy.

A useful alternative is to embed the irregularly shaped fluid domain within a larger parallelepiped computational cell that is subject to periodic boundary conditions. The polymer fluid is confined to the interior domain by applying a "mask" that excludes fluid from the annular region contained between the paral-lelepiped cell and the fluid domain. The mask can consist of a large potential field applied to the annular region, or a constraint (imposed with a Lagrange mul-tiplier field) that the fluid density vanish inside the annulus. By this method, polymers can be confined to domains of arbitrary shape, but pseudo-spectral calculations can still be conducted using a plane wave basis in the larger par-allelepiped cell. There is a modest loss of computational efficiency due to the wasted updates of the field variables in the annular region, but this is usually a small price to pay for the versatility of the masking approach. The recent numer-ical study by Sides and Katsov of helical self-assembly of block copolymer alloys confined to cylindrical pores (see Fig. 5.22) is a nice illustration of the technique (Wu *et al.*, 2004).

With either a potential or density constraint mask, it is important that the inner boundary of the mask confining the fluid domain be diffuse so that the convergence of the spectral solution is not sacrificed. A good rule of thumb is that the mask should have a diffuse interface, e.g. a "tanh" profile, with a width not less than the width of the fluid–fluid interfaces that are to be resolved inside the polymer fluid.

## 4.10   Density functional theory

Density functional theory (DFT) is a formal construct that has general applica-bility to inhomogeneous systems at equilibrium. The formalism is best known in the context of quantum many-body theory, where it has proven to be a powerful tool for electronic structure calculations of large molecules and extended solids (Parr and Yang, 1989). In the statistical mechanics of classical fluids, a similar DFT formalism is useful for studying fluid–vapor interfaces and capillary phe-nomena (Rowlinson and Widom, 1989), as well as liquid–solid transformations, nucleation phenomena, and other materials-related structures and processes (Ox-toby, 2002; Löwen, 2002).

Density functional theory has been broadly applied to atomistic models of inhomogeneous polymer systems, where it has been used to examine the equi-librium structure of polymer surfaces, thin films, and interfaces (Frischknecht *et al.*, 2002). In the case of *mesoscopic* models of inhomogeneous polymers, the DFT formalism is also applicable, but it is an unnecessary addition to the field-theoretic framework that has been outlined in the present chapter. Nevertheless, the language of density functional theory has become a standard one, so it is important to understand the connection with the approach taken in this mono-graph.

The basic strategy underlying the DFT formalism is as follows: (i) impose an inhomogeneous external potential field $J(\mathbf{r})$ that acts independently on each atom of the fluid, (ii) make a Legendre transformation to express the free energy as a functional of the average density created by that potential, and (iii) derive the equilibrium density profile by a *variational principle*. Namely, the equilib-rium density is the field that minimizes the free energy functional, subject to the constraint that the density variations conserve the number of atoms. These three steps are detailed below in the context of the monatomic fluid model of Section 4.1.1.

### 4.10.1  *Monatomic fluid*

Upon imposition of an external potential $J$, the $nVT$ canonical partition function of eqn (4.1) is modified to

$$\mathcal{Z}_C[J] = \frac{1}{n!\lambda_T^{3n}} \int d\mathbf{r}^n \ \exp\left[-\beta U(\mathbf{r}^n) - \int d\mathbf{r} \ J(\mathbf{r})\hat{\rho}(\mathbf{r})\right] \qquad (4.197)$$

where the functional dependence on $J(\mathbf{r})$ is manifest. The expectation value of the atomic density in the presence of the potential is given by

$$\langle \hat{\rho}(\mathbf{r}) \rangle_J = -\frac{\delta \ln \mathcal{Z}_C[J]}{\delta J(\mathbf{r})} \qquad (4.198)$$

Assuming that the right-hand side can be evaluated, this equation expresses the average density as a functional of the field $J$. An important theorem of DFT known as the Hohenberg–Kohn theorem (Hohenberg and Kohn, 1964) states that this nonlinear functional relationship also has an inverse, i.e. to within an additive constant, $J$ is uniquely determined by $\langle \hat{\rho} \rangle_J$.

With this knowledge, it is convenient to introduce a dimensionless free energy functional $F[\langle \hat{\rho} \rangle_J]$, which is a Legendre transform of the Helmholtz free energy $A[J] = -k_B T \ln \mathcal{Z}_C[J]$

$$F[\langle \hat{\rho} \rangle_J] \equiv -\ln \mathcal{Z}_C[J] - \int d\mathbf{r} \ J(\mathbf{r})\langle \hat{\rho}(\mathbf{r}) \rangle_J \qquad (4.199)$$

It is understood that the dependence on $J$ on the right-hand side of this expression is to be replaced by a functional dependence on $\langle \hat{\rho} \rangle_J$ through the application of eqn (4.198). Functional differentiation of eqn (4.199) leads to

$$\frac{\delta F[\langle \hat{\rho} \rangle_J]}{\delta \langle \hat{\rho}(\mathbf{r}) \rangle_J} = -J(\mathbf{r}) \qquad (4.200)$$

We thus conclude that $\delta F/\delta \langle \hat{\rho} \rangle_J = 0$ if no external potential is applied, which corresponds to thermodynamic equilibrium. Indeed, the *equilibrium density profile* $\rho_e(\mathbf{r})$ is obtained by *minimizing* the functional $F$ subject to the constraint that

$$\int d\mathbf{r} \ \langle \hat{\rho}(\mathbf{r}) \rangle_J = n \qquad (4.201)$$

This is accomplished by introducing a *grand potential* $\Omega[\rho]$ according to

$$\Omega[\rho] \equiv F[\rho] - \mu \left[\int d\mathbf{r} \ \rho(\mathbf{r}) - n\right] \qquad (4.202)$$

where $\mu$ plays the role of a Lagrange multiplier. The condition that determines the equilibrium density profile is thus

$$\frac{\delta \Omega[\rho]}{\delta \rho(\mathbf{r})}\bigg|_{\rho=\rho_e} = \frac{\delta F[\rho]}{\delta \rho(\mathbf{r})}\bigg|_{\rho=\rho_e} - \mu = 0 \tag{4.203}$$

The intermediate expression in this equation has a satisfying physical interpretation. The object $\delta F / \delta \rho(\mathbf{r})$ can be viewed as a *local chemical potential* near the point $\mathbf{r}$. Equation (4.203) states that at thermodynamic equilibrium, this local chemical potential is everywhere a constant $\mu$, which we can identify as the bulk chemical potential. Uniformity of chemical potential is of course one of the foundations of fluid phase equilibria.

The above DFT formalism is useful only to the extent that accurate expressions for the "density functional" $F[\rho]$ can be constructed. Unfortunately it is not possible to exactly derive this functional for any non-trivial model of an interacting fluid. Indeed, evaluation of $F$ is a formidable many-body problem in its own right – comparable in difficulty to evaluation of the starting partition function $\mathcal{Z}_C$. Nevertheless, a variety of analytical techniques are available for developing approximate forms for $F$, and these have met with some success in application (Oxtoby, 2002; Löwen, 2002).

### 4.10.2   Polymer solutions: Model A

The formal DFT construct can be combined with the particle-to-field transformation for *any* classical fluid model, atomistic or mesoscopic. We illustrate with the simplest model of a polymer solution – Model A.

Inclusion of an external potential $J(\mathbf{r})$ into the canonical partition function of Model A is accomplished by adding an extra term $U_1$ to the potential energy

$$\beta U_1[\mathbf{r}^{nN}] = \int d\mathbf{r} \; J(\mathbf{r}) \hat{\rho}(\mathbf{r}) \tag{4.204}$$

where the microscopic segment density $\hat{\rho}(\mathbf{r})$ is defined in eqn (4.65). By tracing the effect of this extra term through the particle-to-field transformation, the canonical Model A field theory is modified to

$$\mathcal{Z}_C[J] = \mathcal{Z}_0 \int \mathcal{D}w \; \exp(-H[w, J]) \tag{4.205}$$

where

$$H[w, J] = \frac{1}{2u_0} \int d\mathbf{r} \; [w(\mathbf{r})]^2 - n \ln Q[iw + J] \tag{4.206}$$

Comparison with eqns (4.71) and (4.72) shows that the effect of the external potential $J$ is to shift the argument of the single chain partition function according to $Q[iw] \rightarrow Q[iw + J]$.

The average segment density in the presence of the potential field is given by

$$\langle \hat{\rho}(\mathbf{r}) \rangle_J = -\frac{\delta \ln \mathcal{Z}_C[J]}{\delta J(\mathbf{r})} = \langle \tilde{\rho}(\mathbf{r}; [iw + J]) \rangle_J \qquad (4.207)$$

where $\tilde{\rho}$ is the operator defined in eqn (4.75). The average on the right-hand side of eqn (4.207) denotes a $w$ field average at fixed $J$, i.e.

$$\langle G[w] \rangle_J \equiv \frac{\int \mathcal{D}w \; G[w] \exp(-H[w, J])}{\int \mathcal{D}w \; \exp(-H[w, J])} \qquad (4.208)$$

Equation (4.207) expresses the average density as a nonlinear functional of $J$. This functional dependence is established by solving eqn (3.25) with $w \to iw + J$, evaluating $\tilde{\rho}$ by means of eqn (4.75), and performing the average over the $w$ field fluctuations. These steps cannot be carried out analytically without further approximation, as will be discussed in the next chapter. In order to complete the DFT program, it is further necessary to invert this functional relationship, expressing $J$ in terms of $\langle \hat{\rho} \rangle_J$, and then substitute the result into eqn (4.199) to obtain the density functional $F$. Finally, $F$ is minimized with respect to the density according to eqn (4.203).

It is clear from the structure of the above equations that, unless rather drastic analytical approximations are imposed, the DFT approach must be implemented with computationally demanding numerical methods. Indeed, the iterative numerical inversion of eqn (4.207) to find the potential $J$ for a prescribed density profile is very expensive, especially if solutions are required for many different density profiles. A more direct numerical approach to find the equilibrium density is to cast aside the DFT formalism and evaluate $\rho_e$ according to

$$\rho_e(\mathbf{r}) = \langle \tilde{\rho}(\mathbf{r}; [iw]) \rangle \qquad (4.209)$$

The density operator on the right-hand side of this expression can be evaluated for a particular $w$ field by using the numerical methods of Section 3.6. The ensemble average over the statistically relevant $w$ field variations is then performed by applying numerical techniques that will be described in the remaining two chapters of this book.

## 4.11  Summary

Through a series of examples, the present chapter has illustrated how statistical field theory models can be constructed for a wide variety of classical fluids at equilibrium, both simple and polymeric. Here we recap some of the main results.

- The starting point for building a field theory model of a classical fluid is a "particle model" in which the relevant degrees of freedom in the partition function are atomic or molecular coordinates and the potential energy is expressed as a function of these coordinates. The particle model can be *atomistic* or *mesoscopic*, i.e. coarse-grained to a $\sim 1$ nm scale.

- Hubbard–Stratonovich transformations and related techniques are available for formally and *exactly* converting the partition function of any particle-based model to a statistical field theory. The auxiliary chemical potential "$w$" fields that are introduced in this particle-to-field conversion serve to decouple the interactions among particles, replacing them with independent interactions between each particle and the field variables. Evaluation of the "effective Hamiltonian" $H[w]$ of the resulting field theory thus sidesteps the many-body problem – only the partition function $Q[w]$ of a *single* particle (or polymer) experiencing the auxiliary fields need be computed. It is through this dependence of $H[w]$ on $Q[w]$ that the connection to the subject matter of Chapter 3 is made. Although the particle coordinates have been eliminated in the conversion to a field theory, the remaining $w$ field variables are strongly interacting, so the many-body problem is not actually avoided, but is conveniently reformulated.

- Atomistic models with "hard-core" repulsive interactions generally produce field theories in which fluctuating density *and* chemical potential fields are the relevant degrees of freedom. Mesoscopic models without hard-core interactions can often be converted to simplified field theories in which only fluctuating chemical potential fields appear. Field-theoretic models of charged fluids are characterized by an additional fluctuating electrostatic potential field.

- Field-theoretic models of simple and polymeric fluids are complicated by the fact that the statistical weight $\exp(-H[w])$ is *not positive definite*. This feature is manifest in Models A–K because the effective Hamiltonian $H$ is complex, even for strictly real $w$ fields.

- For each model, *operators* can be defined whose average over the auxiliary field variables yields observables of interest. Such operators, which are functionals of the $w$ fields, can be identified for a wide range of observable quantities including segment and polymer densities, elastic stress, osmotic pressure, and charge density.

- The approach described in the present chapter is flexible and can be used to derive field-theoretic models for a wide variety of polymers and complex fluids. Implicit and explicit solvents, charges, polymer architecture and composition variations, as well as interaction and chain models can be adapted at will, producing field theories that retain the physical content and parameters of the starting particle model. Models can be formulated in a variety of statistical ensembles, including the canonical and grand canonical ensembles.

- Models A–K are intended as simple, generic examples of field-theoretic models for the types of polymeric fluids indicated. In spite of their simplistic assumptions, very little is known about the exact thermodynamic properties of *any* of these models. Even Model A, which has been known for decades, has only recently been studied outside of the mean-field approximation for inhomogeneous situations.

- Models H–K illustrate how various types of *chemical disorder*, quenched and annealed, can be incorporated into the field-based approach. Similar techniques can be applied to construct models for polymer systems exhibiting any well-defined type of chain-length disorder (polydispersity), architectural disorder, or compositional heterogeneity.

- In concentrated polymer systems, boundary layers can arise in which the polymer density varies rapidly near a bounding surface or interface. The use of *effective boundary conditions* can alleviate the need to numerically resolve density and chemical potential fields within such layers, potentially reducing the overall computational effort.

- Density functional theory (DFT) is a framework that can be applied in tandem with the field-theoretic description presented here. While the DFT formalism is analytically intractable for an unapproximated field theory, we shall see that it can be combined with the approximation schemes of Chapter 3 to provide insights into mean-field solutions (Chapter 5). DFT is not recommended for numerical computations with mesoscopic polymer models because it adds an unnecessary computational burden.

# 5

# SELF-CONSISTENT FIELD THEORY

## 5.1  The mean-field approximation

The previous chapter demonstrated how statistical field theories can be constructed from particle-based models of both simple and complex fluids. It is now appropriate to discuss how to analyze such field theories and extract useful information about the structure and thermodynamic properties of polymeric fluids at equilibrium.

The field-theoretic models of Chapter 4 generically express the relevant partition function as a functional integral over one or more chemical potential fields $w(\mathbf{r})$, i.e.

$$\mathcal{Z} = \int \mathcal{D}w \, \exp(-H[w]) \tag{5.1}$$

where $H[w]$ is an effective Hamiltonian that is a non-local functional of the field variables and is generally complex (not strictly real). The form of $H[w]$ depends on the particular interaction and chain model used to construct the field theory and is thus sensitive to polymer architecture, molecular weight, polydispersity, composition, etc. To compute the ensemble average of some observable $G$, one applies the formula

$$\langle G[w] \rangle = \mathcal{Z}^{-1} \int \mathcal{D}w \, G[w] \exp(-H[w]) \tag{5.2}$$

The evaluation of free energies and derivative quantities, through a thermodynamic connection formula to $\mathcal{Z}$, involves the computation of the functional integral given by eqn (5.1). Similarly, by application of eqn (5.2) with $G[w]$ corresponding to an appropriate density operator, the fluid structure is computed as a ratio of two functional integrals.

In practice, for non-trivial models of three-dimensional fluids, *none* of these functional integrals can be evaluated in closed form. There are, however, several options:

1. Generate numerical approximations to the exact field theory.
2. Impose analytical approximations to simplify the field theory, then use *analytical* methods to evaluate quantities of interest.
3. Impose analytical approximations to simplify the theory, then use *numerical* methods to extract information from the simplified theory.

The first strategy, which we shall refer to as the "field-theoretic simulation" (FTS) technique, is the subject of Chapter 6. The second and third approaches are considered in the present chapter.

### 5.1.1  *The mean-field approximation: general considerations*

The most important analytical approximation technique is the *mean-field approximation*, which is also known in the polymer physics literature as *self-consistent field theory (SCFT)* (Edwards, 1965; de Gennes, 1969). This technique is widely used in many physical contexts, perhaps most notably in the theory of phase transformations (Amit, 1984; Parisi, 1988; Goldenfeld, 1992). In the present case, the mean-field approximation amounts to the assumption that a *single* field configuration $w^*(\mathbf{r})$ dominates the functional integrals in eqns (5.1) and (5.2). This field configuration is obtained by demanding that $H[w]$ be stationary with respect to variations in $w(\mathbf{r})$, i.e.

$$\left.\frac{\delta H[w]}{\delta w(\mathbf{r})}\right|_{w=w^*} = 0 \qquad (5.3)$$

Having obtained the "mean-field" potential $w^*(\mathbf{r})$ from this equation, one completes the approximation by imposing

$$\mathcal{Z} \approx \exp(-H[w^*]), \quad \langle G[w]\rangle \approx G[w^*] \qquad (5.4)$$

If $\mathcal{Z}$ represents a canonical partition function, then the Helmholtz free energy follows immediately as $\beta A = -\ln \mathcal{Z} = H[w^*]$.

In the mean-field approximation, all configurations of the potential field are neglected except for the particular configuration $w^*(\mathbf{r})$. For atomic or small molecule fluids, this approximation of neglecting all "field fluctuations" is usually quite poor. This is because the typical coordination number of an atom or small molecule at liquid densities is quite low, $\sim 10$, so large fluctuations occur in the potential experienced by a particle as neighboring particles change positions. Indeed, these strong local field fluctuations are responsible for producing the density correlations that characterize the atomic-scale structure of liquids (Hansen and McDonald, 1986).

The mean-field approximation is also inaccurate in polymer solutions or melts at the *atomic scale* because the coordination number remains small. However, at *mesoscopic scales*, the situation changes qualitatively due to the ability of polymer coils to interpenetrate one another. From a mesoscopic perspective, a useful definition of an effective coordination number in a polymeric fluid is

$$C = \rho_c R_g^3 \qquad (5.5)$$

where $\rho_c = n/V$ is the average number density of chains and $R_g$ denotes the radius of gyration of a polymer. $C$ thus corresponds to the average number of (other) polymer chains that pervade the volume (estimated as $R_g^3$) occupied by one chain of interest. This effective coordination number can be very large for concentrated solutions or melts of polymers of high molecular weight. For example, in a homopolymer melt, $\rho_c = 1/(v_0 N)$, where $v_0$ is the volume occupied by each statistical segment and $N$ is the number of segments per chain. Since polymers are asymptotically ideal in the melt with $R_g = b(N/6)^{1/2}$, it follows that

$C \sim (b^3/v_0)N^{1/2}$. Thus, the effective coordination number grows as the square root of the molecular weight under melt conditions. It follows that at mesoscopic scales, the fluctuations in the environment of each polymer diminish with increasing molecular weight, since the potential field variations are averaged out by contact with increasingly large numbers of surrounding chains. This constitutes the standard argument that the mean-field approximation is accurate for concentrated solutions or melts of high-molecular-weight polymers (de Gennes, 1979).

Implementation of the mean-field approximation requires solving eqn (5.3) to obtain the "mean field" $w^*(\mathbf{r})$. For realistic models, such equations are nonlinear, non-local, and defy closed form analytical solution. Nevertheless, the mean-field equations still represent a great simplification over the difficult functional integrals that comprise the exact field theory. Before we discuss numerical and (approximate) analytical strategies for solving the mean-field equations,[53] it is helpful to address some generic issues related to the analytic structure of the field theory in the complex plane and the existence and significance of multiple solutions of eqn (5.3).

### 5.1.2  *Analytic structure of the field theory*

If the effective Hamiltonian $H[w]$ were strictly real, then the physical solutions of eqn (5.3) would correspond to *local minima* of $H$ and thus minima of the mean-field approximation to the free energy, $\beta A \approx H$. However, the situation is complicated by the complex nature of $H[w]$ that was noted for the models described in Chapter 4. Physically meaningful solutions of eqn (5.3) can lie off the real axis, so that $w^*$ can be purely imaginary or complex. The partition function $\mathcal{Z}$ is real, however, so it follows that $H[w^*]$ must also be real. Similarly, $G[w^*]$ must be real for any real physical observable $G$.

Solutions of eqn (5.3) in the complex plane are commonly referred to as *saddle points* (Bender and Orszag, 1978), by virtue of the analytic structure of $H[w]$ in the vicinity of $w^*$. For this reason, the mean-field approximation is also referred to as the *saddle point approximation* (Amit, 1984; Parisi, 1988). As previously mentioned, another equivalent name for the mean-field approximation in the context of inhomogeneous polymers is *self-consistent field theory (SCFT)* (Edwards, 1965; de Gennes, 1969; de Gennes, 1979; Helfand, 1975; Hong and Noolandi, 1981). This name derives from the notion that the average potential field $w^*(\mathbf{r})$ experienced by a polymer segment at position $\mathbf{r}$ is determined "self-consistently" by eqn (5.3).

For the purpose of implementing the mean-field approximation and understanding when it is valid, it is important to examine the analytic structure of the effective Hamiltonian defining the model. We illustrate with a simple *one-dimensional* "toy" model defined by

---

[53] A different perspective on self-consistent field theory can be found in the excellent reviews by Matsen (2002), Matsen (2005a), and Schmid (1998).

$$\mathcal{Z} = \int_{-\infty}^{\infty} dw \ \exp[-H(w)] \tag{5.6}$$

with effective Hamiltonian

$$H(w) = p\left[iw + \frac{1}{2}w^2\right] \tag{5.7}$$

and where $p > 0$ is a real constant. The path of integration is the real $w$ axis, but due to the explicit factor of $i = \sqrt{-1}$ in the Hamiltonian, $H(w)$ is complex. Nonetheless, the partition function, which can be evaluated exactly to yield $\mathcal{Z} = (2\pi/p)^{1/2}\exp(-p/2)$, is real.

The integrand of eqn (5.6), $\exp[-H(w)]$, is an analytic function of the variable $w$ throughout the complex plane. It follows from Cauchy's theorem of complex integration (Ahlfors, 1979), that the integral along the real axis defining $\mathcal{Z}$ can be deformed onto an arbitrary curve in the complex $w$ plane, $w = w_R + iw_I$, that connects to the real axis for $w_R \to \pm\infty$. One such curve, labelled $\Gamma_1$, is shown in Fig. 5.1. A second curve $\Gamma_2$ is also shown, which connects to the real axis by means of vertical segments at $w_R = \pm X$. The application of Cauchy's theorem thus justifies the alternative expressions

$$\mathcal{Z} = \int_{\Gamma_1} dw \ \exp[-H(w)] = \lim_{X \to +\infty} \int_{\Gamma_2} dw \ \exp[-H(w)] \tag{5.8}$$

Such reexpression of $\mathcal{Z}$ is useful only if the integrals along $\Gamma_1$ or $\Gamma_2$ are easier than the integral along the real axis. The original integral has an *oscillatory* integrand along the integration path $w = w_R$ by virtue of the factor $\exp(-ipw)$. While this is not problematic for the trivial model at hand, oscillatory integrands can lead to considerable numerical difficulties in most realistic cases. One might ask whether it is possible to find a curve $\Gamma$ for which the imaginary part of $H(w)$, the so-called *phase*, $H_I(w_R, w_I)$, is constant along the path of integration. If this were possible, the oscillations would be eliminated. Indeed, such a procedure is the basis of the *steepest descent* (or *saddle point*) technique for the asymptotic analysis of integrals such as eqn (5.8) (Bender and Orszag, 1978).

The steepest descent technique is aided by the knowledge that, at least for large $p$, $p \gg 1$, the integral along a constant phase contour $\Gamma$ is dominated by *local mimima* of the real part of the Hamiltonian, $H_R$. If $H(w)$ is analytic, then it follows from the Cauchy–Riemann equations that these local minima $w^*$ correspond to saddle points where the complex derivative vanishes:

$$\left.\frac{dH(w)}{dw}\right|_{w=w^*} = 0 \tag{5.9}$$

In the case of eqn (5.7), the toy model is seen to possess a single saddle point located on the imaginary axis at $w^* = -i$. It can be shown that at such a second-order saddle point, where the second derivative $H''(w^*)$ is non-vanishing,

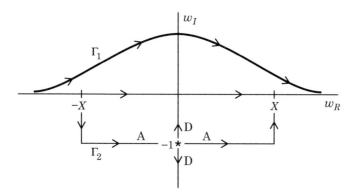

FIG. 5.1. Complex $w$ plane, $w = w_R + iw_I$, indicating paths of integration for the toy model defined by eqns (5.6) and (5.7). The original integration path is along the real axis, $w_I = 0$. Cauchy's theorem permits the integral to be deformed onto an arbitrary contour $\Gamma_1$ that connects to the real axis at $w_R \to \pm\infty$. Similarly, the segmented contour $\Gamma_2$ can be used, provided the limit $X \to +\infty$ is applied. The toy model has a saddle point on the imaginary axis at $w^* = -i$, which is denoted by an asterisk. Constant phase curves passing through the saddle point have $H_I = 0$ and correspond to the lines $w_I = -1$ and $w_R = 0$. Ascent curves are labelled A and descent curves D.

two constant phase curves intersect.[54] The two sections of one of these curves, delineated by the saddle point, correspond to *ascent curves* on which $H_R$ increases away from the saddle point. Similarly, the two sections of the other curve are *descent curves*, along which $H_R$ decreases. For the toy model, $H_I(w_R, w_I) = p\,w_R(1 + w_I)$, so that $H_I = 0$ at the saddle point.[55] The two constant phase curves emanating from $w^* = -i$ thus correspond to $w_R = 0$ and $w_I = -1$. It is easily verified that the two sections of $w_R = 0$ are the descent curves, while $w_I = -1$ generates the two ascent curves. These are shown in Fig. 5.1, where it should be noted that the ascent curves are coincident with the segment of $\Gamma_2$ parallel to the real axis.

The ascent curves are those of primary interest, because they can be connected to the original path of integration by means of the contour $\Gamma_2$ and they produce a finite result for the partition function. The contributions from the contour segments at $w_R = \pm X$ vanish as $\sim \exp(-pX^2/2)$ for $X \to +\infty$, so can be safely neglected. Thus, we can formally reexpress the original integral as an integral over a constant phase path $\Gamma$ parameterized by $w = x - i$ according to

[54]Higher-order saddle points are also possible. For example, at a third-order saddle point, the first non-vanishing derivative is the third, and three constant phase curves intersect.

[55]The observation that $H_I = 0$ at the saddle point is a general result, since $H[w^*]$ must be real for any physical saddle point $w^*$.

$$\mathcal{Z} = \int_{\Gamma} dw \ \exp[-H(w)] = \int_{-\infty}^{\infty} dx \ \exp[-H_R(x, -1)]$$

$$= \int_{-\infty}^{\infty} dx \ \exp[-p(1 + x^2)/2] \tag{5.10}$$

The final integral no longer has an oscillatory integrand, because the phase factor $\exp(iH_I)$ is constant at unity along the integration path. This integral can be analytically evaluated by means of eqn (B.1) to yield the exact result $\mathcal{Z} = (2\pi/p)^{1/2} \exp(-p/2)$. An important observation is that for large $p$, the integral along the constant phase path is dominated by the behavior of $H_R$ very near its local minimum at the saddle point $x = 0$. This provides the basis for developing asymptotic expansions for $p \to +\infty$ in situations where the Laplace-type integral cannot be analytically evaluated (Bender and Orszag, 1978).

There are several important points from this discussion of the one-dimensional "toy" model that can be immediately generalized to a high-dimensional[56] statistical field theory such as eqn (5.1):

- The original path of integration for each degree of freedom of the field $w(\mathbf{r})$ in the defining partition function is along the real axis. Nevertheless, for analytic integrands $\exp(-H[w])$, it is useful, at least conceptually, to deform the integration path onto a constant phase "ascent" contour (surface) that passes through one or more saddle points $w^*(\mathbf{r})$ in the multi-dimensional complex plane. The phase factor $\exp(iH_I[w])$ is constant at unity along such a contour, so that oscillations of the integrand are eliminated.

- By inspection of the models of Chapter 4, it is easily verified that the statistical weights $\exp(-H[w])$ and $\exp(-H_G[w])$ are analytic functionals of $w$. Thus, the deformation described above is conceptually possible for all of the models. It should be noted, however, that only in the grand canonical ensemble is the Hamiltonian itself analytic. In the canonical ensemble, $H[w]$ has branch points that limit the region of analyticity.

- A functional integral deformed onto a constant phase contour has important contributions from saddle point field configurations, $w^*$, which represent mean-field solutions. On the constant phase (ascent) manifold, $H_R[w]$ has local minima at the saddle points. The extent to which one of these saddle point field configurations dominates the integral depends on the value of a "Ginzburg parameter," analogous to $p$ in the toy model. Indeed, it is straightforward to show that the coordination number $C = \rho_c R_g^3$ is the relevant parameter for many of the models described in Chapter 4. *It follows that the mean-field approximation for these models becomes asymptotically exact for $C \to +\infty$.*

---

[56]Strictly speaking, a field theory model is infinite dimensional. However, for numerical purposes the fields are given a discrete representation and the functional integrals are approximated by multiple integrals of finite, but large, dimension.

- It is useful to determine the qualitative location and orientation of a saddle point in the complex plane prior to attempting its computation. In Model A, for example, the requirement that $H[w^*]$ or $H_G[w^*]$ be real implies that any saddle point $w^*(\mathbf{r})$ must be *purely imaginary*. For Models B–E, $w_-^*$ and $w_+^*$ are purely real and purely imaginary, respectively. It is often convenient to compute a purely imaginary saddle point by a relaxation scheme along the imaginary axis, which is a search direction that is orthogonal to the physical path of integration. With such a scheme, it is important to recognize that this is more than likely a *descent* direction for $H_R[w]$, so *one should seek a local maximum, as opposed to a local minimum!*

### 5.1.3  *Multiple solutions*

Equation (5.3) has multiple solutions for most fluid models of interest, corresponding to more than one saddle point. A broad classification scheme identifies such saddle points as being either *homogeneous*, where $w^*(\mathbf{r})$ is independent of position $\mathbf{r}$, or *inhomogeneous*, in which case $w^*(\mathbf{r})$ has explicit position dependence. Homogeneous saddle points can often be determined analytically, while inhomogeneous saddle points normally require numerical methods for accurate evaluation.

In general, one can associate a *pure state* saddle point with each stable or metastable phase of a fluid.[57] For example, models of AB diblock copolymer melts have "pure state" saddle points that can be associated with the disordered phase (D), and the lamellar (L), cylindrical (C), gyroid (G), and spherical (S) mesophases (Matsen and Schick, 1994a). The L, C, G, and S saddle points are inhomogeneous and spatially periodic; the stable S saddle point turns out to have body-centered cubic (bcc) symmetry. Other pure state saddle points are known for Model E, e.g. saddle points with double diamond (DD) and hexagonally perforated layer (HPL) symmetry have been computed (Matsen and Bates, 1996a). Within mean-field theory, these are metastable throughout the parameter space of $\chi N$ and $f$.

Inhomogeneous pure state saddle points need not be spatially periodic. In such cases, the inhomogeneities usually result from boundary conditions applied to the model. For example, Model A, describing a solution of homopolymers in a good solvent, has a unique, non-uniform, pure state saddle point $w^*(z)$ when solved in a confined slit geometry[58] with Dirichlet boundary conditions. The segment density profile corresponding to this potential in the mean-field approx-

---

[57]We distinguish stability/metastability of a fluid phase and a saddle point. A *phase* is deemed "locally stable" if, subject to small displacements, its free energy surface is convex. Otherwise, the phase is said to be unstable. A locally stable phase is deemed "stable" if its free energy is lower than the free energy of all competing phases; otherwise it is "metastable". A *saddle point* is classified similarly, except that the energy functional $H[w]$ is used in place of the free energy. Local stability implies that $H_R$ is convex in the vicinity of the saddle point, provided that field variations are restricted to a constant phase manifold coinciding with a deformed physical integration path.

[58]$z$ is the coordinate normal to the faces of the slit.

imation is given by $\tilde{\rho}(z; [iw^*])$, where $\tilde{\rho}$ is the density operator of eqn (4.75). Numerical examples of such inhomogeneous profiles are shown in Fig. 6.2.

Pure state saddle points are unique, apart from translations and rotations that do not change the energy $H[w^*]$. Even within the class of spatially periodic structures, it is difficult to compute and assess the stability of all pure state saddle points for a specified complex fluid model. In practice we are primarily interested in the saddle points corresponding to phases that have some region of stability in the parameter space of the model. Fortunately the most stable saddle points (i.e. those with lowest $H[w^*]$ values) are usually also those with the largest basin of attraction in the energy landscape, so they can be identified by large-cell simulations, relaxing the $w$ fields from random initial configurations (see Section 5.3.4). More generally, the problem of finding *the* lowest-energy pure state solution of eqn (5.3) for an arbitrary fluid model is an unsolved problem in global optimization (Nocedal and Wright, 1999).

Besides the "pure state" saddle points, one can find inhomogeneous solutions of eqn (5.3) that correspond to *defect states*. These are so named because they reflect topological defects in an otherwise perfect periodic structure. For example, thin films of the cylindrical mesophase of diblock copolymers can possess defects known as *dislocations* (Hammond *et al.*, 2003), similar to dislocations in a two-dimensional hexagonal crystal (Chaikin and Lubensky, 1995; Nelson, 2002). When the cylinders are aligned normal to the plane of the film, such an in-plane dislocation would reflect a neighboring "disclination pair" of cylindrical copolymer aggregates, one with 5 nearest neighbors and one with 7. As shown in Fig. 5.2, besides these disclination pairs, all other cylinders have 6 nearest neighbors. Saddle points corresponding to a defect state such as the dislocation just described are usually metastable, because the pure state saddle point has a lower energy. However, for imposed boundary conditions that are incommensurate with a pure state solution of eqn 5.3, such as the case of crystallization on the surface of a sphere (Nelson, 1983), "defect state" saddle points can become stable. There is also theoretical and experimental support for *glass formation* in systems that possess an exponentially large (in the system size) number of metastable non-periodic defect states (Monasson, 1995; Zhang and Wang, 2005). Upon cooling, such a system can become trapped in one of these metastable states resulting in a type of glass transition.[59]

A third type of saddle point is a *mixed state* that is produced when two or more pure states coexist. Models C and D of a binary homopolymer blend, for example, exhibit a coexistence of two liquid phases, one rich in each of the two homopolymer species, at sufficiently large values of the segmental interaction parameter $\chi_{AB}$. Equation (5.3) thus has mixed state solutions for the canonical ensemble versions of these models, which reflect the coexistence of two

---

[59]In the context of an AB diblock copolymer melt, such a glass transition creates a compositionally disordered, "pattern glass" state. The transition temperature can occur above the glass transition temperatures ($T_g$) of the two blocks, so this not a glass transition in the conventional sense.

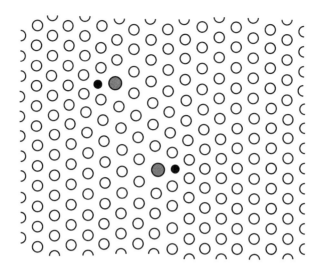

FIG. 5.2. Schematic of a two-dimensional hexagonal lattice of cylinder-forming
block copolymers that contains two dislocations. Each dislocation is itself a
"disclination pair" consisting of a cylindrical micelle with 5 nearest neighbors
(black) adjacent to a micelle with 7 near neighbors (grey). All other cylinders
have 6 neighbors (white). The strain energy associated with the defect leads
to a redistribution of block copolymers between the two micelles involved in
the dislocation. This is known from both experiment and numerical SCFT
calculations to produce a $\sim 20\%$ larger cylinder at the 7-fold disclination and
a $\sim 20\%$ smaller cylinder at the 5-fold disclination (Hammond *et al.*, 2003).

homogeneous phases, differing in composition and separated by an interface. The
geometrical configuration of the interface for a stable "mixed state" saddle point
solution will be dictated by minimum area considerations under the constraints
imposed by species conservation and the applied boundary conditions. Similar to
pure state solutions, mixed state saddle points can have degeneracies associated
with, e.g., uniform translations or rotations. Metastable, *defective*, mixed state
saddle point solutions can also be found where the interfaces separating the pure
state phases are not configured in the ideal topological arrangement.

### 5.1.4 *Homogeneous saddle points*

Homogeneous saddle points can often be located by analytical methods. For
example, eqn (5.3) for the canonical Model A field theory reduces to

$$\frac{1}{u_0}w^*(\mathbf{r}) + i\tilde{\rho}(\mathbf{r};[iw^*]) = 0 \tag{5.11}$$

where the definition of the density operator $\tilde{\rho}(\mathbf{r};[iw])$ in eqn (4.75) has been
applied in constructing the functional derivative. In an unbounded system, or in

a cubic cell subject to periodic boundary conditions, the only physical solution of this equation is homogeneous, i.e. $w^*(\mathbf{r}) = w^*$. It follows from eqns (3.25) and (3.22) that $q(\mathbf{r}, s; [iw^*]) = \exp(-iw^*s)$ and $Q[iw^*] = \exp(-iw^*N)$, and from eqn (4.75) that $\tilde{\rho}(\mathbf{r}; [iw^*]) = nN/V$, which coincides with the average segment density $\rho_0$. Equation (5.11) thus reduces to

$$w^* = -iu_0\rho_0 \tag{5.12}$$

Since both the excluded volume parameter $u_0$ and the average segment density $\rho_0$ are real and positive, the saddle point value of the potential field, $w^*$, is located on the negative *imaginary* axis of the complex $w$ plane. This is similar to the situation with the toy model discussed in Section 5.1.2.

The effective Hamiltonian of the canonical Model A at the homogeneous saddle point is evaluated from eqn (4.72) as $H[w^*] = (1/2)u_0\rho_0^2 V$. It follows that the mean-field approximation to the Helmholtz free energy of the polymer solution is given by

$$A(n, V, T) = A_0 + \frac{k_B T}{2}u_0\rho_0^2 V \tag{5.13}$$

where $A_0 = -k_B T \ln \mathcal{Z}_0$ is the free energy of an ideal gas of non-interacting polymers. In the mean-field approximation, the excess portion of the free energy arises solely from the average interaction energy among the polymer segments, with no account taken for their spatial correlations.

It is instructive to repeat this analysis for the *grand canonical* Model A. The saddle point equation amounts to

$$\left.\frac{\delta H_G[w]}{\delta w(\mathbf{r})}\right|_{w=w^*} = \frac{1}{u_0}w^*(\mathbf{r}) + i\tilde{\rho}_G(\mathbf{r}; [iw^*]) = 0 \tag{5.14}$$

where $\tilde{\rho}_G$ is the segment density operator given in eqn (4.76). For uniform $w^*(\mathbf{r}) = w^*$, one finds that

$$\tilde{\rho}_G(\mathbf{r}; [iw^*]) = zN \exp(-iw^*N) = \frac{\langle n \rangle N}{V} \equiv \rho_0 \tag{5.15}$$

where $\langle n \rangle = zVQ[iw^*] = zV \exp(-iw^*N)$ is the average number of polymers in the volume $V$ and $\rho_0$ is the average segment density. It follows that the mean-field $w^*$ satisfies the transcendental equation

$$iw^* \exp(iw^*N) = u_0 zN \tag{5.16}$$

Because the right-hand side is real and positive, this equation has a unique solution $w^*$ that lies on the negative *imaginary* axis for any polymer activity $z$.

The connection to thermodynamics for the grand canonical Model A relates the *osmotic pressure* $\Pi$ to $\mathcal{Z}_G$ via the formula

$$\beta\Pi = \frac{1}{V} \ln \mathcal{Z}_G(z, V, T)$$

$$\approx -\frac{1}{V}H_G[w^*] = \frac{\rho_0}{N} + \frac{1}{2}u_0\rho_0^2 \tag{5.17}$$

where the mean-field approximation has been invoked in the second line. Thus, we see that the mean-field expression for the osmotic pressure consists of the sum of an ideal gas term proportional to the average density of polymers, $\rho_0/N$, and an interaction term that coincides with the corresponding term in the Helmholtz free energy per unit volume. These are of course well-known results (de Gennes, 1979). Similar expressions can be derived for the homogeneous mean-field solutions of the other models presented in Chapter 4.[60]

## 5.2   Further approximations

Of greater interest are the *inhomogeneous* solutions of the saddle point equation (5.3). Numerical analysis is generally required to develop accurate descriptions of such mean-field solutions. The discussion of numerical methods is deferred to Section 5.3; here we consider further analytical approximations that render the inhomogeneous saddle point equations tractable. These approximations are based on the schemes presented in Section 3.4 for evaluating the statistical properties of a single chain in an external potential.

### 5.2.1   *Weak inhomogenities – the RPA*

An important type of approximation that can be applied in tandem with the mean-field approximation is a weak inhomogeneity expansion. In situations where a solution of eqn (5.3) is nearly uniform, we can write by analogy with eqn (3.113)

$$w^*(\mathbf{r}) = w_0 + \omega^*(\mathbf{r}) \qquad (5.18)$$

where $w_0 \equiv (1/V) \int d\mathbf{r}\, w^*(\mathbf{r})$ is the volume-averaged mean-field potential. Assuming that the deviation $\omega^*(\mathbf{r})$ is everywhere small compared with $w_0$, the procedure of Section 3.4.1 can be followed to develop a weak inhomogeneity expansion. This expansion is commonly referred to in the polymer literature as the *random phase approximation, or RPA* (de Gennes, 1969; de Gennes, 1979). We illustrate it with Model A in the canonical ensemble.

The saddle point equation for the canonical Model A is given in eqn (5.11). Substitution of eqn (5.18) and application of eqns (3.134) and (4.75) lead to the following expansion:

$$u_0^{-1}\omega^*(\mathbf{r}) + \rho_0 N \int d\mathbf{r}'\, g_D(|\mathbf{r}-\mathbf{r}'|)\,\omega^*(\mathbf{r}') + O((\omega^*)^2) = 0 \qquad (5.19)$$

where eqn (5.12) has been used to cancel the leading homogeneous terms and $g_D(r)$ is the Debye function of eqn (3.133). The only solution of eqn (5.19) is $\omega^*(\mathbf{r}) = 0$, which is consistent with our earlier claim that the only saddle point of the canonical Model A in bulk, or in a cell with periodic boundary conditions, is the homogeneous solution $w^*(\mathbf{r}) = w_0$.

---

[60]The homogeneous mean-field solution of Model G for a nematic phase is non-trivial and requires numerical methods to evaluate (Wang and Warner, 1986).

The RPA expansion can also be used to provide insights into the form of the *density functional* $F[\rho]$ that is central to the DFT formalism of Section 4.10. In the mean-field approximation, eqn (4.207) for the canonical Model A reduces to

$$\rho(\mathbf{r}) \approx \tilde{\rho}(\mathbf{r}; [iw^* + J]) \tag{5.20}$$

where we have used the shorthand of $\rho(\mathbf{r})$ in place of $\langle \hat{\rho}(\mathbf{r}) \rangle_J$. This equation determines the average segment density $\rho(\mathbf{r})$ produced by an arbitrary external potential $J(\mathbf{r})$. Furthermore in the mean-field approximation, the partition function of eqn (4.205) reduces to $\mathcal{Z}_C[J] \approx \mathcal{Z}_0 \exp(-H[w^*, J])$. In both of these expressions, the saddle point $w^*(\mathbf{r})$ is determined by

$$\left. \frac{\delta H[w, J]}{\delta w(\mathbf{r})} \right|_{w=w^*} = u_0^{-1} w^*(\mathbf{r}) + i\tilde{\rho}(\mathbf{r}; [iw^* + J]) = 0 \tag{5.21}$$

By choosing $J$ to be weak in amplitude and to have vanishing volume average, i.e. $(1/V) \int d\mathbf{r} \, J(\mathbf{r}) = 0$, the right-hand side of eqn (5.21) can be developed in an RPA expansion analogous to eqn (5.19). At leading order in $J$,

$$\int d\mathbf{r}' \left[ u_0^{-1} \delta(\mathbf{r} - \mathbf{r}') + \rho_0 N g_D(|\mathbf{r} - \mathbf{r}'|) \right] w^*(\mathbf{r}')$$

$$= i\rho_0 N \int d\mathbf{r}' \, g_D(|\mathbf{r} - \mathbf{r}'|) J(\mathbf{r}') + O(J^2) \tag{5.22}$$

Fourier transformation of this result produces the following formula relating $w^*$ to $J$:

$$\hat{w}^*(\mathbf{k}) = \frac{iu_0 \rho_0 N \hat{g}_D(x)}{1 + u_0 \rho_0 N \hat{g}_D(x)} \hat{J}(\mathbf{k}) + O(J^2) \tag{5.23}$$

where $x = k^2 R_g^2$ is a squared wavenumber made dimensionless with the square of the unperturbed radius of gyration, $R_g^2 = Nb^2/6$.

The next step is to use eqns (5.23) and (3.131) to expand the functional $H[w^*, J]$ given in eqn (4.206). This leads to

$$H[w^*, J] = H_0 - \frac{1}{2V} \sum_{\mathbf{k}} \frac{\rho_0 N \hat{g}_D(x)}{1 + u_0 \rho_0 N \hat{g}_D(x)} \hat{J}(\mathbf{k}) \hat{J}(-\mathbf{k}) + O(J^3) \tag{5.24}$$

where $H_0 \equiv (1/2u_0) V w_0^2 + w_0 Nn$ is the homogeneous contribution to the Hamiltonian. The final step necessary to construct the free energy functional $F[\rho]$ is to Legendre transform from $J$ to $\rho$ by means of eqn (4.199), i.e.

$$F[\rho] = -\ln \mathcal{Z}_0 + H[w^*, J] - \int d\mathbf{r} J(\mathbf{r}) \rho(\mathbf{r}) \tag{5.25}$$

This transform requires the expansion of eqn (5.20) to establish the relationship between $J$ and $\rho$. Application of eqn (3.134) leads to

$$\widehat{\Delta\rho}(\mathbf{k}) = -\frac{\rho_0 N \hat{g}_D(x)}{1 + u_0 \rho_0 N \hat{g}_D(x)} \hat{J}(\mathbf{k}) + O(J^2) \tag{5.26}$$

where $\Delta\rho(\mathbf{r}) \equiv \rho(\mathbf{r}) - \rho_0$ is the inhomogeneous part of the monomer density field. Combination of eqns (5.24)–(5.26) produces the desired free energy functional

$$F[\rho] = F_0 + \frac{1}{2V} \sum_{\mathbf{k}} \left( \frac{1}{\rho_0 N \hat{g}_D(x)} + u_0 \right) \widehat{\Delta\rho}(\mathbf{k}) \widehat{\Delta\rho}(-\mathbf{k}) + O(\Delta\rho^3) \tag{5.27}$$

where $F_0$ is the mean-field free energy of the homogeneous fluid.

Equation (5.27) is an expression for the free energy (in units of $k_B T$) of a weakly inhomogeneous polymer solution corresponding to Model A. It is valid to the extent that the mean-field approximation applies and the density inhomogeneities are small. As will be discussed in Chapter 6, the mean-field approximation holds for Model A at sufficiently high concentrations such that

$$C \gg B \equiv \frac{u_0 N^2}{R_g^3} \tag{5.28}$$

where $C = nR_g^3/V$ is the dimensionless chain concentration introduced in eqn (5.5) and $B$ is a dimensionless excluded volume parameter.

One useful application of eqn (5.27) is to assess the *stability* of the homogeneous phase of a polymer solution against small-amplitude density perturbations. Stability can be determined by examining the quadratic coefficient

$$\hat{\Gamma}_2(k) = \frac{1}{\rho_0 N \hat{g}_D(k^2 R_g^2)} + u_0 \tag{5.29}$$

as a function of *wavenumber* $k = |\mathbf{k}|$. Since $\hat{g}_D(x)$ is a monotonically decaying function of $x$, the minimum value of $\hat{\Gamma}_2(k)$ coincides with $k = 0$. The stability limit of the homogeneous phase, or *spinodal*, thus corresponds to

$$\hat{\Gamma}_2(0) = \frac{1}{\rho_0 N} + u_0 = 0 \tag{5.30}$$

It follows that in the mean-field approximation, Model A shows a long-wavelength ($k = 0$) instability when $u_0$ achieves a negative value of $-1/(\rho_0 N)$. This instability produces a macroscopic phase separation into two liquid phases: one rich in polymer; the other solvent-rich. The mean-field spinodal is properly located by eqn (5.30) in spite of the fact that Model A is ill defined for negative values of the excluded volume parameter $u_0$, which correspond to poor solvent conditions. To study the emergence and properties of the new phases that arise in crossing the spinodal boundary, however, it is necessary to switch to Model B.

The RPA expansion can be used to investigate the stability of the homogeneous phase for many of the other models described in Chapter 4. In general, one obtains an expression analogous to eqn (5.27), i.e.

$$F[\rho] = F_0 + \frac{1}{2V} \sum_{\mathbf{k}} \hat{\Gamma}_2(k) \widehat{\Delta\rho}(\mathbf{k}) \widehat{\Delta\rho}(-\mathbf{k}) + O(\Delta\rho^3) \qquad (5.31)$$

but where the form of $\hat{\Gamma}_2(k)$ depends on the model. For Model B, one obtains

$$\hat{\Gamma}_2(k) = v_0 \left( \frac{1}{\phi_{P0} N \hat{g}_D(k^2 R_g^2)} + \frac{1}{\phi_{S0}} - 2\chi_{PS} \right) \qquad (5.32)$$

where $\phi_{P0} \equiv n_P N v_0 / V$ is the average polymer volume fraction, $\phi_{S0} = n_S v_0 / V = 1 - \phi_{P0}$ is the average solvent volume fraction, and $v_0$ is the volume per polymer segment and solvent molecule. In Model B, $\Delta\rho(\mathbf{r})$ can be interpreted as either $\Delta\rho_P(\mathbf{r})$ or $\Delta\rho_S(\mathbf{r})$, because it follows from incompressibility that $\Delta\rho_P(\mathbf{r}) = -\Delta\rho_S(\mathbf{r})$. For small polymer volume fractions, $\phi_{P0} \ll 1$, eqn (5.32) leads to a stability threshold that coincides with eqn (5.30), provided we make the correspondence $u_0 = v_0(1 - 2\chi_{PS})$. The spinodal boundary to liquid–liquid phase separation obtained by setting $\hat{\Gamma}_2(0) = 0$ for Model B, however, differs from the Model A spinodal at higher polymer concentrations. Indeed, the Model B expression coincides with the spinodal of the familiar Flory–Huggins lattice theory of polymer solutions (de Gennes, 1979).

A final example corresponds to an incompressible AB diblock copolymer melt. In a classic paper (Leibler, 1980), Leibler worked out the RPA expansion of Model E to fourth order in the density inhomogeneity, $\Delta\rho \equiv \Delta\rho_A = -\Delta\rho_B$. For equal statistical segment lengths, $b \equiv b_A = b_B$, the quadratic coefficient in the expansion has the form

$$\hat{\Gamma}_2(k) = \frac{v_0}{N} \left[ \gamma(k^2 R_g^2, f) - 2\chi_{AB} N \right] \qquad (5.33)$$

where $R_g^2 = N b^2 / 6$ is the unperturbed radius of gyration of a copolymer and the function $\gamma(x, f)$ is defined by

$$\gamma(x, f) = \frac{\hat{g}(1, x)}{\hat{g}(f, x)\hat{g}(1 - f, x) - (1/4)[\hat{g}(1, x) - \hat{g}(f, x) - \hat{g}(1 - f, x)]^2} \qquad (5.34)$$

In this expression, $\hat{g}(f, x)$ is the modified Debye function

$$\hat{g}(f, x) \equiv \frac{2}{x^2}[fx + \exp(-fx) - 1] \qquad (5.35)$$

At fixed block copolymer composition $f$ (volume fraction of the A block), the function $\gamma(x, f)$ has an $f$-dependent minimum at a *non-zero* value of $x$, $x_m(f)$, given by

$$\left. \frac{\partial \gamma(x, f)}{\partial x} \right|_{x=x_m} = 0 \qquad (5.36)$$

In the case of a symmetric diblock copolymer with $f = 1/2$, $x_m(1/2) = 3.785$. Thus, the most unstable density mode $\widehat{\Delta\rho_A}(\mathbf{k})$ in a homogeneous block copolymer melt has a non-zero wavenumber $k = [x_m(f)]^{1/2}/R_g$, corresponding to a wavelength of $\lambda = 2\pi R_g/[x_m(f)]^{1/2}$. This finite length scale instability signals the onset of a periodic mesophase with a wavelength close to $\lambda$.

The spinodal, or stability limit, of the homogeneous phase of a diblock copolymer melt towards microphase separation is obtained from the relation $\gamma(x_m, f) - 2\chi_{AB}N = 0$. This produces a curve in the plane of $\chi_{AB}N$ versus $f$ shown in Fig. 5.3 that delineates regions where, in the mean-field approximation, the homogeneous (disordered) diblock melt is either stable or unstable. At $f = 1/2$, we observe the famous result of Leibler that the stability limit of a symmetric diblock copolymer melt is $\chi_{AB}N = 10.495$. In mean-field theory, the order–disorder transition (ODT) between the disordered phase and the various ordered mesophases proves to be a first-order phase transition at all compositions other than $f = 1/2$, so higher terms in the RPA expansion are needed to precisely distinguish the ODT and spinodal curves, as well as to determine the symmetry of the ordered mesophases that emerge upon passing through the ODT (Leibler, 1980). In practice, such higher-order RPA analysis becomes very tedious and is limited to weakly ordered mesophases – the so-called *weak segregation limit*. Computational techniques have advanced to the stage where a direct numerical attack on the full mean-field equations, the subject of Section 5.3, is generally advised for the investigation of mesophase formation in block copolymer melts and solutions.

It is important to understand the physical origins of the distinct linear response properties of polymer solutions and block copolymer melts. A polymer solution resists inhomogeneities because there is a conformational entropy penalty associated with organizing chains to support a non-uniform segment density profile. It follows from eqn (5.29) that this free energy penalty rises monotonically with the wavenumber $k$ of the inhomogeneity and grows asymptotically as $k^2$ for $k \gg R_g^{-1}$. The same physics applies for block copolymer melts as can be deduced from the behavior of eqn (5.33) for large $k$. However, a block copolymer melt cannot sustain arbitrarily long-wavelength (small $k$) composition fluctuations, because this would entail stretching individual copolymer chains to separate A segments from B segments over large distances. This stretching results in a conformational entropy penalty that grows according to eqn (5.33) as $k^{-2}$ for $k \to 0$. Thus, a collective manifestation of the chemical bond linking the two blocks of each copolymer is that a preferred wavenumber, $k_m = \sqrt{x_m}/R_g$, exists for composition fluctuations in a copolymer melt. This preferred wavenumber reflects a conformational entropy compromise that is sensitive to copolymer molecular weight and average composition, and more generally, to block copolymer architecture. Although the length scale set by $k_m$ is relevant to the lattice constant of the periodic mesophases formed right at the ODT, upon entering the ordered region of the phase diagram, the lattice constant evolves in a way that cannot be predicted by RPA analysis (Almdal *et al.*, 1990; Matsen and Bates, 1996*b*).

The quadratic coefficient $\hat{\Gamma}_2(k)$ in the RPA expansion is reciprocally related to the *structure factor* $S(k)$ of the homogeneous fluid phase defined by

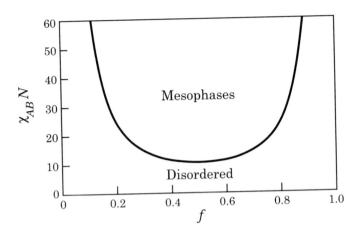

FIG. 5.3. Spinodal curve for an incompressible AB diblock copolymer melt described by Model E. The region denoted "Disordered" indicates values of block segregation strength $\chi_{AB}N$ and composition $f$ for which the homogeneous, compositionally disordered phase is stable to small-amplitude composition perturbations. The region denoted "Mesophases" corresponds to $\chi_{AB}N$ and $f$ such that the disordered phase is unstable to small-amplitude, finite-wavelength composition perturbations, signaling the onset of microphase separation. In the mean-field approximation, the first-order ODT curve is everywhere close to the spinodal shown, and the two curves merge at $f = 1/2$, $\chi_{AB}N = 10.495$, where the transition is predicted to be second order.

$$S(k) = V^{-1} \int d\mathbf{r} \int d\mathbf{r}' \, e^{-i\mathbf{k}\cdot(\mathbf{r}-\mathbf{r}')} \langle [\hat{\rho}(\mathbf{r}) - \rho_0][\hat{\rho}(\mathbf{r}') - \rho_0]\rangle_0 \qquad (5.37)$$

where the subscript 0 on the average denotes a field-free, $J = 0$, ensemble average over chain conformations in the *interacting* system.[61] The structure factor is relevant to radiation scattering experiments carried out with neutrons, x-rays, or light (Hansen and McDonald, 1986). The formal connection between $\Gamma_2(k)$ and $S(k)$ is established through the application of linear response theory, which for Model A implies that

$$\langle \hat{\rho}(\mathbf{r}) \rangle_J = \rho_0 - \int d\mathbf{r}' \, \langle \hat{\rho}(\mathbf{r})\hat{\rho}(\mathbf{r}')\rangle_0 J(\mathbf{r}') + O(J^2) \qquad (5.38)$$

This formula, which is derived by a straightforward perturbation expansion in $J$, assuming that $\int d\mathbf{r} \, J(\mathbf{r}) = 0$, shows that the static linear response function is proportional to the density–density correlation function in the absence of the field, i.e.

[61] Note that the subscript 0 has a different interpretation here than in Chapter 2.

$$\frac{\delta \langle \hat{\rho}(\mathbf{r}) \rangle_J}{\delta J(\mathbf{r}')} \bigg|_{J=0} = -\langle \hat{\rho}(\mathbf{r}) \hat{\rho}(\mathbf{r}') \rangle_0 \qquad (5.39)$$

If we invoke eqn (4.200) from our discussion of density functional theory, it follows that

$$\frac{\delta^2 F[\langle \hat{\rho} \rangle_J]}{\delta \langle \hat{\rho}(\mathbf{r}) \rangle_J \delta \langle \hat{\rho}(\mathbf{r}') \rangle_J} = -\frac{\delta J(\mathbf{r})}{\delta \langle \hat{\rho}(\mathbf{r}') \rangle_J} \qquad (5.40)$$

Combining eqns (5.39) and (5.40) leads to

$$\frac{\delta^2 F[\rho]}{\delta \rho(\mathbf{r}) \delta \rho(\mathbf{r}')} \bigg|_{J=0} = [\langle \hat{\rho}(\mathbf{r}) \hat{\rho}(\mathbf{r}') \rangle_0]^{-1} \qquad (5.41)$$

where we have restored the simplified notation $\rho \equiv \langle \hat{\rho} \rangle_J$ of the present section, and the superscript $-1$ on the right-hand side denotes a functional inverse in the sense of eqn (C.29). Finally, in Fourier space,

$$\frac{\partial^2 F[\rho]}{\partial \widehat{\Delta \rho}(\mathbf{k}) \partial \widehat{\Delta \rho}(-\mathbf{k})} \bigg|_{\Delta \rho = 0} = \frac{\hat{\Gamma}_2(k)}{V} = \frac{1}{V S(k)} \qquad (5.42)$$

which establishes the desired connection $S(k) = 1/\hat{\Gamma}_2(k)$. This is a general result for fluctuations in the disordered phase, although the RPA formulas (5.29), (5.32), and (5.33) rely on the mean-field approximation. These expressions have all been well tested in scattering experiments (des Cloizeaux and Jannink, 1990; Bates and Hartney, 1985).

As a final note, we point out an *inconsistency* that arises in the mean-field approximation. Direct application of eqn (4.78) with the mean-field expression $\langle w(\mathbf{r}) w(\mathbf{r}') \rangle \approx w^*(\mathbf{r}) w^*(\mathbf{r}')$ leads to $S(k) = u_0^{-1}$ for the disordered phase of Model A. This sensible result that the disordered phase is structureless when $w$ field fluctuations are neglected is consistent with a strict mean-field approximation. In contrast, the formula $S(k) = 1/\hat{\Gamma}_2(k)$ derived above, when combined with eqn (5.29), makes a very different statement about density fluctuations in the disordered phase of Model A. Such inconsistencies are well-known in mean-field approximations to interacting field theories (Amit, 1984; Goldenfeld, 1992). In an exact calculation the two approaches would of course have to provide the same result. However, use of the RPA expression for $\hat{\Gamma}_2(k)$ in the exact fluctuation formula $S(k) = 1/\hat{\Gamma}_2(k)$ actually provides a result that goes beyond strict mean-field theory and incorporates $w$ field fluctuations at a Gaussian level. This point will be discussed further in Chapter 6.

### 5.2.2 *Slow gradients*

Another useful approximation scheme that can be combined with the mean-field approximation is the slow gradient expansion of Section 3.4.2. Here the basic assumption is that the self-consistent field $w^*(\mathbf{r})$ and the associated segment density $\langle \hat{\rho}(\mathbf{r}) \rangle$ are both slowly varying on length scales comparable to the polymer radius of gyration $R_g$. Again, we illustrate this with Model A and focus on the construction of a density functional $F[\rho]$. As shown by Tang and Freed (1991), the *grand canonical ensemble* proves most convenient for developing such a gradient expansion.

The starting point is the analogous grand canonical version of the mean-field expression (5.20), i.e.

$$\rho(\mathbf{r}) = \tilde{\rho}_G(\mathbf{r}; [\mu^*]) \tag{5.43}$$

where we have defined $\mu^*(\mathbf{r}) \equiv iw^*(\mathbf{r}) + J(\mathbf{r})$ and again use the shorthand $\rho(\mathbf{r})$ in place of $\langle \hat{\rho}(\mathbf{r}) \rangle_J$. Combining this result with eqns (3.157) and (4.76) leads to[62]

$$\rho(\mathbf{r}) = zNe^{-N\mu^*(\mathbf{r})} \left\{ 1 + R_g^2 \left( \frac{N^2}{6} |\nabla \mu^*|^2 - \frac{N}{3} \nabla^2 \mu^* \right) + \ldots \right\} \tag{5.44}$$

where the neglected terms are fourth order and higher in gradients of $\mu^*$. Equation (5.44) expresses $\rho$ as a gradient expansion in $\mu^*$. This functional relation can be inverted to express $\mu^*$ in gradients of $\rho$, i.e.

$$
\begin{aligned}
\mu^*(\mathbf{r}) = {} & -\frac{1}{N} \ln[\rho(\mathbf{r})/zN] \\
& -\frac{R_g^2}{6N[\rho(\mathbf{r})]^2} |\nabla \rho|^2 + \frac{R_g^2}{3N\rho(\mathbf{r})} \nabla^2 \rho + \ldots
\end{aligned} \tag{5.45}
$$

The mean-field equations for Model A in the grand canonical ensemble imply that $iw^*(\mathbf{r}) = u_0 \rho(\mathbf{r})$. Thus an expansion of $J = \mu^* - iw^* = \mu^* - u_0 \rho$ in gradients of $\rho$ is obtained by subtracting $u_0 \rho$ from both sides of eqn (5.45).

The next step is to develop an expansion of the grand canonical Hamiltonian

$$H_G[w^*, J] = \frac{1}{2u_0} \int d\mathbf{r} \, [w^*(\mathbf{r})]^2 - zV \, Q[\mu^*] \tag{5.46}$$

in gradients of $J$ and hence gradients of $\rho$. Use of eqns (3.156) and (5.45) leads to

$$H_G[w^*, J] = -\int d\mathbf{r} \left( \frac{1}{N} \rho + \frac{u_0}{2} \rho^2 - \frac{R_g^2}{3N} \nabla^2 \rho + \ldots \right) \tag{5.47}$$

Finally, to complete the calculation, we affect the Legendre transform from a functional of $J$ to a functional of $\rho$ by means of

$$F[\rho] = H_G[w^*, J] - \int d\mathbf{r} \, J(\mathbf{r})\rho(\mathbf{r}) \tag{5.48}$$

Substitution of eqns (5.45) and (5.47) into eqn (5.48) produces

$$F[\rho] = \int d\mathbf{r} \left( \frac{1}{N} \rho \ln \rho + \frac{u_0}{2} \rho^2 + \frac{b^2}{36\rho} |\nabla \rho|^2 + \ldots \right) \tag{5.49}$$

where terms linear in $\int d\mathbf{r} \, \rho(\mathbf{r})$ have been dropped because they can be absorbed into the reference chemical potential.

---

[62]The original unscaled variables have been used in eqn (3.157), corresponding to the replacements $\xi_w \to 1$, $\epsilon_s \to R_g^2$, and $W \to Nw$.

Equation (5.49) is a well-known result (Tang and Freed, 1991) that expresses the free energy (in units of $k_B T$) of a polymer solution described by Model A as a functional of the polymer segment density $\rho(\mathbf{r})$. The first term on the right-hand side describes the translational entropy of the polymers, while the second term captures the segment–segment interactions mediated by the solvent. The third "square gradient" or "Lifshitz entropy" term is a long-wavelength approximation to the conformational entropy penalty created by a density inhomogeneity. Equation (5.49) relies on both the mean-field approximation and the slow gradient expansion, so it is valid only for density variations that everywhere satisfy $|\nabla\rho|/\rho \ll R_g^{-1}$. An important distinction from the RPA free energy functional, eqn (5.27), is that eqn (5.49) is not restricted to inhomogeneities that are weak in amplitude, only long in wavelength. In regions where they are both valid, the two expressions coincide. For example, substitution of $\rho(\mathbf{r}) = \rho_0 + \Delta\rho(\mathbf{r})$ into eqn (5.49) and expansion to quadratic order in $\Delta\rho$ produces an expression of the form of eqn (5.31), but with a quadratic coefficient given by

$$\hat{\Gamma}_2(k) = \frac{1}{\rho_0 N}\left(1 + \frac{1}{3}k^2 R_g^2 + ...\right) + u_0 \tag{5.50}$$

This equation, however, is consistent with the small $k$ (long wavelength) expansion of the RPA formula (5.29) to $O(k^2 R_g^2)$.

### 5.2.3  Ground state dominance

The ground state dominance approximation of Section 3.4.3 is a third useful technique that can be combined with the mean-field approximation. This combination is appropriate when polymers of high molecular weight are bound to regions with dimensions small compared to $R_g$. We introduce this subject by returning to the now familiar problem of deriving a free energy functional for Model A in the canonical ensemble.

The mean-field eqns (5.20) and (5.21) for the canonical Model A are conveniently reexpressed as

$$\rho(\mathbf{r}) = \tilde{\rho}(\mathbf{r}; [iw^* + J]), \quad iw^*(\mathbf{r}) = u_0\rho(\mathbf{r}) \tag{5.51}$$

Equations (3.167) and (4.75) can be further combined in the ground state dominance approximation to produce

$$\rho(\mathbf{r}) = \tilde{\rho}(\mathbf{r}; [iw^* + J]) = nN[\psi(\mathbf{r})]^2 \tag{5.52}$$

The ground state eigenfunction $\psi(\mathbf{r})$ in this equation satisfies eqn (3.160) with $w \to iw^* + J$, i.e.

$$\frac{b^2}{6}\nabla^2\psi(\mathbf{r}) = [iw^*(\mathbf{r}) + J(\mathbf{r}) - \Lambda]\psi(\mathbf{r}) \tag{5.53}$$

where $\Lambda$ is the ground state eigenvalue. It follows that the external field $J$ can be expressed as

$$J(\mathbf{r}) = \frac{b^2}{6\psi(\mathbf{r})} \nabla^2 \psi(\mathbf{r}) - u_0 \rho(\mathbf{r}) + \Lambda \tag{5.54}$$

Substitution of this result into eqns (4.206) and (5.25) results in the following density functional:

$$F[\rho] = \int d\mathbf{r} \left( -\frac{nNb^2}{6} \psi \nabla^2 \psi + \frac{u_0}{2} \rho^2 - \Lambda \rho \right) \tag{5.55}$$

where the term $-n \ln Q[iw^* + J]$ has been neglected because it is smaller by $O(1/N)$ than the terms shown.

The remaining task is to reexpress the term in eqn (5.55) involving $\psi$ as a functional of $\rho$. Differentiation of eqn (5.52) produces $|\nabla \psi|^2 = |\nabla \rho|^2/(4nN\rho)$, which upon substitution into eqn (5.55) and integration by parts, leads to the desired free energy functional

$$F[\rho] = \int d\mathbf{r} \left( \frac{b^2}{24\rho} |\nabla \rho|^2 + \frac{u_0}{2} \rho^2 - \Lambda \rho \right) \tag{5.56}$$

In this final expression, the term proportional to $\Lambda$ has been retained. Alternatively, $\Lambda$ could be absorbed into the definition of the Lagrange multiplier (chemical potential) $\mu$ that is used in eqn (4.202) to construct the grand potential $\Omega[\rho]$ and enforce the constraint $\int d\mathbf{r} \, \rho(\mathbf{r}) = nN$.

The ground state free energy functional given by eqn (5.56) is strikingly similar to the functional $F_2[\rho]$ introduced in eqn (3.169). Indeed, the substitution $w \to iw^*/2 = u_0 \rho/2$ in $F_2[\rho]$ leads immediately to eqn (5.56).[63] We observe from this equation that the free energy of an inhomogeneous polymer solution is approximated in the ground state as the sum of a Lifshitz entropy term and a mean-field segment–segment interaction term. The Lifshitz entropy reflects the conformational entropy penalty associated with the inhomogeneous density profile $\rho(\mathbf{r})$.

Equation (5.56) is also similar in form to the free energy functional (5.49) derived by applying the slow gradient expansion. The translational entropy term $(\rho/N) \ln \rho$ is missing in the ground state expression, as is appropriate when $N \to \infty$, or $\xi/R_g \to 0$, where $\xi$ is the width of the region of polymer localization. In addition, the "square gradient coefficient" of the Lifshitz entropy term is given as $1/36$ in eqn (5.49) and as $1/24$ in eqn (5.56). The origin of this discrepancy can be traced to the fact that the first expression is valid for slow density variations on the scale of $R_g$ and the second for rapid variations on the $R_g$ scale. Indeed, we have already seen that for *small-amplitude* density inhomogeneities, the $1/36$ coefficient in the slow gradient expansion (5.49) is consistent with the RPA expansion (5.27) for $kR_g \ll 1$. In the opposite limit of rapid density variations, i.e. $kR_g \gg 1$, the RPA expansion reproduces the $1/24$ coefficient observed

---

[63]The factor of $1/2$ is necessary to cancel the double counting of pair interactions in converting the single-chain functional $F_2$ to the many-chain functional $F$.

in eqn (5.56). In other words, the RPA and the ground state dominance approximations are consistent for density variations that are simultaneously small in amplitude and rapid on the scale of $R_g$. The ground state expression (5.56), however, is not restricted to weak inhomogeneities.

As a second application of the ground state dominance approximation, we turn to consider the classic problem of the *symmetric polymer–polymer interface* first solved by Helfand and Tagami (Helfand and Tagami, 1971). Their analytical solution was developed for an incompressible homopolymer blend model that is a special case of Model C with equal statistical segment lengths and polymerization indices, i.e. $b_A = b_B \equiv b$ and $N_A = N_B \equiv N$. The solution relies on both the mean-field and the ground state dominance approximations, which are satisfied if $\chi_{AB} \ll 1$ and $N \gg 1$ such that $\chi_{AB}N \gg 1$. This is the so-called *strong-segregation limit*, where the system consists of two coexisting phases that are nearly pure in polymers A and B, respectively, separated by a narrow interface of width $\xi$ much less than the radius of gyration $R_g = b(N/6)^{1/2}$.

The situation considered by Helfand and Tagami corresponds to a flat interface, located at $z = 0$, separating a pure phase of A homopolymer at $z \to +\infty$ from a pure phase of B homopolymer at $z \to -\infty$. The mean-field equations correspond to

$$\left. \frac{\delta H[w_{\pm}]}{\delta w_+(z)} \right|_{w_{\pm}=w_{\pm}^*} = i\rho_0[\phi_A(z; [w_A^*]) + \phi_B(z; [w_B^*]) - 1] = 0 \qquad (5.57)$$

$$\left. \frac{\delta H[w_{\pm}]}{\delta w_-(z)} \right|_{w_{\pm}=w_{\pm}^*} = \rho_0 \left[ (2/\chi_{AB})w_-^*(z) - \phi_A(z; [w_A^*]) + \phi_B(z; [w_B^*]) \right] = 0 \quad (5.58)$$

where $w_A = iw_+ - w_-$ and $w_B = iw_+ + w_-$ are the chemical potentials conjugate to the A and B monomer species and the $\phi_K$ fields are *local volume fractions* of type $K$ monomers ($K = A$ or $B$) defined by

$$\phi_K(\mathbf{r}; [w_K]) \equiv \tilde{\rho}_K(\mathbf{r}; [w_K])/\rho_0 \qquad (5.59)$$

The first mean-field equation, eqn (5.57), expresses the incompressibility condition that the volume fractions of the two species must sum to unity at each position $z$. The second saddle point equation, eqn (5.58), shows that the exchange chemical potential field, $w_-$, is related to the A species volume fraction in the mean-field approximation by $w_-^*(z) = \chi_{AB}[\phi_A(z; [w_A^*]) - 1/2]$.

Equations (5.57) and (5.58) are a formidable set of equations by virtue of the nonlinear dependence of the volume fractions $\phi_K(z; [w_K^*])$ on the saddle point fields $w_{\pm}^*(z)$. However, upon imposing the ground state dominance approximation, the volume fractions simplify to

$$\phi_K(z; [w_K^*]) \approx [\psi_K(z)]^2 \qquad (5.60)$$

in which the ground state eigenfunctions $\psi_K(z)$ satisfy

$$\frac{b^2}{6}\frac{d^2}{dz^2}\psi_A(z) = \left[iw_+^*(z) - w_-^*(z) - \Lambda_A\right]\psi_A(z) \tag{5.61}$$

and

$$\frac{b^2}{6}\frac{d^2}{dz^2}\psi_B(z) = \left[iw_+^*(z) + w_-^*(z) - \Lambda_B\right]\psi_B(z) \tag{5.62}$$

The ground state eigenvalues $\Lambda_K$ appearing in these equations are chosen as described below. We have also adopted a different normalization of the eigenfunctions than in eqn (3.161); namely, $\int d\mathbf{r}\psi_K^2 = V\phi_{K0}$, where $\phi_{K0}$ is the average volume fraction of type $K$ polymer segments contained in the system.

A single planar interface between the two homopolymer phases can be imposed by applying suitable boundary conditions. Specifically, if the coexisting phases are pure, as is asymptotically the case for $\chi_{AB}N \to \infty$, then $\psi_A(+\infty) = \psi_B(-\infty) = 1$ and $\psi_A(-\infty) = \psi_B(+\infty) = 0$. The following gradient conditions also apply for $K = $ A and B: $\psi_K'(\pm\infty) = \psi_K''(\pm\infty) = 0$ . For eqns (5.61) and (5.62) to be consistent with these boundary conditions, it must be the case that $\Lambda_A = \Lambda_B \equiv \Lambda$ and $iw_+(\pm\infty) - \chi_{AB}/2 - \Lambda = 0$. A convenient, but arbitrary, choice of the bulk pressure field is $w_+(\pm\infty) = 0$, which implies that $\Lambda = -\chi_{AB}/2$. Consequently, eqns (5.61) and (5.62) reduce to

$$\frac{b^2}{6}\frac{d^2}{dz^2}\psi_A(z) = \left[iw_+^*(z) + \chi_{AB}\,\phi_B(z)\right]\psi_A(z) \tag{5.63}$$

and

$$\frac{b^2}{6}\frac{d^2}{dz^2}\psi_B(z) = \left[iw_+^*(z) + \chi_{AB}\,\phi_A(z)\right]\psi_B(z) \tag{5.64}$$

Thus, the mean potential field felt by the A polymer segments is the sum of a "pressure" term, $iw_+^*(z)$, and an interaction term, $\chi_{AB}\,\phi_B(z)$, describing local contacts with type B segments. Similarly, the B segments experience the same pressure field, but an interaction potential $\chi_{AB}\,\phi_A(z)$ proportional to the local volume fraction of type A segments.

Helfand and Tagami showed that the five equations consisting of eqns (5.57), (5.60) for $K = $ A and B, (5.63), and (5.64) possess an analytical solution for the planar boundary conditions described above. The solution corresponds to (Helfand and Tagami, 1971)

$$\phi_A(z) = 1 - \phi_B(z) = \frac{1}{1 + \exp(-z/\xi)} \tag{5.65}$$

$$\psi_K(z) = [\phi_K(z)]^{1/2} \tag{5.66}$$

and

$$iw_+^*(z) = -3\chi_{AB}\,\phi_A(z)\phi_B(z) \tag{5.67}$$

where $\xi$ is a measure of the *interfacial width*, defined by

$$\xi = \frac{b}{2(6\chi_{AB})^{1/2}} \tag{5.68}$$

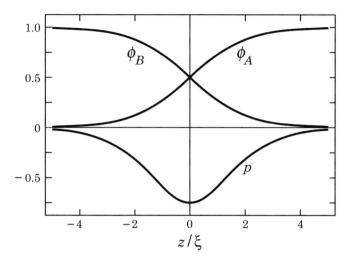

FIG. 5.4. The Helfand–Tagami solution for the interfacial profiles of a symmetric polymer–polymer interface. The volume fractions of the type A and B
polymer segments, respectively, are shown as the curves marked $\phi_A$ and $\phi_B$.
The total segment volume fraction is everywhere unity. The curve labelled $p$
denotes the pressure field $p(z) \equiv iw_+^*(z)/\chi_{AB}$.

The equilibrium volume fraction profiles $\phi_A(z)$ and $\phi_B(z)$ corresponding to
the Helfand–Tagami solution are sketched in Fig. 5.4. The interfacial width $\xi$
corresponds to the width of the region over which type A and B segments in-
termix. This intermixing is opposed by the A–B contact energy, parameterized
by $\chi_{AB}$, and is favored by the conformational entropy gain provided by a diffuse
interface. The pressure field $iw_+^*(z)$ is seen to be negative and localized in the
interfacial region. This field serves to attract polymer segments to the interface
and thereby maintain a uniform total segment density at all positions $z$.

In Section 3.4.3, it was shown that the validity of the ground state dominance approximation relies on the range of a localizing potential being much
smaller than $R_g$. The localizing potential in the present example is the pressure field $iw_+^*(z)$, which binds not entire chains, but loops of chain segments
that form the interface. The range of the pressure field is of the same order as
the width of the interface, $\xi$, and this range meets the necessary criterion of
$\xi/R_g = 1/[2(\chi_{AB}N)^{1/2}] \ll 1$, provided that $\chi_{AB}N \gg 1$. Similarly, the validity
of the mean-field approximation requires that $C \gg 1$, where $C$ is the coordination number defined in eqn (5.5). Because $C \sim \rho_0 b^3 N^{1/2}$ in a molten blend, this
condition is satisfied provided that the A and B polymers are both of high molecular weight, i.e. $N \gg 1$. Finally, for the Helfand–Tagami solution to be valid, the
interfacial width must not fall below mesoscopic scales. In other words, it must
be the case that $\xi \gg b$ or, equivalently, $\chi_{AB} \ll 1$. We have thus argued that

the Helfand–Tagami solution is accurate, provided that the conditions $\chi_{AB} \ll 1$, $N \gg 1$, and $\chi_{AB}N \gg 1$ are all met.

It is important to note that eqn (5.68) describes the "intrinsic" interfacial width. Experiments on fluid–fluid interfaces typically measure an "extrinsic" interfacial width that includes long-wavelength capillary wave undulations of the interface and thus exceeds the intrinsic width (Weeks, 1977; Huse *et al.*, 1985; Binder *et al.*, 2001). Such capillary wave undulations can be viewed as a special type of field fluctuation that is particular to the layer geometry.[64]

The *interfacial tension* of a symmetric polymer–polymer interface is readily obtained from the Helfand–Tagami solution by noting that the tension is the surface excess Helmholtz free energy per unit area of interface (Chandler, 1987) and that in the mean-field approximation, $A(n_A, n_B, V, T) \approx k_B T \, H[w_+^*, w_-^*]$. From eqn (4.98) it thus follows that the interfacial tension $\gamma$ is given by (Helfand and Tagami, 1971)

$$\gamma = k_B T \rho_0 \int_{-\infty}^{\infty} dz \left( \frac{1}{\chi_{AB}} [w_-^*(z)]^2 - i w_+^*(z) - \frac{1}{4} \chi_{AB} \right)$$

$$= -k_B T \rho_0 \int_{-\infty}^{\infty} dz \left( \chi_{AB} \, \phi_A(z) \phi_B(z) + i w_+^*(z) \right)$$

$$= 2 k_B T \rho_0 \chi_{AB} \int_{-\infty}^{\infty} dz \, \phi_A(z) \phi_B(z)$$

$$= k_B T b \rho_0 (\chi_{AB}/6)^{1/2} \tag{5.69}$$

The interfacial tension of strongly segregated polymer–polymer interfaces is therefore independent of polymer molecular weight and scales as the square root of the Flory interaction parameter $\chi_{AB}$.

A simple scaling argument can be used to better understand the physical origins of eqns (5.68) and (5.69). As shown in Fig. 5.5, the interfacial region can be envisioned as a mixture of "loops" of A polymer segments and "loops" of B polymer segments. The interfacial width $\xi$ characterizes the thickness of the region over which the loops mix with each other. Consider that a typical loop contains $g$ statistical segments. The value of $g$ can be estimated by equipartition; namely, a loop represents a penetration of a chain into a dissimilar polymer phase and this penetration will occur only to the extent that the energy cost is $O(k_B T)$. The energy cost of a loop with $g$ segments is $\sim g \chi_{AB} k_B T$, so we arrive at the conclusion that $g \sim 1/\chi_{AB}$, which is a large number of statistical segments for $\chi_{AB} \ll 1$. Since the forces acting on each loop are associated only with the weak thermal energy scale, the width of the interface can be estimated as the size of an unperturbed polymer with $g$ segments, i.e. $\xi \sim b g^{1/2} \sim b/\chi_{AB}^{1/2}$. This scaling is

---

[64]The reader should note that the condition of finite, but large, $C$ implies that the $w_\pm$ field fluctuations are weak at wavelengths comparable to $R_g$. Soft modes can arise on longer length scales in special circumstances, such as the case of undulation modes of a planar interface or a lamellar mesophase, which have little to do with the "field fluctuations" that we normally associate with a breakdown of the mean-field approximation.

evidently consistent with eqn (5.68). Finally, the interfacial tension can be estimated as the surface excess free energy density arising from A–B contacts in the interfacial region, $\sim k_B T \rho_0 \chi_{AB}$, times the thickness of that region, $\xi$, resulting in the scaling formula $\gamma \sim k_B T \rho_0 b \chi_{AB}^{1/2}$. This result is evidently consistent with eqn (5.69).

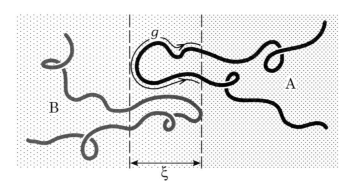

FIG. 5.5. Schematic of chain conformations in a polymer–polymer interface. A typical polymer chain passing through the interface contributes one or more "loops" with $g$ statistical segments. The type A loops and type B loops mix throughout the interfacial region of width $\xi$.

### 5.2.4  *Strong stretching*

Another powerful analytical technique is the combination of the strong stretching (or classical) approximation described in Section 3.4.4.2 with the mean-field approximation. A useful context to illustrate the method is the solution polymer brush described by Model K. At sufficiently high surface density of tethered chains, $\sigma$, the polymers in the brush are strongly stretched and the classical approximation can be applied.

In the mean-field approximation, the (real) potential $\mu(z) \equiv iw^*(z)$ and the average segment density $\rho(z) \equiv \tilde{\rho}(z; [iw^*])$ are related at each position $z$ in the brush by

$$\mu(z) = u_0 \rho(z) \tag{5.70}$$

where the coordinate system of Fig. 4.7 is applied. The classical path $z(s)$ for a chain that originates at the grafting surface $z(0) = 0$ and terminates at position $z(N) = z_0$ satisfies eqn (3.194), i.e.

$$\frac{3}{b^2} \frac{d^2}{ds^2} z(s) = \frac{d\mu(z(s))}{dz(s)} \tag{5.71}$$

where dimensional units have been restored. Because all chains in the brush have $N$ statistical segments, irrespective of their free end positions $z_0$, it can be argued

(Milner *et al.*, 1988) that this "equal time" constraint implies that the classical approximation to $\mu(z)$ must be a harmonic potential, i.e. $\mu(z) = C_1 - C_2 z^2$. With such a harmonic potential, eqn (5.71) can be solved subject to the conditions $z(0) = 0$, $z(N) = z_0$, and $[dz(s)/ds]_{s=N} = 0$ (no tension on the free chain end) to yield

$$z(s) = z_0 \sin(\pi s/(2N)) \tag{5.72}$$

for the classical path, and

$$C_2 = \frac{3\pi^2}{8b^2 N^2} \tag{5.73}$$

for the quadratic coefficient. The second coefficient, $C_1$, is determined by the conservation of segments relation

$$\sigma N = \int_0^h dz\, \rho(z) = u_0^{-1} \int_0^h dz\, (C_1 - C_2 z^2) \tag{5.74}$$

where $h = (C_1/C_2)^{1/2}$ is the thickness of the polymer brush, defined by $\rho(h) = 0$. The condition (5.74) amounts to

$$C_1 = \frac{\sigma u_0 N}{h} + \frac{C_2 h^2}{3} \tag{5.75}$$

By combining these results, we obtain a well-known result for the equilibrium brush thickness (Milner *et al.*, 1988)

$$h = \left(\frac{4}{\pi^2}\right)^{1/3} (\sigma u_0)^{1/3} b^{2/3} N \tag{5.76}$$

A criterion for the validity of this formula is that the polymers are stretched to a height that greatly exceeds their unperturbed dimensions, $h/R_g \gg 1$. This amounts to the condition $(\sigma u_0/b)^{1/3} N^{1/2} \gg 1$, or equivalently that $(\sigma R_g^2) B \gg 1$, where $B$ is the dimensionless excluded volume parameter defined in eqn (5.28). The parameter $B$ will be seen in Chapter 6 to play an important role in assessing the strength of excluded volume effects in polymer solutions.

The equilibrium segment density profile in the combined mean-field, strong stretching approximation, $(u_0/C_1)\rho(z) = 1 - (z/h)^2$, is plotted in Fig. 5.6. This parabolic profile closely mirrors the full mean-field segment density distribution at high grafting density (Milner, 1990). However, as qualitatively indicated by the dashed curves in the figure, the strong stretching approximation is not a uniformly valid approximation to the mean-field profile and deviates significantly from the (numerical) mean-field segment density distribution both very near the surface and close to the outer edge of the brush. The existence of a depletion boundary layer near $z = 0$ was already discussed in Section 4.9.4. Indeed, the full mean-field solution of Model K with a Dirichlet condition applied to the propagators at $z = 0$ exhibits a sharp boundary layer, as sketched in the dashed curve of Fig. 5.6. The steepness of this density rise increases with increasing grafting density $\sigma$. As is clear from the figure, the strong stretching approximation is applicable only outside of the boundary layer, where the potential $\mu(z)$ is slowly varying, and is consistent with an effective Neumann boundary condition at $z = 0$. Similarly, at the outer edge of the brush, numerical solutions of the mean-field equations show a smooth decay of the density profile (Milner, 1990), in contrast to the parabolic profile which abruptly terminates at $z = h$. This outer boundary layer arises from a breakdown of the strong stretching assumption near the edge of the brush; in this region many polymer conformations contribute to the average density profile. Matsen has recently made similar comparisons between the strong stretching approximation and numerical mean-field solutions for a "dry" (i.e. melt) polymer brush and for the interface between a dry brush and a chemically identical homopolymer melt (Matsen, 2004; Matsen, 2005a).

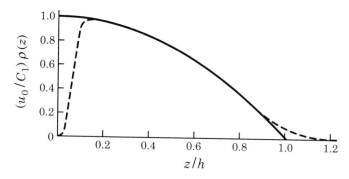

FIG. 5.6. Parabolic segment density profile $\rho(z)$ obtained by combining the mean-field and strong stretching approximations for Model K of a polymer brush. At high grafting densities, the parabolic profile (solid) faithfully reproduces the full mean-field segment distribution, except in boundary layers indicated schematically as dashed curves. Near the surface, the classical approximation imposes an effective Neumann boundary condition and ignores a narrow depletion layer. At the outer edge of the brush, the strong stretching approximation breaks down, resulting in a smoothing of the profile.

Both the strong stretching formulas and the full mean-field solutions for Model K suffer from the neglect of $w$ field fluctuations. These field fluctuations are strongest at short wavelengths below the average spacing between grafting points, $\sim \sigma^{-1/2}$, and generate a local excluded volume swelling of the brush that is not captured by mean-field theory. Such excluded volume effects can be approximately incorporated into the mean-field theory by applying "blob" arguments (Milner $et\ al.$, 1988; Netz and Schick, 1998), but a full treatment requires the numerical techniques that are described in Chapter 6.

In closing this section we note that a powerful $electrostatic\ analogy$ for the classical approximation was identified by Semenov (Semenov, 1985). This analogy has proved very useful for a variety of calculations on inhomogeneous polymers where the assumption of strong chain stretching can be justified (Semenov, 1985; Zhulina *et al.*, 1992; Fredrickson *et al.*, 1992).

## 5.3   Numerical methods

It should be apparent from the above discussion that the available analytical techniques for studying the mean-field equations are rather crude tools that rarely provide a complete picture of the solution. Hence, numerical solutions are often desired. In the present section we discuss strategies for implementing numerical self-consistent field theory ("numerical SCFT") in a computationally efficient manner.

It is helpful at this stage to review the task of solving the mean-field equations. For a specified field theory model with effective Hamiltonian $H[w]$, we seek saddle point field configurations $w^*(\mathbf{r})$ that satisfy the stationary condition eqn (5.3). In the case of the canonical Model A this amounts to

$$\left.\frac{\delta H[w]}{\delta w(\mathbf{r})}\right|_{w=w^*} = \frac{1}{u_0}w^*(\mathbf{r}) + i\tilde{\rho}(\mathbf{r};[iw^*]) = 0 \qquad (5.77)$$

For the purpose of generating numerical solutions to this equation, the field $w^*(\mathbf{r})$ is approximated by an $M$-component column vector $\mathbf{w}^* = (w_1^*, w_2^*, ..., w_M^*)^T$, representing, e.g., values of the field at collocation points. With such a finite representation, eqn (5.77) can be viewed as a set of $M$ simultaneous nonlinear equations to be solved for the components of $\mathbf{w}^*$.

The *nonlinearity* and *non-locality* of the above equations is manifest in the functional dependence of the density operator $\tilde{\rho}$ on the field variables $\mathbf{w}^*$. The density operator is related locally at each collocation point $\mathbf{r}$ to the propagator $q(\mathbf{r}, s; [i\mathbf{w}^*])$ through the bilinear formula eqn (4.75). A further nonlinearity in this expression is provided by the normalizing factor of $1/Q[i\mathbf{w}^*]$. The non-locality of the theory is restricted to the Fokker–Planck equation (3.25), which connects $\mathbf{w}^*$ to the propagator $q$. It is only in solving eqn (3.25) that boundary conditions need to be imposed.

Section 3.6 described numerical methods that can be used for the efficient computation of propagators, partition functions, and density operators, *given* a prescribed finite representation of the field $\mathbf{w}^*$. It was observed that, at least for simple geometries, pseudo-spectral techniques provide a powerful and flexible platform for evaluating these objects. Here we focus on the remaining task necessary to generate a numerical SCFT solution, namely how to adjust the components of $\mathbf{w}^*$ in order to satisfy eqn (5.77) to within a prescribed accuracy.

### 5.3.1   *Nonlinear equation approach*

One strategy for solving the SCFT equations is to view them as a set of *nonlinear equations* to be solved for the $M$-component numerical approximation $\mathbf{w}^*$ to the continuous field $w^*(\mathbf{r})$. For simplicity, we restrict our discussion to the canonical

Model A and employ collocation at spatial grid points to construct the discrete approximation $\mathbf{w}^*$. On physical grounds, the mean-field solution $\mathbf{w}^*$ is expected to be purely imaginary so that $H[\mathbf{w}^*]$ is real. Thus, it is convenient to seek a *purely real* $M$-vector $\boldsymbol{\mu}$, representing a numerical approximation to the field $\mu(\mathbf{r}) \equiv iw^*(\mathbf{r})$. Equation (5.77) determining $\boldsymbol{\mu}$ is equivalent to the nonlinear equation set

$$\frac{1}{u_0}\mu_l - \tilde{\rho}(\mathbf{r}_l; [\boldsymbol{\mu}]) = 0, \quad l = 1, 2, ..., M \tag{5.78}$$

where the $\mathbf{r}_l$ represent $M$ collocation points in the spatial domain at which the various fields are evaluated. In matrix notation, this equation set can be written compactly as

$$\mathbf{F}(\boldsymbol{\mu}) \equiv \frac{1}{u_0}\boldsymbol{\mu} - \boldsymbol{\rho}(\boldsymbol{\mu}) = 0 \tag{5.79}$$

where $\boldsymbol{\rho}$ is a $M$-vector with components $\tilde{\rho}(\mathbf{r}_l; [\boldsymbol{\mu}])$.

A standard method for solving such a set of nonlinear equations is the *Newton–Raphson* technique (Press *et al.*, 1992), which relies on the Taylor expansion

$$\mathbf{F}(\boldsymbol{\mu} + \delta\boldsymbol{\mu}) = \mathbf{F}(\boldsymbol{\mu}) + \mathbf{J} \cdot \delta\boldsymbol{\mu} + O(\delta\boldsymbol{\mu}^2) \tag{5.80}$$

where $\delta\boldsymbol{\mu}$ is some small departure from the desired solution and $J_{ij} \equiv \partial F_i/\partial\mu_j$ is the *Jacobian matrix*. The Newton–Raphson scheme proceeds by neglecting the $O(\delta\boldsymbol{\mu}^2)$ terms and selecting $\delta\boldsymbol{\mu}$ so that $\mathbf{F}(\boldsymbol{\mu} + \delta\boldsymbol{\mu}) = 0$. This results in the algorithm

$$\mathbf{J}^j \cdot \delta\boldsymbol{\mu}^j = -\mathbf{F}^j \tag{5.81}$$

where the superscript $j$ denotes the $j$th iteration of the procedure. Current gradient and function information (i.e. at the point $\boldsymbol{\mu}^j$) are used to evaluate $\mathbf{J}^j$ and $\mathbf{F}^j$. Equation (5.81) represents a linear equation set that can be solved for the "Newton step" $\delta\boldsymbol{\mu}^j$ by using standard techniques such as LU decomposition. The new value of $\boldsymbol{\mu}$ is obtained from this solution by means of

$$\boldsymbol{\mu}^{j+1} = \boldsymbol{\mu}^j + \delta\boldsymbol{\mu}^j \tag{5.82}$$

When it converges, the Newton–Raphson technique enjoys a super-linear rate of convergence to a solution of high numerical accuracy. Unfortunately, it is often difficult to initialize the algorithm with an initial guess $\boldsymbol{\mu}^0$ that is sufficiently good for convergence to be obtained. Globally convergent variants are available, however, which replace the full Newton step of eqn (5.82) with a fractional Newton step defined by $\boldsymbol{\mu}^{j+1} = \boldsymbol{\mu}^j + \lambda^j\delta\boldsymbol{\mu}^j$ in which $\lambda^j \in (0, 1]$. The parameter $\lambda^j$ is chosen by a backtracking line search strategy (Press *et al.*, 1992) to ensure that the functional

$$\mathcal{F}(\boldsymbol{\mu}) = \frac{1}{2}\mathbf{F}(\boldsymbol{\mu}) \cdot \mathbf{F}(\boldsymbol{\mu}) \tag{5.83}$$

decreases during each Newton iteration. The functional $\mathcal{F}(\boldsymbol{\mu})$ has a local minimum with $\mathcal{F} = 0$ for $\boldsymbol{\mu}$ satisfying eqn (5.79), so the sequence of partial Newton steps ensures that $\boldsymbol{\mu}^j$ evolves towards a solution for almost any starting guess $\boldsymbol{\mu}^0$. In the rare case where this globally convergent Newton technique fails, the failure is due to the algorithm landing on a local minimum of $\mathcal{F}$ that is not a solution of eqn (5.79). Normally, restarting the algorithm from a new starting point $\boldsymbol{\mu}^0$ will resolve the problem.

The above scheme can generate highly accurate numerical solutions to the SCFT equations, but it is expensive to implement. Evaluation of the function $\mathbf{F}(\boldsymbol{\mu})$ involves calculation of the density operator $\boldsymbol{\rho}(\boldsymbol{\mu})$, which was shown to require $O(N_s M \log_2 M)$ operations by means of the pseudo-spectral algorithms of Section 3.6. As will be discussed below, numerical evaluation of the Jacobian matrix $\mathbf{J}(\boldsymbol{\mu})$ requires $M$ times this effort, or $O(N_s M^2 \log_2 M)$ operations. Most expensive is the solution of the linear equation set eqn (5.81), which requires $O(M^3)$ operations due to the fact that the Jacobian is not sparse. The cost of implementing a Newton–Raphson scheme, with or without a global convergence wrapper, is clearly prohibitive for a high-resolution SCFT calculation with $M \sim 10^6$ spatial degrees of freedom.

A more practical class of numerical schemes are the so-called *quasi-Newton* methods, which approximate the Jacobian at each step in a way that lowers the overall operation count of the algorithm. The most successful of these techniques is *Broyden's method* (Dennis and Schnabel, 1996), which can also be embedded in a globally convergent strategy. In Broyden's method, the Jacobian matrix $\mathbf{J}^j$ in eqn (5.81) is replaced by an approximate Jacobian $\mathbf{B}^j$ and is updated in $O(M^2)$ operations according to

$$\mathbf{B}^{j+1} = \mathbf{B}^j + \frac{(\delta \mathbf{F}^j - \mathbf{B}^j \cdot \delta \boldsymbol{\mu}^j) \otimes \delta \boldsymbol{\mu}^j}{\delta \boldsymbol{\mu}^j \cdot \delta \boldsymbol{\mu}^j} \tag{5.84}$$

where $\delta \mathbf{F}^j \equiv \mathbf{F}^{j+1} - \mathbf{F}^j$ and the symbol $\otimes$ denotes a direct matrix product. Remarkably, because of the special form of eqn (5.84), the QR decomposition can be used in place of LU decomposition to affect the update of the solution of eqn (5.81) (with $\mathbf{J} \to \mathbf{B}$) between the $j$th and $(j+1)$th iteration using only $O(M^2)$ operations (Press *et al.*, 1992). To start Broyden's scheme, an initial approximation to the Jacobian $\mathbf{B}^0$ is required. Here there are two options, both requiring roughly $O(N_s M^2)$ operations: (i) compute $\mathbf{B}^0$ by evaluating analytical expressions for the Jacobian (see below) using a pseudo-spectral numerical scheme, or (ii) compute $\mathbf{B}^0$ by combining function evaluations with a finite difference approximation. Overall, Broyden's scheme reduces the cost per quasi-Newton step to $O(M^2)$ operations and yet, in globally convergent form, enjoys the same excellent performance as the full Newton–Raphson method.

It is useful to discuss the nature of the Jacobian that enters the Newton and quasi-Newton approaches to solving the SCFT equations. Returning to the continuum theory, the Jacobian corresponds to a two-point function defined by

$$J(\mathbf{r}, \mathbf{r}'; [\mu]) = -\frac{\delta^2 H[\mu]}{\delta \mu(\mathbf{r}) \delta \mu(\mathbf{r}')} \tag{5.85}$$

For the canonical Model A, the right-hand side of this expression becomes

$$J(\mathbf{r}, \mathbf{r}'; [\mu]) = \frac{1}{u_0} \delta(\mathbf{r} - \mathbf{r}') + n \frac{\delta^2 \ln Q[\mu]}{\delta \mu(\mathbf{r}) \delta \mu(\mathbf{r}')} \qquad (5.86)$$

and the second derivative of $\ln Q[\mu]$ was identified in eqn (3.72) as the connected density–density correlation function of a polymer experiencing the potential $\mu(\mathbf{r})$. This object can be numerically evaluated by means of eqn (3.73) in $O(N_s M^2 \log_2 M)$ operations, which is the cost of evaluating the propagator $g(\mathbf{r}, \mathbf{r}', s; [\mu])$ using a pseudo-spectral scheme with $M$ spatial collocation points and $N_s$ chain contour steps.

The Jacobian of eqn (5.86) is a complicated object that is non-local in both real and reciprocal space. However, in certain limiting situations it simplifies considerably. For example, in the case of a potential $\mu(\mathbf{r})$ that is only *weakly inhomogeneous*, eqn (3.136) can be used to approximate the Jacobian by

$$J(\mathbf{r}, \mathbf{r}'; [\mu]) \approx \frac{1}{u_0} \delta(\mathbf{r} - \mathbf{r}') + \rho_0 N \, g_D(|\mathbf{r} - \mathbf{r}'|) \qquad (5.87)$$

where $\rho_0 = nN/V$ is the average segment density. The Debye function $g_D(r)$ decays on the scale of $R_g$, as shown in Fig. 3.9, so the Jacobian is spatially non-local on the same length scale for weakly inhomogeneous $\mu(\mathbf{r})$. However, because of the translational invariance of eqn (5.87), the Fourier transform of the Jacobian is diagonal in the weak inhomogeneity limit. The close connection between the fast gradient expansion and the RPA expansion identified in Section 3.4.4.1 can be used to further argue that eqn (5.87) is also asymptotically correct for potentials $\mu(\mathbf{r})$ that are not necessarily weak in amplitude, but are *rapidly varying*. The latter observation suggests that in reciprocal space, the highest wavevector modes of the Jacobian are *nearly diagonal*. Indeed, this characteristic has been used in conjunction with quasi-Newton schemes to improve the performance of spectral SCFT codes (Morse, 2005).

Overall, quasi-Newton schemes such as Broyden's method are a good match to fully spectral algorithms for solving the SCFT equations. In the semi-discrete approach of Matsen and Schick (Matsen and Schick, 1994a), i.e. eqn (3.228), of order $M^3$ operations are required at each iteration to solve the diffusion equation and evaluate $\mathbf{F}(\boldsymbol{\mu})$. Thus, the $O(M^2)$ cost of a quasi-Newton step is negligible in the overall scheme. Even if the semi-implicit algorithm of eqn (3.233) is substituted for eqn (3.228), so that the operation count of evaluating $\mathbf{F}(\boldsymbol{\mu})$ is reduced to $O(M^2 N_s)$, the cost of the diffusion equation solution is still limiting in each quasi-Newton step.

Quasi-Newton methods are primarily useful for one-dimensional SCFT problems and unit-cell calculations in higher dimensions that exploit maximum symmetry. In the case of large-cell calculations in two and three dimensions, where typically $M \sim 10^4 - 10^6$, the $O(M^2)$ operation count for each quasi-Newton step is prohibitive. For such problems, a field update scheme with "ideal" $O(M)$ or "nearly ideal" $O(M \log_2 M)$ scaling is essential.

### 5.3.2    *Optimization approach*

In the previous section, the SCFT equations were viewed as a set of simultaneous nonlinear equations in the $\boldsymbol{\mu}$ field variables. An alternative strategy is to exploit the fact that the equilibrium mean-field condition corresponds to a saddle point of the complex Hamiltonian $H[w]$, as per the discussion of Section 5.1.2. Depending on how the saddle point is approached in the complex $w$ plane and the nature of the model, it will correspond to a local minimum, maximum, or saddle of the real functional $H_R[w]$. With this knowledge, SCFT can be viewed as a *nonlinear optimization problem* – namely the problem of finding complex field configurations $w(\mathbf{r})$ that are extrema of the functional $H_R[w]$. We shall see that this perspective leads to efficient $O(M \log_2 M)$ SCFT algorithms.

### 5.3.2.1    *Continuous steepest descent*

There are many powerful techniques available for solving nonlinear optimization problems (Dennis and Schnabel, 1996; Nocedal and Wright, 1999; Press *et al.*, 1992). A simple but effective strategy in the present context is to affect a *continuous steepest descent* on the energy landscape. The idea is to introduce a "fictitious" time variable $t$ and evolve the system continuously at each $t$ in the direction of the gradient of the functional to approach a local minimum or maximum. For example, in the case of the canonical Model A, the analytic structure of $H[w]$ is similar to that depicted in Fig. 5.1 for the one-dimensional "toy" model. The saddle point $w^*(\mathbf{r})$ is purely imaginary, and the imaginary axis coincides with a *descent direction* for each component of the field. As a result, $H_R[w]$ has a local *maximum* at $w = w^*$ as $w$ is varied along the imaginary axis. A continuous steepest "ascent" scheme is thus

$$\frac{\partial}{\partial t}\mu(\mathbf{r}, t) = \lambda \frac{\delta H[\mu]}{\delta \mu(\mathbf{r}, t)}$$

$$= \lambda \left( -\frac{1}{u_0}\mu(\mathbf{r}, t) + \tilde{\rho}(\mathbf{r}; [\mu(t)]) \right) \tag{5.88}$$

where $\lambda > 0$ is a real relaxation parameter, $\mu(\mathbf{r}, t) \equiv iw(\mathbf{r}, t)$ is a purely real potential field, and $H[\mu]$ assumes only real values, so $H[\mu] = H_R[\mu]$. Absorbing $\lambda$ into the time scale and collocating eqn (5.88) at $M$ points in the spatial domain leads to the corresponding semi-discrete equation set

$$\frac{d}{dt}\boldsymbol{\mu}(t) = -\frac{1}{u_0}\boldsymbol{\mu}(t) + \boldsymbol{\rho}(\boldsymbol{\mu}(t)) \tag{5.89}$$

These equations evidently have eqn (5.79) as an equilibrium condition. Indeed, it is *only* such steady states that are of physical interest, but eqn (5.89) provides a convenient strategy for "relaxing" an arbitrary initial configuration $\boldsymbol{\mu}(0)$ towards a local solution. The reader should note that in integrating eqn (5.89), *stability* of the algorithm is much more important than *accuracy*. Indeed, we wish to obtain the steady state solution as quickly as possible; the fictitious dynamical trajectory is of little interest!

A simple and useful way to integrate eqn (5.89) is with an *explicit, forward Euler scheme*:

$$\mu^{j+1} = \mu^j + \Delta t \left( -\frac{1}{u_0} \mu^j + \rho(\mu^j) \right) \tag{5.90}$$

where a superscript $j$ again denotes the $j$th iterate and $\Delta t$ is the time step. Provided that $\Delta t$ is small enough, eqn (5.90) will converge for virtually any initial condition $\mu^0$. In this respect, continuous steepest descent algorithms are superior to Newton and quasi-Newton schemes. Moreover, each potential field update requires only that the density operator $\rho(\mu^j)$ be evaluated, which amounts to $O(N_s M \log_2 M)$ operations by the pseudo-spectral Algorithm 3.1. Thus, the explicit Euler scheme nearly meets the ideal of $O(M)$ scaling with the number of spatial degrees of freedom.

The principal weakness of eqn (5.90) is that it is *unstable* for time steps exceeding some threshold $\Delta t_m$, so $\Delta t$ must be set just below this value. Often $\Delta t_m$ is extremely small for saddle points with rapidly varying inhomogeneities, in which case a large number of very tiny steps are required to obtain a solution. Moreover, it is difficult to anticipate in advance how $\Delta t_m$ will vary with the parameters of a problem, so finding an appropriate value of $\Delta t$ amounts to trial and error.

A significantly better relaxation algorithm can be constructed by utilizing analytic information about the form of the density operator. In particular, the weak inhomogeneity expansion of eqn (3.134) can be combined with eqn (4.75) to obtain the linear expression

$$\tilde{\rho}(\mathbf{r}; [\mu]) = \rho_0 \left[ 1 - N \int d\mathbf{r}' \, g_D(|\mathbf{r} - \mathbf{r}'|) \Delta \mu(\mathbf{r}') + ... \right] \tag{5.91}$$

where $\rho_0 = nN/V$ is the average segment density for Model A and $\Delta \mu(\mathbf{r}) \equiv \mu(\mathbf{r}) - (1/V) \int d\mathbf{r} \, \mu(\mathbf{r})$ is the inhomogeneous part of $\mu(\mathbf{r})$. We further recall from Section 3.4.4.1 that eqn (5.91) also captures the leading term of a fast gradient expansion for rapidly varying $\mu(\mathbf{r})$. A useful *semi-implicit* relaxation algorithm is obtained by forward-differencing the time derivative on the left-hand side of eqn (5.89), treating the term $(1/u_0)\mu$ implicitly, and adding and subtracting eqn (5.91) to the right-hand side at time steps $j+1$ and $j$, respectively (Ceniceros and Fredrickson, 2004):

$$\frac{\mu^{j+1} - \mu^j}{\Delta t} = -\rho_0 N \, \mathbf{g}_D * \Delta \mu^{j+1} - \frac{1}{u_0} \mu^{j+1} + \rho(\mu^j) + \rho_0 N \, \mathbf{g}_D * \Delta \mu^j \tag{5.92}$$

In the above, $\mathbf{g}_D * \Delta \mu$ denotes a discrete convolution obtained by collocation of $\int d\mathbf{r}' \, g_D(|\mathbf{r} - \mathbf{r}'|) \Delta \mu(\mathbf{r}')$ at $M$ spatial grid points. Equation (5.92) can be efficiently solved for $\mu^{j+1}$ by applying a single FFT–inverse-FFT pair, requiring $O(M \log_2 M)$ operations. It is thus only slightly more expensive to implement than the explicit scheme (5.90). Upon Fourier transformation, eqn (5.92) corresponds to the update scheme[65]

$$\hat{\mu}_{\mathbf{k}}^{j+1} = \frac{[1 + \Delta t \rho_0 N \hat{g}_D(k^2 R_g^2)]\hat{\mu}_{\mathbf{k}}^j + \Delta t \, \hat{\rho}_{\mathbf{k}}(\boldsymbol{\mu}^j)}{1 + \Delta t[\rho_0 N \hat{g}_D(k^2 R_g^2) + 1/u_0]} \qquad (5.93)$$

for wavevectors $\mathbf{k} \neq 0$, and $\hat{\mu}_0^{j+1} = u_0 \rho_0 V$ for $\mathbf{k} = 0$. In implementing this scheme, it is important to note that the $k$-dependent coefficients can be precomputed once and then applied repeatedly for each iteration $j$.

The semi-implicit relaxation scheme (5.93) can be viewed as an explicit method but with a wavenumber-dependent time step

$$\tau(k) = \Delta t/(1 + \Delta t[\rho_0 N \hat{g}_D(k^2 R_g^2) + 1/u_0]) \qquad (5.94)$$

that relaxes high $k$ modes faster than small $k$ modes. In practice, this has the effect of dramatically improving the stability of the algorithm in comparison with the Euler scheme and permitting higher accuracy solutions to be obtained, especially those with sharp features (Ceniceros and Fredrickson, 2004). An interesting observation is that for large $\Delta t$, $\tau(k)$ approaches $1/[\rho_0 N \hat{g}_D(k^2 R_g^2) + 1/u_0]$, which is the inverse of the approximate Jacobian (5.87). Thus, eqn (5.93) can be interpreted as a type of quasi-Newton scheme at small scales in which the Jacobian is approximated by a diagonal form in Fourier space that coincides with the transform of (5.87). Equation (5.93) is the best continuous steepest ascent algorithm to date for finding mean-field solutions of the canonical Model A.

For the *grand canonical* Model A, only slight changes are necessary to implement the above relaxation schemes. The *explicit* Euler scheme is

$$\boldsymbol{\mu}^{j+1} = \boldsymbol{\mu}^j + \Delta t \left( -\frac{1}{u_0} \boldsymbol{\mu}^j + \boldsymbol{\rho}_G(\boldsymbol{\mu}^j) \right) \qquad (5.95)$$

where the $M$-vector $\boldsymbol{\rho}_G(\boldsymbol{\mu})$ is obtained by evaluating the grand canonical density operator of eqn (4.76) at $M$ collocation points. The *semi-implicit* relaxation algorithm corresponds to solving

$$\hat{\mu}_0^{j+1} = \frac{\hat{\mu}_0^j + \Delta t \, \hat{\rho}_{G,0}(\boldsymbol{\mu}^j)}{1 + \Delta t/u_0} \qquad (5.96)$$

for the $\mathbf{k} = 0$ Fourier component of the field, and

$$\hat{\mu}_{\mathbf{k}}^{j+1} = \frac{[1 + \Delta t \alpha^j N \hat{g}_D(k^2 R_g^2)]\hat{\mu}_{\mathbf{k}}^j + \Delta t \, \hat{\rho}_{G,\mathbf{k}}(\boldsymbol{\mu}^j)}{1 + \Delta t[\alpha^j N \hat{g}_D(k^2 R_g^2) + 1/u_0]} \qquad (5.97)$$

for the $\mathbf{k} \neq 0$ modes, where $\alpha^j \equiv zN \exp(-\hat{\mu}_0^j N/V)$.

---

[65]Carets denote discrete Fourier transforms of the corresponding $M$-vectors.

Similar continuous steepest descent/ascent schemes can be devised for the other models described in Chapter 4. For example, the relevant relaxation equations for an incompressible blend described by the canonical Model C are

$$\frac{\partial}{\partial t}\mu_+(\mathbf{r}, t) = \lambda_+ \frac{\delta H[\mu_+, \mu_-]}{\delta\mu_+(\mathbf{r}, t)}$$
$$= \lambda_+ [\tilde{\rho}_A(\mathbf{r}; [w_A(t)]) + \tilde{\rho}_B(\mathbf{r}; [w_B(t)]) - \rho_0] \qquad (5.98)$$

$$\frac{\partial}{\partial t}\mu_-(\mathbf{r}, t) = -\lambda_- \frac{\delta H[\mu_+, \mu_-]}{\delta\mu_-(\mathbf{r}, t)} = -\lambda_- [(2\rho_0/\chi_{AB})\mu_-(\mathbf{r}, t)$$
$$- \tilde{\rho}_A(\mathbf{r}; [w_A(t)]) + \tilde{\rho}_B(\mathbf{r}; [w_B(t)])] \qquad (5.99)$$

where $\mu_+ \equiv iw_+$, $\mu_- \equiv w_-$, $w_A \equiv \mu_+ - \mu_-$, and $w_B \equiv \mu_+ + \mu_-$. The relaxation coefficients $\lambda_+$ and $\lambda_-$ are real and positive and can be chosen independently. It is important to note in the above equations that the relaxation is *up* the gradient in the $\mu_+$ (pressure) coordinate and *down* the gradient in the $\mu_-$ (exchange potential) coordinate. This mixed continuous ascent/descent scheme is necessary because the (real) functional $H[\mu_+, \mu_-]$ is a saddle surface in the vicinity of the real SCFT field configurations, $\mu_\pm^*(\mathbf{r})$.

The *explicit Euler algorithm* for integrating the collocated Model C equations amounts to

$$\boldsymbol{\mu}_+^{j+1} = \boldsymbol{\mu}_+^j + \Delta t \lambda_+ \left[ \boldsymbol{\rho}_A(\mathbf{w}_A^j) + \boldsymbol{\rho}_B(\mathbf{w}_B^j) - \rho_0 \mathbf{e} \right] \qquad (5.100)$$

$$\boldsymbol{\mu}_-^{j+1} = \boldsymbol{\mu}_-^j - \Delta t \lambda_- \left[ \boldsymbol{\rho}_B(\mathbf{w}_B^j) - \boldsymbol{\rho}_A(\mathbf{w}_A^j) + \frac{2\rho_0}{\chi_{AB}} \boldsymbol{\mu}_-^j \right] \qquad (5.101)$$

where $\mathbf{e}$ is a $M$-vector with 1 in each element. As in the case of Model A, this scheme is conditionally stable for time steps $\Delta t$ smaller than some threshold $\Delta t_m$. A much better semi-implicit algorithm can be constructed by considering the following weak inhomogeneity expansion of the thermodynamic forces

$$\frac{\delta H}{\delta\mu_+(\mathbf{r})} = -\rho_0 \int d\mathbf{r}' \, \{[\phi_{A0}N_A \, g_A(|\mathbf{r} - \mathbf{r}'|) + \phi_{B0}N_B \, g_B(|\mathbf{r} - \mathbf{r}'|)]\Delta\mu_+(\mathbf{r}')$$
$$+ [\phi_{B0}N_B \, g_B(|\mathbf{r} - \mathbf{r}'|) - \phi_{A0}N_A \, g_A(|\mathbf{r} - \mathbf{r}'|)]\Delta\mu_-(\mathbf{r}')\}$$
$$+ O(\Delta\mu_\pm^2) \qquad (5.102)$$

$$\frac{\delta H}{\delta\mu_-(\mathbf{r})} = \rho_0(\phi_{B0} - \phi_{A0}) + (2\rho_0/\chi_{AB})\mu_-(\mathbf{r})$$
$$- \rho_0 \int d\mathbf{r}' \, \{[\phi_{B0}N_B \, g_B(|\mathbf{r} - \mathbf{r}'|) - \phi_{A0}N_A \, g_A(|\mathbf{r} - \mathbf{r}'|)]\Delta\mu_+(\mathbf{r}')$$
$$+ [\phi_{A0}N_A \, g_A(|\mathbf{r} - \mathbf{r}'|) + \phi_{B0}N_B \, g_B(|\mathbf{r} - \mathbf{r}'|)]\Delta\mu_-(\mathbf{r}')\}$$
$$+ O(\Delta\mu_\pm^2) \qquad (5.103)$$

where $\phi_{A0}$ and $\phi_{B0} = 1 - \phi_{A0}$ are the average volume fractions of type A and B segments, respectively, and $g_A(r)$ and $g_B(r)$ are the corresponding Debye functions for A and B homopolymers. Ceniceros and Fredrickson (2004) pointed out that certain terms in the above expressions are destabilizing and should not be included in an implicit scheme. They further argued that the pressure-like potential $\mu_+$ is stiffest, and suggested that it be updated first in the manner of Gauss–Seidel. This leads to the following *semi-implicit Seidel (SIS)* relaxation scheme:[66]

$$
\frac{\mu_+^{j+1} - \mu_+^j}{\Delta t} = - \rho_0(\phi_{A0} N_A \, \mathbf{g}_A + \phi_{B0} N_B \, \mathbf{g}_B) * \Delta \mu_+^{j+1}
$$
$$
+ \rho_A(\mu_+^j - \mu_-^j) + \rho_B(\mu_+^j + \mu_-^j) - \rho_0 \mathbf{e}
$$
$$
+ \rho_0(\phi_{A0} N_A \, \mathbf{g}_A + \phi_{B0} N_B \, \mathbf{g}_B) * \Delta \mu_+^j \qquad (5.104)
$$

$$
\frac{\mu_-^{j+1} - \mu_-^j}{\Delta t} = -\frac{2\rho_0}{\chi_{AB}} \mu_-^{j+1} + \rho_A(\mu_+^{j+1} - \mu_-^j) - \rho_B(\mu_+^{j+1} + \mu_-^j) \qquad (5.105)
$$

Equations (5.104) and (5.105) can be solved sequentially by using FFT–inverse-FFT pairs in $O(M \log_2 M)$ operations. The evaluation of the density operators is most limiting – *two* solutions of the diffusion eqn (3.25) are required per iteration to evaluate the operators appearing in the equations. Thus, each iteration of the SIS algorithm requires $2 \times O(N_s M \log_2 M)$ operations, which is twice the cost of the explicit Euler scheme.[67]

In practice, the SIS scheme significantly outperforms the Euler method, even when the added computational burden is factored in (Ceniceros and Fredrickson, 2004). Figure 5.7 shows the error[68] versus iteration number for a one-dimensional SCFT calculation of a symmetric Model C blend ($N \equiv N_A = N_B$, $\phi_{A0} = 0.5$, $\chi_{AB} N = 4$). Both explicit ($\Delta t = 1.5$) and SIS ($\Delta t = 50.0$) algorithms[69] were implemented pseudo-spectrally with $M = 256$ plane waves and $N_s = 256$ contour steps, applying periodic boundary conditions to a box of size $L = 10 R_g$. The SIS scheme is seen to decrease the error to a level of $O(10^{-14})$ approximately 6.5 times faster than the Euler method, using a factor of 12 fewer iterations.

The volume fraction profiles, $\phi_K(z) = \tilde{\rho}_K(z)/\rho_0$, and self-consistent potential profiles, $\mu_\pm(z)$, resulting from the above optimization of Model C with the SIS algorithm are shown in Fig. 5.8. The periodic boundary conditions are seen to force two interfaces that are similar in form to the analytic Helfand–Tagami solution plotted in Fig. 5.4. An explanation for the observed numerical "stiffness" of the pressure field $\mu_+(z)$ is that it develops sharp attractive cusps centered on

[66]For simplicity, we have set $\lambda_+ = \lambda_- = 1$, although this may not be the optimal choice.

[67]Eric Cochran has reported that in some circumstances a better balance of cost to performance is obtained by evaluating the density operators in eqn (5.105) at time step $j$, thus avoiding a second solution of the diffusion equation. He also notes that the addition of a pair of terms $a\mu_-^j - a\mu_-^{j+1}$ with $a > 0$ to the right-hand side of eqn (5.105) can improve stability at large $\chi_{AB} N$.

[68]The "error" is defined as the $l_1$ norm of the sum of the derivatives $\delta H/\delta\mu_+$ and $\delta H/\delta\mu_-$.

[69]The time steps of the two algorithms were selected for optimum performance. In the case of the explicit scheme, this corresponds to a value of $\Delta t$ just below the stability threshold $\Delta t_m$.

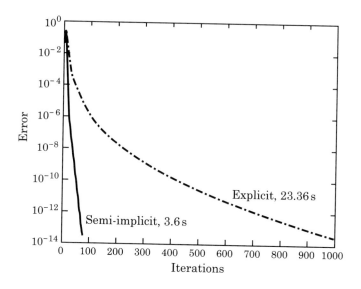

FIG. 5.7. Comparison of the performance of the explicit Euler scheme (dashed–dotted curve) and SIS scheme (solid curve) for a symmetric Model C homopolymer blend with $\chi_{AB}N = 4$, $\phi_{A0} = 0.5$, and $L = 10R_g$. The error shown is the $l_1$ norm of the thermodynamic forces. Reproduced from Ceniceros and Fredrickson (2004).

the interfaces, just as in the Helfand–Tagami solution. In contrast, the exchange potential $\mu_-(z)$ is a smoother function, qualitatively similar to the volume fractions $\phi_A(z) = 1 - \phi_B(z)$. The cusps become narrower and deeper with increasing segregation strength, so the pressure coordinate becomes increasingly stiff as $\chi_{AB}N$ is raised. By damping the high $k$ modes of $\mu_+$ with essentially a fully implicit scheme, the SIS algorithm is able to counteract this stiffness and perform well at large $\chi_{AB}N$. Indeed, the performance of the Euler method deteriorates rapidly with increasing $\chi_{AB}N$, while SIS algorithm continues to be effective.

With slight modifications, the explicit Euler algorithm of eqns (5.100) and (5.101) and the SIS algorithm of eqns (5.104) and (5.105) can be extended to the other models of Chapter 4 that have a pressure-like field $w_+$ and an exchange potential $w_-$. This includes Models B, C, D, E, and J in both the canonical and grand canonical ensembles. Model G also falls in this class, since the effective Hamiltonian has a Hessian matrix[70] with eigenvalues of mixed sign when expressed in terms of the fields $iw$ and $\rho$ and evaluated at a saddle point. Similarly, the explicit and semi-implicit relaxation algorithms of eqns (5.90) and (5.92) are immediately applicable to Models A, H, I, and K. Model F also belongs to this class, since the Hamiltonian expanded in the real fields $iw$ and $i\Phi$

---

[70]The Hessian matrix, or matrix of second derivatives of the energy functional, is simply minus the Jacobian matrix defined by eqn (5.85).

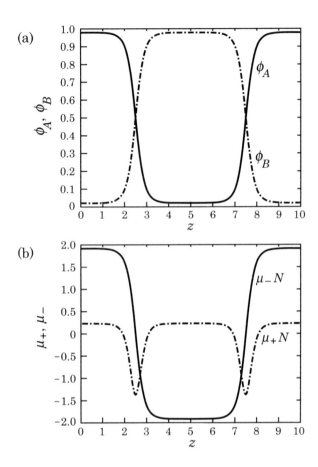

FIG. 5.8. Volume fraction and self-consistent potential profiles for a symmetric Model C homopolymer blend with $\chi_{AB}N = 4$, $\phi_{A0} = 0.5$, and $L = 10R_g$. (a) Volume fractions $\phi_A(z)$ and $\phi_B(z)$. (b) Scaled potentials $\mu_+(z)N$ and $\mu_-(z)N$. Error $= 10^{-12}$. Reproduced from Ceniceros and Fredrickson (2004).

has a Hessian matrix with all negative eigenvalues. In other words, both $iw$ and $i\Phi$ are "pressure-like" in Model F.

In devising the above relaxation schemes, we have used information about the analytic structure of the model Hamiltonian to identify pressure-like coordinates that are to be relaxed up the gradient (to a local maximum) and exchange potential-like coordinates that are to be relaxed down the gradient (to a local minimum). It is important to note that an equivalent approach is to extend the field variables to the complex plane and to always relax *down the complex gradient*. For example, in the case of the canonical Model A, we can relax in the complex $w$ plane to a saddle point solution according to

$$\frac{\partial}{\partial t}w(\mathbf{r}, t) = -\frac{\delta H[w]}{\delta w(\mathbf{r}, t)} = -\frac{1}{u_0}w(\mathbf{r}, t) - i\tilde{\rho}(\mathbf{r}; [iw(t)]) \qquad (5.106)$$

This corresponds to a pair of continuous steepest descent equations for the real and imaginary parts of the complex potential field $w = w_R + iw_I$:

$$\frac{\partial}{\partial t}w_R(\mathbf{r}, t) = -\frac{1}{u_0}w_R(\mathbf{r}, t) + \tilde{\rho}_I(\mathbf{r}; [iw(t)]) \qquad (5.107)$$

$$\frac{\partial}{\partial t}w_I(\mathbf{r}, t) = -\frac{1}{u_0}w_I(\mathbf{r}, t) - \tilde{\rho}_R(\mathbf{r}; [iw(t)]) \qquad (5.108)$$

where the complex density operator $\tilde{\rho}$ has been decomposed according to $\tilde{\rho} = \tilde{\rho}_R + i\tilde{\rho}_I$. For Model A, eqn (5.108) is equivalent to eqn (5.88); in the latter we assumed that $w$ is purely imaginary and made the replacement $\mu \equiv iw = -w_I$. Equation (5.107) is trivially satisfied if $w$ is purely imaginary since then $w_R$ and $\tilde{\rho}_I$ both vanish.

Schemes such as eqn (5.88), which relax SCFT solutions along either the real or imaginary axis of the complex $w$ plane, are obviously less expensive than eqn (5.106), which relaxes both the real and imaginary components of $w$ and thereby requires twice the storage and computational effort. Nevertheless, eqn (5.106) has the advantage of elegance and simplicity, which may outweigh the computational disadvantage, especially when the location and orientation of saddle points in the complex plane are difficult to ascertain. We shall see in Chapter 6 that eqn (5.106) is also a starting point for developing stochastic "complex Langevin" simulation techniques that relax the mean-field approximation.

### 5.3.2.2 Conjugate gradient methods

An optimization expert might wonder why we have devoted so much space to continuous steepest descent (or ascent), which is not considered to be a state-of-the-art numerical optimization technique. The "best" optimization algorithms fall in the class of *conjugate gradient* methods or *variable metric* methods (Press *et al.*, 1992; Dennis and Schnabel, 1996; Nocedal and Wright, 1999), which enjoy super-linear convergence and will locate the minimum of a convex $M$-dimensional quadratic form in $O(M)$ iterations. The variable metric (or "quasi-Newton") methods, however, can be immediately discounted for large-cell SCFT calculations, since they require $O(M^2)$ of storage and $O(M^2)$ operations per iteration.

Nonlinear conjugate gradient techniques are a more likely candidate for high-resolution SCFT, since they require only $O(M)$ storage and operations per iteration. Unfortunately conjugate gradient methods have a serious limitation in the present application; namely their restriction to locating minima and maxima, but *not* saddles. As discussed in the previous section, many of the important polymer field theory models of interest, such as Models B, C, D, E, G, and J, have mean-field solutions that are saddle points of real energy functionals $H[\mu_+, \mu_-]$. In general, conjugate gradient methods are inapplicable to any model with a mixture of repulsive and attractive interactions, because these invariably lead to a saddle character for the functional $H$ in the relevant field variables.

A possible resolution of this difficulty is to return to the functional $\mathcal{F}$ introduced in eqn (5.83). In the case of Model C, for example, $\mathcal{F}$ can be expressed in the continuum description as

$$\mathcal{F}[\mu_+, \mu_-] = \frac{1}{2} \int d\mathbf{r} \left[ \left( \frac{\delta H[\mu_+, \mu_-]}{\delta \mu_+(\mathbf{r})} \right)^2 + \left( \frac{\delta H[\mu_+, \mu_-]}{\delta \mu_-(\mathbf{r})} \right)^2 \right] \qquad (5.109)$$

and has minima with $\mathcal{F} = 0$ that correspond to solutions of the mean-field (SCFT) equations. In addition, $\mathcal{F}$ is convex in the vicinity of these minima, so conjugate gradient methods can in principle be applied. Unfortunately, the functional $\mathcal{F}$ has some undesirable properties that essentially rule out its use for the Model C class of problems. The first is that $\mathcal{F}$ can possess other local minima that do not correspond to solutions of the SCFT equations. More significant is the fact that gradients of $\mathcal{F}$ are prohibitively expensive to compute, requiring $O(N_s M^2 \log_2 M)$ operations. Finally, we have observed that $\mathcal{F}$ confers unphysical stability to the homogeneous saddle point, to an extent that it is difficult to avoid converging this uninteresting solution.

Without some fundamental breakthrough, it appears that conjugate gradient techniques can only be applied to finding SCFT solutions for the class of models that include Models A, F, H, I, and K. Here we outline the technique for the canonical Model A.

In the discrete representation of the $\mu$ field, the problem of locating a SCFT solution of Model A corresponds to that of minimizing the functional $G(\boldsymbol{\mu}) \equiv -H(\boldsymbol{\mu})$. The nonlinear conjugate gradient technique requires that gradients, defined by

$$g_l(\boldsymbol{\mu}) = \frac{\partial G(\boldsymbol{\mu})}{\partial \mu_l}, \quad l = 1, 2, ..., M \qquad (5.110)$$

be computable. For Model A, the gradient $M$-vector $\mathbf{g}(\boldsymbol{\mu})$ coincides with the vector $\mathbf{F}(\boldsymbol{\mu})$ defined in eqn (5.79) and can be computed pseudo-spectrally in $O(N_s M \log_2 M)$ operations. A conjugate gradient algorithm starts with an initial field configuration $\boldsymbol{\mu}^0$, an initial gradient $\mathbf{g}^0 \equiv \mathbf{g}(\boldsymbol{\mu}^0)$, and an initial "direction" of the negative gradient, $\mathbf{h}^0 \equiv -\mathbf{g}^0$. These are updated according to the following series of steps for $j = 0, 1, 2, ...$ until convergence to a desired tolerance is obtained:

1. Obtain a new field configuration according to $\boldsymbol{\mu}^{j+1} = \boldsymbol{\mu}^j + \lambda^j \mathbf{h}^j$, where $\lambda^j$ is determined by a *line minimization*, i.e. from

$$\min_\lambda G(\boldsymbol{\mu}^j + \lambda \mathbf{h}^j)$$

2. Compute a new gradient according to $\mathbf{g}^{j+1} = \mathbf{g}(\boldsymbol{\mu}^{j+1})$.

3. Update the "direction" by means of $\mathbf{h}^{j+1} = -\mathbf{g}^{j+1} + \gamma^j \mathbf{h}^j$, where $\gamma^j$ is given alternatively by the *Fletcher–Reeves* scheme (Press *et al.*, 1992)

$$\gamma^j = \frac{\mathbf{g}^{j+1} \cdot \mathbf{g}^{j+1}}{\mathbf{g}^j \cdot \mathbf{g}^j}$$

or the *Polak–Ribiere* algorithm

$$\gamma^j = \frac{(\mathbf{g}^{j+1} - \mathbf{g}^j) \cdot \mathbf{g}^{j+1}}{\mathbf{g}^j \cdot \mathbf{g}^j}$$

Theoretically, either variant of the above nonlinear conjugate gradient scheme will outperform continuous steepest descent, or steepest descent with line minimization, because optimization directions are chosen to be approximately non-interfering or "conjugate" to directions already used in previous iterations. However, for the method to work well, line minimizations must be performed to a reasonable degree of precision, and in practice this requires about 10 function evaluations per line minimization. In the present SCFT context, function evaluations cost $O(N_s M \log_2 M)$ operations, and gradient evaluations are about 2 times that operation count. Thus, each conjugate gradient step is roughly 5 times as expensive as a relaxation step using an algorithm of the previous section, i.e. eqn (5.90) or (5.92). Nevertheless, for difficult problems with sharp features in the potential field $\mu(\mathbf{r})$ and for high-accuracy calculations, the superior convergence properties of the conjugate gradient technique can pay dividends. In tests on a homopolymer model similar to Model A, Ceniceros and Fredrickson (2004) found that the conjugate gradient method, implemented with either the Fletcher–Reeves or Polak–Ribiere formulas, outperformed the Euler and SIS relaxation schemes by up to a factor of 4 in CPU time.

5.3.2.3 *Other optimization methods*    There are a variety of other *local* optimization strategies that have been applied to converge solutions of the SCFT equations near to some initial guess. Among these are Picard iteration procedures (Shull, 1993; Shi and Noolandi, 1994; Shi *et al.*, 1996; Drolet and Fredrickson, 1999) and Anderson mixing schemes (Thompson *et al.*, 2004). In our experience, none of these schemes are as effective and robust as the algorithms described above. A distinct class of techniques are *global* optimization methods that address the problem of finding the lowest energy configuration among a large number of competing metastable states. Such methods, which include simulated annealing and spectral filtering, are discussed in Section 5.3.4.

### 5.3.3 *Multilevel embedding*

The nonlinear multigrid technique is a powerful way to accelerate the numerical solution of field equations by propagating approximate solutions between fine and coarse spatial grids (Brandt, 1977; Briggs, 1987). A full implementation requires both coarse-to-fine and fine-to-coarse transformations and a relaxation scheme that rapidly damps high-wavenumber components of the error. While a full multigrid scheme for solving the SCFT equations has not yet been achieved, a first step in this direction is the coarse-to-fine *multilevel embedding algorithm* introduced by Ceniceros and Fredrickson (2004). The basic idea is to begin the computation of a mean-field solution on a coarse grid, where the cost per iteration is low, and then gradually increase the resolution of the solution by alternate steps of interpolating to a finer grid and relaxing the interpolated solutions. If designed properly, such an algorithm will result in a significantly reduced CPU time to obtain a solution of desired accuracy, because many of the iterations will have been carried out using grids coarser than the final resolution. It is important to note that the multilevel embedding strategy can be combined with *any* of the relaxation schemes discussed above, including the explicit Euler, SIS, and conjugate gradient methods.

The multilevel embedding scheme requires that a sequence of computational grids be defined corresponding to different numbers of spatial collocation points $M$ and chain contour steps $N_s$. We denote by $G^l = (M^l, N_s^l)$ the number of grid points and contour steps to be used at level $l = 0, 1, 2, ..., N_L$, where $N_L + 1$ is the total number of levels. For a pseudo-spectral implementation, a convenient choice of $M^l$ and $N_s^l$ corresponds to powers of two, so that

$$G^l = (2^l M^0, 2^l N_s^0), \quad l = 0, 1, ..., N_L \tag{5.111}$$

The multilevel embedding algorithm proceeds as follows:

1. Input an initial guess for the potential $\mu$ at the coarsest level, $\mu^0$.
2. For $l = 0, 1, ..., N_L - 1$, carry out the following two steps:
   (a) *Relax* the field on grid $G^l$ using an appropriate relaxation scheme, e.g. the SIS algorithm for Model C, for a total of $m_l$ iterations

$$\mu^{l+1/2} = R_l[\mu^l]$$

   (b) *Interpolate* the resulting relaxed field $\mu^{l+1/2}$ at level $l$ to the next finer level

$$\mu^{l+1} = I_l[\mu^{l+1/2}]$$

3. Starting with $\mu^{N_L}$, relax at the finest-level grid $G^{N_L}$ until the desired accuracy is met.

To complete the definition of the algorithm, it is necessary to specify the relaxation and interpolation operators $R_l$ and $I_l$. $R_l$ is fixed by a choice for the relaxation scheme and the number of relaxation steps at level $l$, $m_l$. For good performance, $m_l$ should decrease with $l$ so that most of the relaxation is done on the coarsest grids. A simple choice that appears to be effective is

$$m_l = \frac{m_0}{2^l}, \quad l = 0, 1, ..., N_L - 1 \tag{5.112}$$

The interpolation to a finer grid is conveniently done in Fourier space. With this choice, $I_l$ is defined by packing the lowest-frequency components of the discrete

Fourier transform of the vector $\boldsymbol{\mu}^{l+1}$, $\hat{\boldsymbol{\mu}}^{l+1}$, with the discrete transform of the vector $\boldsymbol{\mu}^{l+1/2}$. The remaining high-frequency components of $\hat{\boldsymbol{\mu}}^{l+1}$ are padded with zeros and the transform inverted to obtain $\boldsymbol{\mu}^{l+1}$.

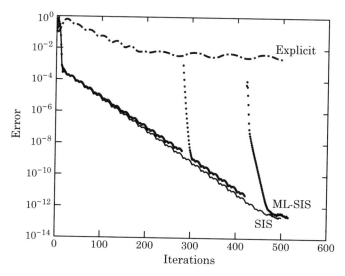

FIG. 5.9. Error versus relaxation step for a symmetric Model C homopolymer blend with $\chi_{AB}N = 8$, $\phi_{A0} = 0.5$, and $L = 20R_g$. The curves labelled Explicit, SIS, and ML-SIS correspond, respectively, to field relaxation on a fine grid ($M = N_s = 256$) using the explicit Euler scheme of eqns (5.100) and (5.101), relaxation on the same fine grid using the SIS scheme of eqns (5.104) and (5.105), and a 3-level embedding with SIS relaxation using $M = N_s = 64$, 128, and 256 grids. Reproduced from Ceniceros and Fredrickson (2004).

Ceniceros and Fredrickson (2004) have applied the above multilevel embedding scheme in combination with the SIS field relaxation algorithm to develop SCFT solutions for Models C and E in one dimension. An impressive acceleration over pure SIS relaxation on the finest grid was observed for both models. For example, Fig. 5.9 shows the error versus iteration number of the relaxation scheme for a Model C calculation with $\chi_{AB}N = 8$, $\phi_{A0} = 0.5$, and $L = 20R_g$. Three grid levels were used with a coarse grid of $G^0 = (64, 64)$ and a fine grid of $G^2 = (256, 256)$. The figure shows the error for explicit Euler stepping on the finest grid ($\Delta t = 0.25$), SIS stepping on the finest grid ($\Delta t = 50.0$), and 3-level embedding with SIS relaxation (ML-SIS). The steep jumps in the error for the ML-SIS curve are associated with the two interpolation events; however, the error rapidly decays back to its original level after a few subsequent iterations. Indeed, after 500 iterations, the errors of the SIS and ML-SIS are comparable. However, the ML-SIS scheme achieved this accuracy using only 6.05 s of CPU time, whereas the SIS stepping on the finest grid required 22.88 s. Thus, in this example, a reduction in computation time of more than a factor of 3 was achieved by multilevel embedding. It is likely that even greater benefits can be obtained by optimizing the choice of $m_l$. These benefits should be most pronounced in large three-dimensional SCFT calculations.

### 5.3.4 *Metastable states*

In Section 5.1.3 we discussed the nature of the multiple solutions exhibited by the mean-field equations. Saddle points corresponding to pure states, defect states, and mixed states are all possible in a given model, and these may be further characterized as stable, metastable, or unstable. It is typically difficult to compute unstable saddle points and they are of limited interest, so we consider them no further.

For a given set of model parameters, the *stable* saddle point, namely the one with the smallest value of $H[w^*]$, is normally the solution of interest. This saddle point corresponds to a field configuration that in the mean-field approximation is lowest in free energy and thus corresponds to the equilibrium configuration of the fluid system. This saddle point is the global minimum of $H[w]$ on the physical $H_I = 0$ ascent manifold. Other saddle points corresponding to local minima with larger values of $H[w^*]$ are thus metastable states of the system. As the model parameters change, two saddle points can exchange stability; namely, a saddle point that was once stable can become metastable, and a new saddle point can become the stable state.

The numerical SCFT algorithms of the previous sections will take an initial guess for the potential field and evolve it to a nearby local minimum of $H = H_R$ on the physical ascent manifold (which might correspond to a local maximum on the search manifold). As such, the algorithms locate stable and metastable states indiscriminately. Since a typical model will possess many metastable states, especially when defect states are considered, how can they be avoided in a numerical SCFT calculation? There is no simple answer to this question that will apply in every situation, so it is useful to gain some insight with a few numerical examples. We begin by discussing the case of a large-cell SCFT simulation.

Figure 5.10 shows an example of the result of a large-cell, two-dimensional SCFT calculation for Model E of an incompressible diblock copolymer melt. The model parameters were set to $\chi_{AB}N = 20.0$ and $f = 0.3$, and 350 field iterations were applied using the SIS relaxation scheme of eqns (5.104)–(5.105). Periodic boundary conditions were imposed on the rhombus-shaped cell shown with side length $L = 32R_g$. Solutions of the diffusion equation were generated pseudo-spectrally using $M = 512^2$ plane waves and collocation on a $512 \times 512$ uniform grid. The cell was chosen in shape and size to be commensurate with 64 primitive unit-cells of the cylindrical mesophase, which is known to be the stable phase for these model parameters. Most importantly, the simulation was started from a *random initial field configuration* created by calls to a random number generator to produce random values of the fields $\mu_+$ and $\mu_-$ at the $512^2$

collocation points.

FIG. 5.10. Segment density configuration (A segments dark, B segments light) for an incompressible diblock copolymer melt (Model E with $b_A = b_B$) resulting from a large-cell SCFT calculation started from random initial conditions. Parameters are $\chi_{AB}N = 20.0$, $f = 0.3$, $L = 32R_g$, and $M = 512^2$. Figure courtesy of E. Cochran.

Figure 5.10 shows a density configuration consistent with the cylindrical mesophase. However, numerous defects are evident in the hexagonal lattice and further field iterations are unable to relax the defects. Thus, we conclude that the configuration obtained is a metastable defect state. Repeated SCFT calculations of this type, but using different random initial conditions, will nearly always return a defective hexagonal lattice. The number, type, and placement of defects in the cell, however, depend on the particular random initial condition that was used to start the simulation.

In stark contrast to Fig. 5.10 is the SCFT result shown in Fig. 5.11, which corresponds to a perfect lattice consisting of 64 primitive cells of the cylindrical mesophase. The parameters used in the simulation were the same as before, but the simulation was started from a "deterministic" initial condition consisting of a single harmonic of the appropriate wavenumber directed along each of the two primitive reciprocal lattice vectors of the rhombic cell. The defect-free structure shown in Fig. 5.11 was fully converged to an error of $1 \times 10^{-3}$ in 30 SIS iterations, the same error achieved in 350 iterations for the randomly initialized structure shown in Fig. 5.10. After 100 iterations, the error in the defect-free configuration of Fig. 5.11 was reduced to $1 \times 10^{-8}$. Thus, by "seeding" a SCFT simulation with an initial condition that has the appropriate symmetry, it is possible to obtain a pure state saddle point that in the present example is the stable state. Moreover, we see that such a seeding procedure can dramatically reduce the computational cost of obtaining a defect-free configuration.

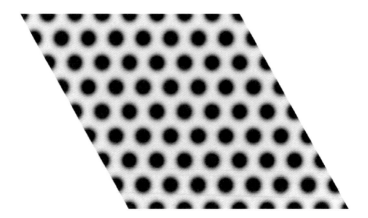

FIG. 5.11. Segment density configuration (A segments dark, B segments light) for an incompressible diblock copolymer melt (Model E) resulting from a large-cell SCFT calculation started from a *deterministic initial condition* that retained only a single harmonic directed along the two primitive reciprocal lattice vectors. Parameters are the same as in Fig. 5.10. Figure courtesy of E. Cochran.

A satisfying feature of the above example is that by starting a large-cell simulation from random initial conditions, a mesophase with the hexagonal *local order* of the stable phase was obtained, albeit a defective mesophase lacking long-range order. Thus, we have some confidence that in tackling a new model where the symmetries of the competing phases are not known in advance, large-cell SCFT calculations launched from random initial conditions can provide insights into the structure of the most stable phases. In the event that two or more structures, distinct by local order or topology, are found in simulations initiated from different random configurations, comparison of the energy values $H[w^*]$ can help establish which is the most stable local arrangement of the fluid. One can then apply one of the procedures described below to obtain a defect-free SCFT solution, which normally corresponds to the stable equilibrium state in the mean-field approximation.

If pure state solutions are desired, effective techniques are needed for removing defects and thereby converting defect states into pure states. Such procedures can be applied in tandem or in alternation with iterative relaxation of the SCFT equations. The following are among the options available:

- *Annealing.* A simple, but useful strategy for removing defects is to gradually increase one or more parameters that control the strength of the inhomogeneities in the fluid during the course of a simulation. In Models C–E for example, the Flory parameter $\chi_{AB}$ can be rescaled by a "coupling parameter" $\lambda$ that is increased linearly or nonlinearly from 0 to 1 while iterating the SCFT equations (Sides and Fredrickson, 2004). This procedure is helpful because the gradual entry into the ordered phases generates large nuclei that have time to grow together in relatively defect-free configurations. It is rarely possible to eliminate all defects by this technique, but the overall defect population is often significantly reduced. A related, but more powerful, global optimization technique of *simulated annealing* has yet to be implemented in large-cell SCFT simulations. In this approach (Kirkpatrick, 1984; Otten and van Ginneken, 1989), a Metropolis or similar Monte Carlo procedure is used to affect barrier crossing while a temperature-like control variable is slowly lowered. The idea behind simulated annealing is to avoid getting stuck in one of the many (higher-energy) local minima while making progress towards locating the global minimum, even if it means that the energy is increased in some iterations. For implementation in the present context, a functional that is convex in the vicinity of the SCFT solutions is required. A good candidate for the models in the Model C class is $G[\mu_-] = H[\mu_+^*[\mu_-], \mu_-]$, i.e. the functional of $\mu_-$ created by replacing $\mu_+$ by its "partial" saddle point configuration $\mu_+^*[\mu_-]$ at a fixed value of $\mu_-$. This partial saddle point can be determined at each $\mu_-$ configuration by a few iterations of one of the relaxation algorithms described in Section 5.3.2.1 to solve $\delta H/\delta\mu_+ = 0$.[71]

- *Zone refining.* In this approach, named by analogy with a familiar metallurgical process, one sweeps a spatial region or "zone" of high temperature (or low segregation strength) through a defective configuration sufficient to cause local melting. As the zone passes a location with defects, the local microstructure melts and then reforms in the wake of the zone, ideally with less defects. In trials with a diblock copolymer melt (Model E), where the zone was a planar slab spanning two of the three dimensions of the simulation cell, we have observed that the method is very effective at creating defect-free mesophases (Sides *et al.*, 2005a). However, there appears to be an undesirable sensitivity to the shape of the zone. In planar zone sweeps, a tendency to recrystallize a lamellar phase was observed, even in situations where the lamellar phase was metastable, rather than stable.

- *Spectral filtering.* A powerful method for removing defects that imparts less of a bias towards particular morphologies is a recently developed spectral filtering technique (Sides *et al.*, 2005a). This strategy takes advantage of the observation noted in the context of Fig. 5.11 that defect-free structures can very quickly be relaxed if a proper "seed" configuration can be supplied as an initial condition. The filtering technique proceeds as follows:

    1. Relax the $\boldsymbol{\mu}$ fields for some number of iterations $m_1$, starting from a random initial condition in a large cell with appropriate boundary conditions.
    2. Perform a fast Fourier transform (FFT) on the resulting configuration $\boldsymbol{\mu}$ and identify the spectral components with the largest amplitudes.

---

[71]The partial saddle point approximation is discussed in more detail in Section 6.5.2.

Apply a suitable filter to remove (set to zero) all components of $\hat{\boldsymbol{\mu}}$ with amplitudes smaller than some threshold $f_1$.

3. Repeat for $j = 2, 3, \ldots$ the steps of relaxing for $m_j$ iterations and filtering out Fourier components with amplitudes less than $f_j$ until a satisfactory defect-free structure has been obtained.

In this filtering algorithm, it is important to choose $f_j$ to be a decreasing function of $j$, so that $\lim_{j \to \infty} f_j = 0$. The essence of the technique is to allow some local order to initially form in the fluid, filter out all but the principal harmonics, and then seed the subsequent relaxation with these dominant modes. In successive steps, the filtering threshold is continuously lowered until it disappears entirely. This procedure has the virtue of not requiring one to guess the symmetry and type of structure before starting the simulation. In preliminary trials, we have found this spectral filtering technique to be robust in both two and three dimensions for a variety of morphologies (Sides *et al.*, 2005a).

- *Unit-cell calculations.* Another strategy for developing pure state SCFT solutions is to carry out unit-cell, rather than large-cell simulations. As discussed in Section 3.6.5, unit-cell calculations attempt to converge a SCFT solution representing one primitive unit-cell of a periodic mesophase structure. A disadvantage of the method is that a reasonably accurate initial guess for the cell shape and size is required, along with a suitable initial condition for the potential field. However, if a large-cell simulation is run in advance, then Fourier analysis can be used to deduce the strongest harmonics and thus extract sufficient information about the local order to initiate a unit-cell calculation. Alternatively, if experimental data is available on the scale and symmetry of the fluid structure, this can serve to define an initial configuration for the fields and cell. If the shape and size of the unit-cell is properly configured, a rough guess for the potential will normally converge rapidly to a defect-free pure state. In some cases, pure states can be obtained from random initial conditions for the potential. However, a more reliable approach is to "seed" the cell with an initial potential constructed from a superposition of the leading harmonics for the structure of interest. Figure 5.11 provided an example of this in a large-cell context; a corresponding example of a unit-cell calculation is shown in Fig. 5.12, where a unit-cell of the gyroid phase in an AB diblock copolymer melt was converged to an error of $1 \times 10^{-5}$ in 60 SIS field iterations starting from an initial guess consisting of the leading harmonics of the Ia$\bar{3}$d cubic space group.

It is evident from the above discussion that procedures for eliminating defect states are currently in a state of development and that the knowledge in this area falls more in the realm of "art" than science. Even if a defect-free SCFT solution can be produced, there is no definitive test that such a solution is the stable state. Metastability can only be strictly ruled out by exhaustively comparing the free

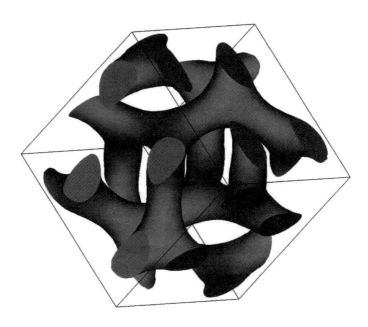

FIG. 5.12. Iso-volume fraction surface corresponding to $\phi_A(\mathbf{r}) = 0.5$ for the gyroid phase of an incompressible diblock copolymer melt (Model E with $b_A = b_B$). The unit-cell SCFT calculation was started from a deterministic initial condition that retained the leading harmonics of the Ia$\bar{3}$d space group. Parameters are $L_1 = L_2 = L_3 = 7.7R_g$, $\chi_{AB}N = 15$, $f = 0.39$, and $M = 64^3$. Figure courtesy of E. Cochran.

energy of the computed structure with the free energies of all possible other structures that could exist in the fluid. This is normally not feasible. However, one can be confident that a structure is very likely stable (in the mean-field approximation) if it is reproducibly obtained from large-cell SCFT simulations starting from different random seeds *and* if it has the lowest free energy of all other structures obtained by repeating this procedure many times and by using several different field relaxation algorithms.

In some circumstances, defect states are intrinsically of interest. Defects are ubiquitous in experimental studies of mesostructured complex fluids, particularly when steps are not taken to eliminate them by annealing or other procedures. An important question is whether defect populations obtained from large-cell SCFT simulations correspond to those observed in experiment and are sensitive to the numerical algorithm applied. Another interesting topic is the dependence of the type and concentration of defects on quenched or annealed chemical disorder present in the fluid. We are aware of only a few studies that have touched on such issues (Hammond *et al.*, 2003; Sides and Fredrickson, 2004).

### 5.3.5  *Variable cell shape methods*

In the previous sections, algorithms were presented for relaxing the potential field variables in both large-cell and unit-cell SCFT calculations. Throughout this discussion, however, it was assumed that the computational cell was fixed in size and shape. In many circumstances, particularly when simulating the spatially periodic mesophases of block copolymer systems, one would like to adjust the cell shape and volume as part of the calculation so as to ensure that the final structure does not have residual stresses imposed by the computational cell. In such instances, it is convenient to continuously adjust the shape and size of the cell, while executing simultaneous potential updates, in order to approach a local minimum of the free energy. Moreover, with such a capability it is straightforward to impose an external stress on the computational cell and thereby examine the *quasi-static mechanical response* of a structured complex fluid phase.

The concept of variable cell size simulations is familiar in the field of particle-based simulations (Allen and Tildesley, 1987; Frenkel and Smit, 1996), where constant pressure molecular dynamics (MD) and Monte Carlo (MC) simulations are now routine. This notion was extended in a powerful way by Parrinello and Rahman (1981) and Ray and Rahman (1984) to allow both cell shape and size to fluctuate during the course of a simulation and in response to external stresses. The Parrinello–Rahman–Ray formalism has proven especially useful in studying phase transformations in crystalline materials and in extracting the linear and nonlinear elastic properties of crystalline and amorphous solids.

In the realm of inhomogeneous polymers, the strategy of minimizing the free energy with respect to cell dimensions has been employed in both spectral and real space unit-cell SCFT calculations (Matsen and Schick, 1994*a*; Bohbot-Raviv and Wang, 2000). Recently, a perturbation scheme for relaxing cell size *and* shape simultaneously with the fields in fully spectral unit-cell calculations was devised (Tyler and Morse, 2003*a*). A related methodology was used to investigate the linear elastic properties of various ordered block copolymer mesophases (Tyler and Morse, 2003*b*). In the large-cell context, Thompson *et al.* (2004) reported the use of *applied strain* SCFT simulations to extract the linear elastic moduli of lamellar multiblock copolymers. Most recently, Barrat *et al.* (2005) developed an *applied stress* formalism that represents an extension of the Parrinello–Rahman–Ray formalism to polymer field theories. The latter is particularly amenable to the pseudo-spectral numerical methods advocated here and is a powerful tool for obtaining stress-free SCFT configurations, as well as for exploring the linear and nonlinear (quasi-static) mechanical response of inhomogeneous polymers. In the following discussion, we closely follow the notions of Parrinello–Rahman–Ray and Barrat *et al.* (2005).

In formulating a variable cell shape scheme, it is necessary to parameterize the shape and volume of the cell. A particularly convenient choice is a parallelepiped

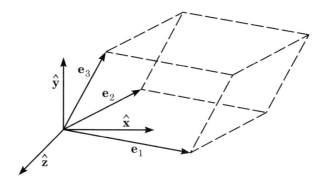

FIG. 5.13. Parallelepiped computational cell used for variable shape simulations. The sides of the cell are defined by three, linearly independent vectors $\mathbf{e}_1$, $\mathbf{e}_2$, and $\mathbf{e}_3$.

spanned by three (in general non-orthogonal) linearly independent vectors $\mathbf{e}_1$, $\mathbf{e}_2$, $\mathbf{e}_3$. With reference to Fig. 5.13, the Cartesian coordinates of any point $\mathbf{r}$ within the cell can be expressed as

$$\mathbf{r} = \mathbf{h} \cdot \mathbf{x} \tag{5.113}$$

where $\mathbf{x}$ is a "cell-scaled" position vector whose components lie in $[0, 1]$. The $3 \times 3$ *cell shape matrix* $\mathbf{h} = (\mathbf{e}_1, \mathbf{e}_2, \mathbf{e}_3)$ has as its $j$th column the Cartesian components of $\mathbf{e}_j$. A similar $2 \times 2$ cell shape matrix can be defined for the purpose of two-dimensional simulations. For example, the shape matrix corresponding to the oblique cell shown in Fig. 3.16 is

$$\mathbf{h} = \begin{pmatrix} L_1 & L_2 \cos \alpha \\ 0 & L_2 \sin \alpha \end{pmatrix} \tag{5.114}$$

Evidently, the length $L_j = |\mathbf{e}_j|$ of the $j$th side of the cell multiplies each term in the $j$th column of $\mathbf{h}$. The cell volume $V$ is given as the determinant of $\mathbf{h}$,

$$V = \det \mathbf{h} \tag{5.115}$$

Thus, we see that the tensor $\mathbf{h}$ contains full information about the shape *and* volume of a parallelepiped cell. An important reason for restricting possible cell shapes to the continuous family of parallelepipeds is that, as described in Section 3.6.5, efficient pseudo-spectral techniques can be used to solve the relevant Fokker–Planck or Chapman–Kolmogorov equations.

Besides $\mathbf{h}$, another important tensor is the *metric tensor* $\mathbf{g}$ defined by[72]

$$\mathbf{g} \equiv \mathbf{h}^T \mathbf{h} \tag{5.116}$$

---

[72]The tensors $\mathbf{h}$ and $\mathbf{g}$ should not be confused with the numerical $M$-vectors introduced in Section 5.3.2.2.

This tensor is useful in transforming dot products and derivatives from the original Cartesian system to cell-scaled coordinates, e.g.

$$\mathbf{r} \cdot \mathbf{r}' = \mathbf{x} \cdot \mathbf{g} \cdot \mathbf{x}', \quad \nabla^2 = \mathbf{g}^{-1} : \nabla_{\mathbf{x}} \nabla_{\mathbf{x}} \tag{5.117}$$

In the Parrinello–Rahman–Ray approach, a new ensemble is applied: the so-called $n\boldsymbol{\tau}T$ ensemble, where $\boldsymbol{\tau}$ is the *thermodynamic tension tensor*, also known as the Piola–Kirchhoff second stress tensor (Wallace, 1973). In this ensemble, the system fluctuates in shape and volume, but at fixed number of particles $n$ and temperature $T$. The thermodynamic tension is conjugate to the strain tensor $\boldsymbol{\epsilon}$ in the sense that the elementary work done on the system is

$$\delta W = V_0 \, \boldsymbol{\tau} : \boldsymbol{\epsilon} \tag{5.118}$$

where $V_0 = \det \mathbf{h}_0$ is the volume of the system in some reference configuration described by a cell shape matrix $\mathbf{h}_0$. The appropriate strain tensor $\boldsymbol{\epsilon}$ to apply in eqn (5.118) is the nonlinear Lagrangian expression

$$\boldsymbol{\epsilon} \equiv \frac{1}{2} \left[ (\mathbf{h}_0^T)^{-1} \mathbf{h}^T \mathbf{h} \, (\mathbf{h}_0)^{-1} - \mathbf{1} \right] = \frac{1}{2} \left[ (\mathbf{h}_0^T)^{-1} \mathbf{g} \, (\mathbf{h}_0)^{-1} - \mathbf{1} \right] \tag{5.119}$$

which relates the way in which a vector in the reference configuration $\mathbf{h}_0$ transforms to a vector $\mathbf{h}$ in the deformed configuration (Landau and Lifshitz, 1986). In the $n\boldsymbol{\tau}T$ ensemble, $\boldsymbol{\tau}$ is viewed as an *external* tension that is applied to the simulation cell, and by prescription of the components of this tensor field, the cell will achieve some average shape and size dictated by $\langle \mathbf{h} \rangle$. The tension can be related to the more familiar externally applied Cauchy stress $\boldsymbol{\sigma}^{ex}$ by the formula (Ray and Rahman, 1984)

$$\boldsymbol{\sigma}^{ex} = \frac{V_0}{V} \, \mathbf{h} \, (\mathbf{h}_0)^{-1} \boldsymbol{\tau} \, (\mathbf{h}_0^T)^{-1} \mathbf{h}^T \tag{5.120}$$

In the case of a purely isotropic external stress, $\boldsymbol{\sigma}^{ex} = P\mathbf{1}$, where $P$ is the pressure, the $n\boldsymbol{\tau}T$ ensemble reduces to the familiar $nPT$ isobaric-isothermal ensemble.

Any of the polymer field theory models described in Chapter 4 can be reformulated in the $n\boldsymbol{\tau}T$ ensemble. This reformulation simply amounts to introducing cell-scaled variables according to eqn (5.113) and augmenting the statistical weight by the Boltzmann factor $\exp(-\beta V_0 \boldsymbol{\tau} : \boldsymbol{\epsilon})$. In the case of Model A, for example, the $n\boldsymbol{\tau}T$ partition function is given by

$$\mathcal{Z}_\tau(n, \boldsymbol{\tau}, T) = \int d\mathbf{h} \int \mathcal{D}w \, \exp(-H_\tau[w, \mathbf{h}]) \tag{5.121}$$

with the effective Hamiltonian

$$H_\tau[w, \mathbf{h}] = \frac{\det \mathbf{h}}{2u_0} \int d\mathbf{x} \, [w(\mathbf{x})]^2 + \beta V_0 \boldsymbol{\tau} : \boldsymbol{\epsilon} - n \ln Q[iw, \mathbf{g}]$$
$$- n \ln \det \mathbf{h} + \ln n! \tag{5.122}$$

which depends on both potential field $w$ and shape $\mathbf{h}$ variables. The last two terms in the Hamiltonian arise from the ideal gas partition function $\mathcal{Z}_0$ appearing in eqn (4.71). The reduced partition function $Q[iw, \mathbf{g}]$ for a polymer chain experiencing the imaginary field $iw(\mathbf{x})$ in a cell of shape $\mathbf{g} = \mathbf{h}^T \mathbf{h}$ is defined by the ratio of path integrals

$$Q[iw, \mathbf{g}] = \frac{\int \mathcal{D}\mathbf{x}\, e^{-\frac{3}{2b^2} \int_0^N ds\, \frac{d\mathbf{x}}{ds} \cdot \mathbf{g} \cdot \frac{d\mathbf{x}}{ds} - \int_0^N ds\, iw(\mathbf{x}(s))}}{\int \mathcal{D}\mathbf{x}\, e^{-\frac{3}{2b^2} \int_0^N ds\, \frac{d\mathbf{x}}{ds} \cdot \mathbf{g} \cdot \frac{d\mathbf{x}}{ds}}} \tag{5.123}$$

and is computed according to

$$Q[iw, \mathbf{g}] = \int d\mathbf{x}\, q(\mathbf{x}, N; [iw, \mathbf{g}]) \tag{5.124}$$

The propagator $q$ satisfies the anisotropic diffusion equation

$$\frac{\partial}{\partial s} q(\mathbf{x}, s) = \frac{b^2}{6} \mathbf{g}^{-1} : \nabla_{\mathbf{x}} \nabla_{\mathbf{x}} q(\mathbf{x}, s) - iw(\mathbf{x}) q(\mathbf{x}, s) \tag{5.125}$$

subject to the initial condition $q(\mathbf{x}, 0; [iw, \mathbf{g}]) = 1$. The reader should note a close analogy between eqns (5.125) and (3.32).

In the mean-field approximation, we wish to solve the simultaneous equations

$$\left. \frac{\delta H_\tau[w, \mathbf{h}]}{\delta w(\mathbf{x})} \right|_{w^*, \mathbf{h}^*} = 0, \quad \left. \frac{\partial H_\tau[w, \mathbf{h}]}{\partial \mathbf{h}} \right|_{w^*, \mathbf{h}^*} = 0 \tag{5.126}$$

at fixed $n$, $\tau$, and $T$ for the self-consistent potential $w^*(\mathbf{x})$ and the optimum box shape and size $\mathbf{h}^*$. For the purpose of satisfying these equations, it is convenient to have an analytical expression for the partial derivative $\partial H_\tau / \partial \mathbf{h}$. It can be shown by explicit differentiation that (Barrat $et$ $al.$, 2005)

$$\frac{\partial H_\tau[w, \mathbf{h}]}{\partial \mathbf{h}} \mathbf{h}^T = \mathbf{1} \left( \frac{V}{2u_0} \int d\mathbf{x}\, [w(\mathbf{x})]^2 - n \right)$$
$$+ \beta V_0 \mathbf{h} \left( \mathbf{h}_0^{-1} \boldsymbol{\tau} (\mathbf{h}_0^T)^{-1} + \mathbf{\Sigma}[iw, \mathbf{g}] \right) \mathbf{h}^T \tag{5.127}$$

where $\mathbf{\Sigma}$ is a symmetric tensor defined by

$$\mathbf{\Sigma}[iw, \mathbf{g}] \equiv -\frac{2nk_B T}{V_0} \frac{\partial \ln Q[iw, \mathbf{g}]}{\partial \mathbf{g}}$$
$$= \frac{3nk_B T}{b^2 V_0} \left( \int_0^N ds \left\langle \frac{d\mathbf{x}(s)}{ds} \frac{d\mathbf{x}(s)}{ds} \right\rangle_{[iw, \mathbf{g}]} \right.$$
$$\left. - \int_0^N ds \left\langle \frac{d\mathbf{x}(s)}{ds} \frac{d\mathbf{x}(s)}{ds} \right\rangle_{[0, \mathbf{g}]} \right) \tag{5.128}$$

and where $\langle ... \rangle_{[iw, \mathbf{g}]}$ denotes the single-chain average defined by eqn (3.47).

It is convenient to define an *internal stress operator* by the expression

$$\tilde{\boldsymbol{\sigma}}[iw, \mathbf{h}] \equiv \frac{V_0}{V} \mathbf{h} \, \boldsymbol{\Sigma}[iw, \mathbf{g}] \mathbf{h}^T$$

$$= \frac{n}{V} \int d\mathbf{r} \, \left( \langle \hat{\boldsymbol{\sigma}}(\mathbf{r}) \rangle_{[iw, \mathbf{h}]} - \langle \hat{\boldsymbol{\sigma}}(\mathbf{r}) \rangle_{[0, \mathbf{h}]} \right) \quad (5.129)$$

where we have recalled the definition of the microscopic, single-chain stress operator $\hat{\boldsymbol{\sigma}}(\mathbf{r})$ given by eqn (3.28).[73] Thus, the internal stress is simply the volume average of the local stress contributed by each of the $n$ chains. The contribution from the term $\langle \hat{\boldsymbol{\sigma}}(\mathbf{r}) \rangle_{[0, \mathbf{h}]}$ in eqn (5.129) proves to be *isotropic*, so the anisotropic contribution to the internal stress arises solely from the remaining term. With the above internal stress definition and recalling the definition (5.120) of the external stress, one sees that eqn (5.127) can be reexpressed as

$$\frac{\partial H_\tau[w, \mathbf{h}]}{\partial \mathbf{h}} \mathbf{h}^T = 1 \left( \frac{V}{2u_0} \int d\mathbf{x} \, [w(\mathbf{x})]^2 - n \right) + \beta V \left( \boldsymbol{\sigma}^{ex} + \tilde{\boldsymbol{\sigma}}[iw, \mathbf{h}] \right) \quad (5.130)$$

The equilibrium condition (5.126) is thus achieved when the sum of the isotropic terms in eqn (5.130) and the isotropic part of $\tilde{\boldsymbol{\sigma}}$ balance the externally imposed pressure,[74] and when the anisotropic (deviatoric) parts of the external stress and the internal stress exactly cancel. In most calculations, we are only interested in the anisotropic part of $\tilde{\boldsymbol{\sigma}}$. This can be computed efficiently from a solution of the diffusion equation (5.125) by means of the following formula derived by combining eqns (5.129) and (3.79):

$$\tilde{\boldsymbol{\sigma}}[iw, \mathbf{h}] = \frac{nb^2 k_B T}{3VQ[iw, \mathbf{g}]} \int d\mathbf{x} \int_0^N ds$$

$$\times q(\mathbf{x}, s; [iw, \mathbf{g}]) \, (\mathbf{h}^T)^{-1} \nabla_\mathbf{x} \nabla_\mathbf{x} \, \mathbf{h}^{-1} \, q(\mathbf{x}, N - s; [iw, \mathbf{g}]) \quad (5.131)$$

The anisotropic part of the tensor $\boldsymbol{\Sigma}$ is given by a similar formula:

$$\boldsymbol{\Sigma}[iw, \mathbf{g}] = \frac{nb^2 k_B T}{3V_0 Q[iw, \mathbf{g}]} \int d\mathbf{x} \int_0^N ds$$

$$\times q(\mathbf{x}, s; [iw, \mathbf{g}]) \, \mathbf{g}^{-1} \nabla_\mathbf{x} \nabla_\mathbf{x} \, \mathbf{g}^{-1} \, q(\mathbf{x}, N - s; [iw, \mathbf{g}]) \quad (5.132)$$

Both of these internal stress expressions can be evaluated pseudo-spectrally in $O(N_s M \log_2 M)$ operations using $M$ collocation points in the scaled cell.

By means of the above formulas it is possible to devise a simple continuous steepest descent scheme to optimize the cell size and shape, described by $\mathbf{h}$,

---

[73] Although we referred to $\hat{\boldsymbol{\sigma}}(\mathbf{r})$ as a microscopic *elastic* stress in Chapter 3, in a field theory model where the segmental interactions are purely local (delta-function-like) the polymers provide the only venue for transmitting stress; e.g., see Section 4.5.2 of Doi and Edwards (1986). Thus, the average of $\hat{\boldsymbol{\sigma}}(\mathbf{r})$ for a many-chain model has both elastic *and* osmotic contributions.

[74] In Model A, the relevant pressure is the osmotic pressure $\Pi$.

while simultaneously relaxing the potential field $w(\mathbf{r})$ to its saddle point value $w^*(\mathbf{r})$. Specifically, because we seek a local minimum of $H_\tau$ with respect to $\mathbf{h}$, it is appropriate to relax down the gradient according to

$$\frac{d}{dt}\mathbf{h}(t) = -\lambda_h \frac{\partial H_\tau[w(t),\mathbf{h}(t)]}{\partial \mathbf{h}(t)} \tag{5.133}$$

where $\lambda_h > 0$ is an adjustable relaxation rate parameter. Equation (5.133) can be integrated in concert with eqn (5.88) to evolve $\mathbf{h}$ and $\mu = iw$ towards a mean-field solution.

In other instances it is useful to relax the cell shape during an SCFT calculation, but at *fixed* cell volume $V = \det \mathbf{h}$. A suitable continuous descent scheme for this purpose is the fictitious dynamics (Barrat *et al.*, 2005)

$$\frac{d}{dt}\mathbf{h}(t) = -\lambda_h\,\mathbf{h}\mathbf{D}\frac{\partial H_\tau[w(t),\mathbf{h}(t)]}{\partial \mathbf{h}(t)} \tag{5.134}$$

In this equation, $\mathbf{D}$ is an anisotropic projection operator whose action on an arbitrary tensor $\mathbf{m}$ is a traceless tensor, i.e.

$$\mathbf{D}\mathbf{m} = \mathbf{m} - \frac{1}{3}\mathrm{Tr}(\mathbf{m})\,\mathbf{1} \tag{5.135}$$

The presence of this operator in eqn (5.134) ensures that $\mathbf{h}^{-1}d\mathbf{h}/dt$ is traceless, which in turn maintains $\det \mathbf{h} = V$ as a constant of the motion because of the identity

$$\frac{d}{dt}\ln \det \mathbf{h} = \mathrm{Tr}\left(\mathbf{h}^{-1}\frac{d}{dt}\mathbf{h}\right) \tag{5.136}$$

In evaluating the right-hand side of eqn (5.134), it should be noted that only the deviatoric components of the stress are required, because

$$\mathbf{D}\frac{\partial H_\tau}{\partial \mathbf{h}}\mathbf{h}^T = \beta V \mathbf{D}(\boldsymbol{\sigma}^{ex} + \tilde{\boldsymbol{\sigma}}) \tag{5.137}$$

Model A is of limited interest for variable cell shape methods because it possesses only a trivial homogeneous saddle point in a simulation cell with periodic boundary conditions. However, the above scheme can be applied to any *compressible* model of a polymer solution or melt, such as Models D, F, G, H, and I. In contrast, the *incompressible* Models B, C, E, and J require a slightly different formulation. For such models, a pressure (isotropic contribution to $\boldsymbol{\sigma}^{ex}$) is not externally applied but is computed internally (through $w_+$) to maintain constant segment density. Since the density is held constant, the system volume $V$ and the number of molecules $n$ are not independent variables, so care must be exercised in forming derivatives with respect to the cell shape tensor $\mathbf{h}$.

There are two principal types of variable cell shape simulations for incompressible systems. One type restricts the shape tensor $\mathbf{h}$ to a constant volume

$V = V_0 = \det \mathbf{h} = \det \mathbf{h}_0$ manifold, which implies that the total number of molecules $n$ is also maintained constant. The derivative formulas (5.127) and (5.130) are thus appropriate, and a suitable relaxation scheme for the cell shape $\mathbf{h}$ at constant volume is given by eqn (5.134). This type of simulation is appropriate for examining the linear or nonlinear quasi-static stress–strain behavior of incompressible models. Starting with a stress-free initial configuration $\mathbf{h}_0$, an external tension $\tau$ is applied and the relaxation scheme (5.134) is used to obtain an equilibrated cell shape $\mathbf{h}$. Equations (5.119) and (5.120) can then be evaluated to obtain the desired relationship between the applied stress $\sigma^{ex}$ and the measured strain $\epsilon$.

A second type of variable cell shape SCFT simulation for an incompressible system allows the cell volume to change, but changes the number of molecules proportionally in order to maintain constant segment density. Such simulations are appropriate for adjusting the cell size and shape in order to obtain a stress-free configuration of the system. This is a particularly useful capability to have when performing unit-cell SCFT calculations of block copolymer mesophases. For such purposes, it is necessary to minimize not the *extensive* energy $H[w_\pm, \mathbf{h}]$ with respect to $\mathbf{h}$, but rather the *intensive* energy density, $H_V \equiv H/V$. As an example, the energy density of Model E for an incompressible diblock copolymer melt is given by[75]

$$H_V[w_\pm, \mathbf{h}] = \rho_0 \int d\mathbf{x} \left[ (1/\chi_{AB}) w_-^2 - i w_+ \right] - \frac{\rho_0}{N} \ln Q[w_A, w_B, \mathbf{g}]$$
$$+ \frac{\rho_0}{N} [\ln(\rho_0/N) - 1] \tag{5.138}$$

where the last two terms are the ideal gas contributions and we recall the definition of the complex fields $w_A, w_B \equiv i w_+ \mp w_-$, which are purely real at a saddle point. The single-chain partition function $Q$ in this expression can be computed by means of eqns (3.99) and (3.104) in cell-scaled form, analogous to eqns (5.124) and (5.125). Upon inspecting eqn (5.138), it is evident that the only contribution to the derivative of $H_V$ with respect to $\mathbf{h}$ at constant total segment density $\rho_0$ comes from the $\ln Q$ term. Thus, by comparison with eqn (5.127), the derivative is given by

$$\frac{\partial H_V[w_\pm, \mathbf{h}]}{\partial \mathbf{h}} = \beta \, \mathbf{h} \, \mathbf{\Sigma}[w_A, w_B, \mathbf{g}] \tag{5.139}$$

where $\mathbf{\Sigma}$ is computed by means of a formula analogous to eqn (5.132)

$$\mathbf{\Sigma}[w_A, w_B, \mathbf{g}] = \frac{n k_B T}{3 V Q[w_A, w_B, \mathbf{g}]} \int d\mathbf{x} \int_0^N ds \, [b(s)]^2 \, q(\mathbf{x}, s; [w_A, w_B, \mathbf{g}])$$
$$\times \mathbf{g}^{-1} \nabla_{\mathbf{x}} \nabla_{\mathbf{x}} \mathbf{g}^{-1} q_c(\mathbf{x}, N - s; [w_A, w_B, \mathbf{g}]) \tag{5.140}$$

---

[75] In deriving this formula, we have introduced cell-scaled coordinates according to eqn (5.113) and the replacement $n \to \rho_0 V/N$ is made.

In this expression, $b(s)$ is the species-dependent statistical segment length defined by eqn (3.95) and $q$ and $q_c$ are the propagators of Section 3.3.2 expressed in cell-scaled coordinates. Thus, a suitable scheme to relax the cell shape *and* size to achieve a stress-free configuration of an incompressible block copolymer melt is

$$\frac{d}{dt}\mathbf{h} = -\lambda_h \frac{\partial H_V}{\partial \mathbf{h}} = -\lambda_h \beta\, \mathbf{h}\, \boldsymbol{\Sigma}[w_A, w_B, \mathbf{g}] \qquad (5.141)$$

A simple forward Euler discretization can be used to approximate the time derivative on the left-hand side. According to this equation, when the equilibrium cell configuration $\mathbf{h}^*$ is achieved, the internal stress $\tilde{\sigma}$ vanishes, because $\tilde{\sigma} = \mathbf{h}\boldsymbol{\Sigma}\mathbf{h}^T = 0$. In concert with the cell relaxation, the fields $w_A$ and $w_B$ (or equivalently $\mu_\pm$) can be relaxed using the Euler or SIS schemes of Section 5.3.2.1. The parameters $\lambda_\pm$ and $\lambda_h$ are to be adjusted for stability and optimal performance. In our experience this requires evolving the cell shape and volume more slowly than the potential fields.

Variable cell shape simulations of Model E at constant volume, both with and without imposed external stress, have been recently reported (Barrat *et al.*, 2005). An example of an SCFT simulation of the second type, namely one to relax the cell shape and volume at zero external stress, is shown in Fig. 5.14. The figure illustrates how an initial square cell configuration of an AB diblock copolymer melt in the cylindrical morphology evolves under the dynamics of eqn (5.141) to the equilibrium rhombic unit-cell with the optimal cell volume.

 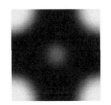 

FIG. 5.14. Evolution of a diblock copolymer morphology under zero applied stress. The calculation for Model E assumed that $b_A = b_B$ and applied the following parameters: $\chi_{AB}N = 20$, $f = 0.3$, $M = 64 \times 64$, $L_1 = L_2 = 4.25R_g$. The angle $\alpha$ of the shape tensor in eqn (5.114) was initialized to $91°$, and the initial potential field configuration was proportional to the leading harmonic of P4 tetragonal symmetry. The three panels from left to right correspond to the A species volume fraction after 10, 30, and 100 potential field and shape tensor updates, respectively, to achieve an error of $1 \times 10^{-3}$. At 200 iterations, the error was $1 \times 10^{-5}$. Figure courtesy of E. Cochran.

### 5.3.6 *Mean-field dynamics*

It is important to emphasize that the "fictitious" dynamics introduced in Section 5.3.2.1 to relax potential fields is not a physically realistic dynamics for either solutions or melts of polymers. Although the time-dependent properties of inhomogeneous polymers are not a focus of this monograph, it is worth commenting briefly on how SCFT relaxation algorithms can be adapted to at least qualitatively capture the kinetic pathway of a complex fluid towards its equilibrium state. Such adaptations are often referred to as *dynamical mean-field theory*, *dynamical SCFT* (Yeung and Shi, 1999; Reister *et al.*, 2001; Hasegawa and Doi, 1997; Müller and Schmid, 2005), or as popularized by Fraaije and coworkers (Fraaije, 1993; Fraaije *et al.*, 1997; Maurits and Fraaije, 1997), *dynamic density functional theory (DDFT)*.

For simplicity, the subject of mean-field dynamics will be introduced in the context of Model A. It is helpful to return to the original form of Model A, described by the two-field $(\rho, w)$ theory summarized in eqns (4.69) and (4.70). The mean-field approximation for this theory corresponds to

$$\left.\frac{\delta H[\rho, w]}{\delta \rho(\mathbf{r})}\right|_{w^*, \rho^*} = -iw^*(\mathbf{r}) + u_0 \rho^*(\mathbf{r}) = 0 \qquad (5.142)$$

$$\left.\frac{\delta H[\rho, w]}{\delta w(\mathbf{r})}\right|_{w^*, \rho^*} = i[\tilde{\rho}(\mathbf{r}; [iw^*]) - \rho^*(\mathbf{r})] = 0 \qquad (5.143)$$

Of interest are physically motivated relaxation schemes that will evolve a system with prescribed values of $\rho(\mathbf{r})$ and $w(\mathbf{r})$ to the equilibrium state $(\rho^*, w^*)$ described by these saddle point equations. Since the density is a *conserved field*, i.e. $\int d\mathbf{r}\rho(\mathbf{r}) = nN$, diffusion limitations will cause the longest-wavelength Fourier components of $\rho(\mathbf{r})$ to be slowly relaxing variables. Indeed, the components of $\hat{\rho}(\mathbf{k})$ with $k = |\mathbf{k}| \to 0$ will relax more slowly than the stress relaxation time of the fluid, so that, at least for these modes, viscoelastic coupling effects can be neglected (Milner, 1993). This suggests the following simple dynamics scheme, whereby the density field relaxes by diffusion and the potential field is adiabatically slaved to the instantaneous density:

$$\frac{\partial}{\partial t}\rho(\mathbf{r}, t) = \Lambda \nabla^2 \frac{\delta H[\rho, \mu]}{\delta \rho(\mathbf{r}, t)} = \Lambda \nabla^2 [u_0 \rho(\mathbf{r}, t) - \mu(\mathbf{r}, t)] \qquad (5.144)$$

The Onsager kinetic coefficient $\Lambda$ $(> 0)$ appearing in this equation defines an effective diffusion coefficient. At each time $t$, the *real* potential field $\mu(\mathbf{r}, t) \equiv iw(\mathbf{r}, t)$ is obtained by applying the local equilibrium condition

$$\frac{\delta H[\rho, \mu]}{\delta \mu(\mathbf{r}, t)} = \tilde{\rho}(\mathbf{r}; [\mu(t)]) - \rho(\mathbf{r}, t) = 0 \qquad (5.145)$$

to determine $\mu$ as a functional of the instantaneous $\rho$. In short, the above scheme assumes that: (i) the field $\rho$ constitutes the slowest relaxing variables in a polymer solution, (ii) all other degrees of freedom can be slaved to $\rho$, including $\mu$, and (iii) one is only interested in asking detailed questions about the lowest-frequency and longest-wavelength dynamical response of the fluid. With this dynamics scheme, polymer-specific kinetic factors, such as degree of entanglement, molecular weight and architecture, monomeric friction coefficients, etc., only contribute through their influence on the kinetic coefficient $\Lambda$. Formal methods have been developed for extending such field equations to describe higher-frequency and shorter-wavelength relaxation phenomena (Zwanzig, 1961; Ottinger, 1998; Kawasaki and Sekimoto, 1987), but these have proved to be very difficult to implement in practice.

The mean-field dynamics defined by eqns (5.144) and (5.145) is the original DDFT scheme devised by Fraaije and collaborators (Fraaije, 1993; Fraaije *et al.*, 1997). This scheme has met with some success in comparison with experiments conducted under weakly non-equilibrium conditions over long times (Knoll *et al.*, 2004). Unfortunately, it suffers from the following computational drawback: at each time step in the integration of eqn (5.144), it is necessary to iteratively solve the nonlinear eqn (5.145) to obtain the potential $\mu$ consistent with the instantaneous density $\rho$. In practice, 5 to 10 iterations per time step are required to solve[76] eqn (5.145) to reasonable accuracy, each iteration requiring an independent solution of the diffusion equation (3.25) and evaluation of the density operator $\tilde{\rho}$. Thus, the DDFT is a factor of 5–10 times more expensive per time step than the relaxation schemes described in Section 5.3.2.1. A more efficient "$\rho$-explicit" alternative to eqns (5.144)–(5.145) (Fraaije and Sevink, 2003) is obtained by eliminating eqn (5.145) and replacing $\mu(\mathbf{r}, t)$ on the right-hand side of eqn (5.144) by $u_0\tilde{\rho}(\mathbf{r}; [u_0\rho(t)])$, i.e.

$$\frac{\partial}{\partial t}\rho(\mathbf{r}, t) = \Lambda\nabla^2 u_0\{\rho(\mathbf{r}, t) - \tilde{\rho}(\mathbf{r}; [u_0\rho(t)])\} \tag{5.146}$$

For a prescribed density field $\rho(\mathbf{r}, t)$, the right-hand side of this equation can be evaluated using only a *single* solution of the diffusion equation, amounting to $O(N_s M \log_2 M)$ operations with a pseudo-spectral technique. The $\rho$-explicit scheme is thus no more expensive per time step than the non-conserved relaxation dynamics of eqn (5.88). Moreover, the time integration can be done semi-implicitly, analogous to eqn (5.92), by extracting the linear contribution to the $\tilde{\rho}$ term.

Both the DDFT and $\rho$-explicit algorithms conserve segment density and have an equilibrium consistent with eqns (5.142) and (5.143). Their dynamical trajectories appear to be qualitatively similar, although to our knowledge no quantitative comparison study has been performed. Since neither scheme appears to be theoretically advantaged, we favor use of the more efficient eqn (5.146). It is interesting to note that in the derivation of the $\rho$-explicit scheme, the fields $\mu$ and $\rho$ were assumed to be constrained according to the equilibrium condition $\mu = u_0\rho$ at all times. This relationship can be used to rewrite eqn (5.146) as the following "$\mu$-explicit" dynamics:

---

[76]The relaxation schemes of Section 5.3.2.1 are useful for this "target homopolymer density" problem of finding the equilibrium potential field consistent with a prescribed segment density (Ceniceros and Fredrickson, 2004).

$$\frac{\partial}{\partial t}\mu(\mathbf{r}, t) = \Lambda_\mu \nabla^2 \left\{ \frac{1}{u_0}\mu(\mathbf{r}, t) - \tilde{\rho}(\mathbf{r}; [\mu(t)]) \right\}$$

$$= -\Lambda_\mu \nabla^2 \frac{\delta H[\mu]}{\delta\mu(\mathbf{r}, t)} \qquad (5.147)$$

where $\Lambda_\mu \equiv u_0^2 \Lambda$ and $H[\mu]$ is the effective Hamiltonian for the simplified version of Model A given in eqn (4.72). The $\mu$-explicit scheme, sometimes referred to as *external potential dynamics (EPD)* (Maurits and Fraaije, 1997), can thus be viewed as a simple modification of the non-conserved relaxation equation (5.88) in which conservation on $\mu$ (and thus $\rho$) is imposed by the replacement $\lambda \rightarrow -\Lambda_\mu \nabla^2$.

Dynamical equations similar to eqns (5.144) and (5.145) can be easily constructed for all the models described in Chapter 4. However, the models that impose an incompressibility constraint on the total species density, e.g. Models B, C, and E, cannot be reduced to $\rho$-explicit or $\mu$-explicit form. This computational inconvenience is created by the fact that in such models there is one more independent potential field than density field. The "extra" potential field can be identified with the pressure $\mu_+ = iw_+$, which in a physically realistic dynamics should be chosen to enforce the incompressibility constraint along the entire dynamical trajectory. For example, in the case of the binary blend Model C, a simple conserved dynamics scheme is

$$\frac{\partial}{\partial t}\mu_-(\mathbf{r}, t) = \lambda_- \nabla^2 \frac{\delta H[\mu_+, \mu_-]}{\delta\mu_-(\mathbf{r}, t)} = \lambda_- \nabla^2 \left[ (2\rho_0/\chi_{AB})\mu_-(\mathbf{r}, t) \right.$$

$$\left. - \tilde{\rho}_A(\mathbf{r}; [w_A(t)]) + \tilde{\rho}_B(\mathbf{r}; [w_B(t)]) \right] \qquad (5.148)$$

where $w_{A/B} = \mu_+ \mp \mu_-$ and $\mu_+$ is chosen at each $t$ to enforce incompressibility, i.e.

$$\frac{\delta H[\mu_+, \mu_-]}{\delta\mu_+(\mathbf{r}, t)} = \tilde{\rho}_A(\mathbf{r}; [w_A(t)]) + \tilde{\rho}_B(\mathbf{r}; [w_B(t)]) - \rho_0 = 0 \qquad (5.149)$$

Because eqn (5.149) must be iteratively solved at each time step to obtain $\mu_+$ corresponding to the instantaneous value of $\mu_-$, this modified EPD strategy suffers from the same computational disadvantage as the dual density–potential dynamics of eqns (5.144) and (5.145). One possible way to minimize the computational cost of solving eqn (5.149) is to attempt to satisfy it by means of the non-conserved relaxation dynamics of eqn (5.98) with a large value of the relaxation coefficient $\lambda_+$. Semi-implicit time stepping of this equation by means of eqn (5.104) in concert with explicit or semi-implicit time stepping of eqn (5.148) might be accurate enough that only two or three solutions of the diffusion equation would be required per time step.

A discussion of related mean-field dynamics schemes can be found in the review article by Müller and Schmid (2005). A major concern about all such strategies is that they neglect interactions between the density degrees of freedom in the fluid and the *momentum density*, which is another conserved field whose small $k$ Fourier components relax slowly. These "hydrodynamic mode couplings" are known to be crucially important to dynamic critical phenomena (Hohenberg and Halperin, 1977) in both simple and complex fluids and to polymer-specific kinetic processes such as viscoelastic phase separation and shear enhancement of concentration fluctuations (Helfand and Fredrickson, 1989; Onuki, 1990; Milner, 1993; Doi and Onuki, 1992; Ottinger and Grmela, 1997; Tanaka, 1993; Araki and Tanaka, 2001). In descriptions of the latter phenomena, couplings between the local polymer stress, i.e. the operator given in eqn (3.79), and the species and momentum densities are also invoked. A framework for mean-field dynamics that includes such couplings and is consistent with the spirit and notations of this monograph has been described by Fredrickson (2002).

## 5.4 Applications and examples

We conclude this chapter on self-consistent field theory by providing some examples of numerical results that were obtained by applying the methods outlined in the previous sections. Since hundreds, and perhaps thousands, of SCFT calculations have been reported in the literature, a comprehensive overview is not attempted.

### 5.4.1 Unit-cell calculations

The most significant numerical SCFT results of the past decade are the unit-cell calculations of Matsen, Schick, and collaborators for the ordered mesophases of block copolymers and their alloys. Whereas earlier investigators had made circular (or spherical) approximations to the unit-cells of block copolymer mesophases for the purpose of reducing the dimensionality of the calculations (Helfand and Wasserman, 1976; Whitmore and Vavasour, 1992; Shull, 1993; Shi and Noolandi, 1994), the Matsen–Schick spectral method (Matsen and Schick, 1994a) created the opportunity to perform high-accuracy unit-cell calculations for both simple and complex morphologies. This capability has led to mean-field phase diagrams that have proved influential in clarifying the universal features of block copolymer self-assembly, as well as in guiding the experimental design of copolymers and alloys of practical significance. These results have been reviewed by Matsen and Schick (Matsen and Schick, 1996; Matsen, 2002; Matsen, 2005b) and in the monograph of Hamley (1998), so our discussion will be limited to a few highlights.

The recent appearance of extensive numerical SCFT results for block copolymers originated from the development (Matsen and Schick, 1994a) of the fully spectral technique for solving the diffusion equation discussed in Section 3.6.2. Matsen and Schick utilize a plane wave basis that takes full advantage of the symmetry relations characteristic of a particular space group and converge the SCFT equations by means of the quasi-Newton scheme of Broyden discussed in Section 5.3.1. In this technique, a space group is selected at the outset, and one has to optimize the dimensions of the unit-cell as part of the calculation. Critical for starting the quasi-Newton scheme is a reasonable initial guess for the Fourier components of the potential fields and an initial approximation for the Jacobian matrix. Because the computational cost per field iteration with $M$ basis functions is $O(M^3)$, calculations using the Matsen–Schick method are practically restricted to $M \lesssim O(10^2)$. This implies a restriction to studies of the *weak* or *intermediate* segregation regimes, since the interfaces between microdomains sharpen at strong segregation (large Flory $\chi$ parameters) and many plane waves are then necessary to accurately resolve the fields. Indeed, the Helfand–Tagami theory of Section 5.2.3 indicates that the interfacial width scales as $\xi \sim \chi^{-1/2}$ for large $\chi$, so a doubling of spatial resolution is required for every 4-fold increase in $\chi$. In the case of a lamellar phase, such a change in $\chi$ would require increasing $M$ by a factor of 2. However, in the case of a complex three-dimensional structure, such as the gyroid phase, $M$ must be increased by a factor of $2^3 = 8$ to affect a doubling of resolution in all three directions.

In spite of these limitations, Matsen and Schick have produced an impressive collection of results, beginning with the mean-field phase diagram of a conformationally symmetric, incompressible AB diblock copolymer melt (Matsen and Schick, 1994$a$; Matsen and Bates, 1996$b$), which corresponds to Model E with $b_A = b_B$ and $\chi \equiv \chi_{AB}$. Figure 5.15 shows this phase diagram in the coordinates of $\chi N$ versus $f$, the average volume fraction of the A block. The disordered-to-ordered mesophase envelope, the so-called order–disorder transition (ODT), is very close to the spinodal boundary deduced by the RPA approximation and shown in Fig. 5.3.[77] Unlike the RPA, the numerical SCFT calculations utilize many harmonics in describing the periodic mesophases and can accurately compute mean-field free energies. The phase boundaries indicated in Fig. 5.15 were obtained by comparing the free energies of the disordered (homogeneous), lamellar, cylindrical, spherical, and bicontinuous gyroid (Ia$\bar{3}$d) phases indicated, as well as other periodic mesophases that did not prove to be stable throughout the $(\chi N, f)$ plane. Schematic drawings of some of these structures are shown in Fig. 5.16. The danger in such calculations is that one might miss an important structure by not including it in the free energy competition. Indeed, Tyler and Morse (2005) recently demonstrated a very small region of stability of a non-cubic bicontinuous network phase (Fddd space group), which has led to the first modification of the famous Matsen–Bates diblock copolymer phase diagram after nearly a decade.

The Matsen–Schick unit-cell calculations have proved to be not only of theoretical interest, but also very influential in the experimental polymer physics community for sorting out the relative stability and structural identification of various block copolymer mesophases. For example, the ordered bicontinuous double diamond (OBDD) phase (Pn$\bar{3}$m space group) (Thomas *et al.*, 1986), which was postulated for some time to be a stable mesophase in AB diblock copolymers,

---

[77]The order–disorder phase boundaries shown in Figs. 5.3 and 5.15 are identical at $f = 0.5$ where the phase transition is predicted by SCFT to be second order (continuous). At other compositions, the first-order ODT occurs at a slightly lower value of $\chi N$ than the spinodal value.

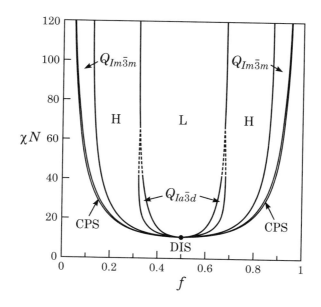

FIG. 5.15. Mean-field phase diagram for conformationally symmetric diblock melts. Phases are labeled L (lamellar), H (hexagonal cylinders), $Q_{Ia\bar{3}d}$ (bicontinuous Ia$\bar{3}$d gyroid), $Q_{Im\bar{3}m}$ (bcc spheres), CPS (close-packed spheres), and DIS (disordered). Dashed lines denote extrapolated phase boundaries, and the dot denotes the mean-field critical point. Reproduced from Matsen and Bates (1996$b$).

was found to be metastable throughout the $(\chi N, f)$ plane in SCFT calculations (Matsen and Schick, 1994$a$; Matsen and Bates, 1996$b$). In part, this theoretical finding encouraged the experimental community to go back and correct the assignment of OBDD in certain samples to the Ia$\bar{3}$d gyroid structure and conclude that the latter structure is the predominant bicontinuous network phase in diblock copolymers (Hajduk *et al.*, 1994; Schulz *et al.*, 1994; Hajduk *et al.*, 1995). Similarly, the hexagonally perforated lamellar phase (HPL) found in experiments was originally believed to be stable (Khandpur *et al.*, 1995), but unit cell SCFT results indicating metastability (Matsen and Bates, 1996$b$) and other theoretical input (Qi and Wang, 1997) were influential in convincing the experimental community that the HPL phase is only a long-lived metastable state (Hajduk *et al.*, 1997).

The dashed lines in Fig. 5.15 indicating an extrapolated pinch-off of the bicontinuous gyroid phase at large $\chi N$ reflect the difficulty of applying the fully spectral method at increasing segregation strength where large numbers of basis functions are needed. Indeed, recent calculations by Cochran using the more efficient pseudo-spectral techniques described in this monograph appear to contradict the loss of gyroid stability at large $\chi N$ (Cochran *et al.*, 2005).

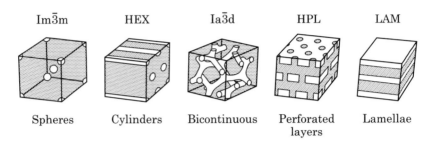

| Im$\bar{3}$m | HEX | Ia$\bar{3}$d | HPL | LAM |
|---|---|---|---|---|
| Spheres | Cylinders | Bicontinuous | Perforated layers | Lamellae |

FIG. 5.16. Schematic of the four stable ordered mesophases observed in experiments on diblock copolymer melts and the metastable structure HPL (hexagonally perforated lamellae). Reproduced from Khandpur *et al.* (1995).

The phase diagram of Fig. 5.15 is one of the remarkable successes of self-consistent field theory. The universal character of this phase diagram, with the relative placement of phases indicated, has been confirmed in experimental studies on several distinct diblock copolymer systems (Thomas *et al.*, 1988; Bates *et al.*, 1994; Khandpur *et al.*, 1995; Schulz *et al.*, 1996; Almdal *et al.*, 1996; Zhao *et al.*, 1996; Hillmyer *et al.*, 1996). For example, an experimental phase diagram for the most commonly studied system of polystyrene-b-polyisoprene (PS-PI) is shown in Fig. 5.17. Comparison with the theoretical mean-field phase diagram of Fig. 5.15 reveals striking qualitative similarities. However, there are quantitative differences that can be traced to factors that could be addressed in SCFT by refining the model, and factors that are associated with non-mean-field (fluctuation) effects. In the first category are statistical segment length differences between PS and PI (so-called "conformational asymmetry") and composition dependence of the Flory $\chi$ parameter, both of which are easily relaxed in SCFT (Matsen and Bates, 1997). These factors are responsible for the asymmetry of the experimental phase diagram on either side of the composition $f = 1/2$. In the second category are thermally excited composition fluctuations, which tend to stabilize the disordered phase over the ordered phases. The effect is to shift the experimental ODT envelope to larger $\chi N$ than the SCFT prediction, to destroy the mean-field critical point at $f = 1/2$, and to open up order–disorder boundaries between the disordered phase and non-spherical ordered phases (Fredrickson and Helfand, 1987). The theoretical treatment of such field fluctuation effects will be discussed in Chapter 6. Aside from these considerations, it is clear that numerical SCFT does a remarkable job of capturing the complexity of self-assembly in diblock copolymer melts.

Matsen, Schick, and coworkers have extended their unit-cell calculations to flexible AB block copolymer melts of a wide variety of types and architectures. These include linear $(AB)_n$ multiblocks (Matsen and Schick, 1994b), $(AB)_n$ star blocks (Matsen and Schick, 1994c), ABA triblock copolymers (Matsen and Thompson, 1999; Matsen, 2000), and $A_2B_2$ star block copolymers (Matsen and Gardiner, 2000). Numerical SCFT results for $AB_n$ "mikto-arm" star

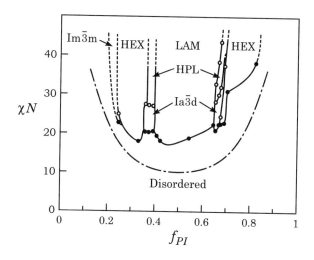

Fig. 5.17. Phase diagram for PS-PI diblock copolymers. Open and filled circles represent the order–order (OOT) and order–disorder (ODT) transitions, respectively, determined from rheological measurements. The dash-dotted curve is the mean-field spinodal for the ODT shown in Fig. 5.3. Solid curves have been drawn to delineate the different phases observed but might not correspond to precise phase boundaries. Reproduced from Khandpur *et al.* (1995).

block copolymer melts have also been recently reported (Grason and Kamien, 2004). Some unit-cell calculations have been carried out for *solutions* of diblock copolymers (Naughton and Matsen, 2002), although a much more extensive body of work on copolymer solutions can be found in the literature on lattice SCFT (Linse, 1993; Noolandi *et al.*, 1996; Svensson *et al.*, 1999).[78]

In addition to bulk copolymers, *thin films* of block copolymers have been studied by means of columnar unit-cell calculations with Dirichlet boundary conditions applied at the film surfaces and periodic conditions imposed in-plane (Matsen, 1997a; Matsen, 1998a; Pickett and Balazs, 1997; Wang *et al.*, 2003). The use of effective boundary conditions of the type described in Section 4.9 should further enhance the performance of spectral and pseudo-spectral algorithms in such situations.

Another area where unit-cell SCFT calculations have made a significant impact is in the phase behavior of block copolymer *alloys*. In blends of block copolymers with homopolymers or with other block copolymers, a bewildering array of ordered mesophases can appear, as well as regions of multiple phase coexistence. Coexisting phases can include one or more homogeneous liquid phases, and

---

[78]The lattice SCFT technique (Scheutjens and Fleer, 1979; Fleer *et al.*, 1993) is difficult to apply in non-cubic computational cells, so accurate unit-cell calculations of complex phases are problematic.

mesophases of arbitrary symmetry. Matsen showed that the grand canonical ensemble is most convenient for studying such alloys by numerical SCFT because coexistence regions (mixed state saddle points) can be easily bracketed (Matsen, 1995$a$). Using this technique, he performed an extensive set of calculations for binary blends of AB diblock copolymers with A homopolymers of varying molecular weight (Matsen, 1995$a$). Numerical SCFT results for blends of two AB diblock copolymers differing in composition and/or molecular weight have also been reported (Matsen, 1995$b$; Matsen and Bates, 1995; Shi and Noolandi, 1994; Shi and Noolandi, 1995). Janert and Schick (1997) conducted similar grand canonical unit-cell calculations to map out a series of ternary phase diagrams for AB diblock copolymers blended with both A and B homopolymers.

Figure 5.18 shows an example of a complex phase diagram that was obtained from unit-cell SCFT calculations for a block copolymer/homopolymer alloy. The figure, corresponding to an incompressible AB + A blend with $\chi N = 11$ and equal degrees of polymerization (Matsen, 1995$a$), shows regions of pure mesophases and pure homogeneous liquid phases, and regions of two-phase coexistence between the various types of phases. Remarkably, the rich features of such mean-field phase diagrams are manifested in experiments on model block copolymer/homopolymer blends (Winey $et\ al.$, 1992; Koizumi $et\ al.$, 1994). The phase diagrams have also provided qualitative guidance in developing templating strategies for creating nanostructured thermoset materials (Hillmyer $et\ al.$, 1997; Lipic $et\ al.$, 1998).

A topic of contemporary interest is the equilibrium phase behavior of $ABC$ $block\ copolymers$ (Bates and Fredrickson, 1999; Sioula $et\ al.$, 1998; Abetz and Goldacker, 2000; Goldacker $et\ al.$, 1999; Huckstadt $et\ al.$, 2000; Auschra and Stadler, 1993; Mogi $et\ al.$, 1992). The phase diagrams of such systems are of daunting complexity and the experimental design space is very large. As a result, the most successful groups working in the area look to numerical SCFT as a complement to experimental investigations. Some experimentalists are relying on published SCFT unit-cell calculations (Matsen, 1998$b$), while others are collaborating with theoreticians (Shefelbine $et\ al.$, 1999), or are exploiting available SCFT packages (Fraaije $et\ al.$, 2004; Doi, 2003). Defect states prove difficult to avoid in large-cell SCFT simulations of ABC block copolymers, presumably because the competing interactions in such systems creates frustration and a rough energy landscape. The prevalence of complex network phases that are nearly degenerate in energy (Epps $et\ al.$, 2004; Tyler and Morse, 2005) makes such large-cell simulations particularly challenging. For these reasons, high-accuracy, unit cell SCFT is an especially valuable tool for studying the self-assembly behavior of ABC block copolymers.

Numerical SCFT calculations have been applied to a wider range of the models described in Chapter 4 than just flexible block copolymers and their alloys. For example, unit-cell SCFT results have been presented for semiflexible, or "rod-coil" diblock copolymers (Matsen, 1996; Matsen and Barrett, 1998; Düchs and Sullivan, 2002), based on the wormlike chain model of Section 3.1.3. In such

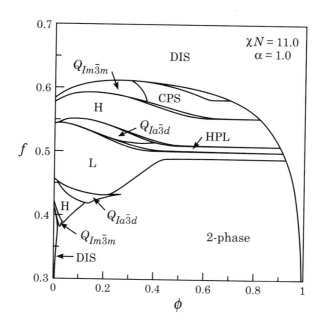

FIG. 5.18. Binary phase diagram of a melt blend of AB diblock and A homopolymer at $\chi N = 11$ and equal polymerization indices ($\alpha = 1$) plotted in
terms of the homopolymer volume fraction $\phi$ and the A segment volume fraction of the diblock, $f$. For clarity, only the largest biphasic region is labelled.
Reproduced from Matsen (1995a).

systems, nematic interactions of the type discussed in the context of Model G are
incorporated in the interaction model, along with the isotropic segment interactions (Flory $\chi$ parameter) and incompressibility constraint normally included
in Model E. An interesting feature of such models is the interplay between the
tendencies for nematic and smectic ordering. Other types of systems that have
been studied by means of unit-cell SCFT, or closely related lattice SCFT, are
polymer brushes consisting of mixtures of two different types of grafted chains
(a generalization of Model K) (Müller, 2002), grafted block copolymers (Zhulina
*et al.*, 1996), polyelectrolyte solutions and alloys (generalizations of Model F)
(Gottschalk *et al.*, 1998; Shi and Noolandi, 1999; Li and Schick, 2000; Schick
and Li, 2000), polyelectrolyte brushes (Shusharina and Linse, 2001), and block
copolyelectrolytes (Wang *et al.*, 2004).

### 5.4.2  *Large-cell calculations*

Large-cell numerical SCFT was pioneered by Fraaije and coworkers in the framework of the dynamic density functional theory (DDFT) described in Section 5.3.6
(Fraaije, 1993; Fraaije *et al.*, 1997). Fraaije's group continues to be active in
this area and has produced a large body of work on the application of DDFT

to block copolymers both at equilibrium and under shear (Fraaije *et al.*, 1997; Zvelindovsky *et al.*, 1998), morphology evolution in reacting polymer blends (Maurits *et al.*, 1999), dynamics of copolymer solutions (van Vlimmeren *et al.*, 1999), liquid crystalline polymers (Hamm *et al.*, 2001; Hamm *et al.*, 2003), block copolyelectrolytes (Kyrylyuk and Fraaije, 2004), as well as other topics.[79] A commercial SCFT code based on the DDFT scheme, MesoDyn, has also been developed (Fraaije *et al.*, 2004).

If kinetic pathways are *not* of intrinsic interest, we recommend that the DDFT field relaxation strategies discussed in Section 5.3.6 be avoided in large-cell simulations. By conserving species density, the DDFT approach suffers from diffusion limitations that slow the relaxation of the largest structures in a fluid, including macrophases and defect structures in mesophases. The numerical methods described in Section 5.3.2 do not suffer this limitation and should converge faster in situations where only stable and metastable equilibrium morphologies are desired.

An advantage of large-cell SCFT is that, unlike unit-cell approaches, it is not necessary to guess the symmetry and lattice constant of a periodic mesophase in advance of starting a calculation. Instead, the potential fields within a large cell can be initialized from a random seed and the structures evolved as part of iterating the fields to satisfy the SCFT equations. As exemplified by Fig. 5.10, this procedure inevitably leads to metastable defect states, rather than pure states. However, by analyzing the local order produced in such a simulation, it is straightforward to re-initialize and converge a unit-cell or multiple unit-cells of the identified periodic structure. A second potential advantage of large-cell SCFT simulations is that by such means it is possible to generate large populations of defect states, which may themselves be of intrinsic interest. Finally, there are situations in which spatially periodic mesophases are not present, yet the fluid is structured in a non-trivial way, possibly due to defects or applied boundary conditions. Large-cell SCFT is easily adapted to such situations.

The extensive DDFT work of Fraaije and collaborators has provided insight into the capabilities of large-cell SCFT simulations. The notion that such simulations could be used for "computational-combinatorial screening" of complex polymer phase diagrams was presented by Drolet and Fredrickson (1999) and Bohbot-Raviv and Wang (2000). Following up on this work, Drolet and Fredrickson (2001) illustrated how large-cell SCFT can be used to computationally optimize a property of interest – the practical toughness of amorphous ABABA pentablock copolymers. The advent of pseudo-spectral techniques (Rasmussen and Kalosakas, 2002) for efficient computation of chain propagators and density operators, along with new parallelization strategies (Sides and Fredrickson, 2003), field relaxation schemes (Ceniceros and Fredrickson, 2004), and variable

---

[79]The reader should note that the typical implementation of DDFT employs a discrete Gaussian chain with a small number of beads, i.e. $N \sim 10$, in contrast to the continuous Gaussian chain used by Matsen and Schick and applied throughout this book. In addition, the DDFT calculations utilize a compressible interaction model similar to Model D.

box shape methods (Barrat *et al.*, 2005), should further serve to advance the capabilities and potential of large-cell SCFT.

Figure 5.19 provides an example of how large-cell SCFT simulations can be used in a discovery mode for exploring the self-assembly and phase behavior of a complex polymer formulation (Sides and Fredrickson, 2003). Each panel corresponds to an independent simulation for an incompressible blend of ABA triblock copolymer[80] with A homopolymer at the indicated value of the homopolymer volume fraction, $\phi$, and A content of the triblock, $f$. The simulations were started from random initial conditions and the potential fields were relaxed for a total of 1000 field iterations. In these simulations two types of parallelization were employed: the assignment of 4 processors per panel to affect a domain decomposition as described in Section 3.6.6, and a trivial parallelization in which different panels, corresponding to different points in the $(\phi, f)$ parameter space, were assigned their own processors. By such means and with access to a distributed memory cluster with a modest number of nodes, it is possible to very rapidly get a global map of a phase diagram in a multi-component system. Such calculations do not determine precise phase boundaries, but allow for a coarse determination of major phases and features of a phase diagram. For example, in the case of the results shown in Fig. 5.19 we observe regions of pure cylindrical and lamellar mesophases at low homopolymer fraction $\phi$, broad regions of two-phase coexistence between a homopolymer-rich disordered phase and a lamellar or cylindrical mesophase at higher $\phi$, and ultimately a homogeneous disordered phase for large $\phi$ and $f$, where the two components are most miscible. These same features are manifest in a much more quantitative way in the AB + A phase diagram shown in Fig. 5.18, but the latter required a tedious series of unit-cell calculations to locate precise phase boundaries.

This example demonstrates a nice synergy between unit-cell and large-cell SCFT. When tackling a new model with a large parameter space, it is useful to first scope out the general features of the phase diagram with large-cell simulations carried out in parallel. Regions of parameter space or phases of particular interest that have emerged from the large-cell "prospecting" can then be studied in quantitative detail by means of unit-cell calculations or higher-resolution large-cell calculations. For example, Fig. 1.6 corresponds to a very high-resolution ($M = 1024^2 \approx 1.05 \times 10^6$, $L = 256R_g$) large-cell SCFT simulation that was performed to zoom in on the $\phi = 0.2$, $f = 0.3$ panel shown in Fig. 5.19.

It is important to point out that the model used to obtain the results shown in Figs. 5.19 and 1.6 was formulated in the *canonical ensemble* and resembles the canonical versions of Models C and E. In large-cell simulations of such models, coexisting phases are readily apparent within the computational cell. For example, Fig. 1.6 shows a coexistence between (light) droplets of a disordered phase rich in the A homopolymer and a cylindrical mesophase rich in the ABA

---

[80]The two A blocks of the triblock are assumed to be of equal length and constitute a volume fraction $f$ of the entire molecule.

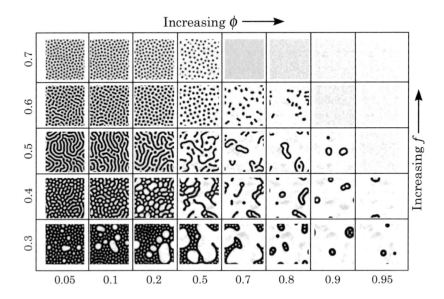

FIG. 5.19. Array of morphologies in the $(\phi, f)$ parameter space obtained from large-cell SCFT simulations of an incompressible blend of ABA triblock and A homopolymer at $\chi N_{ABA} = 32$ with homopolymer polymerization index half that of the triblock ($\alpha \equiv N_A/N_{ABA} = 0.5$). Each panel resulted from a SCFT simulation run on 4 processors with $M = 128^2$ ($L = 32R_g$) and started from a random initial condition. Light regions have the largest total A segment density. Reproduced from Sides and Fredrickson (2003).

triblock copolymer. The configuration is clearly not at equilibrium – further field relaxation produces coalescence of the light droplets, presumably evolving the system towards a lowest free energy configuration of one macroscopic droplet of the disordered phase in contact with the ordered mesophase. Such mixed states obtained from large-cell simulations of models in the canonical ensemble are useful in demonstrating that two or more phases coexist, but are not particularly helpful in determining the width of the coexistence regions. As in the case of unit-cell SCFT (Matsen, 1995a), a more straightforward and accurate determination of phase boundaries in systems exhibiting phase coexistence is obtained from large-cell simulations in the *grand canonical ensemble*. In grand canonical SCFT, two-phase windows are collapsed, so mixed states can be more easily avoided. The discontinuous jump in composition associated with a transition between pure states and driven by a change in chemical potential (i.e. activity) provides a direct measure of the width of a coexistence region.

Large-cell simulations provide a convenient method for studying *defects* in inhomogeneous polymer phases, including the periodic mesophases of block copolymers. Simulations launched from random initial conditions produce defective mesophases, such as the configuration shown in Fig. 5.10 for the cylindrical

phase of an incompressible diblock copolymer melt (Model E). In that configuration one can identify dislocations consisting of 5–7 disclination pairs of the
type illustrated in Fig. 5.2. The 5-coordinated cylinders in Fig. 5.10 are smaller
in diameter than the average cylinder and the 7-coordinated cylinders are larger
than average because chains are exchanged between the cylindrical aggregates in
the disclination pair to reduce the core energy of the defect. This chain exchange
and local equilibration is implicit in SCFT and compares favorably in quantitative terms with defect structures observed in scanning force microscopy of thin
block copolymer films (Hammond et al., 2003).

Lamellar mesophases of block and graft copolymers are analogous to the
smectic A phase of liquid crystals,[81] so lamellae-forming copolymers reveal line
defects, such as dislocations and focal conics, and wall defects, such as kink grain
boundaries, that are familiar in smectics (de Gennes and Prost, 1993). These
same defects appear in large-cell SCFT simulations of lamellar block copolymers.
For example, Fig. 5.20 shows a defective lamellar phase resulting from a large
(square) cell simulation of Model E with $f = 0.5$, $\chi N = 25.1$ subject to periodic
boundary conditions and initialized from a random seed. This two-dimensional
density configuration exhibits a complicated array of various types of defects,
including dislocations and chevron and omega kink boundaries, all of which have
been observed in experiment (Gido and Thomas, 1994).

If the structure and energetics of an isolated defect are of interest, it is possible through the application of suitable boundary conditions to converge a single
defect in a large-cell SCFT simulation. Matsen (1997b) provided a beautiful illustration of this technique in the context of kink grain boundaries in lamellar
block copolymers. He used fully spectral large-cell SCFT calculations of Model
E to study the transition between chevron and omega boundaries as a function
of the angle $\theta$ between the lamellae normals of two opposing grains, the segregation strength, and compositional and conformational asymmetries. The basis
functions employed were an orthonormal set of eigenfunctions of the operator
$\nabla^2$ consistent with the reflection and rotation symmetries necessary to impose
an isolated kink boundary. As shown in Fig. 5.21 and in agreement with experiment (Gido and Thomas, 1994), Matsen's results demonstrate that chevron
kink boundaries transform to omega boundaries upon increasing $\theta$ above 90°.
Similar SCFT calculations have been reported for the other two independent
types of grain boundaries in lamellar block copolymers; twist grain boundaries
(Duque and Schick, 2000) and T-junctions (Duque et al., 2002). A comprehensive understanding of grain boundary free energies in lamellar systems is thus
in place. It is important to emphasize the advantage of numerical SCFT over
phenomenological approaches that employ approximate Ginzburg–Landau-type
energy functionals (Netz et al., 1997). Such methods cannot be used to study the

---

[81]This analogy applies only to the compositional order present in the two types of systems.
The strong orientational order characteristic of the molecules in a smectic A phase is not
exhibited by the segments of a lamellar block copolymer.

FIG. 5.20. Segment density configuration for an incompressible, symmetric diblock copolymer melt (Model E with $b_A = b_B$). Parameters are $\chi_{AB}N = 25.1$, $f = 0.5$, $L = 51.2R_g$, and $M = 512^2$. Figure courtesy of A. Bosse.

influence of changing microscopic parameters such as segment length asymmetry, $b_A \neq b_B$, block composition $f$, or copolymer architecture.

As a second example of how large-cell SCFT simulations can be used in tandem with appropriate boundary conditions to study non-trivial self-assembly, we note the recent work of Wu *et al.* (2004) who showed that simulations of AB diblock/A homopolymer blends exhibiting a hexagonal mesophase in the bulk produce chiral morphologies when confined to a narrow cylindrical pore. The spontaneous symmetry breaking to complex helical structures, some of which are illustrated in Fig. 5.22, in spite of no chirality in the underlying chain model, is a surprising observation. Moreover, the phenomenon is apparently universal because it has been experimentally observed in two very dissimilar polymer systems (Wu *et al.*, 2004; Xiang *et al.*, 2005).

The simulation results shown in Fig. 5.22 also provide a nice illustration of the "masking" technique described in Section 4.9.5. In particular, the cylindrical simulation cell used by Wu *et al.* (2004) was created by masking a cylinder within a slightly larger tetragonal computational cell. This approach allowed for efficient pseudo-spectral calculations utilizing plane waves and periodic boundary conditions in the larger cell.

A few large-cell simulations have been performed for models with quenched or annealed disorder. Fredrickson and Sides (2003) examined the interfacial properties of a polydisperse version of the blend Model C, using the Model H description of quenched chain length disorder and a Shulz molecular weight distribution. The

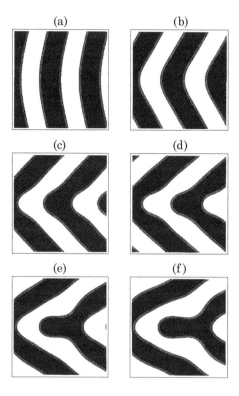

FIG. 5.21. Kink boundary profiles at angles (a) $\theta = 30°$, (b) $\theta = 60°$, (c) $\theta = 90°$, (d) $\theta = 100°$, (e) $\theta = 110°$, and (f) $\theta = 120°$ for a lamellar phase of Model E with $\chi N = 20$, $f = 0.5$, and $b_A = b_B$. The first two profiles, (a) and (b), are chevron boundaries, the second two, (c) and (d), are intermediate boundaries, and the last two, (e) and (f), are omega boundaries. Figure reproduced from Matsen (1997$b$).

structure and phase behavior of AB diblock copolymer melts with monodisperse A blocks and polydisperse B blocks (Shulz distribution) were subsequently studied by Sides and Fredrickson (2004) using two-dimensional large-cell SCFT. In the latter investigation, which employed a modified version of Model E, polydispersity was observed to produce several effects including: a shift in the ODT to lower $\chi$, i.e. the ordered phases are stabilized by polydispersity; a shift in the compositions of phase boundaries driven by the release of stretching energy in the polydisperse B domains; and a stabilization of defect structures relative to the monodisperse case. The stabilization of defects is an interesting phenomenon that occurs because a polydisperse diblock copolymer melt contains a continuum of diblocks of different compositions that can position themselves to lower the energy of a defect. This redistribution of chains of different composition can be studied quantitatively in large-cell simulations by evaluating partial density

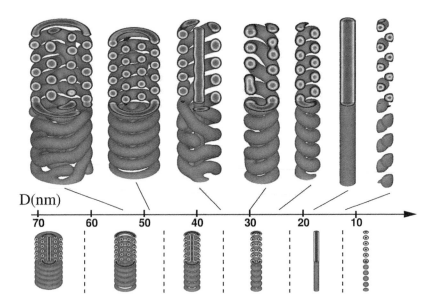

FIG. 5.22. Confined self-assembly in cylindrical pores with varying cylinder diameter (D) for an incompressible AB diblock + A homopolymer blend. In the SCFT simulations shown, boundary conditions that favor wetting by A segments were imposed at the cylinder walls and periodic boundary conditions were applied in the axial direction. Two-dimensional SCFT calculations assuming axial symmetry are also shown (bottom row). In tight confinement, hexagonal packing of the cylindrical micelles running parallel to the pore axis would impose high stress on the system. This stress can be relieved by arranging the system into a set of concentric shells. In each shell the cylinders bend and tilt with respect to the pore axis, thus forming helices or donuts. Figure reproduced from Wu *et al.* (2004).

operators such as the operator given by eqn (4.145). Figure 5.23 shows how the total B segment density in a defective morphology for a polydisperse diblock melt can be decomposed into the contributions to the density from diblocks of varying length (and thus composition). The shortest B blocks are seen to be concentrated within the A–B interfaces of the mesophase structure, while the longest B blocks tend to populate the continuous B domain (light) surrounding the discrete A domains (dark).

## 5.5 Summary

The above examples have served to illustrate that self-consistent field theory is a powerful tool for investigating the equilibrium self-assembly and thermodynamic properties of inhomogeneous polymers. We conclude this chapter by recapping some of the main points:

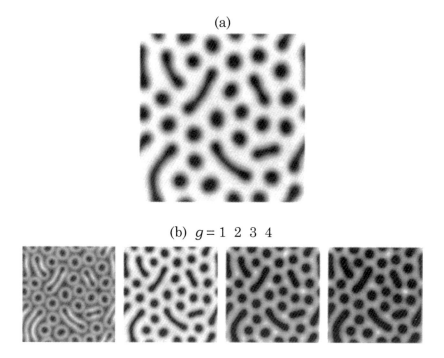

(a)

(b) $g = 1$  2  3  4

FIG. 5.23. Two-dimensional plots of the B segment density for an AB diblock copolymer melt (incompressible Model E, $b_A = b_B$) in which the B block is polydisperse with a Shulz distribution. Parameters are: $\chi N_n = 18$, where $N_n$ is the number average polymerization index for the overall chains, $f = 0.33$, $L = 26R_g$, and $I^B_{pdi} = 2.0$ is the polydispersity index for the B block. (a) Total density of B segments (light regions have high B density). (b) Separate contributions to the B segment density from chains with B blocks having increasing lengths indexed by $g = 1, 2, 3, 4$ (corresponding to abscissas in a 4-point Gauss–Laguerre quadrature). Figure reproduced from Sides and Fredrickson (2004).

- Self-consistent field theory (SCFT) is a type of mean-field theory that can be derived for any field theory model of a simple or complex fluid, including the models of Chapter 4, by invoking a saddle point approximation to the functional integral that defines the model. The saddle point approximation, which is synonymous with the mean-field approximation, amounts to the assumption that a single potential field configuration $w(\mathbf{r}) = w^*(\mathbf{r})$ provides the dominant contribution to the functional integral. This field configuration $w^*$ is the so-called *mean field* or *self-consistent field*.

- The mean-field potential $w^*$ is determined by eqn (5.3), which states that the effective Hamiltonian of the model, $H[w]$, is stationary with respect to field variations when $w = w^*$. In the canonical ensemble, $H[w^*]$ corresponds to the mean-field approximation to the Helmholtz free energy in units of $k_B T$.

- Because of the complex (not strictly real) nature of the Hamiltonian for most models, the mean-field condition corresponds to a vanishing of the complex derivative $\delta H / \delta w$, which implies that $w^*$ is a *saddle point* that may be located off the real axis in the complex plane. The qualitative location of physical saddle points, i.e. gross location within the complex plane, can often be determined by invoking the fact that $H[w^*]$ (and thus the free energy) must be real. Such information can be useful in developing efficient numerical methods for computing saddle points.

- Saddle points (SCFT solutions) $w^*(\mathbf{r})$ can be classified as being *homogeneous* or *inhomogeneous*. Homogeneous saddle points can normally be determined analytically, while numerical methods are usually required to determine inhomogeneous SCFT solutions.

- Saddle points can be further classified as corresponding to *pure states*, *defect states*, or *mixed states*. Pure states can be associated with the pure phases that are present in a fluid system. Defect states are pure phases that contain topological defects. It is usually the case that defect states are metastable and pure states are stable. When two pure phases coexist, the corresponding saddle point is referred to as a mixed state.

- It is sometimes useful to combine *additional approximations* with the mean-field approximation so that analytical progress can be made. The weak inhomogeneity expansion (RPA), slow gradient expansion, ground state dominance approximation, and strong stretching approximation are among the schemes described in Section 3.4 that can be combined with the mean-field approximation. Section 5.2 details some useful analytical results that are obtained by applying these approximation schemes to simplify the SCFT equations.

- A variety of numerical methods have been developed for solving the full SCFT equations. We favor the continuous descent/ascent techniques of Section 5.3.2.1 for relaxing the field variables, combined with a pseudo-spectral evaluation of the relevant chain propagators and density operators. This combination provides both spectral accuracy and high efficiency, and is suitable for acceleration by parallelization (Section 3.6.6) or multilevel embedding (Section 5.3.3).

- Numerical SCFT simulations can be divided into two broad categories: *unit-cell* and *large-cell* calculations. Unit-cell calculations resolve one primitive cell of the Bravais lattice for a spatially periodic mesophase. Large-cell simulations are not restricted to periodic mesophases, and are particularly useful for examining defect states and mixed states. Both types of SCFT simulations can benefit from *variable cell shape methods* in which the size and shape of the computational cell is adjusted as part of the calculation, e.g. to obtain a stress-free configuration of the system. Variable cell shape methods also allow for SCFT calculations of the *quasi-static mechanical response* of a polymeric fluid to applied stresses or strains.

- If kinetic pathways towards the mean-field equilibrium state are of interest, approximate dynamics schemes can be applied that conserve mass and allow for diffusive relaxation of the potential and species density fields. Results from such *dynamical SCFT* simulations should be interpreted with caution, however, because hydrodynamic and elastic stress couplings must often be included for a quantitative description of even the lowest-frequency dynamics of a polymeric fluid.

# 6

# BEYOND MEAN-FIELD THEORY

In the previous chapter we saw that the mean-field approximation can be used to gain significant insights into the structure and thermodynamic properties of inhomogeneous polymers. For situations in which this approximation, i.e. self-consistent field theory (SCFT), is accurate, it is unlikely that the more advanced techniques described in the present chapter will be required. Such situations include polymer alloys and block copolymers of high molecular weight or large strong segregation strength, molten polymer brushes, and highly concentrated polymer solutions, among others. There are, however, quite a few circumstances in which SCFT is inaccurate, or even qualitatively incorrect (de Gennes, 1979). These include

- polymer solutions in the dilute or semidilute regimes
- polymer blends near a critical point, second-order phase transition, or weakly first-order phase transition
- microemulsion and micellar phases of polymers, copolymers, and surfactants
- block copolymers near their order–disorder transition (ODT)
- polyelectrolyte solutions in various concentration regimes

In such situations, the functional integral defining the partition function of the relevant field theory model, e.g. eqn (5.1), is not entirely dominated by the mean-field configuration $w^*(\mathbf{r})$, but field configurations $w(\mathbf{r})$ close to and possibly far from the saddle point can make important contributions. The study of these *field fluctuation effects* is the subject of the present chapter.

## 6.1 Analytical methods

When faced with a field-theoretic model of a polymeric fluid, it is important to establish for what range of parameters, if any, the mean-field approximation is valid. Returning to the discussion surrounding the one-dimensional toy model of eqn (5.6), we observed that the model contained a parameter $p$ that multiplied all the terms in the Hamiltonian $H(w)$ of eqn (5.7). By deforming the integral defining $\mathcal{Z}$ for the model onto a constant phase ascent path $\Gamma$ that passes through the saddle point $w^*$, it was shown that the integral is dominated by the saddle point asymptotically for $p \to +\infty$. This asymptotic method of analysis, the so-called *steepest descent*[82] or *saddle point* technique (Bender and Orszag, 1978), serves to demonstrate that the mean-field approximation, $\mathcal{Z} \approx \exp[-H(w^*)]$,

---

[82]The relevant technique for the present situation is one of *steepest ascent*.

is asymptotically exact for $p \to +\infty$. Furthermore, the steepest ascent analysis is not restricted to models with a quadratic Hamiltonian, but is more broadly applicable. A criterion for the validity of the mean-field approximation for any such toy model with $p$ multiplying all terms in the Hamiltonian is thus $p \gg 1$.

The field-theoretic models of Chapter 4 are of course not one-dimensional, but are defined by functional integrals that are effectively of infinite dimension. Nevertheless, the same steepest ascent analysis can be adapted to establish the conditions for which the mean-field approximation (SCFT) is expected to be valid. For the purpose of illustration, it is convenient to return to the canonical Model A field theory for a bulk three-dimensional polymer solution subject to periodic boundary conditions. As was discussed in Section 5.1.4, the relevant saddle point for this situation is homogeneous.

### 6.1.1  *Gaussian fluctuations about a homogeneous state: canonical ensemble*

Our starting point for a steepest ascent analysis of the canonical Model A defined by eqns (4.71) and (4.72) is to rescale positions and fields according to

$$\mathbf{x} \equiv \mathbf{r}/R_{g0}, \quad W(\mathbf{x}) \equiv Nw(\mathbf{r}) \tag{6.1}$$

where $R_{g0} = b(N/6)^{1/2}$ is the unperturbed radius of gyration. By applying these definitions and recalling the dimensionless chain concentration and excluded volume parameters[83] $C$ and $B$ introduced in the previous chapter,

$$C \equiv \frac{nR_{g0}^3}{V}, \quad B \equiv \frac{u_0 N^2}{R_{g0}^3} \tag{6.2}$$

one can reexpress the canonical Model A as

$$\mathcal{Z}_C = \mathcal{Z}_0 \int \mathcal{D}W \, \exp(-H[W]) \tag{6.3}$$

$$H[W] = \frac{1}{2B} \int d\mathbf{x} \, [W(\mathbf{x})]^2 - C\widetilde{V} \ln Q[iW] \tag{6.4}$$

where $\widetilde{V} = V/R_{g0}^3$ is a dimensionless system volume. The single-chain partition function $Q[iW]$ is computed with these scalings according to $Q[iW] = (1/\widetilde{V}) \int d\mathbf{x} \, q(\mathbf{x}, 1)$, where $q(\mathbf{x}, t; [iW])$ satisfies

$$\frac{\partial}{\partial t} q(\mathbf{x}, t; [iW]) = \nabla_{\mathbf{x}}^2 \, q(\mathbf{x}, t; [iW]) - iW(\mathbf{x}) q(\mathbf{x}, t; [iW]) \tag{6.5}$$

subject to $q(\mathbf{x}, 0; [iW]) = 1$. In the above equation, the contour variable $t \equiv s/N$ has been scaled to $t \in [0, 1]$.

---

[83]The parameter $B$ is proportional to the dimensionless expansion parameter used in perturbation studies of the excluded volume effect (Yamakawa, 1972). In de Gennes' monograph (de Gennes, 1979), this parameter is denoted $\zeta$.

By means of the above scalings, it is apparent that the canonical Model A depends on the parameters $C$ and $B$, along with the scaled volume $\widetilde{V}$. In the thermodynamic limit, the intensive properties of the model are thus a function of only $C$ and $B$, so the model can be considered a *two-parameter* model. For the present case of periodic boundary conditions, the only saddle point is homogeneous and is located according to eqn (5.11) along the negative imaginary axis at

$$W^* = -iCB \tag{6.6}$$

Thus, the analytic structure of Model A appears to be much like the situation depicted in Fig. 5.1 for the toy model of eqn (5.6). Because the integrand $\exp(-H[W])$ is an analytic functional of the field $W$ throughout the complex $W$ plane, it is possible to deform the contour of integration in eqn (6.3) onto a new contour that is parallel to the real axis and passes through the saddle point, cf. the contour $\Gamma_2$ in Fig. 5.1. This deformation is affected by the simple uniform shift in field to $\Omega(\mathbf{x}) = W(\mathbf{x}) + iCB$, which transforms the field theory to

$$\mathcal{Z}_C = \mathcal{Z}_0 \exp(-H[W^*]) \int \mathcal{D}\Omega \ \exp(-H_S[\Omega]) \tag{6.7}$$

$$H_S[\Omega] = \frac{1}{2B} \int d\mathbf{x} \ [\Omega(\mathbf{x})]^2 - iC \int d\mathbf{x} \ \Omega(\mathbf{x}) - C\widetilde{V} \ln Q[i\Omega] \tag{6.8}$$

where $H[W^*] = (1/2)\widetilde{V}C^2B$. The path of integration in eqn (6.7) is along the real axis for each component of the field $\Omega(\mathbf{x})$, and the saddle point of the transformed theory defined by $\delta H_S/\delta\Omega|_{\Omega^*} = 0$ is now at the origin, $\Omega^* = 0$.

Because $C$ multiplies all the terms in the effective Hamiltonian of eqn (6.8), except for the first term, it would appear that $C$ plays the role of the parameter $p$ in the toy model. Namely, we expect that $C$ is the relevant *Ginzburg parameter* (Ginzburg, 1960; Goldenfeld, 1992) that dictates the validity of the mean-field approximation. This would imply that mean-field theory is exact for $C \rightarrow +\infty$. To verify this expectation, it is necessary to perform a *local analysis* and expand the effective Hamiltonian of the theory $H_S[\Omega]$ about the homogeneous saddle point at $\Omega = 0$. This step will serve three purposes: (i) to confirm that the shift in field from $W$ to $\Omega$ has deformed the integral onto a constant phase ascent path passing through the saddle point, (ii) to confirm that $C$ is the proper Ginzburg parameter, and (iii) to obtain a first correction to the mean-field approximation associated with Gaussian field fluctuations about the saddle point.

The local analysis required to analyze eqn (6.7) in the vicinity of the homogeneous saddle point amounts to a Taylor expansion of the effective Hamiltonian $H_S[\Omega]$ in powers of $\Omega$ about $\Omega^* = 0$. The first two contributions to $H_S[\Omega]$ in eqn (6.8) are quadratic and linear in $\Omega$, respectively, so it remains to expand the functional $\ln Q[i\Omega]$. The expansion of $Q[i\Omega]$ in powers of $\Omega$, however, is nothing more than the weak inhomogeneity expansion of Section 3.4.1, so eqns (3.131) and (3.132) can be immediately applied. The Fourier representation of eqn (3.131) proves most convenient, which in the present notation becomes

$$Q[i\Omega] = e^{-i\hat{\Omega}(0)/\widetilde{V}} \left[ 1 + \frac{(i)^2}{2\widetilde{V}^2} \sum_{\mathbf{k}} \hat{g}_D(k^2)\hat{\Omega}(\mathbf{k})\hat{\Omega}(-\mathbf{k}) + O(i\hat{\Omega}^3) \right] \qquad (6.9)$$

where

$$\hat{\Omega}(\mathbf{k}) = \int_{\widetilde{V}} d\mathbf{x} \ \exp(-i\mathbf{k} \cdot \mathbf{x})\Omega(\mathbf{x}) \qquad (6.10)$$

denotes the Fourier coefficient of $\Omega(\mathbf{x})$ with wavevector $\mathbf{k}$. For simplicity, the spatial domain is taken to be a cube of dimensionless volume $\widetilde{V}$. It should be noted that the $\mathbf{k} = 0$ term is to be omitted in the sum over reciprocal lattice vectors in eqn (6.9), since the $\mathbf{k} = 0$ (homogeneous) mode has been extracted in the prefactor. Upon combining eqns (6.8) and (6.9) one finds that the terms linear in $\Omega$ exactly cancel, which leads to the following harmonic expansion of $H_S[\Omega]$ about the saddle point:

$$H_S[\Omega] = \frac{1}{2\widetilde{V}} \sum_{\mathbf{k}} \hat{\gamma}_2(k)\hat{\Omega}(\mathbf{k})\hat{\Omega}(-\mathbf{k}) + O(i\hat{\Omega}^3) \qquad (6.11)$$

where

$$\hat{\gamma}_2(k) = \begin{cases} B^{-1}, & k = 0 \\ B^{-1} + C\,\hat{g}_D(k^2), & k \neq 0 \end{cases} \qquad (6.12)$$

For real values of the field $\Omega(\mathbf{x})$, it follows from eqn (6.10) that $\hat{\Omega}(\mathbf{k})\hat{\Omega}(-\mathbf{k})$ is real. Furthermore, provided the solvent quality is good so that $B > 0$, $\hat{\gamma}_2(k)$ is real and positive definite. Thus, eqn (6.11) shows that to $O(\hat{\Omega}^2)$ about the saddle point, the deformed integration path is a constant phase path with vanishing imaginary part of the Hamiltonian, $H_{SI} = 0$. The real part of the Hamiltonian, $H_S = H_{SR}$, has a local minimum along the path at the saddle point, $\hat{\Omega}(\mathbf{k}) = 0$, for each Fourier mode $\mathbf{k}$. We have thus succeeded in deforming the functional integral of eqn (6.3) onto a contour in the complex plane that *to quadratic order about the homogeneous saddle point is a constant phase ascent path*.

By truncating the Taylor expansion of eqn (6.11) at quadratic order and substituting the result into eqn (6.7), a Gaussian functional integral is obtained. In the Fourier representation of the field $\Omega(\mathbf{x})$, eqn (6.11) shows that each mode is independent, so the integrals over the Fourier amplitudes[84] can be factored for

---

[84]Here we interpret the functional integral over $\Omega(\mathbf{x})$ spectrally as a product of ordinary integrals over the real and imaginary parts of $\hat{\Omega}(\mathbf{k})$ for each $\mathbf{k}$, cf. eqn (C.26). A subtle point is that only half of the reciprocal space should be included, because $\Omega(\mathbf{x})$ is real, and thus $\hat{\Omega}(\mathbf{k})$ and $\hat{\Omega}(-\mathbf{k})$ are complex conjugates (Goldenfeld, 1992).

each wavevector $\mathbf{k}$ and evaluated by means of eqn (B.1). The following expression for the Helmholtz free energy, $\beta A = -\ln \mathcal{Z}_C$, is thus obtained:

$$\beta A(n, V, T) = \beta A_0 + \frac{1}{2} BC^2 \widetilde{V} + \frac{1}{2} \sum_{\mathbf{k}} \ln \left( \frac{\hat{\gamma}_2(k)}{2\pi \widetilde{V}} \right) \tag{6.13}$$

where $\beta A_0 \equiv -\ln \mathcal{Z}_0$. The three terms on the right-hand side of this equation correspond, respectively, to free energy contributions from the translational entropy of an ideal gas of non-interacting polymers, mean-field interactions, and Gaussian field fluctuations.

A convenient reference state of free energy is one in which the excluded volume interaction is turned off, $u_0 = 0$. An "excess" Helmholtz free energy with respect to this non-interacting state is defined as

$$\beta A^{ex}(n, V, T) \equiv \beta A(n, V, T) - \beta A(n, V, T)|_{B \to 0+}$$

$$= \frac{1}{2} BC^2 \widetilde{V} + \frac{1}{2} \sum_{\mathbf{k}} \ln \left[ B\hat{\gamma}_2(k) \right] \tag{6.14}$$

The fluctuation term in this expression can be analyzed by converting the sum to an integral, as is appropriate in the thermodynamic limit (see Appendix A). It follows that

$$\frac{1}{2} \sum_{\mathbf{k}} \ln \left[ B\hat{\gamma}_2(k) \right] = \frac{\widetilde{V}}{2} \frac{1}{\widetilde{V}} \sum_{\mathbf{k}} \ln \left[ B\hat{\gamma}_2(k) \right]$$

$$\approx \frac{\widetilde{V}}{2} \frac{1}{(2\pi)^3} \int d\mathbf{k} \, \ln \left[ B\hat{\gamma}_2(k) \right]$$

$$\approx \frac{\widetilde{V}}{4\pi^2} \int_0^\Lambda dk \, k^2 \ln \left[ 1 + BC \, \hat{g}_D(k^2) \right] \tag{6.15}$$

In the last line of this expression, we have introduced a high wavenumber cutoff, $\Lambda$, in the integral over Fourier modes. Such a step is necessary because the integrand scales as $k^0$ for $k \to \infty$ and so the integral over all of reciprocal space does not exist. This is an example of an *ultraviolet divergence* in a field theory. Ultraviolet divergences arise when the highest Fourier modes of field fluctuations are not sufficiently damped for free energies or other physical quantities to exist. This is an *artifact* of the coarse-grained field-theoretic models that we employ; in an atomistic description of a fluid with realistic hard-core interactions at small scales, the amplitudes of high $k$ fluctuation modes are strongly suppressed and no such divergences arise.

There are several approaches that can be taken to cure or *regularize* a field theory of ultraviolet divergences. These include the addition of extra gradient terms to the effective Hamiltonian to smoothly damp the high $k$ modes, dimensional regularization schemes, and the simple spherical cutoff approach that we adopted above (Amit, 1984; Zee, 2003). In the spherical cutoff scheme, all field fluctuations with wavevectors $\mathbf{k}$ of magnitude $k = |\mathbf{k}| > \Lambda$ are removed. Thus the retained modes fill the sphere $k \leq \Lambda$ in reciprocal space. Physically, $\Lambda^{-1}$ should correspond to a small length scale comparable to the size of a polymer segment or solvent molecule in the fluid. It is important to recognize that when a thermodynamic quantity such as $A^{ex}$ depends on the value of a cutoff parameter, then the quantity is sensitive to microscopic details not included in the field-theoretic description. However, an analysis of fluctuation corrections for other macroscopic properties or observables may not require the introduction of a high $k$ cutoff. Such properties are deemed *universal* and can be computed in a self-contained way without regularization.

As an example of a universal quantity, we consider the osmotic pressure, $\Pi = -(\partial A/\partial V)_{n,T}$. This can be expressed in dimensionless form as

$$\beta R_{g0}^3 \Pi = -\left(\frac{\partial \beta A}{\partial \widetilde{V}}\right)_{n,T} = C - \left(\frac{\partial \beta A^{ex}}{\partial \widetilde{V}}\right)_{n,T} \tag{6.16}$$

Substitution of eqns (6.14) and (6.15) leads to

$$\beta R_{g0}^3 \Pi = C + \frac{1}{2}BC^2$$
$$- \frac{1}{4\pi^2}\int_0^{\Lambda} dk\, k^2 \left\{ \ln[1 + BC\hat{g}_D(k^2)] - \frac{BC\hat{g}_D(k^2)}{1 + BC\hat{g}_D(k^2)} \right\} \tag{6.17}$$

The fluctuation integral in this expression converges even for $\Lambda \to \infty$, so we can remove the cutoff in the upper limit. The dominant contribution to the integral arises from dimensionless wavenumbers $k \sim (BC)^{1/2}$ that are large compared with unity for $BC \gg 1$. For this situation, the Debye function can be approximated by $\hat{g}_D(k^2) \approx 2/k^2$ and the resulting integral evaluated analytically (Edwards, 1966; Doi and Edwards, 1986). The result is

$$\beta R_{g0}^3 \Pi = C + \frac{1}{2}BC^2 - \frac{1}{24\pi}(2BC)^{3/2} \tag{6.18}$$

The final term in this expression, representing the contribution of Gaussian field fluctuations, is small compared to the mean-field term $(1/2)BC^2$, provided that

$$C \gg B \tag{6.19}$$

This defines the so-called *concentrated regime* of polymer solution behavior (Doi and Edwards, 1986). Condition (6.19) also verifies our expectation that $C$ plays the role of a Ginzburg parameter for the canonical Model A. When $C \to \infty$ at fixed $B$, the mean-field term provides the dominant contribution to the osmotic pressure and field fluctuation effects are asymptotically negligible.

Another useful way to express the fluctuation term in eqn (6.18) is through the introduction of the *Edwards correlation length* $\xi_E$,

$$\xi_E \equiv R_{g0}(2BC)^{-1/2} = \frac{b}{2\sqrt{3}(\rho_0 u_0)^{1/2}} \tag{6.20}$$

where $\rho_0 = nN/V$ is the average density of polymer segments. As will be demonstrated below, $\xi_E$ represents the length scale over which segment density fluctuations are correlated in a concentrated polymer solution. The osmotic pressure formula (6.18) can be rewritten with the use of eqn (6.20) as

$$\Pi = k_B T \left( \frac{\rho_0}{N} + \frac{1}{2} u_0 \rho_0^2 - \frac{1}{24\pi \xi_E^3} \right) \tag{6.21}$$

The final fluctuation term in this equation can interpreted by means of the following scaling argument (de Gennes, 1979). Because segment density fluctuations are correlated on a length scale of order $\xi_E$, regions ("blobs") of characteristic volume $\xi_E^3$ fluctuate independently. By equipartition, each such region has an energy of order the thermal energy $k_B T$. It follows that the fluctuation contribution to the osmotic pressure, which has units of energy density, scales as $k_B T/\xi_E^3$ in the concentrated regime.

The interpretation of $\xi_E$ as the correlation length for segment density fluctuations is easily established. The structure factor $S(k)$, defined by eqn (5.37), can be evaluated by combining eqns (4.77) and (4.78) to obtain[85]

$$S(k) = u_0^{-1} - \frac{1}{u_0^2 V} \langle \hat{\omega}(\mathbf{k}) \hat{\omega}(-\mathbf{k}) \rangle \tag{6.22}$$

where $\hat{\omega}(\mathbf{k})$ is the Fourier transform of $w(\mathbf{r}) - \langle w(\mathbf{r}) \rangle$. The average over $w$ fluctuations in this expression can be rewritten as an average over fluctuations in the scaled field $\Omega$ by means of the relation $\hat{\omega}(\mathbf{k}) = (R_{g0}^3/N)\hat{\Omega}(\mathbf{k}R_{g0})$. Furthermore, combining eqns (6.11) and (B.15) produces the following Gaussian fluctuation formula

$$\langle \hat{\Omega}(\mathbf{k})\hat{\Omega}(-\mathbf{k}) \rangle \equiv \frac{\int \mathcal{D}\Omega \, \exp(-H_S[\Omega])\hat{\Omega}(\mathbf{k})\hat{\Omega}(-\mathbf{k})}{\int \mathcal{D}\Omega \, \exp(-H_S[\Omega])} = \frac{\tilde{V}}{\hat{\gamma}_2(k)} \tag{6.23}$$

Substitution of this result into eqn (6.22) leads to

$$S(k) = \frac{\rho_0 N \hat{g}_D(k^2 R_{g0}^2)}{1 + BC \, \hat{g}_D(k^2 R_{g0}^2)} \tag{6.24}$$

In the concentrated regime, it is permissible to make the replacement $\hat{g}_D(k^2 R_{g0}^2) \approx 2/(kR_{g0})^2$, in which case eqn (6.24) reduces to

$$S(k) \approx \frac{12\rho_0 b^{-2}}{k^2 + \xi_E^{-2}} \tag{6.25}$$

This Lorentzian expression for the structure factor implies a pair correlation function of segment density that decays exponentially with a decay (correlation) length $\xi_E$ (Doi and Edwards, 1986), i.e.

$$\langle \hat{\rho}(\mathbf{r})\hat{\rho}(0) \rangle - \rho_0^2 = \frac{3\rho_0}{\pi b^2 \, r} \exp(-r/\xi_E) \tag{6.26}$$

---

[85] Dimensional units and wavevectors have been restored in this expression.

The above calculation also confirms our statement in Section 5.2.1 that the RPA expression for the structure factor is consistent with a Gaussian treatment of potential field fluctuations. Indeed, substitution of eqn (6.24) into the linear response formula (5.42), $\hat{\Gamma}_2(k) = 1/S(k)$, exactly reproduces eqn (5.29).

Another quantity of interest is the *chemical potential* $\mu$, which is evaluated in the Gaussian approximation by explicit differentiation of eqn (6.13):

$$\beta\mu \equiv \left(\frac{\partial\beta A}{\partial n}\right)_{V,T} = \beta\mu_0 + BC + \frac{1}{2\tilde{V}} \sum_k \frac{\hat{g}_D(k^2)}{\hat{\gamma}_2(k)} \tag{6.27}$$

where $\beta\mu_0 = (\partial\beta A_0/\partial n)_{V,T}$. The fluctuation term in eqn (6.27) is again ultraviolet divergent, but a simple subtraction to reference the infinite dilution state of swollen coils with excluded volume ($C \to 0$, $B > 0$) serves to regularize the chemical potential:

$$\beta\tilde{\mu}_E \equiv \beta\mu - \beta\mu(C \to 0) = BC - \frac{B^2C}{2\tilde{V}} \sum_k \frac{[\hat{g}_D(k^2)]^2}{1 + BC\,\hat{g}_D(k^2)} \tag{6.28}$$

The fluctuation sum in this expression can be converted to a convergent integral in the thermodynamic limit. In the concentrated regime, where the approximation $\hat{g}_D(k^2) \approx 2/k^2$ can be applied, the integral can be analytically evaluated to obtain

$$\beta\tilde{\mu}_E = BC - \frac{1}{2\sqrt{2}\pi}(B^3C)^{1/2} \tag{6.29}$$

This expression for the "excess" chemical potential again shows that the validity of the Gaussian approximation is restricted to the concentrated regime, $C \gg B$, where the (final) fluctuation term makes a small correction to the mean-field contribution $BC$.

### 6.1.2 Gaussian fluctuations about a homogeneous state: grand canonical ensemble

It is instructive to repeat the above study of Gaussian fluctuations in the *grand canonical ensemble*. The analysis in this situation proceeds somewhat differently than in the canonical ensemble, where the $\mathbf{k} = 0$ mode of the fluctuating field was seen to be decoupled from the finite $\mathbf{k}$ modes. This proves not to be the case in the grand canonical ensemble.

By introducing the scalings of eqn (6.1) and the dimensionless excluded volume parameter $B$ defined in eqn (6.2), one can express the grand canonical Model A as

$$\mathcal{Z}_G = \int \mathcal{D}W \, \exp(-H_G[W]) \tag{6.30}$$

$$H_G[W] = \frac{1}{2B} \int d\mathbf{x} \, [W(\mathbf{x})]^2 - \tilde{z}\tilde{V}\, Q[iW] \tag{6.31}$$

where $\tilde{z} \equiv z R_{g0}^3$ is a dimensionless polymer activity. Rather than expanding the effective Hamiltonian $H_G[W]$ about the homogeneous saddle point $W(\mathbf{x}) = W^*$, which satisfies

$$iW^* = B\tilde{z}\exp(-iW^*), \tag{6.32}$$

it is useful to expand it about a homogeneous field $W_0$ that corresponds to the volume average of the fluctuating field $W(\mathbf{x})$,

$$W_0 \equiv \frac{1}{\widetilde{V}} \int d\mathbf{x}\, W(\mathbf{x}) \tag{6.33}$$

So defined, $W_0$ retains the character of a fluctuating variable that in general has an expectation value different from $W^*$. Next, we introduce a fluctuating field defined by

$$\Omega(\mathbf{x}) \equiv W(\mathbf{x}) - W_0 \tag{6.34}$$

It is important to note that $\Omega(\mathbf{x})$ has a vanishing $\mathbf{k} = 0$ Fourier mode due to the property that $\int d\mathbf{x}\, \Omega(\mathbf{x}) = 0$. Substitution of eqn (6.34) into eqns (6.30) and (6.31) leads to

$$\mathcal{Z}_G = \int_{-\infty}^{\infty} dW_0 \int \mathcal{D}\Omega\; \exp(-H_{GS}[W_0, \Omega]) \tag{6.35}$$

$$H_{GS}[W_0, \Omega] = \frac{\widetilde{V}}{2B} W_0^2 + \frac{1}{2B}\int d\mathbf{x}\, [\Omega(\mathbf{x})]^2 - \tilde{z}\widetilde{V}\exp(-iW_0)\, Q[i\Omega] \tag{6.36}$$

where it is understood that the integral over $\Omega(\mathbf{x})$ is restricted to a function space in which $\int d\mathbf{x}\, \Omega = 0$.

To quadratic order in $\Omega$, the single-chain partition function $Q[i\Omega]$ can expanded according to eqn (3.131) as

$$Q[i\Omega] = 1 + \frac{(i)^2}{2\widetilde{V}^2}\sum_{\mathbf{k}} \hat{g}_D(k^2)\hat{\Omega}(\mathbf{k})\hat{\Omega}(-\mathbf{k}) + O(i\hat{\Omega}^3) \tag{6.37}$$

where $\hat{\Omega}(\mathbf{k})$ again denotes the $\mathbf{k}$th Fourier coefficient of $\Omega(\mathbf{x})$ and $\hat{\Omega}(0) = 0$. Substitution of eqn (6.37), truncated at quadratic order, into eqns (6.35) and (6.36) produces a set of independent Gaussian integrals for the $\mathbf{k} \neq 0$ modes of $\hat{\Omega}(\mathbf{k})$. These can be evaluated with the result

$$\mathcal{Z}_G = \int_{-\infty}^{\infty} dW_0\; \exp[S_G(W_0)] \tag{6.38}$$

where

$$S_G(W_0) = -\frac{\widetilde{V}}{2B}W_0^2 + \tilde{z}\widetilde{V}\exp(-iW_0) - \frac{1}{2}\sum_{\mathbf{k}} \ln[\hat{s}_2(k^2, W_0)] \tag{6.39}$$

and it is understood that the sum excludes the term $\mathbf{k} = 0$. The function $\hat{s}_2(k^2, W_0)$ is defined by

$$\hat{s}_2(k^2, W_0) \equiv 1 + B\tilde{z}\exp(-iW_0)\,\hat{g}_D(k^2) \tag{6.40}$$

and we note that a spurious term proportional to $\sum_{\mathbf{k}} \ln(2\pi\tilde{V}B)$ has been extracted from $S_G(W_0)$ since it is independent of $W_0$. This is analogous to the subtraction used in eqn (6.14).

The integral in eqn (6.38) can be asymptotically evaluated in the thermodynamic limit by the steepest descent method, since the scaled volume $\tilde{V}$ multiplies each term[86] in $S_G$. The saddle point value of $W_0$, $W_0^*$, satisfies the following equation:

$$iW_0^* = BC_0 - \frac{B^2 C_0}{4\pi^2}\int_0^\Lambda dk\,\frac{k^2\,\hat{g}_D(k^2)}{1 + BC_0\,\hat{g}_D(k^2)} \tag{6.41}$$

where a dimensionless "concentration" $C_0$ has been defined by

$$C_0 \equiv \tilde{z}\exp(-iW_0^*) \tag{6.42}$$

Thus, asymptotically for $\tilde{V} \to \infty$ we have

$$\mathcal{Z}_G \approx \exp[S_G(W_0^*)] \tag{6.43}$$

in which eqns (6.39) and (6.41) can be combined to obtain

$$S_G(W_0^*) = \tilde{V}\left(\frac{iW_0^*}{B} - \frac{(W_0^*)^2}{2B}\right)$$
$$- \frac{\tilde{V}}{4\pi^2}\int_0^\Lambda dk\,k^2\left\{\ln[1 + BC_0\hat{g}_D(k^2)] - \frac{BC_0\hat{g}_D(k^2)}{1 + BC_0\hat{g}_D(k^2)}\right\} \tag{6.44}$$

The fluctuation integral in eqn (6.44) is evidently the same as that in eqn (6.17) and converges for $\Lambda \to \infty$. For $BC_0 \gg 1$, the integral can be analytically evaluated and combined with the grand canonical connection formula (5.17) to obtain for the osmotic pressure

$$\beta R_{g0}^3 \Pi \equiv \frac{1}{\tilde{V}}\ln\mathcal{Z}_G = \frac{1}{\tilde{V}}S_G(W_0^*)$$
$$= \frac{iW_0^*}{B} - \frac{(W_0^*)^2}{2B} - \frac{1}{24\pi}(2BC_0)^{3/2} \tag{6.45}$$

---

[86]Note that there is effectively a factor of $\tilde{V}$ in front of the fluctuation sum because $\sum_{\mathbf{k}} \to [\tilde{V}/(2\pi)^3]\int d\mathbf{k}$ in the thermodynamic limit.

The remaining step is to express $W_0^*$ and $C_0$ in terms of the dimensionless chain concentration $C$ defined by

$$C \equiv \frac{\langle n \rangle R_{g0}^3}{V} = \tilde{z} \langle Q[iW] \rangle \qquad (6.46)$$

which is analogous to the definition of $C$ applied in eqn (6.2) for the canonical ensemble. Substitution of eqns (6.34) and (6.37) into $Q[iW]$ in eqn (6.46) leads to

$$C = C_0 - \frac{BC_0}{4\pi^2} \int_0^\Lambda dk \, \frac{k^2 \, \hat{g}_D(k^2)}{1 + BC_0 \, \hat{g}_D(k^2)} \qquad (6.47)$$

Combining this equation with eqn (6.41) produces

$$iW_0^* = BC \qquad (6.48)$$

which can be used to eliminate $W_0^*$ from the first two terms of eqn (6.45). Equation (6.48) shows that the saddle point shift of the $\mathbf{k} = 0$ mode in the grand canonical ensemble, when expressed in terms of $C$, has the same form as eqn (6.6) for the canonical ensemble. Finally, asymptotic analysis of the integral in eqn (6.47) for $C_0 \to \infty$ shows that

$$C = C_0 \left[ 1 + O(\Lambda^3/C_0) \right] \qquad (6.49)$$

so that $C_0$ can be replaced by $C$ in eqn (6.45). The osmotic pressure formula of eqn (6.18) is immediately recovered.

### 6.1.3  Validity of the Gaussian approximation

The above analysis of Gaussian fluctuations in the canonical and grand canonical ensembles has shown that $C$, suitably defined, serves as a Ginzburg parameter that can be used to assess the accuracy of the mean-field approximation. The mean-field approximation, i.e. SCFT, has been shown to be asymptotically exact for Model A in the limit $C \to \infty$, and the Gaussian fluctuation formulas were argued to be accurate for polymer solutions in the *concentrated regime*, i.e.

$$C \gg C^{**} \qquad (6.50)$$

where $C^{**} = B$. For polymers of high molecular weight in a good solvent, $B \sim (u_0/b^3)N^{1/2} \gg 1$, so both $B$ and $C$ must be large compared with unity for these conditions to be met.

Unfortunately, the Gaussian fluctuation formulas do not apply outside of the concentrated regime, namely to the *dilute* and *semi-dilute* regimes, which are at least as important in practice (de Gennes, 1979). In the present notation, the *semi-dilute regime* corresponds to dimensionless concentrations in the range

$$C^* \ll C \ll C^{**} \qquad (6.51)$$

where $C^*$ is the so-called *overlap concentration*. At $C^*$, as depicted in Fig. 6.1, individual polymer coils swollen to a characteristic size $R_g \sim b_s N^\nu$ given by eqn (2.2) are packed to a volume fraction of unity, so that the average segment concentration inside a coil is the same as the overall segment concentration. This leads to the condition $nR_g^3/V \sim 1$, or

$$C^* = \left(\frac{R_{g0}}{R_g}\right)^3 \sim B^{-3(2\nu-1)} \sim B^{-0.528} \tag{6.52}$$

where we have distinguished the radii of gyration of swollen and ideal chains and have applied the asymptotic dependence of the swelling ratio $R_g/R_{g0}$ on the parameter $B$ for $B \to \infty$. At high molecular weight, $B \gg 1$, the semi-dilute regime can evidently be very broad when expressed in the concentration variable $C$. The *dilute regime* is correspondingly

$$C \ll C^* \tag{6.53}$$

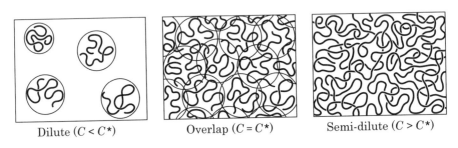

| Dilute ($C < C^*$) | Overlap ($C = C^*$) | Semi-dilute ($C > C^*$) |

FIG. 6.1. Typical polymer conformations in the dilute and semi-dilute regimes for flexible polymers dissolved in a good solvent. At overlap, $C = C^*$, the polymer coils pack to fill space with unit volume fraction. The magnification is higher in the semi-dilute panel, so the polymers appear longer.

In the dilute and semi-dilute regimes, the Gaussian treatment of fluctuations breaks down entirely. The functional integrals that define the partition functions $\mathcal{Z}_C$ and $\mathcal{Z}_G$ for Model A are no longer dominated by field configurations close to the saddle point, but by a much larger region of configuration space. Indeed, for $C \ll C^{**}$, the field fluctuations make contributions to the solution structure and thermodynamics that overwhelm the mean-field contributions. In the case of the semi-dilute regime, for example, scaling arguments can be used to argue that the osmotic pressure is given by the formula (de Gennes, 1979)

$$\Pi \sim \frac{k_B T}{\xi^3} \tag{6.54}$$

where $\xi$ is the correlation length for segment density fluctuations. This is similar in form to the Gaussian fluctuation term in eqn (6.21) and can be derived by a similar "blob" argument, but the semi-dilute correlation length $\xi$ has a different scaling[87] than the Edwards length $\xi_E$,

$$\xi \sim R_{g0} \left(B^{2\nu-1}C^\nu\right)^{-\frac{1}{3\nu-1}} \sim R_{g0}B^{-0.230}C^{-0.770} \qquad (6.55)$$

In dimensionless form, eqn (6.54) becomes

$$\beta R_{g0}^3 \Pi \sim \left(\frac{R_{g0}}{\xi}\right)^3 \sim \left(B^{2\nu-1}C^\nu\right)^{\frac{3}{3\nu-1}} \sim B^{0.691}C^{2.31} \qquad (6.56)$$

which should be compared with eqn (6.18) for the concentrated regime. These semi-dilute scaling results cannot be reproduced by a Gaussian treatment of fluctuations but require non-perturbative analytical techniques, such as the renormalization group discussed in Section 6.1.5, or the numerical "field-theoretic simulation" methods described later in this chapter.

Similar studies of Gaussian fluctuations about homogeneous saddle points have been performed for several of the models described in Chapter 4, including the polymer blend Model C (de Gennes, 1977; Joanny, 1978; Binder, 1984; de la Cruz et al., 1988; Wang, 2002) and the block copolymer Model E (Fredrickson and Helfand, 1987). This work has shown that a parameter similar to $C$ plays the role of the Ginzburg parameter in these models, so that the mean-field approximation (SCFT) is asymptotically valid for $C \to \infty$. However, for the incompressible melt models, $n/V = 1/(v_0N)$, where $v_0$ is the volume per segment, so $C \sim (b^3/v_0)N^{1/2} \gg 1$ at high molecular weight.[88] Thus, the mean-field approximation has a much broader range of validity for polymer melts than for solutions. However, even in the case of polymer melts of high molecular weight, the mean-field approximation can fail catastrophically in the close vicinity of a critical point (e.g. the critical consolute point of Model C) (Goldenfeld, 1992; Wang, 2002), a multicritical point such as a Lifshitz point (Bates et al., 1997), or any second-order or weakly first-order phase transition such as the block copolymer order–disorder transition (Fredrickson and Helfand, 1987). In these situations, the breakdown of the mean-field approximation is restricted to a region of reduced temperature, $\tau = |T - T_t|/T_t$, where $T_t$ is the temperature of the relevant phase transition, that vanishes as $C \to \infty$. For example, the width of the non-mean-field region for a symmetric polymer blend is (de Gennes, 1977) $\tau \sim C^{-2}$, while $\tau \sim C^{-2/3}$ for symmetric diblock copolymer melts (Fredrickson and Helfand, 1987) and for asymmetric blends near the spinodal (Binder, 1984; Wang, 2002).

---

[87]The scaling formula (6.55) results from the ansatz that $\xi$ is independent of $N$ in the semi-dilute regime, except through its dependence on the average segment density $\rho_0 = nN/V$.

[88]The Ginzburg parameter is sometimes written in a form proportional to polymer molecular weight, $\bar{N} \equiv Nb^6/v_0^2$. This parameter is related to $C$ by $C = 6^{-3/2}\bar{N}^{1/2}$.

### 6.1.4 Gaussian fluctuations about an inhomogeneous state

An analysis of Gaussian field fluctuations can also be performed for the more interesting and typical case of an *inhomogeneous* mean-field solution. The methodology for such analysis of polymer field theories was first described by Shi, Noolandi, and coworkers (Yeung *et al.*, 1996; Shi *et al.*, 1996; Laradji *et al.*, 1997*a*; Laradji *et al.*, 1997*b*) and was recast in a form more consistent with the present notation in a recent review (Müller and Schmid, 2005). We discuss the topic here in the simplest context – the canonical Model A.

When solved in a confined geometry, e.g. a slit with Dirichlet boundary conditions applied, Model A exhibits inhomogeneous saddle points. Figure 6.2 shows numerical examples of the reduced mean-field density profiles, $\phi(z) = \tilde{\rho}(z; [iW^*])/\rho_0$, corresponding to such saddle points for a slit width of $L/R_{g0} = 1$. The profiles are sensitive only to the combined variable $BC$ and for $BC \gg 1$ rise from the walls to a plateau on a scale that corresponds to the Edwards correlation length, $\xi_E = R_{g0}/\sqrt{2BC}$. For more general cases of three-dimensional confinement by Dirichlet boundaries, the relevant saddle points $W^*(\mathbf{x})$ have a more complicated position dependence, but we expect the same qualitative behavior – namely a "healing" of the density profile on the scale of $\xi_E$ normal to the boundaries. Such inhomogeneous saddle points are also pure imaginary, just as in the homogeneous case, cf. eqn (6.6).

Given an inhomogeneous saddle point $W^*(\mathbf{x})$, the role of Gaussian fluctuations can be studied by a shift of integration contour in eqn (6.3) and expansion of $H[W]$ to quadratic order about $W^*$. As in Section 6.1.1, the appropriate shift of contour is to a new integration path for the field

$$\Omega(\mathbf{x}) \equiv W(\mathbf{x}) - W^*(\mathbf{x}) \tag{6.57}$$

that is parallel to the real axis. In a discrete representation of $\Omega(\mathbf{x})$, this means that the value of the field at each grid point $\mathbf{x_l}$ is integrated over an independent parallel path that is offset from the real axis by an amount $-W^*(\mathbf{x_l})$. The aggregate of such integrals over all grid points constitutes the discrete approximation to the functional integral over $\Omega$. The expansion of $H[W]$ to quadratic order amounts to a functional Taylor expansion, cf. eqn (C.6),

$$H[W] = H[W^*] + \int d\mathbf{x} \left. \frac{\delta H[W]}{\delta W(\mathbf{x})} \right|_{W^*} \Omega(\mathbf{x})$$
$$+ \frac{1}{2!} \int d\mathbf{x} \int d\mathbf{x'} \left. \frac{\delta^2 H[W]}{\delta W(\mathbf{x}) \delta W(\mathbf{x'})} \right|_{W^*} \Omega(\mathbf{x})\Omega(\mathbf{x'}) + O(\Omega^3) \tag{6.58}$$

The linear term in this expression vanishes identically because of the definition of $W^*$. It follows that substitution of eqn (6.58) into eqn (6.3) leads to a transformed field theory of exactly the same form as eqn (6.7), but where

$$H_S[\Omega] = \frac{1}{2!} \int d\mathbf{x} \int d\mathbf{x'} \, \gamma_2(\mathbf{x}, \mathbf{x'}) \, \Omega(\mathbf{x})\Omega(\mathbf{x'}) + O(\Omega^3) \tag{6.59}$$

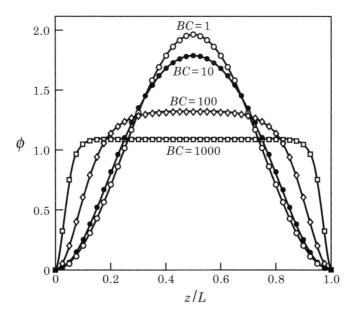

FIG. 6.2. Mean-field reduced density profiles for Model A solved in a slit geometry. The points are numerical SCFT results for $BC = 1$ (open circles),
$BC = 10$ (filled circles), $BC = 100$ (open diamonds), and $BC = 1000$ (open
squares). The solid lines represent fits to an analytical expression based on the
ground state dominance approximation. Reproduced from Alexander-Katz
*et al.* (2003).

and $\gamma_2(\mathbf{x}, \mathbf{x}')$ is defined by

$$\gamma_2(\mathbf{x}, \mathbf{x}') = \frac{1}{B}\,\delta(\mathbf{x} - \mathbf{x}') - C\widetilde{V}\,\left.\frac{\delta^2 \ln Q[iW]}{\delta W(\mathbf{x})\delta W(\mathbf{x}')}\right|_{W^*} \qquad (6.60)$$

The local form of the Hamiltonian $H_S[\Omega]$ is thus very similar to that in the
homogeneous saddle point case, cf. eqn (6.11). However, when the saddle point
$W^*(\mathbf{x})$ is inhomogeneous, the function $\gamma_2(\mathbf{x}, \mathbf{x}')$ is *not translationally invariant*,
i.e. $\gamma_2(\mathbf{x}, \mathbf{x}') \neq \gamma_2(|\mathbf{x} - \mathbf{x}'|)$, so $H_S[\Omega]$ is not diagonal in a Fourier representation.
Nevertheless, upon introducing a field $Y(\mathbf{x}) = iW(\mathbf{x})$, which is *real* at the saddle
point, we see that eqn (6.60) can be reexpressed as

$$\gamma_2(\mathbf{x}, \mathbf{x}') = \frac{1}{B}\,\delta(\mathbf{x} - \mathbf{x}') + C\widetilde{V}\,\left.\frac{\delta^2 \ln Q[Y]}{\delta Y(\mathbf{x})\delta Y(\mathbf{x}')}\right|_{Y^*} \qquad (6.61)$$

which shows that $\gamma_2$ is *real*. This function can be identified as a dimensionless
version of the *Jacobian* $J(\mathbf{r}, \mathbf{r}'; [\mu])$ introduced in eqn (5.86). Thus, $\gamma_2(\mathbf{x}, \mathbf{x}')$ can
be evaluated by means of eqn (3.73) in $O(N_s M^2 \log_2 M)$ operations, using a

pseudo-spectral scheme with $M$ spatial collocation points and $N_s$ contour steps. In such a discrete representation, $\gamma_2$ is a *real, symmetric $M \times M$ matrix*[89] that is ordinarily not sparse. However, as described in the context of eqn (5.87), the Jacobian matrix becomes diagonal dominant for the large $\mathbf{k}$ modes in a Fourier representation, so this might be used to computational advantage.

Provided that the eigenvalues of $\gamma_2$ are all positive, which would imply a *locally stable* saddle point, the Gaussian fluctuation integral of eqn (6.7) can be performed analytically, cf. eqn (B.11), with the result[90]

$$\beta A(n, V, T) = \beta A_0 + H[W^*] + \frac{1}{2} \ln \left( \frac{(\Delta x)^{6M} \det \gamma_2}{(2\pi)^M} \right) \qquad (6.62)$$

where $(\Delta x)^3 = \widetilde{V}/M$ is the volume per grid point in the discrete representation of the field. As in the case of a homogeneous system, it is convenient to reference the free energy to the non-interacting state with $B = 0$. This leads to an excess free energy given by

$$\beta A^{ex}(n, V, T) = H[W^*] + \frac{1}{2} \ln \det[(\Delta x)^3 B \gamma_2] \qquad (6.63)$$

which is analogous to eqn (6.14) for the homogeneous case.

Equation (6.63) exhibits the same ultraviolet divergence as eqn (6.15) when the collocation mesh is refined, i.e. $\Delta x \to 0$, $M \to \infty$, subject to $(\Delta x)^3 M = \widetilde{V}$. However, other derivative quantities such as $\Pi$ and $\tilde{\mu}_E$ are ultraviolet convergent, as in the homogeneous case. It should be noted that such expressions must be evaluated numerically, and the operation count can be significant, scaling as $O(M^3)$ for large $M$.

A quantity of particular interest is the structure factor (scattering function) $S(\mathbf{k})$ that describes the anisotropic nature of field fluctuations about the saddle point. In dimensional units, $S(\mathbf{k})$ is defined by an expression analogous to eqn (5.37):

$$S(\mathbf{k}) = V^{-1} \int d\mathbf{r} \int d\mathbf{r}'\, e^{-i\mathbf{k}\cdot(\mathbf{r}-\mathbf{r}')} \langle [\hat{\rho}(\mathbf{r}) - \langle \hat{\rho}(\mathbf{r}) \rangle][\hat{\rho}(\mathbf{r}') - \langle \hat{\rho}(\mathbf{r}') \rangle] \rangle \qquad (6.64)$$

Substituting eqns (4.77) and (4.78) into this equation and introducing dimensionless variables leads to

$$S(\mathbf{q}) = \frac{N^2}{VB} \int d\mathbf{x} \int d\mathbf{x}'\, e^{-i\mathbf{q}\cdot(\mathbf{x}-\mathbf{x}')} [\delta(\mathbf{x} - \mathbf{x}') - B^{-1} \langle \Omega(\mathbf{x})\Omega(\mathbf{x}') \rangle] \qquad (6.65)$$

where $\mathbf{q} = R_{g0}\mathbf{k}$ is a dimensionless wavevector. If we invoke the present Gaussian treatment of fluctuations, described by the Hamiltonian of eqn (6.59), the correlation function appearing in this expression is given by (cf. eqn (B.15))

---

[89]Following our earlier convention, boldface is used when referring to $\gamma_2$ as a matrix.

[90]Note that the fluctuation term can be written in a form similar to the corresponding term in eqn (6.13) by means of the identity $\ln \det \gamma_2 = \mathrm{Tr} \ln \gamma_2$.

$$\langle \Omega(\mathbf{x})\Omega(\mathbf{x}')\rangle = \gamma_2^{-1}(\mathbf{x}, \mathbf{x}') \tag{6.66}$$

where $\gamma_2^{-1}$ is the functional inverse of $\gamma_2$ in the sense of eqn (C.29). Combining these results, we finally obtain

$$S(\mathbf{q}) = \frac{N^2}{VB} \int d\mathbf{x} \int d\mathbf{x}' \, e^{-i\mathbf{q}\cdot(\mathbf{x}-\mathbf{x}')}[\delta(\mathbf{x} - \mathbf{x}') - B^{-1}\gamma_2^{-1}(\mathbf{x}, \mathbf{x}')] \tag{6.67}$$

For a general inhomogeneous saddle point, this expression can be evaluated numerically by inverting the matrix $\gamma_2$ in the discrete representation and performing a numerical quadrature. In the case of a homogeneous saddle point, eqn (6.67) reduces to eqn (6.24). Provided that the confinement length scale $L$ responsible for creating the inhomogeneity in Model A is large compared to the Edwards length $\xi_E$, it can be shown that the applicability of the above expressions is again limited to $C \gg C^{**} = B$, i.e. to the concentrated regime.

Laradji et al. (Laradji et al., 1997a; Laradji et al., 1997b) have applied similar methods to study Gaussian fluctuation effects in diblock copolymer melts, namely the canonical Model E with $b_A = b_B$. As in a local analysis about the homogeneous (disordered phase) saddle point, $C \sim N^{1/2}$ is the relevant Ginzburg parameter the governs the strength of anisotropic field fluctuations about the inhomogeneous saddle points (periodic SCFT mesophase solutions) of Model E. Thus, at high molecular weight the fluctuations are weak, except in the close vicinity of the order–disorder transition (ODT). By considering the sign of the eigenvalues of the matrix corresponding to $\gamma_2$, Laradji et al. were able to examine the local stability of the SCFT solutions for the mesophases shown in Fig. 5.16. Moreover, their analysis has provided insights into kinetic pathways between the various ordered mesophases and especially near the L–H boundaries in Figs. 5.15 and 5.17, where the "complex" Ia3̄d and HPL phases are experimentally observed (Khandpur et al., 1995; Tyler and Morse, 2005).

### 6.1.5 *Renormalization techniques*

In the previous sections we have seen that it is possible to compute corrections to mean-field (SCFT) results for the thermodynamic properties of polymeric fluids that arise from Gaussian field fluctuations. Such computations require a local analysis to quadratic order around the relevant saddle point that can usually be performed *analytically* in the case of a homogeneous saddle point, or *numerically* for an inhomogeneous saddle point. The "fluctuation corrections" so obtained are useful in establishing the range of validity of the mean-field approximation, but are also of interest in the interpretation of radiation scattering experiments, assessing the stability of SCFT solutions, and examining kinetic pathways between solutions of different symmetry.

Unfortunately there are a number of situations in which the field fluctuations are so strong that a Gaussian treatment is inadequate or breaks down entirely. Several of these were listed at the beginning of the present chapter. For example, in the case of Model A we have seen that Gaussian fluctuations make a small contribution to thermodynamic properties, relative to the mean-field contribution, only if $C \gg C^{**} = B$. For smaller values of $C$, the Gaussian fluctuation formulas are both unjustified and inaccurate. Thus, the important semi-dilute and dilute polymer solution regimes are inaccessible by this theoretical framework. Similarly, the Gaussian treatment of fluctuations fails catastrophically for Model C in the vicinity of its critical consolute point and for Model E near the order–disorder phase boundary (ODT). For these and many other situations, we require a more powerful theoretical tool capable of dealing with field fluctuations that are large in amplitude, or strongly correlated with a divergent correlation length.

Indeed, such a tool exists for analyzing field theories that exhibit strong field fluctuations – the *renormalization group (RG)* (Wilson and Kogut, 1974; 't Hooft and Veltman, 1972; Amit, 1984; Goldenfeld, 1992; Zee, 2003). This technique evolved from pioneering work in the fields of high energy and condensed matter physics (specifically the subject of critical phenomena), but has since become a standard field-theoretic tool that has been applied in many disciplines. In the context of polymer physics, excellent introductions to the subject of the renormalization group are available, including the monographs by de Gennes (1979), Freed (1987), and des Cloizeaux and Jannink (1990).

The basic idea underlying "renormalization" and the renormalization group is to study in a systematic way how a field theory model is changed when certain degrees of freedom are eliminated by integrating them out of the defining functional integral. In the case of polymer field theories, this step normally amounts to a type of *coarse-graining*, because it is the finest scale (or largest wavenumber $k$) field fluctuations that are removed. Since the fine scale modes are coupled to coarser fluctuation modes, when the finest modes are removed, the remaining modes interact with each other differently. This can be quantified by shifting or *renormalizing* the interaction parameters that enter the original "bare" field theory model. For example, in the case of Model A, the excluded volume parameter $u_0$ describes the strength of segment–segment repulsions. In a RG analysis, $u_0$ is allowed to change or be "renormalized" as the highest $k$ fluctuation modes are integrated out of the model.

6.1.5.1   *The Hartree approximation*   Before discussing the full RG theory, it is useful to consider a simple example of renormalization in the context of Model A. This example will serve to illustrate the power of renormalization techniques, while taking advantage of the Gaussian fluctuation analysis that has already been performed.

For simplicity, we restrict attention to the canonical Model A in a three-dimensional bulk system and return to eqns (6.17) and (6.18) for the osmotic pressure that resulted from a local analysis about the homogeneous saddle point. One way of reinterpreting the fluctuation term in eqn (6.17) is to use it to "renormalize" the dimensionless excluded volume parameter $B$ appearing in the mean-field term $(1/2)BC^2$. The renormalized parameter, $B_r$, will be seen to depend on both $B$ and $C$. To affect this renormalization, we simply rewrite eqn (6.17) as

$$\beta R_{g0}^3 \Pi = C + \frac{1}{2} B_r C^2 \qquad (6.68)$$

where

$$B_r \equiv B - \frac{1}{2\pi^2 C^2} \int_0^\Lambda dk \, k^2 \left\{ \ln[1 + BC\hat{g}_D(k^2)] - \frac{BC\hat{g}_D(k^2)}{1 + BC\hat{g}_D(k^2)} \right\} \qquad (6.69)$$

This expression for $B_r$ implies that the effective interaction between polymer segments is altered from the mean-field interaction $B$ due to correlation effects treated at the level of Gaussian fluctuations. The formula, however, is only valid for $C > B$, i.e. in the concentrated regime, because otherwise $B_r$ could change sign and the osmotic pressure could become negative – an unphysical result.

We noticed previously when evaluating the integral in eqn (6.17) that the dominant contribution arises from field fluctuation modes with dimensionless wavenumbers $k \sim (BC)^{1/2}$, or physical length scales comparable to the Edwards' correlation length $\xi_E$. A more sophisticated approximation for the osmotic pressure can be obtained if we allow this length scale to be determined *self-consistently* as part of the calculation. Namely, we replace $B$ with $B_r$ inside the fluctuation integral of eqn (6.69), which amounts to the assumption that the dominant length scale for segment concentration fluctuations is the renormalized correlation length

$$\xi_r \equiv R_{g0}(2B_r C)^{-1/2} \qquad (6.70)$$

Upon evaluating the integral with this replacement, one obtains

$$B_r = B - \frac{2^{3/2}}{12\pi} B_r^{3/2} C^{-1/2} \qquad (6.71)$$

For specified values of $B$ and $C$, this equation is a cubic equation in $B_r^{1/2}$ that can be solved analytically. The full expression for the physical root will not be given here; instead, we restrict our attention to its asymptotic behavior in various regimes of polymer concentration. In the *concentrated regime*, $C \gg C^{**} = B$, the asymptotic behavior is

$$B_r \approx B - \frac{2^{3/2}}{12\pi} B^{3/2} C^{-1/2} \qquad (6.72)$$

which is nothing more than the Gaussian fluctuation result.

In the more interesting cases of the *semi-dilute* and *dilute* regimes, where $C \ll C^{**}$, the two terms on the right-hand side of eqn (6.71) are dominant and lead to

$$B_r \approx 2^{1/3}(3\pi)^{2/3} B^{2/3} C^{1/3} \qquad (6.73)$$

Substitution of this result into eqn (6.70) produces the semi-dilute scaling prediction

$$\xi_r \sim R_{g0} B^{-1/3} C^{-2/3} \tag{6.74}$$

which compares reasonably well with the "exact" scaling formula (6.55). More impressive is the result for the osmotic pressure when eqn (6.73) is substituted into eqn (6.68). Upon balancing the ideal gas contribution $C$ against the interaction term $(1/2)B_r C^2$, the following estimate is obtained for the overlap concentration $C = C^*$:

$$C^* \sim B^{-1/2} \tag{6.75}$$

This compares quite favorably with the "exact" eqn (6.52). For $C^* \ll C \ll C^{**}$, the interaction term is dominant over the ideal gas term, so that eqns (6.73) and (6.68) lead to the following *semi-dilute* scaling formula

$$\beta R_{g0}^3 \Pi \sim B^{2/3} C^{7/3} \tag{6.76}$$

which is in remarkably good agreement with the "exact" scaling expression (6.56). Finally, in the *dilute regime*, $C \ll C^*$, the ideal gas term in eqn (6.68) is dominant, and we obtain

$$\beta R_{g0}^3 \Pi \approx C \tag{6.77}$$

This example has shown that the simple process of self-consistently renormalizing the interaction parameter $B$ in our Gaussian fluctuation formulas has transformed an approximation that breaks down outside the concentrated regime into a much more sophisticated approximation. Indeed, the renormalized theory makes a reasonably accurate prediction for the dilute to semi-dilute boundary ($C^*$) and an exact identification of the semi-dilute to concentrated boundary ($C^{**}$). It also makes surprisingly accurate predictions for the scaling exponents of the correlation length and osmotic pressure in the semi-dilute regime. Furthermore, due to the self-consistent nature of the approximation, it is fundamentally *non-perturbative* in character. In other words, these predictions could not be obtained by continuing the perturbation expansion used to derive the Gaussian approximation, i.e. by including anharmonic terms in eqn (6.11), to any *finite* order.

The self-consistent approximation described above is one of the simplest renormalization schemes that is widely applied in field theory. It is commonly referred to as the *Hartree approximation* or the *self-consistent one-loop approximation* (Amit, 1984; Chaikin and Lubensky, 1995).[91] In the case of Model A, a more sophisticated version of the approximation was devised by Edwards and Muthukumar that self-consistently treats both single-chain and collective (many-chain) properties (Edwards, 1975; Muthukumar and Edwards, 1982).

The Hartree approximation has been used to examine fluctuation effects in several of the other models described in Chapter 4. For example, Wang (2002) applied this scheme to study the role of Gaussian concentration fluctuations in

---

[91]The Hartree approximation is sometimes called the random phase approximation, or RPA, but we avoid this terminology to prevent confusion with the use of the term RPA in the polymer context (de Gennes, 1979), cf. Section 5.2.1.

renormalizing the Flory interaction parameter, $\chi_{AB}$, for the binary blend Model C. Useful insights were also gained into the effect of fluctuations on the stability limit (spinodal) of the disordered phase. The approximation, however, fails at the critical composition – incorrectly predicting a first-order phase transition (Amit and Zannetti, 1973), rather than the theoretically expected and experimentally observed second-order transition with Ising exponents (de Gennes, 1977; Bates *et al.*, 1990).

The Hartree approximation also forms the basis for analytical studies of fluctuation effects in block copolymers. Brazovskii (1975) used the technique to analyze a model of weak crystallization that has soft fluctuation modes centered on a spherical shell $|\mathbf{k}| = k_m$ in reciprocal space. Fredrickson and coworkers (Fredrickson and Helfand, 1987; Barrat and Fredrickson, 1991) adapted the Brazovskii analysis to the study of fluctuation effects on the order–disorder transition (ODT) of diblock copolymer melts (Model E with $b_A = b_B$). Their results show that strong field fluctuations with wavevectors nearly isotropically distributed on the shell $k_m$ (given by the Leibler expression (5.36)) act to destroy the mean-field critical point shown at $(f = 1/2, \chi N = 10.495)$ in Fig. 5.15. At this symmetric composition, the ODT becomes a *fluctuation-induced, weakly first-order* phase transition, and the ODT boundary between the disordered and lamellar phases is shifted to larger $\chi N$ according to

$$(\chi N)_{ODT} = 10.495 + 41.022\,\bar{N}^{-1/3} = 10.495 + 6.837\,C^{-2/3} \qquad (6.78)$$

We have expressed this result[92] both in terms of the parameter $C$ employed here and the more conventional (Bates *et al.*, 1988) $\bar{N} = 6^3 C^2$. From eqn (6.78), we see that thermal fluctuations stabilize the disordered phase at finite $C$ and that the mean-field prediction of a continuous transition at $\chi N = 10.495$ is restored for $C \sim N^{1/2} \to \infty$, again confirming the role of $C$ as the relevant Ginzburg parameter. The non-trivial $\sim C^{-2/3}$ scaling of the correction term arises from the self-consistent treatment of fluctuations.

Another important prediction of the Fredrickson–Helfand theory is that the composition fluctuations open up windows of direct first-order transitions in the phase diagram between the disordered phase and ordered phases other than the spherical CPS and Im$\bar{3}$m phases shown in Fig. 5.15. As illustrated in Fig. 6.3 for the case of $C = 68.0$ ($\bar{N} = 10^6$), the Hartree analysis predicts direct disordered-to-lamellae and disordered-to-cylinder transitions at the ODT boundary. More recent theoretical studies have extended these calculations in various ways, including a more sophisticated self-consistent treatment of the fluctuation integrals appearing in the theory (Barrat and Fredrickson, 1991; de la Cruz, 1991; Stepanow, 1995), the treatment of other copolymer architectures and blends

---

[92]It should be noted that the calculation used to obtain this fluctuation correction (Fredrickson and Helfand, 1987) is not fully consistent because the two-field ($\rho, w$) description was employed, e.g. eqn (4.69), but the Hartree analysis was only applied to the $\rho$ integral (the $w$ integral was performed by the saddle point approximation). This inconsistency was addressed by Stepanow (1995).

(Mayes and de la Cruz, 1991; Olmsted and Milner, 1994; Dobrynin and Leibler, 1997; Potemkin and Panyukov, 1998), analysis of complex mesophases such as the Ia$\bar{3}$d (gyroid) phase (Hamley and Podneks, 1997), and the extension to thin film geometries (Milner and Morse, 1996).

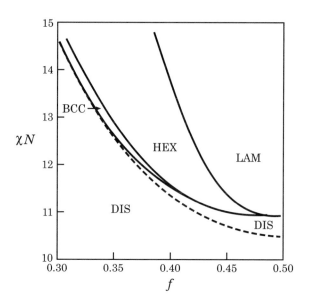

FIG. 6.3. The diblock copolymer phase diagram for $C = 68.0$ ($\bar{N} = 10^6$) derived from the Hartree approximation. The regions labeled LAM, HEX, BCC, and DIS correspond, respectively, to the lamellar, hexagonal cylinder, bcc sphere (Im$\bar{3}$m), and disordered phases depicted in Fig. 5.16. The dashed curve is the mean-field spinodal of Leibler (1980) shown in Fig. 5.3. Reproduced from Fredrickson and Helfand (1987).

The Hartree analysis brings the theoretical phase diagram for diblock copolymer melts, cf. Figs. 5.15 and 6.3, in closer qualitative agreement with experiment, such as the experimental PS-PI phase diagram shown in Fig. 5.17. In addition, the theory has been successfully applied to interpret scattering and rheological data for high molecular weight, nearly symmetric diblock melts (Bates *et al.*, 1988; Bates *et al.*, 1990). However, the approximation as conventionally implemented fails rather dramatically for strongly asymmetric melts (Fredrickson and Helfand, 1987), $f \lesssim 0.3$. This failure can be traced to the local analysis that is performed around the ordered phases to evaluate the fluctuation contribution to the free energy. From the discussion in Section 6.1.4 and the work of Shi, Noolandi and coworkers (Shi *et al.*, 1996; Laradji *et al.*, 1997b), it is apparent that numerical methods are required to properly compute Gaussian fluctuation corrections about an inhomogeneous saddle point solution. In the Brazovskii–

Fredrickson–Helfand (BFH) analysis, the Gaussian fluctuation integral is simplified by ignoring all but leading harmonics in the saddle point solution. While this
is a reasonable procedure near $f = 1/2$, where the weakly ordered lamellar saddle
point is well approximated by a pure sinusoid (at least for $C \gg 1$), the strategy
fails to capture the strong first-order character of the transition for asymmetric
copolymers. A possible resolution of this deficiency would be to implement the
Gaussian fluctuation analysis of Laradji *et al.* (1997*b*) self-consistently to affect
a "proper" Hartree theory. However, this would increase the computational burden of the method to the point that the "field-theoretic simulation" techniques
considered later in this chapter would be more attractive. A final criticism of
the BFH analysis is that it ignores certain ultraviolet divergences that should
properly be absorbed into a renormalization of $\chi = \chi_{AB}$ (Wang, 2002; Kudlay
and Stepanow, 2003). This presumably is responsible for the discrepancy in $\chi$
values obtained by fitting experimental data for diblock copolymers and binary
homopolymer blends (composed of the same A and B monomers) to the BFH
theory and the corresponding theory for Model C (Maurer *et al.*, 1998).

In spite of a growing list of successes, the Hartree approximation has a number of drawbacks, some of which have already been noted. In the application
to Model A, the Hartree scaling predictions for the correlation length and osmotic pressure in the semi-dilute regime, eqns (6.74) and (6.76), are not exact,
cf. eqns (6.55) and (6.56), and it is not obvious how to go about systematically
improving them. Moreover, the theory is not internally consistent, because it
violates the scaling relation (6.54). In other words, eqn (6.76) for the osmotic
pressure does not agree with the expression $\Pi \sim k_B T / \xi_r^3$. For Model C, we have
already mentioned the failure of the Hartree approximation to predict a continuous transition at the critical composition. Probably the most serious drawback,
however, relates to the difficulty of properly implementing the Hartree approximation for inhomogeneous saddle points, as was noted above for Model E. The
necessary local analysis requires numerical methods that can amount to a significant computational burden once the self-consistency condition is implemented.
Worse yet, this analysis must be repeated about *each* inhomogeneous saddle
point that competes in the global phase diagram of the model. A full implementation of the Hartree approximation for a complicated field-theoretic model
with multiple saddle points can thus prove to be tedious and computationally
demanding.

6.1.5.2 *Renormalization group theory*  Renormalization group (RG) theory is
a powerful formalism that in principle allows for systematic computation of scaling exponents and provides theoretical justification for scaling formulas such as
eqn (6.54). It thus resolves some of the shortcomings of the Hartree approximation noted above.

The renormalization group is implemented as a series of "RG transformations", each of which consists of a *coarse-graining* step, immediately followed by

a *change of scale* (Wilson and Kogut, 1974; Goldenfeld, 1992).[93] These steps are performed by eliminating and rescaling degrees of freedom in the partition sum or integral that defines the model. In an RG calculation, the main focus is to study how parameters in the Hamiltonian, and thus the form and strength of interactions among the fluctuating modes, are modified by the application of an RG transformation. The coarse-graining step can be performed in one of several ways. For lattice models, such as Ising models, coarse-graining is normally carried out in real space by averaging adjacent spins or other degrees of freedom into "blocks." An RG transformation based on such an approach is sometimes referred to as *position space RG*. In the case of continuum field theory models, the coarse-graining step is more commonly performed using a Fourier representation of the fields – the so-called *momentum space RG*. In this approach, which will be adopted in the present section, coarse-graining is achieved by integrating out a spherical shell of field fluctuation modes with wavenumbers $\Lambda' < |\mathbf{k}| < \Lambda$ just below some upper cutoff $\Lambda$. The *change of scale* step is performed to maintain a constant grid size if the coarse-graining is performed in real space, or to restore the sphere of retained modes $|\mathbf{k}| \leq \Lambda'$ to its original radius $\Lambda$ if the coarse-graining is done in reciprocal space.

It is the coarse-graining step of an RG transformation that is technically most difficult because the modes to be removed normally interact strongly with the modes retained, so the requisite integrals are not Gaussian. Nevertheless, the beauty of the RG formalism is that it provides justification in some cases[94] for such integrals to be performed using perturbation theory techniques, even though the formalism is capable of generating results that are inherently *nonperturbative* in nature, i.e. beyond the realm of any finite-order extension of the weak amplitude expansion used in Sections 6.1.1-6.1.4. The second rescaling step of the RG transformation is usually straightforward to perform and does not require any difficult calculations.

Upon implementing an RG transformation, a vector of parameters $\mathbf{P} = (P_1, P_2, ...)$ that defines the initial model is transformed into a new vector $\mathbf{P}'$ that characterizes the interactions among field fluctuation modes in the coarse-grained and rescaled model

$$\mathbf{P}' = \mathcal{R}(\mathbf{P}) \tag{6.79}$$

The RG operator $\mathcal{R}$ appearing in this expression defines a nonlinear mapping from $\mathbf{P}$ to $\mathbf{P}'$. Physical *critical points* correspond to *fixed points* $\mathbf{P}_f$ in parameter space such that

$$\mathbf{P}_f = \mathcal{R}(\mathbf{P}_f) \tag{6.80}$$

---

[93]We shall adopt the "Wilson-type" renormalization scheme that is familiar in condensed matter physics. A different RG strategy based on Callan–Symanzik equations (Amit, 1984; Brezin *et al.*, 1976) has its origin in particle physics, but has also been applied to polymers (Freed, 1987; Oono, 1985; des Cloizeaux and Jannink, 1990).

[94]These are problems with so-called "weak-coupling" fixed points.

At such special points, the model is scale invariant, normally because a correlation length has become infinite. *Critical exponents*, which define the scaling behavior of thermodynamic quantities in the close vicinity of the critical point, are related to the *eigenvalues* of the matrix $\mathcal{R}_L$ that is obtained by linearizing eqn (6.79) about $\mathbf{P}_f$, i.e.

$$\delta\mathbf{P}' = \mathcal{R}_L \cdot \delta\mathbf{P} \tag{6.81}$$

where $\delta\mathbf{P} \equiv \mathbf{P} - \mathbf{P}_f$ and $\delta\mathbf{P}' \equiv \mathbf{P}' - \mathbf{P}_f$. In the polymer context, the eigenvalues of $\mathcal{R}_L$ will be seen to be connected to the scaling exponents of properties such as the osmotic pressure and correlation length with respect to variations in polymer concentration and molecular weight.

Renormalization group theory is a complex subject that goes considerably beyond the scope of the current text. Indeed, entire monographs have been written on the application of RG to just the *simplest* of the models described here – the canonical Model A in unbounded space (Freed, 1987; des Cloizeaux and Jannink, 1990). As will be discussed in more detail below, analytical RG calculations are essentially intractable for models with inhomogeneous saddle points, so few results for other models have been reported. For the purpose of illustration, we shall restrict our attention to Model A subject to periodic boundary conditions. The presentation and notations are adapted from the work of Ohta and Nakanishi (Ohta and Nakanishi, 1983; Nakanishi and Ohta, 1985).

To conduct RG calculations, it proves convenient to generalize Model A to $d$ dimensions and to absorb a factor of $3/b^2$ into the contour variable $s$ in the potential energy expression eqn (4.64). The potential energy of this generalized Model A can be written

$$\beta U[\mathbf{r}^{nN_0}] = \frac{1}{2}\sum_{j=1}^{n}\int_0^{N_0} ds \left|\frac{d\mathbf{r}_j(s)}{ds}\right|^2 + \frac{u_0}{2}\int d^d\mathbf{r} \, [\hat{\rho}(\mathbf{r})]^2 \tag{6.82}$$

where $\mathbf{r}_j(s)$ is a $d$-dimensional space curve describing the conformation of the $j$th polymer and $\int d^d\mathbf{r}$ denotes a $d$-dimensional volume integral over a hypercube of volume $V$. Periodic boundary conditions are applied. The parameter $N_0$ is again proportional to the polymer molecular weight, but has "engineering dimensions" of length squared,[95] i.e. $[N_0] = L^2$. The contour variable $s$ has the same dimensions as $N_0$, i.e. $[s] = L^2$, and it follows that $[\hat{\rho}] = L^{2-d}$ and $[u_0] = L^{-\epsilon}$, where $\epsilon \equiv 4 - d$.

It is a simple matter to retrace the particle-to-field transformation that led to the canonical Model A field theory. Starting with the modified potential energy eqn (6.82) we find that

$$\mathcal{Z}_C = \mathcal{Z}_0 \exp(-H[w^*])\int \mathcal{D}\omega \, \exp(-H[\omega]) \tag{6.83}$$

where

---

[95] We adopt the notation $[x]$ to indicate the engineering dimension of a variable $x$.

$$H[\omega] = \frac{1}{2u_0} \int d^d\mathbf{r} \, [\omega(\mathbf{r})]^2 - n \ln Q[i\omega] \qquad (6.84)$$

The single-chain partition function is given as usual by

$$Q[i\omega] = \frac{1}{V} \int d^d\mathbf{r} \, q(\mathbf{r}, N_0; [i\omega]) \qquad (6.85)$$

where the propagator $q$ satisfies

$$\frac{\partial}{\partial s} q(\mathbf{r}, s; [i\omega]) = \left( \frac{1}{2}\nabla^2 - i\omega(\mathbf{r}) \right) q(\mathbf{r}, s; [i\omega]) \qquad (6.86)$$

subject to $q(\mathbf{r}, 0; [i\omega]) = 0$ and periodic boundary conditions. It should be noted that the uniform mode, $w_0 = (1/V) \int d^d\mathbf{r} \, w(\mathbf{r})$, has been extracted so that the functional integral in eqn (6.83) is over a field $\omega(\mathbf{r}) = w(\mathbf{r}) - w_0$ with vanishing spatial integral. As shown in Section 6.1.1, the uniform mode fluctuates independently of the $\mathbf{k} \neq 0$ Fourier components of $w$, and its removal isolates the mean-field contribution to the free energy, $H[w^*] = (1/2)u_0\rho_0^2 V$, which is explicit in eqn (6.83). The field $\omega(\mathbf{r})$ has engineering dimensions $[\omega] = L^{-2}$.

Ohta and Nakanishi (1983) carried out a weak amplitude expansion of the effective Hamiltonian $H[\omega]$ to fourth order in the $\mathbf{k} \neq 0$ Fourier components of $\omega(\mathbf{r})$, denoted by $\hat{\omega}(\mathbf{k})$. This amounts to an extension of eqn (6.11) to include terms up to $O(\hat{\omega}^4)$. The following expression was obtained

$$\begin{aligned}
H[\hat{\omega}] = \frac{1}{2} \int_{\mathbf{k}} &\left( u_0^{-1} + 2c_0\hat{\Gamma}^{(2)}(k) \right) \hat{\omega}(\mathbf{k})\hat{\omega}(-\mathbf{k}) \\
&+ ic_0 \int_{\mathbf{k}} \int_{\mathbf{q}} \hat{\Gamma}^{(3)}(\mathbf{k}, \mathbf{q}) \, \hat{\omega}(\mathbf{k})\hat{\omega}(\mathbf{q})\hat{\omega}(-\mathbf{k} - \mathbf{q}) \\
&+ c_0 \int_{\mathbf{k}} \int_{\mathbf{q}} \int_{\mathbf{p}} \hat{\Gamma}^{(4)}(\mathbf{k}, \mathbf{q}, \mathbf{p}) \, \hat{\omega}(\mathbf{k})\hat{\omega}(\mathbf{q})\hat{\omega}(\mathbf{p})\hat{\omega}(-\mathbf{k} - \mathbf{q} - \mathbf{p}) \\
&- \frac{c_0}{2V} \int_{\mathbf{k}} \int_{\mathbf{q}} \hat{\Gamma}^{(2)}(k)\hat{\Gamma}^{(2)}(q) \, \hat{\omega}(\mathbf{k})\hat{\omega}(-\mathbf{k})\hat{\omega}(\mathbf{q})\hat{\omega}(-\mathbf{q}) + O(\hat{\omega}^5) \quad (6.87)
\end{aligned}$$

where the sums over reciprocal lattice vectors have been converted to integrals according to eqn (A.14), as is appropriate in the large volume limit, and the shorthand $\int_{\mathbf{k}} \equiv (2\pi)^{-d} \int d^d\mathbf{k}$ has been employed. A spherical upper cutoff at $|\mathbf{k}| = \Lambda$ is implied on the wavevector integrals to regularize any ultraviolet divergences. The parameter $c_0 \equiv n/V$ is the average polymer density.

The "vertex functions" $\hat{\Gamma}^{(m)}$ appearing in eqn (6.87) are $m$-point segment density correlation functions for an ideal continuous Gaussian chain. The two-point function $\hat{\Gamma}^{(2)}$ is related to the Debye function $\hat{g}_D$ of eqn (3.130) by

$$\hat{\Gamma}^{(2)}(k) = \frac{N_0^2}{2} \hat{g}_D(k^2 N_0/2) \qquad (6.88)$$

as should be apparent by comparing eqns (6.11) and (6.87). The higher-order vertex functions $\hat{\Gamma}^{(3)}$ and $\hat{\Gamma}^{(4)}$, reflecting three- and four-point segment density correlations, are more complicated functions that are presented in the appendix of Ohta and Nakanishi (1983). Their detailed forms will not be required for the present purposes.

By noting that $\hat{\Gamma}^{(m)} \sim N_0^m$ for small wavevector arguments and that $\hat{\omega}(\mathbf{k})$ can be estimated to be of order $\sim u_0^{1/2} N_0^{d/4}$, Ohta and Nakanishi argued that eqn (6.87) amounts to an expansion in the parameter $(u_0 N_0^{2-d/2})^{1/2}$ at fixed $c_0 u_0 N_0^2$. Recalling the dimensionless variables $B$ and $C$ for the three-dimensional case considered in Section 6.1.1, we see that eqn (6.87) amounts to an expansion in powers of $B^{1/2}$ at fixed $BC$. For $d > 4$, the parameter $u_0 N_0^{2-d/2}$ is small for polymers of high molecular weight, so the expansion is well controlled. In dimensions less than four, however, this parameter diverges for $N_0 \to \infty$, so the expansion breaks down. Nevertheless, we shall see that the perturbation theory can be justified by restricting consideration to dimensions just slightly below four, the "upper critical dimension," so that $\epsilon = 4 - d$ is a small positive parameter. This is the so-called *epsilon expansion* approach of Wilson and Fisher (1972).

The effective Hamiltonian $H$ of the "bare" model described by eqns (6.83) and (6.87) depends on the three parameters $\mathbf{P} = (N_0, u_0, c_0)$, as well as the momentum cutoff $\Lambda$. Upon implementing the first coarse-graining step of a momentum space RG transformation, the resulting effective Hamiltonian $H_1$ contains a new set of parameters $\mathbf{P}_1 = (N_1, u_1, c_1)$ that describes the energy of interaction of the remaining (longer wavelength) modes. The upper momentum cutoff at this stage is $\Lambda'$, which is smaller than $\Lambda$. A convenient choice is

$$\Lambda' = \Lambda/\ell \tag{6.89}$$

where $\ell$ is a real positive number just slightly greater than unity. After the second rescaling step of the RG transformation, the upper cutoff has been restored to $\Lambda$, but the effective Hamiltonian evolves to $H'$, characterized by the parameter set $\mathbf{P}' = (N', u', c')$. We now turn to consider these individual steps and the resulting expressions for the renormalized parameters.

The coarse-graining step is implemented by removing the field fluctuation modes with wavenumbers in the spherical shell $\Lambda/\ell < |\mathbf{k}| < \Lambda$, i.e.

$$\exp(-H_1[\hat{\omega}]) \equiv \left( \prod_{\Lambda/\ell < |\mathbf{k}| < \Lambda} \int d\hat{\omega}(\mathbf{k}) \right) \exp(-H[\hat{\omega}]) \tag{6.90}$$

where it is understood that $H_1[\hat{\omega}]$ depends only on the Fourier modes $\hat{\omega}(\mathbf{k})$, $|\mathbf{k}| < \Lambda/\ell$ that have not been integrated out. The integrals on the right-hand side of eqn (6.90) cannot be performed exactly because $H[\hat{\omega}]$ contains terms beyond quadratic order in $\hat{\omega}(\mathbf{k})$. Nevertheless, the integrals can be approximately evaluated by expanding the integrand in powers of the cubic and quartic terms contained in $H[\hat{\omega}]$ – at each order producing Gaussian integrals that can be evaluated using the formulas of Appendix B. This procedure amounts to a series expansion of $H_1[\hat{\omega}]$ in powers of $\hat{\omega}$ with parameters $\mathbf{P}_1 = (N_1, u_1, c_1)$ expanded in powers of the dimensionless parameter[96] $u_0 N_0^{\epsilon/2}$. The technical details of this calculation will not be presented here but can be found in Ohta and Nakanishi (1983) and in more general references on RG theory (Goldenfeld, 1992; Amit, 1984). Ohta and Nakanishi showed that to leading order in $u_0 N_0^{\epsilon/2}$, the theory is "renormalizable" in the sense that $H_1[\hat{\omega}]$ has the same form as $H[\hat{\omega}]$ given in eqn (6.87), but with $\mathbf{P} \to \mathbf{P}_1$. In particular, the quadratic term in the coarse-grained Hamiltonian is given by

$$H_1[\hat{\omega}] = \frac{1}{2} \int_{\mathbf{k}}^{\Lambda/\ell} \left[ u_1^{-1} + c_1 N_1^2 \, \hat{g}_D(k^2 N_1/2) \right] \hat{\omega}(\mathbf{k}) \hat{\omega}(-\mathbf{k}) + O(\hat{\omega}^3) \qquad (6.91)$$

where the upper limit of the integral is a reminder that the retained modes are those with $|\mathbf{k}| < \Lambda/\ell$. It should be emphasized that we have retained the full form of the Debye function for convenience, although the renormalization conditions are applied only to the $O(k^0)$ and $O(k^2)$ coefficients of this expression as described below. The parameters $\mathbf{P}_1$ in the coarse-grained model $H_1$ are given to first order in $u_0 N_0^{\epsilon/2}$ by

$$u_1 N_1^2 = u_0 N_0^2 \left( 1 - \frac{u_0 N_0^{\epsilon/2} S_d}{2^{1-d/2}(2\pi)^d} \int_{\Lambda R_0/\ell}^{\Lambda R_0} dK \, \frac{K^{d-1}[\hat{g}_D(K^2)]^2}{[1 + c_0 u_0 N_0^2 \, \hat{g}_D(K^2)]^2} \right) \qquad (6.92)$$

$$N_1 = N_0 \left( 1 + \frac{u_0 N_0^{\epsilon/2} S_d}{2^{-(d/2+2)} d (2\pi)^d} \int_{\Lambda R_0/\ell}^{\Lambda R_0} dK \, K^{d+1} V_D(K) \right) \qquad (6.93)$$

$$c_1 = c_0 \qquad (6.94)$$

where $R_0 \equiv (N_0/2)^{1/2}$ is the characteristic size of an ideal polymer and $K \equiv k R_0$ is a dimensionless wavenumber. The quantity $S_d$ denotes the surface area of a unit sphere in $d$ dimensions and $V_D(K)$ is a dimensionless Debye-like function whose full expression is given in the appendix of Ohta and Nakanishi (1983).

The physical interpretation of eqn (6.94) is straightforward – the polymer number density is not influenced by the coarse-graining procedure, which changes neither the number of polymers nor the volume. Equation (6.92) arises from the effect of the eliminated shell of fluctuation modes on the *osmotic compressibility*, $(\partial\Pi/\partial c_0)^{-1}$, which is proportional to the $k \to 0$ limit of the structure factor $S(k)$ defined by eqn (6.22). In the mean-field approximation for the bare model, $\beta \partial\Pi/\partial c_0 \approx 1 + c_0 u_0 N_0^2$, so it is natural that the renormalization is applied to the product $u_0 N_0^2$. Another important observation is that the fluctuation integral in eqn (6.92) has exactly the same form as eqn (6.17) for $\partial\Pi/\partial C$ (in $d = 3$), except that the range of integration spans only the shell of eliminated Fourier modes.

In contrast to the *collective* (many chain) renormalization condition (6.92) derived from the osmotic compressibility, eqn (6.93) results from examining the

---

[96] The smallness of this parameter and therefore the utility of the asymptotic expansion will be subsequently justified.

effect of the eliminated modes on the dimensions of a *single chain* at infinite dilution. Indeed, the $O(k^2)$ expansion coefficient of the single-chain scattering function $\hat{g}_D(k^2 N_0/2)$ is proportional to the squared radius of gyration $R_{g0}^2 = dN_0/6$ of a single chain in the bare model. The eliminated high-$k$ fluctuation modes serve to renormalize the radius of gyration in order to account for short-ranged excluded volume correlations. This effect is manifested in eqn (6.93) as a renormalization of $N_0$ to $N_1$ and, at $O(k^2)$ in eqn (6.91), with a similar replacement in the argument of the Debye function.

The results presented thus far are applicable to any space dimension $d$. However, to control the above expansion in $u_0 N_0^{\epsilon/2}$ for $d < 4$ it is necessary to simultaneously expand about four dimensions, i.e. by making the assumption that the parameter $\epsilon = 4 - d$ is small (Wilson and Fisher, 1972). We therefore expand the right-hand sides of eqns (6.92) and (6.93) in powers of $\epsilon$, using advance knowledge of the fact that the fixed point value of the excluded volume parameter $u_0$ turns out to be $O(\epsilon)$. At leading order in $\epsilon$, the fluctuation integrals can thus be performed in four dimensions. A further simplification results from the fact that the cutoff is microscopic, $\Lambda R_0 \gg 1$, so that the functions $\hat{g}_D$ and $V_D$ can be approximated by their $K \gg 1$ asymptotic forms, $\hat{g}_D(K^2) \approx 2/K^2$ and $V_D(K) \approx 1/K^6$. With these simplifications, eqns (6.92) and (6.93) reduce to

$$u_1 N_1^2 = u_0 N_0^2 \left[1 - 8K_4 u_0 \ln \ell + O(\epsilon^2)\right] \qquad (6.95)$$

$$N_1 = N_0 \left[1 + 4K_4 u_0 \ln \ell + O(\epsilon^2)\right] \qquad (6.96)$$

where $K_4 \equiv S_4/(2\pi)^4 = 1/(8\pi^2)$. It is important to note the appearance of a factor of $\ln \ell$ in both expressions, reflecting a logarithmic sensitivity to the width of the momentum shell that has been removed. Finally, we combine eqns (6.95) and (6.96), again noting that $u_0 \sim O(\epsilon)$, to obtain a separate renormalization condition for $u_0$ alone,

$$u_1 = u_0 \left[1 - 16K_4 u_0 \ln \ell + O(\epsilon^2)\right] \qquad (6.97)$$

so that eqns (6.95) and (6.96) can be equivalently replaced by eqns (6.96) and (6.97).

At this stage, the first coarse-graining step ($H \to H_1$) of the RG transformation is complete; only the *rescaling step* ($H_1 \to H'$) remains. In order to restore the sphere of retained modes to its original radius $\Lambda$, all wavevectors in eqn (6.91) are to be rescaled according to $\mathbf{q} \equiv \ell \mathbf{k}$. However, if we desire an expression for $H'$ that is invariant to the RG transformation, i.e. that has the same form as $H$ given in eqn (6.87) except for the replacement $\mathbf{P} \to \mathbf{P}'$, it is also necessary to rescale the Fourier components of the *field* according to

$$\hat{\psi}(\mathbf{q}) \equiv z_w^{-1} \hat{\omega}(\mathbf{k}) \qquad (6.98)$$

where $z_w$ is a dimensionless $\ell$-dependent rescaling factor. These changes transform the quadratic term in $H_1$, given by eqn (6.91), into the following expression for $H'$:

$$H'[\hat{\psi}] = \frac{1}{2} \int_{\mathbf{q}}^{\Lambda} z_w^2 \ell^{-d} \left[ u_1^{-1} + c_1 N_1^2 \, \hat{g}_D(q^2 \ell^{-2} N_1/2) \right] \hat{\psi}(\mathbf{q}) \hat{\psi}(-\mathbf{q}) + O(\hat{\psi}^3) \quad (6.99)$$

For this expression to have the same form as the quadratic term in $H[\hat{\omega}]$, the following correspondences are necessary:

$$N' = \ell^{-2} N_1 \quad u' = z_w^{-2} \ell^d u_1 \quad c' = z_w^2 \ell^\epsilon c_1 \qquad (6.100)$$

At this juncture we can inject another piece of physics, namely that the wavevector rescaling shrinks the volume by a factor of $\ell^{-d}$ without changing the number of polymers, so that $c' = \ell^d c_1$. The field rescaling factor $z_w$ is thus given by $z_w = \ell^{d-2}$, so that $u' = \ell^\epsilon u_1$. If one combines these results with eqns (6.94), (6.96), and (6.97), it follows that the net effect of the RG transformation $(\mathbf{P} \to \mathbf{P'})$ is summarized to $O(\epsilon)$ by

$$c' = \ell^d c_0 \qquad (6.101)$$

$$r' = \ell^2 r_0 \left[ 1 - 4K_4 u_0 \ln \ell + O(\epsilon^2) \right] \qquad (6.102)$$

$$u' = \ell^\epsilon u_0 \left[ 1 - 16 K_4 u_0 \ln \ell + O(\epsilon^2) \right] \qquad (6.103)$$

where we have defined $r' \equiv 1/N'$ and $r_0 \equiv 1/N_0$.

Equations (6.101)–(6.103) define the way in which the parameters of Model A "flow" under an RG transformation performed close to four dimensions. These discrete flow equations can be converted to *differential flow equations* by introducing a fictitious "time" variable $t$ that describes the progress of successive, continuously applied RG transformations. In the case of the polymer concentration, for example, we write $c' = c(t + dt)$, $c_0 = c(t)$, and take $\ell = \exp(dt) = 1 + dt + O(dt^2)$, so that each coarse-graining transformation shaves a differential shell of width $\Lambda dt$ from the sphere of allowed wavevectors. It follows that eqn (6.101) reduces to the differential equation

$$\frac{d}{dt} c(t) = d \, c(t) \qquad (6.104)$$

Similarly, eqns (6.102) and (6.103) reduce to the differential flow equations

$$\frac{d}{dt} r(t) = 2 \, r(t) - 4K_4 r(t) u(t) \qquad (6.105)$$

$$\frac{d}{dt} u(t) = \epsilon \, u(t) - 16 K_4 [u(t)]^2 \qquad (6.106)$$

The analysis of such RG flow equations is standard (Goldenfeld, 1992). Equation (6.106) has two *fixed points*, $u_f$, which describe values of the excluded volume parameter that are invariant to the renormalization group. The *Gaussian fixed point*, $u_f = 0$, is a physically meaningful critical point only for $d > 4$, so will be considered no further. Of more interest is the *Wilson–Fisher fixed point*

$$u_f = \frac{\epsilon}{16K_4} = \frac{1}{2}\pi^2\epsilon + O(\epsilon^2) \qquad (6.107)$$

which is the physical critical point for $d < 4$. This result confirms our expectation that $u_f \sim O(\epsilon)$ and, for small $\epsilon$, justifies the perturbation expansion in $u_0$ that was used to affect the coarse-graining transformation. The values of $c$ and $r$ at the Wilson–Fisher fixed point are $c_f = r_f = 0$, which correspond, respectively, to zero polymer concentration and infinite polymer molecular weight.

*Scaling exponents* are obtained by linearizing the above RG flow equations about the Wilson–Fisher fixed point, cf. eqn (6.81). By defining a column vector $\delta\mathbf{P}(t) \equiv (\delta r(t), \delta u(t), \delta c(t))^T$, which describes the deviations of the model parameters from the fixed point $\mathbf{P}_f = (0, \pi^2\epsilon/2, 0)^T$, the linearized flow equations can be written in matrix form as

$$\frac{d}{dt}\delta\mathbf{P}(t) = \mathcal{R}_L \cdot \delta\mathbf{P}(t) \qquad (6.108)$$

where the linearized RG transformation matrix $\mathcal{R}_L$ proves to be diagonal:

$$\mathcal{R}_L = \begin{pmatrix} 2 - \epsilon/4 & 0 & 0 \\ 0 & -\epsilon & 0 \\ 0 & 0 & d \end{pmatrix} \qquad (6.109)$$

The eigenvalues of $\mathcal{R}_L$, denoted by $y_j$, can be trivially read off as the diagonal elements, yielding to $O(\epsilon)$: $y_r = 2 - \epsilon/4$, $y_u = -\epsilon$, and $y_c = d$.

Qualitative trajectories of the RG flow equations for $d < 4$ are shown in Fig. 6.4 in the quadrant $u > 0$, $r > 0$. The non-negativity of $r_0$ is implied by the physical constraint that the polymer chain length $N_0 = 1/r_0$ is positive definite. The parameter $u_0$ must also be positive due to the restriction of Model A to *good* solvent conditions, where the pair potential of mean force between polymer segments is repulsive. Solutions of eqns (6.105) and (6.106) starting from points $(r_0, u_0)$ in the positive quadrant have the property that $u, r > 0$ for all $t$.

The connection between the eigenvalues of $\mathcal{R}_L$ and physical *scaling exponents* can be clarified by examining the effect of the RG transformation on quantities such as the free energy density, the osmotic pressure, or the correlation length. For example, the correlation length $\xi$ transforms under the RG according to

$$\xi(r', u', c') = \ell^{-1}\xi(r_0, u_0, c_0) \qquad (6.110)$$

This follows because the coarse-graining step removes fluctuations below the correlation length scale, but has no effect on $\xi$. The rescaling step reduces the correlation length by a factor of $\ell^{-1}$, so this is the net effect of a single RG transformation. Correspondingly, after a series of differential RG transformations, the correlation length transforms as

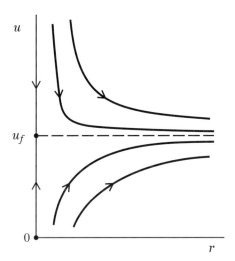

Fig. 6.4. Schematic RG flows in the $u$–$r$ plane produced by integrating eqns (6.105) and (6.106) for various bare model parameters (initial conditions) $(r_0, u_0)$ and $d < 4$. The Wilson–Fisher fixed point is denoted by a solid dot at $u_f$ on the $u$ axis, while the Gaussian fixed point is located at the origin. Qualitatively similar flows are observed in the $u$–$c$ plane, corresponding to solutions of eqns (6.104) and (6.106).

$$\xi(r_0, u_0, c_0) = \exp(t)\, \xi(r(t), u(t), c(t)) \tag{6.111}$$

where $(r_0, u_0, c_0)$ are the initial $(t = 0)$ parameters of the model. If we conveniently choose $u_0 = u_f$, so that $u(t) = u_f$, and replace $r(t)$ and $c(t)$ by the solution of the corresponding flow equations, eqn (6.111) becomes

$$\xi(r_0, u_f, c_0) = \exp(t)\, \xi(r_0 \exp(y_r t), u_f, c_0 \exp(y_c t)) \tag{6.112}$$

Equation (6.112) is a general result that holds for any "time" $t$. One convenient choice is $t = t^*$ so that $r_0 \exp(y_r t^*) = 1$. This corresponds to an integration path of the RG equations that sweeps $r(t)$ from a small initial value $r_0 = 1/N_0$ near the fixed point $(r_f = 0)$ to an $O(1)$ value at $t = t^*$ that is outside of the critical region. Evaluation of eqn (6.112) at $t = t^*$ and substituting $r_0 = 1/N_0$ leads to

$$\xi(N_0^{-1}, u_f, c_0) = N_0^{\nu}\, \xi(1, u_f, c_0 N_0^{d\nu}) \tag{6.113}$$

where

$$\nu \equiv \frac{1}{y_r} = \frac{1}{2}\left(1 + \frac{\epsilon}{8} + O(\epsilon^2)\right) \tag{6.114}$$

is the scaling exponent $\nu$ defined in eqn (2.2) that characterizes the size $R \sim N^{\nu}$ of an isolated polymer chain under good solvent conditions. Indeed, eqn (6.113) predicts that at infinite dilution, $c_0 = c_f = 0$, $\xi \sim R \sim N^{\nu}$. More generally, eqn (6.113) amounts to a derivation of the well known scaling ansatz for dilute and semi-dilute polymer solutions (de Gennes, 1979)

$$\xi = R\, F_\xi(nR^d/V) \qquad (6.115)$$

where $F_\xi(x)$ is a dimensionless scaling function with the property that $F_\xi(0) \sim O(1)$ (dilute solution limit). The behavior of the scaling function for $x \gg 1$ (semi-dilute limit) can be deduced by going back to eqn (6.112) and making a different choice of $t$. Selecting $t = t^*$ such that $r_0 \exp(y_r t^*) = c_0 \exp(y_c t^*)$ leads to

$$\xi(N_0^{-1}, u_f, c_0) = \rho_0^{\nu/(1-\nu d)}\, \xi(N_0^{-1}\rho_0^{1/(1-\nu d)}, u_f, N_0^{-1}\rho_0^{1/(1-\nu d)}) \qquad (6.116)$$

where $\rho_0 \equiv N_0 c_0 = nN_0/V$ is the average number density of polymer segments. Taking $N_0 \to \infty$ at fixed $\rho_0$ thus produces the semi-dilute scaling formula $\xi \sim \rho_0^{-\nu/(\nu d-1)}$, which is consistent with eqn (6.55). It also implies that the scaling function in eqn (6.115) has the asymptotic behavior $F_\xi(x) \sim x^{-\nu/(\nu d-1)}$ for $x \gg 1$.

The power of the RG formalism is now readily apparent and can be summarized by the following discussion points:

- The renormalization group theory provides rigorous theoretical justification for the scaling picture of dilute and semi-dilute solutions, as evidenced by eqns (6.113)–(6.116). This framework is general and transcends approximate strategies for computing RG flow equations, fixed points, eigenvalues, and associated scaling exponents. Based on this formalism, a comprehensive understanding of the equilibrium structure and thermodynamic properties of the homogeneous Model A has been achieved (Ohta and Nakanishi, 1983; Nakanishi and Ohta, 1985; Freed, 1987; Schäfer, 1982; des Cloizeaux and Jannink, 1990).

- Provided that the relevant fixed point is a so-called *weak-coupling* fixed point, approximation strategies can be devised for the systematic analytical computation of *scaling exponents*. For example, in the case of Model A we observed an upper critical dimension of $d = 4$ and found that it was possible to compute the scaling exponents in an asymptotic "epsilon expansion" in powers of $\epsilon = 4 - d$. The fact that $u_f \sim O(\epsilon)$ made it possible to get over the hurdle of the coarse-graining step and justify a bare perturbation expansion in powers of $u_0 N_0^{\epsilon/2}$.

- At $O(\epsilon)$, the scaling exponents so obtained are not very accurate in three dimensions. For example, setting $\epsilon = 1$ in eqn (6.114) predicts that $\nu = 9/16 = 0.562$, while the most accurate estimate in $d = 3$ is $\nu = 0.588$ (Le Guillou and Zinn-Justin, 1977). The latter estimate of $\nu$ was obtained by combining a high-order epsilon expansion with series analysis techniques.

- The observation that the RG-derived scaling exponents are independent of the choice of microscopic cutoff $\Lambda$ provides strong theoretical support to the notion that the scaling behavior of polymer solutions is *universal*. In particular, exponents such as $\nu$ are predicted to be independent of microscopic details including the chemical composition of the polymer and the atomistic nature of the solvent. All such systems and associated models that exhibit a common set of scaling exponents are said to belong to a particular *universality class*. As we have seen, varying the dimensionality of a system changes the universality class, although other model features such as the symmetry group of the effective Hamiltonian and the presence of long-ranged interactions can also lead to different universality classes. In the specific case of polymeric fluids, solvent quality and polymer topology can be additional factors in determining the universality class to which a system belongs.

- The RG theory, in conjunction with the epsilon expansion, can also be used to study the explicit form of *crossover scaling functions*, such as the function $F_\xi(x)$ discussed above. Crossover scaling functions have been computed to $O(\epsilon)$ for properties including the correlation length, osmotic compressibility, radius of gyration, osmotic pressure, and structure factor (Ohta and Oono, 1982; Ohta and Nakanishi, 1983; Nakanishi and Ohta, 1985; Schäfer, 1982). These results have found impressive agreement with experimental measurements on polymer solutions spanning the dilute and semi-dilute regimes (Noda *et al.*, 1981; Noda *et al.*, 1983; Wiltzius *et al.*, 1983).

- Closely related to crossover phenomena are *finite size scaling effects*, which describe the influence of a finite system size $L$ on critical phenomena and scaling behavior. The RG theory has much to say about the mathematical structure of finite size scaling effects and has been used to extrapolate thermodynamic properties deduced from Monte Carlo simulations of small fluid or magnetic systems to systems of infinite extent (Binder and Heermann, 1988; Binder, 1995; Landau and Binder, 2000).

These successes notwithstanding, the application of RG theory to the *inhomogeneous* polymer systems and models that are the focus of this monograph is problematic. Even in the case of Model A, the coarse-graining step of the RG transformation becomes very difficult when imposed boundary conditions create an inhomogeneous saddle point. The technical reason for this difficulty is that translational invariance is lost in the momentum shell integrals indicated in eqn (6.90) when the field fluctuations are about a non-uniform state. Although a perturbation expansion of the vertex functions and parameters of $H_1$ in powers of $u_0 N_0^{\epsilon/2}$ can be formally carried out in such a situation, the fluctuation integrals analogous to those appearing in eqns (6.92) and (6.93) cannot be evaluated analytically without further approximations. This is similar to the situation described in Section 6.1.4, where we discussed Gaussian fluctuations about a non-uniform state. It is possible to take advantage of special features of the saddle point to simplify an RG calculation, e.g. a $w^*(\mathbf{r})$ with weak or slowly varying inhomogeneities, but one is then faced with imposing additional approximations on top of the usual epsilon expansion in order to make analytical headway. *The consequence is that very few analytical RG calculations have been performed about an inhomogeneous state of a polymeric fluid.*

Fortunately, *numerical* simulation techniques such as Monte Carlo simulations can be used to execute RG transformations (Ma, 1976; Swendsen, 1979; Swendsen, 1982), and such *Monte Carlo RG* methods are equally applicable to models with homogeneous and inhomogeneous saddle points. Monte Carlo RG has the merit of allowing direct calculations in the space dimension of interest, e.g. $d = 3$, and thereby avoiding epsilon expansions and other approximations that are necessary in analytical RG calculations. Monte Carlo RG been successfully applied to a wide range of spin models, lattice gauge theories, percolation models, and simple continuum field theories of condensed matter systems, such as the $\phi^4$ model. As will be discussed in the subsequent sections of this chapter, the application of Monte Carlo methods to polymer field theories is complicated by the non-positive-definite nature of the relevant statistical weights, $\exp(-H[w])$. Nevertheless, the specialized Monte Carlo and Langevin techniques that are described in Sections 6.3–6.5 can be adapted to implement *numerical RG* transformations. This topic will be discussed briefly in Section 6.7.

Very few of the other models described in Chapter 4 have been investigated by analytical RG methods. An extended version of Model A in which an additional interaction parameter, $v_0 > 0$, has been introduced to describe the effective *three-body* contribution to the potential of mean force, has been used to study polymer solutions in the *theta point* region (de Gennes, 1979). In this enlarged parameter space $\mathbf{P} = (r_0, u_0, v_0, c_0)$, the theta point corresponds to a tricritical-like fixed point under the RG in which $r_f = c_f = u_f = 0$ and where logarithmic corrections arise in three dimensions from residual three-body correlations. Analytical RG treatments of this model have been presented by Duplantier (1982) and Freed and coworkers (Cherayil *et al.*, 1985; Freed, 1987).

Model C of an incompressible binary polymer blend has not, to our knowledge, been investigated by renormalization group techniques, even in the homogeneous region of the phase diagram. However, the model is believed to belong to the $d = 3$ Ising universality class (Goldenfeld, 1992), so numerous investigators (de Gennes, 1977; de Gennes, 1979; Bates *et al.*, 1990; Hair *et al.*, 1992; Belyakov and Kiselev, 1992; Anisimov *et al.*, 1992) have attempted to approximately map Model C onto the familiar (and much simpler) $\phi^4$ model that belongs to the same universality class and for which extensive RG calculations have been performed (Wilson and Kogut, 1974; Brezin *et al.*, 1976; Amit, 1984; Goldenfeld, 1992). While 3d Ising exponents are indeed observed experimentally in model polymer blends very close to the critical point (Bates *et al.*, 1990; Hair *et al.*, 1992; Chu *et al.*, 1992; Schwahn *et al.*, 1995), the experiments reveal large discrepancies with theoretical predictions for the mean-field to Ising crossover in scaling behavior as the critical point is approached from the disordered phase. It seems likely that the source of this discrepancy is an inadequate treatment of short-wavelength field fluctuations, below the $R_g$ scale, which differ considerably between Model C and the $\phi^4$ model. A more sophisticated treatment of such fluctuations for Model C was presented by Wang (2002), although the Hartee approximation was employed and the focus of the work was nucleation phenomena, rather than critical phenomena.

Model E for diblock copolymer melts has also not been subjected to direct attack using the renormalization group, although a closely related RG calculation was performed by Hohenberg and Swift (1995). These authors investigated the RG flow behavior of a phenomenological model proposed by Brazovskii (Brazovskii, 1975) that takes the form

$$H[\phi] = \int d^d\mathbf{r} \left\{ \frac{\tau_0}{2}\phi^2 + \frac{\lambda_0}{4!}\phi^4 + \frac{\xi_0^4}{2}[(\nabla^2 + k_0^2)\phi]^2 \right\} \qquad (6.117)$$

with bare parameters $(\tau_0, \lambda_0, \xi_0, k_0)$. While similar to the Hamiltonian of the $\phi^4$ model, the lowest-energy field fluctuation modes of eqn (6.117) correspond to those on the *spherical surface* $|\mathbf{k}| = k_0$ in reciprocal space, rather than at the *point* $\mathbf{k} = 0$ in the $\phi^4$ model. As was discussed in Section 6.1.5.1, Brazovskii studied the above model in the Hartree approximation and predicted a fluctuation-induced first-order phase transition from the disordered phase to the lamellar phase. Fredrickson and Helfand (1987) approximately reduced Leibler's RPA-derived free energy functional (Leibler, 1980), eqns (5.31) and (5.33), to the above form, making the correspondence between the parameters of Model E and the parameters of eqn (6.117), and were thereby able to apply Brazovskii's Hartree analysis in the block copolymer context. Subsequently, Hohenberg and Swift (1995) showed that it is possible to adapt a novel momentum-shell RG approach developed by Shankar (1994) for the Fermi liquid to the Brazovskii Hamiltonian, thus enabling a systematic investigation of the model.

The essence of the Shankar approach is to set a momentum cutoff $\Lambda$ centered around the sphere of dominant fluctuation modes so that the retained modes have wavenumbers $k = |\mathbf{k}|$ satisfying $|k - k_0| < \Lambda$. A convenient choice of the cutoff is $\Lambda = k_0$, so that all Fourier modes in a sphere of radius $2k_0$ centered at the origin are initially included in the model. As illustrated in Fig. 6.5, the coarse-graining step of the RG transformation amounts to integrating out narrow inner *and* outer shells of field fluctuation modes such that the band of retained modes is narrowed to $|k - k_0| < \Lambda/\ell$. The rescaling step of $\mathbf{k} \to \ell\mathbf{k}$ and $k_0 \to \ell k_0$ then restores the band to its original thickness $\Lambda$. Renormalization conditions are applied on the sphere at $k = k_0$. Hohenberg and Swift showed that the Brazovskii model is indeed renormalizable by this scheme and that the coarse-grained Hamiltonian flows to a form that is similar in character to the Hartree potential (Fredrickson and Binder, 1989). The Shankar–Hohenberg–Swift RG technique has not yet been directly applied to Model E, but it is likely that such an approach could resolve the spurious ultraviolet divergences that have been noted in the Hartree approximation by Stepanow and coworkers (Stepanow, 1995; Kudlay and Stepanow, 2003).

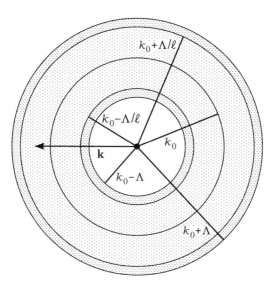

FIG. 6.5. Momentum shells used in the Shankar–Hohenberg–Swift RG scheme. The bare model of eqn (6.117) retains all reciprocal space wavevectors **k** in the spherical shell of thickness $2\Lambda$ centered at $k_0$, i.e. $|k - k_0| < \Lambda$. Coarse graining amounts to the removal of field fluctuation modes at the inner and outer edges of the shell in narrow bands of width $\Lambda(1 - 1/\ell)$. In the rescaling step, all wavevectors are rescaled by a factor of $\ell$, which restores the shell of allowed wavevectors to its original thickness.

### 6.1.6 *Other analytical methods*

There are several other analytical techniques that can be used to approximately treat deviations from mean-field behavior associated with field fluctuations.

#### 6.1.6.1 *Blob scaling methods*
One important approach is the class of "blob" scaling methods popularized by de Gennes (de Gennes, 1979). These techniques are phenomenological rather than from first principles, but are versatile in that they can be applied to both homogeneous and inhomogeneous polymer melts and solutions.

The basic idea behind a blob calculation is to apply scaling arguments, physical intuition, and possibly rigorous RG calculations to deduce the "blob" length scale $\xi$, below which field fluctuations are strongly correlated. For example, $\xi$ corresponds to the segment density correlation length (6.55) for Model A in the semi-dilute regime. Establishing the effective interactions in the model at the blob scale are also necessary for this type of analysis. Given this information, a blob calculation amounts to the assumption that field fluctuations are weak and uncorrelated at and beyond the blob scale, so that *mean-field* expressions can be applied to a suitably coarse-grained and renormalized theory.

An example of such an approach was given in Section 6.1.5.1, where the mean-field expression (6.68) was applied for the osmotic pressure, but with a renormalized interaction parameter $B_r$ computed from the Hartree approximation. A more standard blob scaling argument would precede as follows (de Gennes, 1979). In the *semi-dilute regime*, the blob scale $\xi \sim \rho_0^{-\nu/(3\nu-1)}$ can be viewed as the "mesh size", defining the scale below which strands of polymers are self-avoiding random walks. Beyond this scale, the effects of excluded volume on the conformations of a single chain are screened due to multiple interactions with other strands, and the correlations in segment density are lost. As depicted in Fig. 6.6, it is thus possible to view each chain as a string of blobs, each blob consisting of a strand of $g$ segments of size $\xi$ related by eqn (2.2):

$$\xi = b_s \, g^\nu \tag{6.118}$$

The blob strings associated with all the polymers in a semi-dilute solution evidently pack to fill space. Thus, when coarse-grained to the blob scale, it is possible to view a semi-dilute solution as a *melt* of blob strings. At high molecular weight, so that the number of blobs per chain $N_b \equiv N/g$ is large, mean-field theory can be applied to this melt. For the osmotic pressure, we thus have

$$\beta \Pi \approx \frac{1}{2} u_b \rho_b^2 \tag{6.119}$$

as follows from eqn (5.17), but where $\rho_b \sim 1/\xi^3$ is the number density of blobs and $u_b$ is the excluded volume parameter between pairs of blobs. Guessing that the interactions between blobs are analogous to those between hard spheres of diameter $\xi$, it follows that $u_b \sim \xi^3$, so that $\beta \Pi \sim 1/\xi^3$. This is of course consistent with the scaling ansatz (6.54), which is believed to be an exact result.

Such blob arguments can be extended in many useful ways. For example, the picture of a semi-dilute solution as a melt of blob chains can be combined with the "Flory theorem," which states that the conformational statistics of a chain in a homogeneous melt are asymptotically ideal (cf. Section 2.1), to predict the characteristic size of a polymer chain in the semi-dilute regime (Daoud *et al.*, 1975; de Gennes, 1979). In particular, the size $R$, e.g. the radius of gyration, of a blob string consisting of $N_b$ blobs of size $\xi$ scales as

$$R \sim \xi N_b^{1/2} \sim \rho_0^{-(2\nu-1)/[2(3\nu-1)]} N^{1/2} \tag{6.120}$$

where eqn (6.118) has been applied to evaluate $g$ in the expression $N_b = N/g$. This result, which has been experimentally verified, can also be derived using RG techniques analogous to those described in the previous section for $\xi$. The blob approach, however, is much simpler. With this simplicity comes a *price* – blob arguments cannot be used to derive fundamental exponents such as $\nu$, nor do they provide access to crossover scaling functions and the prefactors in scaling formulas.

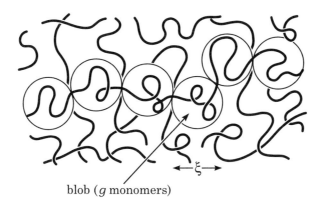

blob ($g$ monomers)

FIG. 6.6. Scaling picture of a semi-dilute solution, where each chain is considered to be a pearl-necklace-like string of blobs, each blob having a diameter $\xi$ and containing $g$ segments. These blob strings fill space to form a melt. Reproduced from de Gennes (1979).

A beautiful example of the application of blob techniques to an *inhomogeneous system* was provided by Broseta *et al.* (1987). These authors considered the equilibrium interfacial properties of a phase-separated blend of two homopolymers (A and B), dissolved in a mutual (non-selective) good solvent S. By adopting a blob picture in the semi-dilute concentration regime where both the A and B chains are envisaged as strings of blobs, Broseta *et al.* argued that the *melt* Model C is applicable, but where the following renormalizations are made ($K = A$ or $B$)

$$b_K \to \xi \sim \rho_0^{-\nu/(3\nu-1)}$$
$$N_K \to N_b \equiv N/g$$
$$\chi_{AB} \to u_{AB} \sim \rho_0^{\mu/(3\nu-1)} \tag{6.121}$$

and the symmetric case of $b_A = b_B$, $N_A = N_B \equiv N$ is assumed. The first renormalization amounts to the replacement of the segment length by the blob diameter, where $\rho_0$ is the total (A plus B) segment number density in the solution. The second expression replaces the number of segments per chain by the number of blobs per chain with $g$ given in eqn (6.118). Finally, the last expression in eqn (6.121) replaces the interaction parameter between segments $\chi_{AB}$ in the melt theory with a renormalized parameter $u_{AB}$ that describes the interactions between dissimilar (A and B) blobs. The quantity $\mu \approx 0.22$ in the scaling expression for $u_{AB}$ is a universal scaling exponent that has been derived on the basis of RG calculations for a *homogeneous* $A + B + S$ solution (Joanny *et al.*, 1984; Kosmas, 1984; Schäfer and Kappeler, 1985). Broseta *et al.* (1987) argued that outside of the critical region, the interfaces created by phase separation of A and B chains in the common good solvent can be described by a mean-field treatment of Model C with the replacements noted above. Specifically, for chains of high molecular weight the Helfand–Tagami ground state analysis is applicable, cf. Section 5.2.3, so that the interfacial width $\xi_I$ and the interfacial tension $\gamma$ follow by applying the renormalizations of eqn (6.121) to eqns (5.68) and (5.69):[97]

$$\xi_I \sim \xi\, u_{AB}^{-1/2} \sim \rho_0^{-0.9} \tag{6.122}$$

$$\gamma \sim \xi^{-2} u_{AB}^{1/2} \sim \rho_0^{1.7} \tag{6.123}$$

Modified versions of these expressions resulting from mean-field calculations to account for finite molecular weight corrections and a slight excess of solvent in the interfacial region have found excellent agreement with experiment (Broseta et al., 1987). Similar blob techniques have been used to adapt mean-field results for Model E to semi-dilute solutions of diblock copolymers (Fredrickson and Leibler, 1989; de la Cruz, 1989) and to correct mean-field/strong-stretching formulas for Model K to account for excluded volume correlations in solution polymer brushes (Milner et al., 1988; Kreer et al., 2004).

6.1.6.2 *Long-wavelength methods*   The blob scaling techniques just described attempt to semi-quantitatively capture the effects of strong field fluctuations at *short* length scales (below the blob size) through phenomenological renormalization of parameters in mean-field expressions. Another class of analytical methods is appropriate when the relevant field fluctuations occur on scales that are *long* compared with the finest inhomogeneous structures in the fluid. These *long-wavelength methods* attempt to phenomenologically embed mean-field results into fluctuation calculations for long-wavelength translational or undulation modes of a mesostructured polymer phase.

An elegant example of such a long-wavelength strategy is the work by Dormidontova and Lodge (2001) and Wang et al. (2005) in describing the phase behavior of highly asymmetric ($f \approx 0.1$) diblock copolymer melts in the vicinity of the order–disorder transition (ODT). Experiments (Schwab and Stühn, 1996; Sakamoto et al., 1997; Kim et al., 1999; Wang et al., 2002; Adams et al., 1996) have conclusively shown the existence of a fluid phase with a disordered (liquid-like) packing of spherical micelles in a narrow temperature range intermediate between the disordered phase (DIS) and the ordered close-packed sphere (CPS) phase shown in Fig. 5.15. The mean-field calculations that underpin this phase diagram for Model E (Matsen and Bates, 1996b) provide no insights into the micellar phase that intervenes between order and disorder. Dormidontova and Lodge (2001) combined a strong-segregation calculation analogous to that of Section 5.2.4 for the free energy of formation of a micelle (Semenov, 1989) with a standard treatment of micellization. The latter allows for equilibrium of micellized and unmicellized block copolymers and includes the translational entropy of both free polymers and micelles. These authors were able to demonstrate the

---

[97]In eqn (5.69) it is also necessary to replace the segment density with the total blob density according to $\rho_0 \to \rho_b \sim 1/\xi^3$.

existence of the micellar phase and a critical micelle temperature, $T_{CM}$, above which the population of micelles diminishes rapidly. In the Dormidontova–Lodge calculation, micelles can be viewed as large-amplitude localized composition variations whose energy is estimated by a mean-field/strong-stretching calculation. Long-wavelength field fluctuations are treated in a phenomenological fashion through the inclusion of the translational entropy of the micelles.

Wang *et al.* (2005) have further embellished this theory by performing a full numerical SCFT calculation to compute the excess free energy of micelle formation. Their work indicates the existence of a precisely defined "micelle dissociation" temperature $T_{MD}$ (greater than $T_{CM}$), above which micelles cannot exist with finite lifetimes. The observed micelles for $T \lesssim T_{CM}$ are therefore *thermally activated*. For asymmetric block copolymers of low molecular weight, i.e. $C \sim N^{1/2}$ not too large, Wang *et al.* further predict that the micellar phase intervenes between the (structureless) disordered and ordered spherical phases. In contrast, at very high molecular weight, they argue that the micellar phase is absent and direct transitions between the disordered phase and the CPS ordered spherical phase should occur. These predictions are consistent with the experiments to date.

It is important to contrast the fluctuation theory of Wang *et al.* (2005) with that of Fredrickson and Helfand (1987) described in Section 6.1.5.1. The Fredrickson–Helfand calculation treats field fluctuations at a self-consistent Gaussian level using the Hartree approach of Brazovskii. This scheme is appropriate when the ordered phases are weakly formed, describable by only a few harmonics, and when the fluctuations about the ordered composition patterns are similarly diffuse. These conditions are approximately met for nearly symmetric block copolymer melts in the vicinity of the ODT, where the lamellar-like fluctuations lack sharp A–B interfaces and are not thermally activated. In contrast, the Brazovskii–Fredrickson–Helfand approach breaks down in asymmetric melts where the ordering occurs at larger $\chi N$, the ODT has a much stronger first-order character, and where many Fourier components are required to describe the ordered phase and the activated field fluctuations. The approach of Wang et al. (and Dormidontova and Lodge) cleverly treats the asymmetric case by applying mean-field theory (numerical SCFT) to address the short length scale degrees of freedom semi-quantitatively, while using physical intuition to associate the lowest-energy fluctuation modes at longer wavelengths with translations of the sphere centers of mass. A more sophisticated treatment would also include undulation modes associated with deformations in the spherical shape of the micelles, similar to the field-theoretic description of nucleation phenomena (Langer, 1980; Gunton and Droz, 1983).

Long-wavelength techniques have also been applied to treat capillary wave undulations of fluid–fluid interfaces (Weeks, 1977; Huse *et al.*, 1985). Capillary waves normally have wavelengths that greatly exceed the "intrinsic width" of the interface, even for polymer–polymer interfaces. Thus, the *local* structure and thermodynamics of the interface can be computed by mean-field techniques, cf.

Section 5.2.3, possibly invoking blob arguments to capture short-ranged excluded volume correlations. In contrast, the long-wavelength undulations, which serve to broaden an interface from its intrinsic width, can be treated by a phenomenological approach utilizing an effective Hamiltonian that contains the mean-field interfacial tension as a parameter (Binder *et al.*, 2001).

### 6.1.6.3 *Liquid state methods: PRISM*

Finally, we should mention *liquid state theory* approaches to studying fluctuation phenomena in polymeric fluids. These methods, most notably the polymer reference interaction site models (PRISM) (Schweizer and Curro, 1994; Schweizer and Curro, 1997), extend well-established liquid state theories for atomic and molecular fluids (Hansen and McDonald, 1986; Chandler, 1982) to the case of polymers. In some ways the PRISM approach is much more ambitious than the other methods that we have described, since the theory attempts to self-consistently treat structure and correlations on length scales ranging from atomic to macroscopic. Moreover, the approach is distinct from the formalism presented here in that the theory is formulated directly for a particle-based model, rather than for a field theory that emerges from a particle-to-field transformation. The PRISM theory relies on a "closure approximation" to produce a set of coupled nonlinear integral equations whose solution yields the structure (pair correlation functions) of a polymeric fluid. Equilibrium properties follow by applying well known connection formulas between structure and thermodynamics (Hansen and McDonald, 1986). Most PRISM calculations are carried out by specifying the *intramolecular* pair correlation functions, usually computed from ideal chain models such as those described in Chapter 2, and then solving the PRISM equations to obtain the *intermolecular* correlations. A more sophisticated type of "self-consistent PRISM" theory attempts to simultaneously compute inter- and intra-molecular correlation functions. It is of interest to note that PRISM and related atomic liquid state theories constitute a sophisticated type of Gaussian field theory for density fluctuations in which a local constraint is imposed to exclude particles from the hard-core regions (Chandler, 1993). This theory is distinct from the unconstrained Gaussian theories for *potential* fluctuations described in Sections 6.1.1 and 6.1.2.

The principal weakness of the PRISM approach lies in the closure approximations, which are deduced using arguments based on atomic scale physics and can prove unreliable in their treatment of long-wavelength fluctuations. The earliest closure approximations, for example, were found to make physically unrealistic predictions for the critical properties of a polymer blend model analogous to Model C. Unfortunately, there is no systematic way to go about improving a specific closure scheme. Nevertheless, the current "molecular closures" have resolved the obvious shortcomings of early approximation schemes (Yethiraj and Schweizer, 1993; Schweizer and Yethiraj, 1993). A further difficulty with the PRISM theory is that the nonlinear equations must be solved numerically, except in special cases where the fluid is homogeneous on average and the "thread limit" is applied.[98] The PRISM equations are not believed to be reliable for an *inhomogeneous* fluid state, such as a block copolymer mesophase. Instead, numerical PRISM calculations for a homogeneous reference state are normally combined with approximate density functional theories (DFT) to address inhomogeneous systems (Donley *et al.*, 1995; Frischknecht *et al.*, 2002). Finally, we note that the PRISM formalism is not easily extended to include three-body interactions.

In spite of these difficulties, Schweizer, Curro and coworkers have made considerable progress in advancing the PRISM theory and testing it against experiment, other theories, and particle-based computer simulations (Schweizer and Curro, 1994; Schweizer and Curro, 1997). The comparisons have been particularly extensive in the case of homogeneous polymer blends (Singh *et al.*, 1995; Schweizer and Singh, 1995), where important insights were obtained regarding the interplay of entropic and enthalpic factors in fluctuation phenomena. Analytical results in the thread limit are particularly useful in this regard. An impressive range of PRISM calculations have been reported for a variety of polymer systems including melts, solutions, block copolymers, polymer-particle composites, liquid crystalline polymers, and strained polymer networks (Guenza and Schweizer, 1997; Schweizer and Curro, 1997; Pickett and Schweizer, 2000; Fuchs and Schweizer, 2002; Hooper *et al.*, 2004; Oyerokun and Schweizer, 2004).

The power of PRISM and PRISM combined with DFT (for inhomogeneous systems) is that these methods can address the local atomic-scale structure of a complex fluid, unlike the coarse-grained field theories that are the focus of this monograph. In some ways, the PRISM theory is analogous to the Hartree theory discussed in Section 6.1.5.1 because it renormalizes interaction parameters such as $\chi$ by including fluctuations of *all* wavelengths. In contrast, the RG theory provides a more measured and controlled approach to renormalization by integrating out successive shells of field fluctuation modes. If the PRISM closure schemes could be trusted at all length scales, numerical implementation issues aside, PRISM would be a near-perfect theory since it would provide a self-contained treatment of structure and thermodynamics from atomic scales upward. However, different closure schemes and thermodynamic connection formulas yield varying and sometimes inconsistent predictions for $R_g$-scale structure and properties (Singh *et al.*, 1995; Wang, 2002). Thus, it is perhaps too much to ask of PRISM to quantitatively describe fluctuation phenomena on mesoscopic and macroscopic length scales, just as it is unreasonable to expect the coarse-grained models and methods of this book to address local liquid structure.

## 6.2   Field-theoretic simulations

It is clear from the above discussion that the available analytical tools for studying fluctuation effects in polymeric fluids are very limited in their capabilities.

---

[98]The "thread limit" corresponds to a situation where the hard-core diameter of the monomers is shrunk to zero, while a fixed monomer volume fraction is maintained.

Virtually all techniques require numerical methods when applied to inhomogeneous systems. If expensive numerical computations are required to analyze an *approximate* theory, such as an implementation of Gaussian fluctuations about an inhomogeneous saddle point (Section 6.1.4) or a corresponding inhomogeneous PRISM + DFT calculation (Section 6.1.6.3), it is worth considering whether it is preferable to attempt a direct numerical attack on the underlying statistical mechanical model *without any simplifying approximations*. In the case of a particle-based model, such a strategy amounts to a *computer simulation*, which can be conducted using a variety of methods including Monte Carlo (MC) and molecular dynamics (MD) techniques (Binder and Heermann, 1988; Binder, 1995; Binder and Ciccotti, 1996; Landau and Binder, 2000; Kotelyanskii and Theodorou, 2004). While less well known, we shall see that it is also possible to carry out a computer simulation of a field theory model. This subject of *field-theoretic simulations (FTS)* (Ganesan and Fredrickson, 2001; Fredrickson *et al.*, 2002) will be the focus of the remainder of the monograph.

The basic idea behind a field-based computer simulation has already been described in Chapter 1 surrounding the discussion of eqns (1.7) and (1.8). Given a field theory model of a simple or complex fluid, such as one of the models described in Chapter 4, it is straightforward to define an associated *discrete field theory model* in which the relevant fields have been given a finite representation so as to render the theory suitable for numerical calculations. Specifically, we might choose to approximate a field $w(\mathbf{r})$ by an $M$-vector $\mathbf{w} = (w_1, w_2, ..., w_M)^T$ whose elements correspond to field values at $M$ collocation points. In the case of Model A, for example, this would lead to the discrete approximation

$$\mathcal{Z}_C = \mathcal{Z}_0 \int \mathcal{D}w \ \exp(-H[w]) \approx \mathcal{Z}_0 \int d\mathbf{w} \ \exp[-H(\mathbf{w})] \qquad (6.124)$$

where in the final expression the partition function is computed as an $M$-dimensional Riemann integral[99] over the components of the field and $H(\mathbf{w})$ denotes the discrete approximation to the effective Hamiltonian given in eqn (4.72). The corresponding ensemble average of some observable $G[w]$, defined according to eqn (4.59), is expressed in the discrete approximation as

$$\langle G[w] \rangle \approx \langle G(\mathbf{w}) \rangle \equiv \frac{\int d\mathbf{w} \ G(\mathbf{w}) \exp[-H(\mathbf{w})]}{\int d\mathbf{w} \ \exp[-H(\mathbf{w})]} \qquad (6.125)$$

The essence of a field-theoretic simulation (FTS) is to devise an efficient numerical strategy whereby eqn (6.125) can be evaluated for an arbitrary observable $G[w]$ and at high field resolution, typically $M \sim 10^4$–$10^7$. Because deterministic quadrature schemes are not feasible for such values of $M$, a *stochastic* integration method such as the Monte Carlo technique must be applied. On the surface, eqn (6.125) appears to have the same form as corresponding expressions for

---

[99]The notation $\int d\mathbf{w}$ denotes an $M$-dimensional integral in which each component of the vector $\mathbf{w}$ is integrated along the real axis.

particle-based fluid models where the **w** variables correspond to either coarse-grained or atomic particle coordinates. It would thus seem that the extensive literature on Monte Carlo (MC) simulations of classical fluids is immediately applicable (Allen and Tildesley, 1987; Frenkel and Smit, 1996; Landau and Binder, 2000). However, there are a number of subtle issues related to the evaluation of eqn (6.125) in the field-theoretic case:

- *Sign problem.* First and foremost, the field theory models constructed in Chapter 4 have a *complex* (not purely real) character associated with the introduction of fields necessary to decouple repulsive interactions. Since all realistic fluid models incorporate repulsive interactions, the complex nature of the effective Hamiltonian is an essential feature of the field-theoretic description. Unfortunately, standard Monte Carlo techniques for evaluating eqn (6.125) require that the "statistical weight" $\exp[-H(\mathbf{w})]$ appearing in numerator and denominator, when suitably normalized, be interpreted as the *probability* of observing a particular discrete field configuration **w**. However, $\exp[-H(\mathbf{w})]$ is not a suitable probability weight in the present situation because it *is not positive semi-definite* for complex $H$. In other words, direct application of MC simulation techniques is thwarted by the fact that the supposed probability weight can be negative in sign and, worse yet, not strictly real. This complication is referred to as the *sign problem* and is familiar in other branches of physics and chemistry, especially in lattice gauge theories, time-dependent quantum chemistry, and correlated electron problems that involve discrete quantum field theories (Montvay and Münster, 1994; Doll and Freedman, 1988; Makri and Miller, 1987; Loh Jr. *et al.*, 1990). Addressing the sign problem is at the heart of the FTS method and will be a central theme in the subsequent sections of this book.

- *Choice of discrete approximation.* Unlike the case of particle simulations, there is considerable flexibility in how we choose to represent a continuous field by a discrete approximation. Since approximation theory provides limited guidance in the case of functional integrals, the best strategy appears to be a discretization scheme that is computationally convenient and that does not introduce any field-dependent "Jacobian" factors in the integration measure. One such scheme is the *lattice field theory* approach, in which fields are inscribed on a regular lattice and gradients are evaluated by using finite difference approximations. With this method, **w** is a vector representing the values of the continuous field $w(\mathbf{r})$ at the $M$ sites of the lattice. However, as was noted in Chapters 3 and 5, *pseudo-spectral* methods are particularly convenient for efficient evaluation of discrete approximations to $H[w]$ and observables such as the density operator $\rho(\mathbf{r}; [iw])$. Indeed, the pseudo-spectral approach is our method of choice because it allows the integrals in eqn (6.125) to be interpreted *either* as $M$-dimensional integrals over the Fourier coefficients of the field, or as $M$-dimensional integrals over the values of the field at $M$ collocation points.

- *Cutoffs and ultraviolet divergences.* In Section 6.1 it was observed that
  continuum field theories can possess ultraviolet divergences that reflect in-
  adequacies of the models below some characteristic length scale. We "regularized" (removed) these divergences in our analytical calculations by introducing a large $k$ cutoff $\Lambda$, which corresponds to a cutoff length scale
  $2\pi/\Lambda$. Such a step is not necessary for a discrete field theory produced
  by either the lattice field theory or pseudo-spectral approach, because an
  ultraviolet cutoff is automatically imposed by the underlying spacing of
  the lattice or collocation grid. Nevertheless, it is important in the course
  of an FTS simulation to test for the dependence of average quantities of
  interest on the grid spacing. Such tests will reveal not only whether sufficient spatial resolution is being applied in a particular situation, but also
  whether the properties of interest are *universal* (not sensitive to atomic
  details neglected in the model).[100]

- *Convergence to the continuum theory.* In the opening paragraph of this
  section, it was stated that the FTS method represents a direct numerical
  attack on an underlying field-theoretic model "without any simplifying
  assumptions." This is true, of course, only to the extent that the numerical
  approximations inherent in the discrete approximation to the theory can
  be neglected. For average properties $\langle G[w] \rangle$ that are universal, it must be
  the case that FTS simulations based on the discrete approximation (6.125)
  will converge to a finite result upon increasing the spatial resolution, i.e.
  taking $M \to \infty$. For non-universal quantities, such convergence will not
  be obtained. It is important to emphasize that the delineation of universal
  and non-universal properties can sometimes be difficult and is highly model
  dependent. For example, a continuum field-theoretic model derived from
  a fully atomistic description of a fluid, e.g. the monatomic Lennard-Jones
  fluid of eqn (1.3), presumably has no ultraviolet divergences. However,
  when a discrete approximation to the theory is simulated using a grid
  spacing that exceeds the atomic size $\sigma$, a strong sensitivity to the lattice
  cutoff will be observed. Nonetheless, as the grid spacing is reduced to well
  below $\sigma$, we expect this sensitivity to be lost and numerical simulations to
  converge to "universal" values for all physical properties of interest.

More details surrounding these issues will be provided in the following sections. However, it should be emphasized that field-theoretic simulations of classical fluids, and especially polymers, are still in a very early stage of development.
Much remains to be understood about all aspects of conducting and interpreting
the results of such simulations.

---

[100]In this and subsequent sections, a "universal" average property $\langle G[w] \rangle$ is one that exhibits
no ultraviolet divergences as the discrete approximation to $w(\mathbf{r})$ is refined.

## 6.3 Monte Carlo simulations

The most straightforward way to address the sign problem inherent in eqn (6.125) is to separate the real and imaginary contributions to the Boltzmann-like factor $\exp[-H(\mathbf{w})]$ by means of the decomposition

$$H(\mathbf{w}) = H_R(\mathbf{w}) + iH_I(\mathbf{w}) \tag{6.126}$$

where subscripts $R$ and $I$ denote the real and imaginary parts of a complex quantity. It should be emphasized that while $H$ is complex, the field $w(\mathbf{r})$ and its discrete representation $\mathbf{w}$ are strictly *real*.[101] For models formulated in the *canonical ensemble*, the presence of the $\ln Q(i\mathbf{w})$ "entropy" term in $H(\mathbf{w})$ makes it necessary to cut the complex $Q$ plane from the origin to render $H$ single valued. The choice of branch cut is arbitrary but should be applied consistently in all expressions involving $H_I$. In contrast, $H_G(\mathbf{w})$ is a single-valued analytic function of $\mathbf{w}$ for models formulated in the *grand canonical ensemble*, so no branch cut is required to affect the decomposition of eqn (6.126).

With the above separation, eqn (6.125) can be reexpressed as

$$\langle G(\mathbf{w}) \rangle = \frac{\int d\mathbf{w}\, G(\mathbf{w}) P(\mathbf{w}) \exp[-iH_I(\mathbf{w})]}{\int d\mathbf{w}\, P(\mathbf{w}) \exp[-iH_I(\mathbf{w})]} \tag{6.127}$$

where $P(\mathbf{w})$ is a real, non-negative function defined by

$$P(\mathbf{w}) \equiv \frac{\exp[-H_R(\mathbf{w})]}{\int d\mathbf{w}\, \exp[-H_R(\mathbf{w})]} \tag{6.128}$$

This function is suitable as a probability weight and is normalized so that $\int d\mathbf{w}\, P(\mathbf{w}) = 1$ (cf. Appendix B). Equation (6.127) can thus be written compactly as

$$\langle G(\mathbf{w}) \rangle = \frac{\langle G(\mathbf{w}) \exp[-iH_I(\mathbf{w})] \rangle_R}{\langle \exp[-iH_I(\mathbf{w})] \rangle_R} \tag{6.129}$$

where $\langle ... \rangle_R$ denotes an average computed with the probability distribution $P(\mathbf{w})$, i.e.

$$\langle O(\mathbf{w}) \rangle_R \equiv \int d\mathbf{w}\, P(\mathbf{w}) O(\mathbf{w}) \tag{6.130}$$

Equation (6.129) is now in a form (de Raedt and Lagendijk, 1981; Lin and Hirsch, 1986) that is suitable for evaluation by means of standard Monte Carlo (MC) simulation techniques. The essence of the MC method for evaluating an average such as eqn (6.130) is to devise a strategy for generating a large number

---

[101] The functional integrals in the models of Chapter 4 are along the real $w$ axis. In a discrete approximation to a field theory, the $M$ integrals over each component of $\mathbf{w}$ are correspondingly along the real axis. For the particular case of a spectral field representation, this analytic structure can be maintained by selecting the real, rather than the complex, form of the Fourier series.

$N_{CR}$ of random field configurations, $\mathbf{w}^j$, $j = 1, 2, ..., N_{CR}$, that have statistical properties described by the probability distribution function $P(\mathbf{w})$. The ensemble average in eqn (6.130) can then be approximated as

$$\langle O(\mathbf{w}) \rangle_R \approx \frac{1}{N_{CR}} \sum_{j=1}^{N_{CR}} O(\mathbf{w}^j) \tag{6.131}$$

and the correspondence becomes exact for $N_{CR} \to \infty$. This MC strategy is known as *importance sampling* (Landau and Binder, 2000) because the random field configurations are not chosen with a uniform distribution, but rather with a distribution corresponding to $P(\mathbf{w})$ appearing in eqn (6.130).

It remains to discuss how the random sequence of field configurations necessary to evaluate the right-hand side of eqn (6.131) can be generated. The MC approach is to invent a "fictitious" dynamics described by a *master equation* (van Kampen, 1981)

$$\frac{\partial}{\partial t} P_j(t) = \sum_{k(\neq j)} [T_{jk} P_k(t) - T_{kj} P_j(t)] \tag{6.132}$$

where $P_j(t)$ is the probability of the system being in the configurational state $j$, i.e. $\mathbf{w}^j$, at time $t$. The matrix $T_{jk}$ appearing in this equation describes the probability per unit time of observing a transition from configurational state $k$ to state $j$. Evidently the first term on the right-hand side of eqn (6.132) corresponds to a gain in probability of state $j$ to due a transition from a different state $k$. The second term reflects a loss of probability in state $j$ due to transitions to other states $k$. For eqn (6.132) to be useful in an MC simulation, it is important that the *steady state* solution of the equation, $P_j$, satisfies $P_j = P(\mathbf{w}^j)$ for the probability distribution $P(\mathbf{w})$ that we wish to sample. This condition provides a constraint on the choice of transition rate matrix $T_{jk}$:

$$\sum_{k(\neq j)} [T_{jk} P_k - T_{kj} P_j] = 0 \quad \forall j \tag{6.133}$$

Equation 6.133 is a necessary, but not sufficient, condition to determine $T_{jk}$. Much of the "art" of MC simulations comes down to a proper choice of this transition rate matrix. A common, but not exclusive, way to satisfy eqn (6.133) is to demand *microscopic reversibility* (or "detailed balance") whereby

$$T_{jk} P_k = T_{kj} P_j \tag{6.134}$$

This condition amounts to the requirement that the forward and reverse probability currents between any two states $j$ and $k$ are in exact balance. Clearly even with microscopic reversibility imposed, the transition rate matrix is not uniquely specified.

### 6.3.1  Metropolis Monte Carlo

The first and still most popular choice of $T_{jk}$ in Monte Carlo simulations is the
*Metropolis form* (Metropolis *et al.*, 1953)

$$T_{jk} = \begin{cases} \alpha \exp(-\Delta E_{jk}), & \Delta E_{jk} > 0 \\ \alpha, & \Delta E_{jk} < 0 \end{cases} \tag{6.135}$$

where $\alpha$ is some attempt frequency for transitions and $\Delta E_{jk} \equiv E_j - E_k$ is the
difference in energy between the new $(j)$ and old $(k)$ states in units of $k_B T$. For
the present case of $P_j = P(\mathbf{w}^j) \propto \exp[-H_R(\mathbf{w}^j)]$, we see that eqns (6.134) and
(6.135) are consistent, provided that

$$\Delta E_{jk} = H_R(\mathbf{w}^j) - H_R(\mathbf{w}^k) \tag{6.136}$$

A *Metropolis algorithm* for generating a sequence of random field configurations $\mathbf{w}$ consistent with the probability distribution $P(\mathbf{w})$ thus proceeds as
follows:

### Metropolis Algorithm 6.1

1. Generate an initial field configuration $\mathbf{w}^0$ of the system corresponding to
   state $j = 0$.
2. Evaluate the energy $E^0 = H_R(\mathbf{w}^0)$ of the initial configuration.
3. For $j = 0, 1, 2, ...$, repeat steps 4–7 below.
4. Generate a "trial" field configuration $\mathbf{w}_{trial}$ by means of

$$\mathbf{w}_{trial} = \mathbf{w}^j + \mathbf{R}$$

   where $\mathbf{R}$ is an $M$-vector of random numbers selected from a *uniform* distribution of real numbers over the interval $[-\Delta, +\Delta]$.
5. Evaluate the energy $E_{trial} = H_R(\mathbf{w}_{trial})$ of the trial configuration and the
   change in energy $\Delta E = E_{trial} - E^j$ from state $j$.
6. Generate a random number $r$ from a *uniform* distribution over $r \in [0, 1]$.
7. If $r \leq \exp(-\Delta E)$, then *accept* the trial configuration and set

$$\mathbf{w}^{j+1} = \mathbf{w}_{trial}, \quad E^{j+1} = E_{trial}$$

   If $r > \exp(-\Delta E)$, then *reject* the trial configuration and set

$$\mathbf{w}^{j+1} = \mathbf{w}^j, \quad E^{j+1} = E^j$$

Steps 6 and 7 are evidently responsible for generating transitions between
configurational states that are consistent with eqn (6.135). After running this
algorithm for a suitable equilibration period, such that the states generated are
characteristic of the steady state distribution $P_j$, one simply collects a *Markov*

*chain* of $N_{CR}$ successive[102] field configurations which are then used to construct averages according to eqn (6.131). These "time averages" are assumed to be equivalent to ensemble averages calculated with the statistical weight $P(\mathbf{w})$ by invoking the ergodic theorem of statistical mechanics (McQuarrie, 1976).

Some comments regarding the implementation of this algorithm are in order. The initial configuration $\mathbf{w}^0$ can be selected in one of several ways. The simplest strategy is to use $M$ calls to a random number generator to initialize the elements of the configuration vector. A more sophisticated approach, which is appropriate for a field theory where the saddle point has been shifted onto the real axis [cf. eqns (6.137)–(6.138) below], is to apply the methods of Section 5.3.2 to obtain a numerical approximation to the saddle point and set $\mathbf{w}^0 = \mathbf{w}^*$. This method starts the simulation from a "highly relevant" field configuration, but is obviously more expensive to implement.

Each energy evaluation in steps 2 and 5 above (for Model A) requires a solution of the diffusion eqn (3.25) and an evaluation of the real part of eqn (4.72) for a pure imaginary field configuration $i\mathbf{w}$. The overall operation count for these steps scales as $O(N_s M \log_2 M)$ using the pseudo-spectral methods of Section 3.6.3. In contrast, steps 4 and 7 require only $O(M)$ operations, so these steps are not limiting. It is important to emphasize that Algorithm 6.1 attempts to update *all* $M$ components of the potential in each pass, whereas it is more common in Monte Carlo simulations of spin models and particle-based fluid models to attempt moves of *single* spins or particles (Frenkel and Smit, 1996; Landau and Binder, 2000). In the present situation the operation count for evaluating the energy is the same regardless of how many components of the vector $\mathbf{w}$ are updated, *so it would be very wasteful to attempt updates of only one field component at a time.*

In addition to an efficient routine for evaluating the real part of $H$, the above algorithm requires a random number generator that can generate real numbers from uniform distributions over the intervals $[-\Delta, \Delta]$ and $[0, 1]$. It is essential that the random numbers applied to different components of $\mathbf{w}$ and in different MC cycles are statistically independent. The real parameter $\Delta > 0$ can be adjusted to tune the performance of the algorithm. Too large a value of $\Delta$ will lead to a small fraction of the trial updates being accepted, i.e. a small "acceptance ratio," and thus slow evolution of the system through field configuration space. In contrast, too small a value of $\Delta$ will produce a large acceptance ratio, but again a slow evolution of the system because the individual moves are very small. The optimal choice of $\Delta$ usually corresponds to a value for which the acceptance ratio is approximately 0.5, i.e. roughly half of the trial moves are accepted and half are rejected over the course of a long simulation.

While Algorithm 6.1 is a reasonable way to produce a chain of states selected from the distribution $P(\mathbf{w}) \sim \exp[-H_R(\mathbf{w})]$, the use of eqn (6.131) to evaluate

---

[102]Note that immediately adjacent field configurations from the algorithm are usually not statistically independent, so it is typical in practice to construct the Markov chain from configurations that are spaced by 10–100 iterations of the Metropolis algorithm.

the averages in the numerator and denominator of eqn (6.129) *proves to be a very poor way of addressing the sign problem*. The difficulty lies in the structure of eqn (6.129). For **w** along the real axis, $H_I$ assumes values that are typically $O(M)$ because of the extensive nature of the continuum energy functional (4.72). As a result, the phase factor $\exp[-iH_I(\mathbf{w})]$ oscillates wildly along the sampling path. Extensive cancellation occurs in averages computed with eqn (6.131) so that both numerator and denominator in eqn (6.129) become very small for large systems, often decaying exponentially with $M$ (Loh Jr. *et al.*, 1990). For even modest values of $M$ it thus becomes very difficult to extract meaningful estimates of $\langle G(\mathbf{w}) \rangle$ from eqn (6.129).

The situation is considerably improved if it is possible to deform the path of functional integration in the field theory from the real axis onto a path that is a *constant phase ascent path* passing through one or more saddle points of interest.[103] For example, the canonical Model A subject to periodic boundary conditions has a single homogeneous saddle point located on the imaginary axis at $w^* = -iu_0\rho_0$, cf. eqn (5.12). A simple shift in field to $\omega(\mathbf{r}) = w(\mathbf{r}) - w^*$ with $\omega$ real serves to deform the path of integration from the real axis $\Gamma_1$ onto a new parallel path $\Gamma_2$ shown in Fig. 6.7 that passes through $w^*$. By analogy with eqns (6.7)–(6.8), Model A is transformed according to

$$\mathcal{Z}_C = \mathcal{Z}_0 \exp(-H[w^*]) \int \mathcal{D}\omega \, \exp(-H_S[\omega]) \qquad (6.137)$$

$$H_S[\omega] = \frac{1}{2u_0} \int d\mathbf{r} \, [\omega(\mathbf{r})]^2 - i\rho_0 \int d\mathbf{r} \, \omega(\mathbf{r}) - n \ln Q[i\omega] \qquad (6.138)$$

where the path of integration of $\omega(\mathbf{r})$ is over the real axis for each field component and the saddle point of the "shifted" Hamiltonian $H_S[\omega]$ is located at the origin, $\omega^*(\mathbf{r}) = 0$. The mean-field interaction energy is again given by $H[w^*] = (1/2)u_0\rho_0^2 V$.

It was demonstrated in Section 6.1.1 that the shifted path $\Gamma_2$ is a constant phase ascent path to quadratic order about the saddle point. Specifically, to $O(\omega^2)$ about the origin, $H_{SR}$ is a convex, positive definite quadratic form and $H_{SI} = 0$. Thus, for field configurations $\omega(\mathbf{r})$ of small amplitude, the phase factor of the shifted theory is approximately given by $\exp(-iH_{SI}[\omega]) \approx 1$ and the problematic phase fluctuations are suppressed. While this is only true to quadratic order, the shifted theory of eqns (6.137)–(6.138) provides a much better basis for carrying out Monte Carlo simulations on Model A than the original formulation given by eqns (4.71)–(4.72).[104] In particular, the counterpart of eqn (6.129) for the discrete shifted theory is

---

[103] At this juncture, the reader should find it helpful to review Section 5.1.2 discussing the analytic structure of the field theory models in Chapter 4.

[104] Similar benefits have been realized by utilizing shifted contours in *ab initio* quantum Monte Carlo simulations of fermions (Baer *et al.*, 1998).

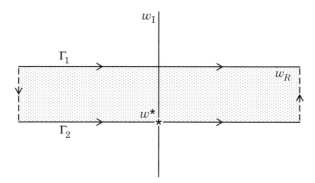

FIG. 6.7. Cauchy's theorem and the analyticity of $\exp(-H[w])$ for Model A in the shaded strip indicated can be used to deform the integration path of the field theory from the real axis $\Gamma_1$ onto a parallel path $\Gamma_2$ passing through the saddle point $w^*$. The connecting dashed contours make negligible contributions if they are applied at $w_R = \pm X$, with $X \to \infty$.

$$\langle G(\boldsymbol{\omega}) \rangle = \frac{\langle G(\boldsymbol{\omega}) \exp[-iH_{SI}(\boldsymbol{\omega})]\rangle_R}{\langle \exp[-iH_{SI}(\boldsymbol{\omega})]\rangle_R} \tag{6.139}$$

where $\boldsymbol{\omega}$ is an $M$-component discrete approximation to the field $\omega$. The notation $\langle ... \rangle_R$ reflects an average with a probability weight $P(\boldsymbol{\omega})$ that is constructed from the real part of the shifted Hamiltonian $H_S[\omega]$, i.e.

$$P(\boldsymbol{\omega}) \equiv \frac{\exp[-H_{SR}(\boldsymbol{\omega})]}{\int d\boldsymbol{\omega} \, \exp[-H_{SR}(\boldsymbol{\omega})]} \tag{6.140}$$

The Metropolis algorithm can again be used to evaluate the averages in the numerator and denominator of eqn (6.139), noting the replacements of $\boldsymbol{\omega}$ for $\mathbf{w}$ and $H_{SR}$ for $H_R$.

Under conditions where the fluctuations in the $\omega$ field are nearly Gaussian, eqn (6.139) does not suffer from the sign problem – in striking contrast to eqn (6.129). Unfortunately, the sign problem has not been completely circumvented by this simple trick, because the $\omega$ field fluctuations are strictly Gaussian only in the concentrated regime for which $C \gg C^{**} = B$, cf. eqn (6.50). As the polymer concentration is decreased to approach the semi-dilute regime, phase oscillations reappear because fluctuations about the saddle point grow in amplitude and the anharmonic terms in $H_S(\boldsymbol{\omega})$ are of increasing importance. Indeed, as the numerical examples provided below indicate, the sign problem is strongly evident in the semi-dilute regime.

Figure 6.8 shows the average sign defined as $\langle \text{sign} \rangle \equiv \langle \exp[-iH_{SI}(\boldsymbol{\omega})]\rangle_R$ determined from Monte Carlo simulations of the canonical Model A with the discrete Gaussian chain substituted for the continuous Gaussian chain. The average sign is plotted as a function of the dimensionless concentration $C$ for three fixed values of the product $BC$, which we recall from eqn (6.20) determines the Edwards correlation length $\xi_E$. The simulations were conducted in three dimensions with periodic boundary conditions and using a collocation grid with $M = 16^3$ points. A discrete Gaussian chain with $N = 100$ springs was employed and the system size was adjusted in the three cases to maintain the grid spacing at a constant fraction of $\xi_E$. The figure shows that the average sign increases monotonically with $C$ at fixed $BC$, approaching unity in the concentrated regime. The deterioration of the average sign upon lowering $C$ is seen in each case to begin in the concentrated regime, $C \gtrsim B$, and occurs over several decades. The shifted path Monte Carlo technique breaks down entirely when the average sign vanishes.

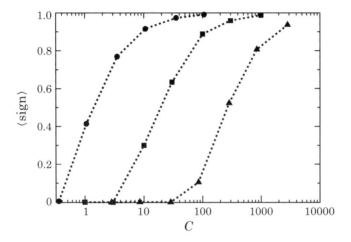

FIG. 6.8. Average sign versus dimensionless concentration $C$ for the canonical Model A at three fixed values of $BC = 5$ (circles), 10 (squares), and 20 (triangles). The simulations were conducted in three dimensions using periodic boundary conditions, discrete Gaussian chains with 100 beads, and $M = 16^3$ collocation points. The system sizes correspond to $L = 1.4142R_{g0}$ ($BC = 5$), $L = 1.0R_{g0}$ ($BC = 10$), $L = 0.7071R_{g0}$ ($BC = 20$). Figure courtesy of K. Katsov.

Figure 6.9 shows the dependence of the average sign on the volume of the system for the canonical Model A at $C = 3$ and $BC = 10$. The Monte Carlo simulations were again conducted in three dimensions using periodic boundary conditions and by applying the discrete Gaussian chain model with $N = 100$ beads. The collocation grid spacing $\Delta x = \Delta y = \Delta z$ was fixed in the simulations at $(1/8)R_{g0}$ and the average sign was calculated for cubic lattices using $M_x$, $M_y$, and $M_z$ grid points in the three orthogonal directions. The dimensionless system volume is given by $\tilde{V} = (1/8)^3 M_x M_y M_z$. Figure 6.9 shows that the average sign decays approximately exponentially with the system volume at fixed spatial resolution. At a particular volume, the average sign is largest for cubic simulation cells with $M_x = M_y = M_z$ and is smallest in tetragonal cells where the lattice is extended in one dimension at the expense of the other two. The worsening of the sign problem with increasing system volume is evidently very limiting – Monte Carlo simulations are practically useful only for small systems, in spite of the contour shift.

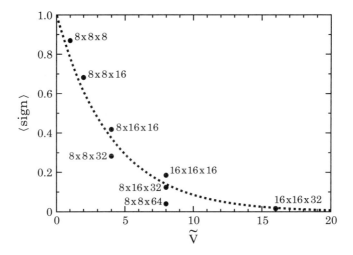

FIG. 6.9. Average sign versus dimensionless system volume $\widetilde{V}$ for the canonical Model A with parameters $C = 3$, and $BC = 10$. Uniform collocation grids were employed with a fixed lattice spacing $\Delta x = (1/8)R_{g0}$ and the number of grid points indicated ($M_x \times M_y \times M_z$). The simulations were conducted in three dimensions using periodic boundary conditions and discrete Gaussian chains with 100 beads. The dashed curve is a guide to the eye. Figure courtesy of K. Katsov.

The above trick of shifting the integration path for each component of a field from the real axis to a *parallel path* passing through a saddle point is helpful in situations other than the homogeneous Model A case just described. For example, the same technique can be effective for treating Model A in a *bound geometry*, but then the necessary shift is an *inhomogeneous* one,

$$\omega(\mathbf{r}) = w(\mathbf{r}) - w^*(\mathbf{r}) \qquad (6.141)$$

where the purely imaginary saddle point $w^*(\mathbf{r})$ must be pre-computed numerically by the methods of Chapter 5 (cf. the slit profiles of Fig. 6.2). The Gaussian analysis of Section 6.1.4 confirms our expectation that the deformed path with $\omega(\mathbf{r})$ strictly real is again a constant phase ascent path with $H_{SI} = 0$ to harmonic order. As in the homogeneous case, however, the sign problem reemerges outside of the concentrated regime, $C < C^{**}$, and for large systems.

Similar contour displacements of the defining functional integrals can be useful in preparing other field-theoretic models for numerical investigation by Monte Carlo simulation. However, many models, including Models B, C, D, E, and J of Chapter 4, possess *multiple* saddle points that can be either pure states or mixed states. For a Monte Carlo simulation to perform well against the sign problem, the contour shift must be onto a constant phase path through the "relevant" saddle point for the specified parameters of the model. Since the relevant saddle point is often not known in advance, it may be necessary to pre-compute a number of saddle points and explore the performance of MC simulations conducted by shifting through each of them. Worse yet, for models in the B–E class, cf. eqn (6.199) for Model C, a deformed path *parallel* to the real axis and passing through a saddle point may not correspond to a constant phase path, even locally. In such a situation, a local harmonic analysis can be used to identify a *non-parallel* constant phase ascent path through the saddle point.[105] However, the extra computational burden and tedium of conducting a local analysis around multiple saddle points of a model renders this approach very unappealing. For such models, we recommend applying an "adaptive" field-theoretic simulation technique, such as the complex Langevin method described in Section 6.4.

### 6.3.2 *Force-bias Monte Carlo*

The Metropolis algorithm is but one of many possible Monte Carlo algorithms that have been devised for sampling states from a real non-negative probability distribution $P(\mathbf{w})$ (Landau and Binder, 2000). If one applies the Metropolis scheme, it is necessary to evaluate the real part of the effective Hamiltonian, $H_R$, for each trial configuration and the imaginary part, $H_I$, for each *successful* configurational update. Another important class of MC algorithms use not only values of the energy at each attempted move, but also instantaneous values of the force (Ceperley *et al.*, 1977; Pangali *et al.*, 1978; Rossky *et al.*, 1978). Such *force-bias* MC algorithms can potentially sample field configuration states more efficiently than Metropolis-type algorithms, but this improved performance comes with the extra computational burden of the force evaluation. It thus remains to be seen whether force-bias schemes can play an important role in FTS simulations.

In the present field-theoretic case, where we wish to sample a distribution $P(\boldsymbol{\omega}) \sim \exp[-H_{SR}(\boldsymbol{\omega})]$ for a theory shifted through the saddle point, the relevant force is

$$\mathbf{F}(\boldsymbol{\omega}) = -\mathrm{Re}\frac{\partial H_S(\boldsymbol{\omega})}{\partial \boldsymbol{\omega}} = -\frac{\partial H_{SR}(\boldsymbol{\omega})}{\partial \boldsymbol{\omega}} \tag{6.142}$$

where Re denotes the operation of taking the real part of a complex function. Specifically, for the canonical Model A in three dimensions, this force is given by

$$\mathbf{F}(\boldsymbol{\omega}) = -(\Delta x)^3 \left[\frac{1}{u_0}\boldsymbol{\omega} - \boldsymbol{\rho}_I(i\boldsymbol{\omega})\right] \tag{6.143}$$

---

[105] Recall from Section 6.1.4 that such an analysis is very expensive for an inhomogeneous saddle point, requiring $O(N_s M^2 \log_2 M)$ operations.

where $(\Delta x)^3$ is the volume of a cell of the collocation grid and $\boldsymbol{\rho}_I(i\omega)$ is a vector consisting of the values of the imaginary part of the density operator $\tilde{\rho}(\mathbf{r}; [i\omega])$ collocated at $M$ grid points. From an existing solution of the complex diffusion equation, the force vector can be evaluated from eqn (4.75) on the collocation grid using an additional $O(N_s M)$ operations.

The variant of force-bias MC that is most suited to field-theoretic simulations is the approach of Rossky *et al.* (1978), which is also referred to as *smart Monte Carlo*. This MC technique was inspired by an earlier simulation strategy, the *Brownian dynamics* method (Ermak, 1975), which is based on the following "Langevin equation" (van Kampen, 1981; McQuarrie, 1976):[106]

$$\frac{d}{dt}\boldsymbol{\omega}(t) = -\lambda \frac{\partial H_{SR}(\boldsymbol{\omega})}{\partial \boldsymbol{\omega}(t)} + \boldsymbol{\eta}(t) \qquad (6.144)$$

A Langevin equation is a type of *stochastic differential equation* that contains a random force term; in this case, an added random $M$-vector $\boldsymbol{\eta}(t)$. The physical content of eqn (6.144) is a relaxational dynamics in which the frictional force proportional to the "velocity" $d\boldsymbol{\omega}/dt$ is balanced by the sum of a deterministic potential force $\mathbf{F} = -\partial H_{SR}/\partial \boldsymbol{\omega}$ and the random force $\boldsymbol{\eta}$. These dynamics should not be interpreted literally in the present context, but as with the master eqn (6.132), should be viewed as an artifice for generating a Markov chain of equilibrium configurational states. The real parameter $\lambda > 0$ is similar to that appearing in the relaxation schemes of Section 5.3.2.1 and can be freely adjusted.

Once the statistical properties of $\boldsymbol{\eta}(t)$ are specified along with an initial condition, the Langevin equation is itself specified. A particularly useful representation of the "solution" of such a stochastic differential equation is an expression for a probability distribution function $P(\boldsymbol{\omega}, t)$, which gives the probability of observing the field configuration state $\boldsymbol{\omega}$ at time $t$. If the random force statistics are simple enough, it is possible to derive an explicit partial differential equation satisfied by $P(\boldsymbol{\omega}, t)$, which is a type of *Fokker–Planck* equation. The most common choice of random force corresponds to "white noise" statistics in which $\boldsymbol{\eta}(t)$ is assumed to be a Gaussian random process (van Kampen, 1981) with mean and covariance given by

$$\langle \eta_j(t) \rangle = 0 \qquad (6.145)$$

$$\langle \eta_j(t)\eta_k(t') \rangle = 2\lambda\, \delta_{jk}\delta(t - t') \qquad (6.146)$$

and where $\eta_j(t)$ denotes the $j$th component of the $M$-vector $\boldsymbol{\eta}(t)$. With this choice, the random force is uncorrelated at different field collocation sites $j \neq k$ at all times and decorrelates in time at a single site instantaneously. The term "white noise" refers to the fact that the Fourier transform (power spectrum) of eqn (6.146) is frequency independent. It is important to note that the noise

---

[106]We note that this is an example of a conventional, or *real*, Langevin equation since the state vector $\boldsymbol{\omega}$ and the effective Hamiltonian $H_{SR}(\boldsymbol{\omega})$ are strictly real. A different type of "complex" Langevin equation will be the subject of Section 6.4.

strength in eqn (6.146) is $2\lambda$, i.e. two times the dissipative coefficient $\lambda$ multiplying the force in eqn (6.144). This relationship, known as a *fluctuation-dissipation theorem*, is necessary to ensure that the Langevin equation will generate the correct equilibrium distribution of field states.

Using methods similar to those described in Appendix D, one can show that eqns (6.144)–(6.146) are equivalent to the Fokker–Planck equation

$$\frac{\partial}{\partial t} P(\boldsymbol{\omega}, t) = \frac{\partial}{\partial \boldsymbol{\omega}} \cdot \boldsymbol{\lambda} \cdot \left[ \frac{\partial}{\partial \boldsymbol{\omega}} + \frac{\partial H_{SR}(\boldsymbol{\omega})}{\partial \boldsymbol{\omega}} \right] P(\boldsymbol{\omega}, t) \tag{6.147}$$

where the $M \times M$ "transport coefficient matrix" $\boldsymbol{\lambda}$ is diagonal with elements $\lambda_{jk} = \lambda \delta_{jk}$. It is easily confirmed that the probability weight $P(\boldsymbol{\omega})$ given in eqn (6.140) is a steady state (equilibrium) solution of this equation. It can also be proven for physically realistic conditions on $H_{SR}$ that this steady state distribution is unique and is approached from an arbitrary initial distribution $P(\boldsymbol{\omega}, 0)$. As a result, time integration of the Langevin eqn (6.144) with noise statistics given by eqns (6.145) and (6.146) is guaranteed, after an equilibration period, to produce a Markov chain of field configuration states that is representative of the probability distribution $P(\boldsymbol{\omega})$. The essence of a Brownian dynamics simulation is to use these states for the purpose of evaluating ensemble averages such as eqn (6.139) as time averages according to eqn (6.131).

A Brownian dynamics algorithm amounts to a time integration scheme for the Langevin eqn (6.144). The simplest scheme is the Euler–Maruyama (EM) algorithm (Higham, 2001; Kloeden and Platen, 1992), which is the analog of forward Euler time stepping of the corresponding deterministic equation. The EM algorithm amounts to

$$\boldsymbol{\omega}^{j+1} = \boldsymbol{\omega}^j + \lambda \Delta t \, \mathbf{F}(\boldsymbol{\omega}^j) + \mathbf{R}^j \tag{6.148}$$

for $j = 0, 1, 2, \ldots$, where a superscript is again used to index successive field configuration states and $\Delta t$ is the time step. The random force vector $\mathbf{R}^j$ at time $t^j$ is formally defined by

$$\mathbf{R}^j = \int_{t^j}^{t^j + \Delta t} ds \, \boldsymbol{\eta}(s) \tag{6.149}$$

and it is easily verified from eqns (6.145) and (6.146) that $\mathbf{R}^j$ is a Gaussian random variable with moments

$$\langle R_n^j \rangle = 0 \tag{6.150}$$

$$\langle R_n^j R_m^k \rangle = 2\lambda \Delta t \, \delta_{jk} \delta_{nm} \tag{6.151}$$

It is important to note from eqn (6.151) that the random force term in eqn (6.148) is characteristically $O(\sqrt{\Delta t})$, while the deterministic force is $O(\Delta t)$.

Given a random number generator capable of generating random numbers with a Gaussian distribution, eqn (6.148) can be readily implemented to conduct a Brownian dynamics simulation. Superficially, this would seem to be a better algorithm than the Metropolis algorithm for sampling states from $P(\boldsymbol{\omega})$ because the moves are biased by the local force in a direction that (for sufficiently small $\Delta t$) tends to lower the energy $H_{SR}(\boldsymbol{\omega})$. Thus, faster equilibration might be possible. In practice, however, eqn (6.148) is not a good algorithm because it suffers from problems of *stability* and *accuracy* unless $\Delta t$ is taken to be very small. While the stability issue can be addressed with a more sophisticated time integration scheme, analogous to the semi-implicit scheme given in eqn (5.92), the accuracy issue is more problematic. Brownian and Langevin dynamics schemes in general suffer from the fact that the error associated with the time integration causes the distribution of states to vary from the target distribution $P(\boldsymbol{\omega})$. This leads to systematic errors in computed averages, in addition to the statistical sampling errors that are inherent in any stochastic simulation technique. In contrast, Monte Carlo algorithms do not suffer from lack of stability or accuracy; their performance reflects only the tradeoff between the simultaneous desire for large trial moves and a high acceptance ratio of attempted moves.

The smart MC algorithm of Rossky *et al.* (1978) is a strategy for preserving the force-bias of the Brownian dynamics method, while embedding the update scheme in a robust MC framework. The basic idea is to employ eqn (6.148) to define a set of transition probabilities $T_{jk}$ that can be used in a MC procedure. However, unlike in Brownian dynamics, where the parameter

$$\Delta \equiv \lambda \Delta t \tag{6.152}$$

must be small to achieve accuracy and stability, we place no *a priori* constraints on the magnitude of $\Delta > 0$. In the smart MC method, a transition from a state $j$ to a state $k$ is defined by an expression analogous to eqn (6.148):

$$\boldsymbol{\omega}^k = \boldsymbol{\omega}^j + \Delta \, \mathbf{F}(\boldsymbol{\omega}^j) + \mathbf{R} \tag{6.153}$$

where $\mathbf{R}$ is a Gaussian random $M$-vector. The normalized probability distribution function of $\mathbf{R}$ is given by

$$\mathcal{P}(\mathbf{R}) = (4\pi\Delta)^{-M/2} \exp[-\mathbf{R} \cdot \mathbf{R}/(4\Delta)] \tag{6.154}$$

which is consistent with eqns (6.150) and (6.151). A matrix of transition probabilities between states $j$ and $k$ can be defined according to the above equations as

$$T_{kj}^* \equiv \frac{\int d\mathbf{R} \, \mathcal{P}(\mathbf{R}) \, \delta(\boldsymbol{\omega}^k - \boldsymbol{\omega}^j - \Delta \, \mathbf{F}(\boldsymbol{\omega}^j) - \mathbf{R})}{\int d\boldsymbol{\omega}^k \int d\mathbf{R} \, \mathcal{P}(\mathbf{R}) \, \delta(\boldsymbol{\omega}^k - \boldsymbol{\omega}^j - \Delta \, \mathbf{F}(\boldsymbol{\omega}^j) - \mathbf{R})} \tag{6.155}$$

Using the properties of the Dirac delta function, one reduces this matrix to

$$T_{kj}^* = (4\pi\Delta)^{-M/2} \exp[-|\boldsymbol{\omega}^k - \boldsymbol{\omega}^j - \Delta \, \mathbf{F}(\boldsymbol{\omega}^j)|^2/(4\Delta)] \tag{6.156}$$

While $T_{kj}^*$ is normalized, it does not satisfy the desired microscopic reversibility condition (6.134). However, the related transition probability matrix $T_{kj}$ defined by

$$T_{kj} \equiv \begin{cases} T_{kj}^*, & T_{jk}^* P_k \geq T_{kj}^* P_j \\ T_{jk}^* P_k / P_j, & T_{jk}^* P_k < T_{kj}^* P_j \end{cases} \tag{6.157}$$

with $P_j = P(\boldsymbol{\omega}^j) \propto \exp[-H_{SR}(\boldsymbol{\omega}^j)]$ does have this characteristic. Indeed, it follows from eqn (6.157) that

$$T_{kj} P_j = \min(T_{kj}^* P_j, T_{jk}^* P_k) = T_{jk} P_k \tag{6.158}$$

The smart Monte Carlo algorithm for a contour-shifted field theory is defined as follows (Rossky *et al.*, 1978):

### Smart Monte Carlo Algorithm 6.2

1. Generate an initial field configuration $\boldsymbol{\omega}^0$ of the system corresponding to state $j = 0$.
2. Evaluate the energy $E^0 = H_{SR}(\boldsymbol{\omega}^0)$ and the force $\mathbf{F}^0 = \mathbf{F}(\boldsymbol{\omega}^0)$ associated with the initial configuration.
3. For $j = 0, 1, 2, ...$, repeat steps 4–7 below.
4. Generate a "trial" field configuration $\boldsymbol{\omega}^k$ by means of

$$\boldsymbol{\omega}^k = \boldsymbol{\omega}^j + \Delta \, \mathbf{F}^j + \mathbf{R}$$

   where $\mathbf{R}$ is an $M$-vector of independent random numbers selected from the Gaussian distribution eqn (6.154).
5. Evaluate the energy $E^k = H_{SR}(\boldsymbol{\omega}^k)$ and the force $\mathbf{F}^k = \mathbf{F}(\boldsymbol{\omega}^k)$ of the trial configuration and the ratio

$$\tau = (T_{jk}^* P_k)/(T_{kj}^* P_j)$$

   by means of eqn (6.156).
6. Generate a random number $r$ from a *uniform* distribution over $r \in [0, 1]$.
7. If $r \leq \tau$, then *accept* the trial configuration and set

$$\boldsymbol{\omega}^{j+1} = \boldsymbol{\omega}^k, \quad E^{j+1} = E^k, \quad \mathbf{F}^{j+1} = \mathbf{F}^k$$

   If $r > \tau$, then *reject* the trial configuration and set

$$\boldsymbol{\omega}^{j+1} = \boldsymbol{\omega}^j, \quad E^{j+1} = E^j, \quad \mathbf{F}^{j+1} = \mathbf{F}^j$$

Again, steps 6 and 7 are responsible for implementing the detailed balance condition, in this case accepting trial moves with a force-biased probability $p_{SMC} = \min(1, T_{jk}^* P_k / T_{kj}^* P_j)$. In contrast, the Metropolis Algorithm 6.1 accepts trial moves with a probability $p_{MC} = \min(1, P_k / P_j)$ that only takes into account the energy difference between states. The parameter $\Delta$ is usually adjusted to achieve an acceptance ratio of approximately 0.5, although the optimal value of $\Delta$ (the one producing the fastest equilibration) is expected to be model dependent.

The computational burden of the smart MC algorithm per iteration is approximately 2 times that of the Metropolis scheme. This follows because the force must be recomputed for each *trial* move and a force evaluation has an incremental cost of $O(N_s M)$ operations – roughly equal to the cost of the diffusion equation solution and energy evaluation.

Figure 6.10 compares the effectiveness of Metropolis and smart MC simulations in sampling the distribution $P(\omega)$. Specifically, the figure shows the normalized energy time correlation function of the shifted canonical Model A defined according to[107]

$$\langle H(0)H^*(t) \rangle \equiv \frac{\langle H_S(0)H_S^*(t) \rangle_R}{\langle H_S(0)H_S^*(0) \rangle_R} \qquad (6.159)$$

The time $t$ in each panel refers to the corresponding number of Metropolis or smart MC steps (attempts to update all sites of the lattice). The simulations were conducted in three dimensions with periodic boundary conditions and using a cubic cell with side length $L = 1R_{g0}$ and $M = 16^3$ collocation points. A discrete Gaussian chain with $N = 100$ beads was adopted and $BC$ was fixed at 10. Figure 6.10 shows that irrespective of the value of $C$, the energy correlations decay on the scale of approximately $10^4$ Metropolis MC steps or 50 smart MC steps. Thus, the force-bias scheme leads to a reduction in the correlation time by a factor of approximately 200 in this example. This correlation time advantage translates directly into a large decrease in computer time (by a factor of approximately 100) when the extra expense of the smart MC algorithm is taken into account. By biasing trial MC moves along the force direction, we see that the smart MC algorithm leads to a more efficient exploration of phase space and a very substantial performance advantage over conventional Metropolis Monte Carlo for field-theoretic simulations.

It is important to emphasize that the correlation time results shown in Fig. 6.10 relate to the efficiency of sampling the real distribution $P(\omega)$, but do not account for the sign problem. Specifically, these correlation times dictate the number of Monte Carlo steps $N_{CR}$ that are required to accurately evaluate the average $\langle G(\omega) \rangle_R$, rather than the number of steps $N_C$ necessary to evaluate $\langle G(\omega) \rangle$ according to eqn (6.139). It is the latter phase factor-weighted average that is the physically significant one. The relationship between $N_C$ and $N_{CR}$ can be shown to be (Kieu and Griffin, 1994)

$$N_C \approx \frac{N_{CR}}{\langle \text{sign} \rangle^2} \qquad (6.160)$$

where $\langle \text{sign} \rangle = \langle \cos H_{SI}(\omega) \rangle_R$ is the average sign. Thus, in the case of $C = 10$, $B = 1$, where the correlation time from smart Monte Carlo is seen from Fig. 6.10 to be on the order of 50 MC steps, we might expect that about $N_{CR} = 5\,000$ steps (100 times the correlation time) would be sufficient to accurately compute

---

[107]The asterisk superscript denotes complex conjugation.

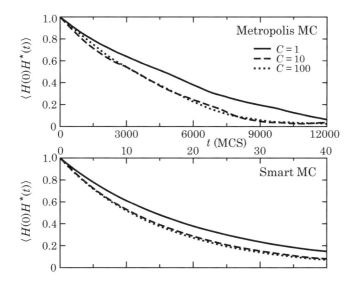

FIG. 6.10. Normalized energy autocorrelation function of the canonical Model A calculated using Metropolis (upper panel) and smart (lower panel) MC simulations. The simulations were conducted on a three-dimensional cubic lattice with side length $L = 1R_{g0}$, $16^3$ sites, and $BC = 10$. Discrete Gaussian chains with 100 beads were employed, and the solid, dashed, and dotted curves correspond to the indicated values of $C$. Time is measured in MC steps (attempts to update the entire lattice). Figure courtesy of K. Katsov.

a $P(\boldsymbol{\omega})$-weighted average $\langle G(\boldsymbol{\omega})\rangle_R$. However, since the average sign in this case is 0.30, a much larger number $N_C \approx 56\,000$ of smart Monte Carlo steps is necessary to evaluate the thermodynamically relevant average $\langle G(\boldsymbol{\omega})\rangle$.

A number of variants of force-bias Monte Carlo methods are possible. A simple, but potentially useful, modification is to replace the scalar $\lambda$ by a positive definite $M \times M$ matrix $\boldsymbol{\lambda}$, both in front of the force in the Langevin eqn (6.144) and in the noise covariance, so that eqn (6.146) becomes

$$\langle \boldsymbol{\eta}(t)\boldsymbol{\eta}(t')\rangle = 2\boldsymbol{\lambda}\,\delta(t-t') \tag{6.161}$$

It can be shown that the Fokker–Planck eqn (6.147) is unchanged by this modification, so $P(\boldsymbol{\omega}) \propto \exp[-H_{SR}(\boldsymbol{\omega})]$ remains the steady state distribution. Algorithm 6.2 is immediately applicable, but with the scalar $\Delta$ replaced by the tensor $\boldsymbol{\Delta} = \boldsymbol{\lambda}\Delta t$, and factors of $1/\Delta$ interpreted as the inverse matrix $\boldsymbol{\Delta}^{-1}$. Most choices of the matrix $\boldsymbol{\lambda}$ produce algorithms that are computationally expensive to implement. However, an important class of matrices correspond to translationally invariant, continuum transport coefficients $\lambda(|\mathbf{r} - \mathbf{r}'|)$, which are diagonal in a Fourier representation. Such "Fourier-accelerated" force-bias algorithms, which apply an optimized wavenumber-dependent time step $\hat{\Delta}(k)$, have proven effective in simulations of quantum field theories (Katz *et al.*, 1988). While Fourier acceleration has not been implemented to date in MC simulations of polymer field theories, similar benefits are anticipated.

### 6.3.3  Hybrid Monte Carlo

Closely related to force-bias and smart Monte Carlo methods are *hybrid Monte Carlo (HMC)* algorithms developed in the context of lattice quantum field theory (Duane *et al.*, 1987). While smart MC was inspired by Brownian dynamics, the HMC technique shares a close connection with the molecular dynamics (MD) simulation method (Allen and Tildesley, 1987; Frenkel and Smit, 1996). The basic idea behind the method is to extend the dimensions of the discrete shifted field theory by introducing a "momentum" field $\boldsymbol{\pi}$ (a $M$-vector) that is conjugate in the classical dynamics sense to the discrete field $\boldsymbol{\omega}$. In particular, one introduces a real Hamiltonian

$$H(\boldsymbol{\omega}, \boldsymbol{\pi}) \equiv \frac{1}{2}\boldsymbol{\pi}^T\boldsymbol{\pi} + H_{SR}(\boldsymbol{\omega}) \tag{6.162}$$

that differs from $H_{SR}$ by the inclusion of a "kinetic energy" term $(1/2)\boldsymbol{\pi}^T\boldsymbol{\pi}$. A molecular dynamics scheme would amount to a time integration strategy for the Hamiltonian dynamics

$$\frac{d}{dt}\boldsymbol{\omega}(t) = \frac{\partial H(\boldsymbol{\omega}, \boldsymbol{\pi})}{\partial \boldsymbol{\pi}(t)} = \boldsymbol{\pi}(t)$$

$$\frac{d}{dt}\boldsymbol{\pi}(t) = -\frac{\partial H(\boldsymbol{\omega}, \boldsymbol{\pi})}{\partial \boldsymbol{\omega}(t)} = \mathbf{F}(\boldsymbol{\omega}(t)) \tag{6.163}$$

where $\mathbf{F}(\boldsymbol{\omega})$ is the force defined by eqn (6.142). Because $H$ is a constant of the motion under such dynamics, a microcanonical MD simulation based on these equations would not generate the desired Boltzmann distribution of states, $P(\boldsymbol{\omega}, \boldsymbol{\pi}) \propto \exp[-H(\boldsymbol{\omega}, \boldsymbol{\pi})]$.

The hybrid MC method is a strategy for ensuring that $P(\boldsymbol{\omega}, \boldsymbol{\pi})$ is recovered at steady state, while avoiding the time discretization errors that are inherent to the MD technique. Starting from some initial field configuration $\boldsymbol{\omega}$, an initial momentum $\boldsymbol{\pi}$ is generated by selecting $M$ independent random numbers from the Gaussian distribution $P(\boldsymbol{\pi}) \propto \exp(-\boldsymbol{\pi}^T\boldsymbol{\pi}/2)$. Given this initial state $(\boldsymbol{\omega}, \boldsymbol{\pi})$, the system is allowed to evolve *deterministically* for $t \in [0, \tau_0]$ according to the MD eqns (6.163). This produces a "trial" configuration $(\boldsymbol{\omega}', \boldsymbol{\pi}')$. The trial configuration is accepted with probability

$$p_{HMC} = \min[1, \exp(-\delta H)] \tag{6.164}$$

where $\delta H \equiv H(\boldsymbol{\omega}', \boldsymbol{\pi}') - H(\boldsymbol{\omega}, \boldsymbol{\pi})$. Duane *et al.* (1987) have proven that this algorithm satisfies detailed balance and when iterated many times generates a Markov chain of states consistent with the distribution $P(\boldsymbol{\omega}, \boldsymbol{\pi}) \propto \exp[-H(\boldsymbol{\omega}, \boldsymbol{\pi})]$.

An interesting feature of the HMC algorithm is that if the time integration during the deterministic relaxation interval $[0, \tau_0]$ could be conducted without error, $\delta H$ would vanish and the acceptance probability would be unity. Instead, the method relies on the time discretization error and the momentum randomization to create ergodicity and evolve the system towards the proper distribution of states. As with any Monte Carlo method, there are no errors incurred by the use of a finite time step, nor is the method subject to stability concerns. One constraint on implementation, however, is that the discrete time algorithm used to integrate eqns (6.163) must preserve exact time reversibility.[108]

In the HMC method there are two parameters that can be tuned to optimize the performance of a simulation. The first is the parameter $\tau_0$ that defines the length of the MD relaxation period between acceptance tests, and the second is the time step $\Delta t \, (< \tau_0)$ used in the numerical solution of eqns (6.163). These parameters are be adjusted to minimize the correlation time of a simulation, while maintaining a high acceptance ratio. The philosophy behind the MD relaxation period is that it provides an efficient way to generate uncorrelated field configuration between acceptance tests, unlike the smart MC method where successive field configurations are normally similar because a small $\Delta$ is required for a high acceptance ratio. Although the HMC method has not yet been applied to polymer field theories, it outperforms other MC methods in simulations of lattice quantum field theories. Thus, HMC would seem to be promising for application in the present context. Kennedy (1999) has provided a comprehensive review of the HMC method and its generalizations and extensions. We note also a recent "Fourier-accelerated" version of HMC (Catterall and Karamov, 2002).

Many other specialized "configurational bias" techniques have been developed for conducting MC simulations of a wide variety of models in statistical physics. In addition, "reweighting" methods, such as umbrella sampling, histogram methods, and multicanonical sampling, have been shown to greatly accelerate MC investigations of important classes of systems, including particle–based models of polymeric fluids. These advanced topics are beyond the scope of the present book, so the interested reader is referred to the extensive literature (Frenkel and Smit, 1996; Landau and Binder, 2000). Undoubtedly such sampling methods, which are enabled by the flexibility of the Monte Carlo technique, will become of increasing significance as the subject of field-theoretic simulations matures.

## 6.4   Complex Langevin simulations

In the previous sections we saw that Monte Carlo simulations, coupled with a shift in contour of the defining functional integrals, can be an effective tool against the sign problem. However, when the fluctuations about the relevant saddle point are strong, when it is not possible to deform onto a constant phase contour in the vicinity of the saddle point, or when the system size is large, then the sign problem becomes significant and the contour-shifted MC approach breaks down. Unfortunately these conditions are more the rule than the exception in multicomponent polymer solutions and alloys. A second major drawback of

---

[108]The "leap-frog" algorithm, for example, has this characteristic.

the contour-shifted MC technique is that it is not *adaptive*, in the sense that a separate contour displacement is required for every saddle point that competes in the parameter space of a model. It may be necessary to pre-compute all of the saddle points and to run a number of unsuccessful MC simulations until a sampling path is identified that is effective against the sign problem.

In contrast, the *complex Langevin (CL)* simulation technique described in the present section is fully adaptive and requires no advance computation of saddle points. The method is subject to the same stability and accuracy concerns that were discussed in the context of Brownian dynamics simulations, but we shall see that these can be adequately addressed. Indeed, the CL technique is the most versatile and powerful field-theoretic simulation tool that we have to date for fighting the sign problem in a broad range of classical statistical field theory models.

The theory underlying the CL simulation method is discussed in some detail in Appendix D. Here we provide a brief synopsis of the theory and focus primarily on numerical implementation issues. It is again convenient to introduce the subject in the context of the *discrete* canonical Model A, cf. eqn (6.124):

$$\mathcal{Z}_C \approx \mathcal{Z}_0 \int d\mathbf{w} \ \exp[-H(\mathbf{w})] \qquad (6.165)$$

where the most useful discrete approximation $\mathbf{w}$ to the real field $w(\mathbf{r})$ is an $M$-vector consisting of field values at $M$ collocation points within the simulation cell. Thus, the integral in eqn (6.165) should be interpreted as an $M$-dimensional integral over these grid point field values, each integral taken along the real axis. When the Fourier coefficients of an $M$-term spectral approximation are required, a discrete Fourier transform can be performed on the vector $\mathbf{w}$ to produce a coefficient vector $\hat{\mathbf{w}}$.

In the discrete representation, the ensemble average of some observable $G(\mathbf{w})$, defined by eqn (6.125), can be expressed as

$$\langle G(\mathbf{w}) \rangle = \int d\mathbf{w} \ G(\mathbf{w}) P_c(\mathbf{w}) \qquad (6.166)$$

where $P_c(\mathbf{w})$ is a "complex probability" density defined by

$$P_c(\mathbf{w}) = \frac{\exp[-H(\mathbf{w})]}{\int d\mathbf{w} \ \exp[-H(\mathbf{w})]} \qquad (6.167)$$

As discussed in Appendix D, $P_c(\mathbf{w})$ is not a true probability distribution function because it is not positive semi-definite for complex $H(\mathbf{w})$. Nevertheless, $P_c(\mathbf{w})$ is the relevant normalized weight function that is to be applied in computing averages with a discrete field theory.

In the complex Langevin (CL) method (Parisi, 1983; Klauder, 1983), the real $M$-vector $\mathbf{w}$ is extended to a complex $2M$-vector $\mathbf{z} = \mathbf{w} + i\mathbf{v}$, and eqn (6.166) is rewritten as a $2M$-dimensional integral over the *whole complex plane*:

$$\langle G(\mathbf{w}) \rangle = \int d\mathbf{w} \int d\mathbf{v}\, G(\mathbf{w} + i\mathbf{v}) P(\mathbf{w}, \mathbf{v}) \tag{6.168}$$

In this expression, $P(\mathbf{w}, \mathbf{v})$ is a *real, non-negative* probability density. The equivalence of eqns (6.166) and (6.168) relies on the existence of a "real" probability density $P(\mathbf{w}, \mathbf{v})$ such that the "complex" probability can be expressed as

$$P_c(\mathbf{w}) = \int d\mathbf{v}\, P(\mathbf{w} - i\mathbf{v}, \mathbf{v}) \tag{6.169}$$

If such a distribution $P(\mathbf{w}, \mathbf{v})$ can be found for a given model Hamiltonian, eqn (6.168) provides a powerful route to circumventing the sign problem. Namely, one approximates the average according to

$$\langle G(\mathbf{w}) \rangle \approx \frac{1}{N_C} \sum_{j=1}^{N_C} G(\mathbf{z}^j) \tag{6.170}$$

where $\mathbf{z}^j = \mathbf{w}^j + i\mathbf{v}^j$, for $j = 1, 2, ..., N_C$ are random points in the $2M$-dimensional complex plane selected from the real distribution $P(\mathbf{w}, \mathbf{v})$. It is important to note that in contrast to eqn (6.129), there is no complex phase factor $\exp[-iH_I]$ appearing in either of eqns (6.168) and (6.170). Thus, the problematic phase and sign oscillations are avoided. Although the CL strategy requires a doubling of the dimensions of phase space (from $M$ to $2M$) in comparison with the Monte Carlo methods of Section 6.3, this computational price is usually more than compensated by the avoidance of the sign problem.

For the above scheme to be implemented it is necessary to verify the existence of a real distribution $P(\mathbf{w}, \mathbf{v})$ satisfying eqn (6.169) and to devise a numerical algorithm for importance sampling of this distribution. While necessary and sufficient conditions have been identified for the existence of $P$ given a complex Hamiltonian $H$ (Salcedo, 1997; Weingarten, 2002), these conditions are difficult to verify in practice. Nevertheless, we shall see that a numerical simulation of the CL equations provides a practical strategy for accomplishing both tasks. The *"standard" complex Langevin (CL) equations* are as follows (Parisi, 1983; Klauder, 1983):

$$\frac{d}{dt}\mathbf{w}(t) = -\lambda \operatorname{Re}\left[\frac{\partial H(\mathbf{z})}{\partial \mathbf{z}(t)}\right] + \boldsymbol{\eta}(t)$$
$$\frac{d}{dt}\mathbf{v}(t) = -\lambda \operatorname{Im}\left[\frac{\partial H(\mathbf{z})}{\partial \mathbf{z}(t)}\right] \tag{6.171}$$

where Re and Im denote the operations of taking the real and imaginary parts, respectively, of a complex function. The real positive parameter $\lambda$ is a scalar relaxation coefficient similar to that appearing in the Brownian dynamics of eqn (6.144), and $\boldsymbol{\eta}(t)$ is a real $M$-vector representing a random force. The random force is taken to have Gaussian white noise statistics identical to eqns (6.145)–(6.146)

$$\langle \boldsymbol{\eta}(t) \rangle = 0$$
$$\langle \boldsymbol{\eta}(t) \boldsymbol{\eta}(t') \rangle = 2\lambda \, \mathbf{1} \, \delta(t - t') \tag{6.172}$$

where $\mathbf{1}$ is the $M \times M$ unit tensor.

The CL equations (6.171) are a set of $2M$ stochastic differential equations describing the evolution of field configurations in the extended complex space of $\mathbf{z}(t) = \mathbf{w}(t) + i\mathbf{v}(t)$. These equations are similar to the "real" Langevin equations (6.144) that form the basis of the Brownian dynamics simulation technique. However, there is a marked *asymmetry* of the CL equations; the noise is added only to the equation for the real component $\mathbf{w}(t)$ and is omitted from the equation for $\mathbf{v}(t)$. This asymmetry is necessary to preserve the analytic structure of the starting model eqn (6.165) which has $\mathbf{w}$ real, but $H(\mathbf{w})$ complex. A second important feature of the CL equations is that when the random force term $\boldsymbol{\eta}(t)$ is removed, the equations have steady state solutions that correspond to *saddle points* of the model. Indeed, without the noise, eqns (6.171) are a discrete version of the relaxation scheme (5.106).

With the noise included, eqns (6.171) have an interesting physical interpretation. The second of the two equations relaxes the imaginary component towards an instantaneous value of $\mathbf{v}$ such that the "phase," i.e. the imaginary part of $H(\mathbf{z})$, is nearly constant with respect to variations in $\mathbf{w}$:

$$\text{Im} \left( \frac{\partial H(\mathbf{z})}{\partial \mathbf{z}} \right) = \frac{\partial}{\partial \mathbf{w}} H_I(\mathbf{w}, \mathbf{v}) \approx 0 \tag{6.173}$$

Thus, the second equation causes the dynamical trajectory to drift towards a *constant phase contour*, while the first CL equation drives a stochastic Brownian dynamics along that contour. The constant phase contours that dominate the CL trajectories are ascent paths that pass through the most relevant saddle points for a specified set of model parameters. It should be clear that the CL equations are fully *adaptive* because no saddle points or contours need be specified or computed in advance.

By following the procedure described in Appendix D, eqns (6.171) can be used to derive a *Fokker–Planck equation* for a probability distribution function $P(\mathbf{w}, \mathbf{v}, t)$ characterizing the CL stochastic process. This real distribution function describes the probability density of observing a field configuration $\mathbf{z} = \mathbf{w} + i\mathbf{v}$ at time t. In the long time limit, the function $P(\mathbf{w}, \mathbf{v}) = \lim_{t \to \infty} P(\mathbf{w}, \mathbf{v}, t)$, *if it exists*, is the steady state distribution function appearing in eqns (6.168)–(6.169). The Fokker–Planck equation can be written

$$\frac{\partial}{\partial t} P(\mathbf{w}, \mathbf{v}, t) = -\lambda \frac{\partial}{\partial \mathbf{w}} \cdot [\mathbf{F}_R(\mathbf{w}, \mathbf{v}) P(\mathbf{w}, \mathbf{v}, t)] - \lambda \frac{\partial}{\partial \mathbf{v}} \cdot [\mathbf{F}_I(\mathbf{w}, \mathbf{v}) P(\mathbf{w}, \mathbf{v}, t)]$$
$$+ \lambda \frac{\partial}{\partial \mathbf{w}} \cdot \frac{\partial}{\partial \mathbf{w}} P(\mathbf{w}, \mathbf{v}, t) \tag{6.174}$$

where $\mathbf{F}_R$ and $\mathbf{F}_I$ are the real and imaginary components of the complex "force" defined by

$$\mathbf{F}(\mathbf{z}) \equiv -\frac{\partial H(\mathbf{z})}{\partial \mathbf{z}} = -(\Delta x)^3 \left[ \frac{1}{u_0} \mathbf{z} + i\boldsymbol{\rho}(i\mathbf{z}) \right] \qquad (6.175)$$

the final expression being specific to Model A. The complex $M$-vector $\boldsymbol{\rho}(i\mathbf{z})$ corresponds to the values of the complex segment density operator $\tilde{\rho}(\mathbf{r}; [iz])$ collocated at $M$ grid points, and $(\Delta x)^3$ denotes the volume associated with each grid point.

A proof that steady state solutions of eqn (6.174) satisfy the necessary condition (6.169) is outlined in Appendix D. It follows that *if* the Fokker–Planck equation for the CL process has a steady state solution $P(\mathbf{w}, \mathbf{v})$, then the ensemble averages given in eqns (6.166) and (6.168) are equivalent. In such a situation, time integration of the CL eqns (6.171) can be used to generate a Markov chain of states $\mathbf{z}^j$, representative of $P(\mathbf{w}, \mathbf{v})$, and the chain further used to approximate ensemble averages by time averages along the CL trajectory according to eqn (6.170). This procedure is the essence of the *complex Langevin (CL) simulation* technique.

The missing theoretical element is a proof that eqn (6.174) has a steady state. To date, there is no such proof, although Gausterer and Lee (Gausterer and Lee, 1993; Lee, 1994) have proven the following theorem that can be used *a posteriori* to confirm the validity of results obtained from a CL simulation:

**Theorem 6.1** *If expectation values for analytic observables $G(\mathbf{z})$, computed as averages over states obtained by integrating eqns (6.171), become time independent over the course of a complex Langevin simulation, then these necessarily converge to correct values consistent with the ensemble average (6.166).*

In practice, we have not observed a failure of CL simulations to converge for any of the models described in Chapter 4. Thus, at least for the classical statistical field theory models relevant to inhomogeneous polymeric fluids, the CL simulation technique appears to have a sound theoretical basis.

It is important to note that the CL method presupposes that the effective Hamiltonian of the model $H(\mathbf{w})$ is an *analytic* function of the extended variable $\mathbf{z}$, so that the complex force $\mathbf{F} = -\partial H/\partial \mathbf{z}$ exists. This condition is met for all models in Chapter 4 formulated in the grand canonical ensemble. In the *canonical ensemble*, the situation is complicated by the fact that the "entropy" term $\ln Q(i\mathbf{z})$ appearing in the model Hamiltonians has branch points at zeros of $Q(i\mathbf{z})$. Normally, however, the complex $Q$ plane can be cut so that $H(\mathbf{z})$ is analytic in the cut plane, and branch points and cut lines are not traversed during a CL simulation. Since the statistically relevant CL trajectories tend to be localized near saddle points, a qualitative understanding of the location of saddle points in the complex plane is a useful prerequisite to conducting a CL simulation on a model formulated in the canonical ensemble.

One unfortunate feature of the CL technique is that it relies on accurate time integration of the stochastic complex Langevin eqns (6.171). We saw in

Section 5.3.2.1 that deterministic versions of these equations have severe *stability* limitations when applying explicit time integration methods. The CL equations have an even greater propensity for stability problems, because the random forces destroy the smoothness of the simulated fields. Thus, a stable and robust time stepping scheme is an *essential* element of a CL simulation. This appears not to be widely recognized; some of the reported "failures" of CL simulations likely originate from inadequate numerical methods, rather than from any theoretical deficiency of the technique (Lin and Hirsch, 1986; Klauder and Petersen, 1985).

In the relaxation algorithms of Section 5.3.2.1 for numerically computing saddle points (mean-field solutions), our concern was simply on stability because the *accuracy* of time integration was not relevant to the accuracy of the final steady state solution. This is not the case for Brownian and Langevin dynamics schemes, where there are *systematic errors* in computed averages associated with the use of a finite time step. These errors can be reduced by employing a high-order time integration scheme, but often identifying such a scheme that is also very stable and computationally efficient can be difficult. It is unfortunate that a closed form analytic solution has not been found for the steady state solution $P(\mathbf{w}, \mathbf{v})$ of the CL Fokker–Planck equation. If such a solution were in hand, the CL dynamics could be replaced by a "complex" force-bias Monte Carlo procedure similar to that described in Section 6.3.2. The focus would then be on optimizing the design and acceptance of trial moves, rather than addressing stability and accuracy concerns.

The simplest time integration scheme for eqns (6.171) is the Euler–Maruyama (EM) algorithm (Higham, 2001; Kloeden and Platen, 1992) already discussed in the context of Brownian dynamics. This explicit, forward-Euler-type scheme amounts to

$$\mathbf{w}^{j+1} = \mathbf{w}^j + \lambda \Delta t \, \mathbf{F}_R(\mathbf{w}^j, \mathbf{v}^j) + \mathbf{R}^j$$
$$\mathbf{v}^{j+1} = \mathbf{v}^j + \lambda \Delta t \, \mathbf{F}_I(\mathbf{w}^j, \mathbf{v}^j) \tag{6.176}$$

where $\mathbf{R}^j$ is the Gaussian random force defined by eqns (6.149)–(6.151). Successive values of this force can be obtained from a random number generator capable of reproducing a normal distribution (Press *et al.*, 1992). The evaluation of the deterministic force $\mathbf{F}(\mathbf{z}^j)$ requires, for the canonical Model A according to eqn (6.175), a solution of the complex diffusion eqn (3.25) with $w \rightarrow i\mathbf{z}^j$ and evaluation of the complex density operator $\rho(i\mathbf{z}^j)$ using eqn (4.75) with $w \rightarrow \mathbf{z}^j$. Both tasks can be performed pseudo-spectrally in $O(N_s M \log_2 M)$ operations as described in Sections (3.6.3) and (5.3.2.1).

A CL simulation is performed with the EM algorithm by iterating eqns (6.176) for $j = 0, 1, 2, ...$, starting from some initial configuration $\mathbf{z}^0$, to generate a sequence of field configuration states $\mathbf{z}^j = \mathbf{w}^j + i\mathbf{v}^j$. A preliminary "equilibration run" is normally used to bring the stochastic process to a steady state prior to the "production run," which serves to generate the Markov chain that is applied in eqn (6.170). Standard tests for equilibration (Frenkel and Smit, 1996; Landau and Binder, 2000) are especially important to perform in the case of CL simulations, due to the reliance of the method on numerical evidence that a stationary state has been achieved (cf. Theorem 6.1).

There are several sources of error in a CL simulation, including:

- *Time discretization error.* This is the error in the expectation value of an observable associated with the discrete time approximation to the stochastic differential equation.[109] In the Euler–Maruyama scheme, the time discretization error is $O(\Delta t)$ and can be reduced by decreasing the size of the time step. Alternatively, the error can be reduced by switching to a higher-order algorithm.

- *Spatial discretization error.* This is the error associated with approximating a complex continuous field $z(\mathbf{r})$ by a complex $M$-vector $\mathbf{z}$. This error is closely related to the ultraviolet divergences discussed in Section 6.1 and can only be defined for observable quantities that are *universal* as described in Section 6.2. In contrast, non-universal quantities exhibit divergences as the spatial resolution is refined. We shall defer further discussion of this topic of *lattice cutoff effects* to Section 6.6.

- *Sampling error.* This is the error associated with the stochastic nature of the simulation method, namely the approximation of the ensemble average (6.168) by the "time" average (6.170). The sampling error is $O(N_C^{-1/2})$ and can thus be reduced by increasing the length of the Markov chain (length of simulation run) used to compute expectation values.

- *Random number bias.* This reflects errors associated with the numerical algorithm used to generate random numbers. These errors, which are manifested as weak correlations in supposedly independent random variables, can be significant in long simulations. Thus it is important to test the quality of the random numbers produced by a given algorithm and, ideally, verify that simulation results are insensitive to the particular random number generator applied.

- *Roundoff error.* This error is associated with accumulated floating point roundoff inherent to digital computation.

Of these five sources of error, the roundoff error is usually the least important to address in CL simulations, because the random number bias is typically more limiting in long runs that aim to achieve high accuracy (Komori *et al.*, 1994). Upon verifying that random number bias is not playing a role in a simulation, the primary focus of error analysis is thus usually on "systematic" discretization and "statistical" sampling errors.

The biggest drawback of the Euler–Maruyama algorithm is not accuracy, but rather *stability*. As with the forward Euler scheme (5.90) for the deterministic eqn (5.89), the EM algorithm (6.176) is only conditionally stable. The instability

---

[109]The relevant time discretization error for field-theoretic simulations relates to convergence in the *weak* sense, rather than *strong* or pathwise convergence of the stochastic process (Kloeden and Platen, 1992).

threshold in the stochastic CL case can be severe depending on the model, so that very small time steps $\Delta t$ must be employed. This is presumably a consequence of the loss of smoothness of the instantaneous fields. An obvious remedy is to switch to an implicit or semi-implicit scheme, either of which is known to be effective in improving the stability of numerical solutions to stochastic differential equations (Kloeden and Platen, 1992). A fully implicit algorithm is not practical because of the expense of the force computation in the present context. However, a suitable semi-implicit scheme can be fashioned after the relaxation algorithm (5.92) by extracting the linear part of the deterministic force. In particular, for the canonical Model A, a useful *semi-implicit CL algorithm* is (Lennon *et al.*, 2005)

$$
\begin{aligned}
\mathbf{w}^{j+1} = {} & \mathbf{w}^j + \mathbf{R}^j - \lambda \Delta t (\Delta x)^3 \\
& \times \left[ \frac{1}{u_0} \mathbf{w}^{j+1} - \boldsymbol{\rho}_I(i\mathbf{z}^j) + \rho_0 N \, \mathbf{g}_D * (\Delta \mathbf{w}^{j+1} - \Delta \mathbf{w}^j) \right] \\
\mathbf{v}^{j+1} = {} & \mathbf{v}^j - \lambda \Delta t (\Delta x)^3 \\
& \times \left[ \frac{1}{u_0} \mathbf{v}^{j+1} + \boldsymbol{\rho}_R(i\mathbf{z}^j) + \rho_0 N \, \mathbf{g}_D * (\Delta \mathbf{v}^{j+1} - \Delta \mathbf{v}^j) \right] \qquad (6.177)
\end{aligned}
$$

where $\Delta \mathbf{z} = \Delta \mathbf{w} + i\Delta \mathbf{v}$ is the discrete approximation to the *inhomogeneous part* of the complex field, i.e. $\Delta z(\mathbf{r}) = z(\mathbf{r}) - V^{-1} \int d\mathbf{r}\, z(\mathbf{r})$, and $\mathbf{g}_D * \Delta \mathbf{w}$ denotes a discrete convolution sum obtained by collocating the convolution integral $\int d\mathbf{r}'\, g_D(|\mathbf{r} - \mathbf{r}'|)\Delta w(\mathbf{r}')$. As with eqn (5.92), this semi-implicit algorithm can be efficiently solved for $\mathbf{z}^{j+1}$ by applying a discrete Fourier transform pair and with a single evaluation of the complex density operator $\boldsymbol{\rho}(i\mathbf{z}^j)$.

While the semi-implicit CL algorithm (6.177) has the same accuracy as the EM scheme, namely time discretization errors that are $O(\Delta t)$ in the weak sense, it has dramatically improved stability, which allows for larger time steps. Indeed, this is currently the algorithm of choice for complex Langevin simulations of polymer field theories.

Figure 6.11 compares the normalized autocorrelation function of the energy $H_S(\boldsymbol{\omega})$ for the shifted canonical model A, evaluated both by smart Monte Carlo (SMC) and complex Langevin (CL) simulations. The simulation conditions were the same as in Fig. 6.10; namely, a cubic lattice was applied with side length $L = 1R_{g0}$, $16^3$ grid points, and $BC = 10$. The time $t$ is measured in numbers of Monte Carlo steps and CL time steps, respectively, for the SMC and CL simulations. The parameter $\lambda \Delta t$ in the CL simulations was fixed at $0.1\Delta$, where $\Delta$ is the corresponding trial step size in SMC. This choice, with $\Delta$ selected to produce an acceptance ratio of 0.5 in SMC, results in an error of less than 1% in the average energy computed using CL.

For the case of $C = 100$, Fig. 6.11 shows that the energy autocorrelation function has a correlation time (fit to a single exponential) of $\tau_{CL} = 48$ deduced from CL and $\tau_{SMC} = 15$ from SMC. The larger correlation time in the CL case is due to the small value of $\lambda \Delta t$ (in comparison with $\Delta$) necessary for accurate

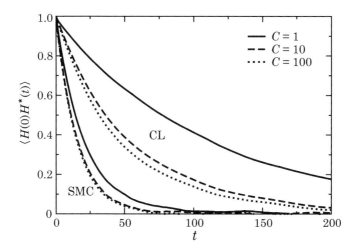

FIG. 6.11. Normalized energy autocorrelation function of the canonical Model A calculated using SMC and CL simulations. The simulations were conducted using the same conditions as in Fig. 6.10, notably at fixed $BC = 10$ and the three $C$ values 1, 10, and 100. Time is measured in SMC steps and CL time steps for the two techniques. Figure courtesy of K. Katsov.

time integration of the CL equations. However, we recall from eqn (6.160) that the average sign factors into the efficiency of the SMC technique for computing physical averages and that $\tau_{SMC}$ ignores the role of the rapidly oscillating phase factor that must be included in such averages. In contrast, the CL technique completely avoids the sign problem. Thus, to put the techniques on equal footing, it is appropriate to compare $\tau_{CL}$ not with $\tau_{SMC}$, but rather with a modified SMC correlation time that accounts for the sign problem inherent to Monte Carlo methods

$$\tilde{\tau}_{SMC} = \frac{\tau_{SMC}}{\langle \text{sign} \rangle^2} \qquad (6.178)$$

For the case of $C = 100$, where $\langle \text{sign} \rangle = 0.89$, it follows that $\tilde{\tau}_{SMC} = 19$, so SMC is clearly favored over CL for this high polymer concentration example.

The situation changes dramatically, however, when $C$ is lowered into the semi-dilute and dilute regimes. For the case of $C = 10$, where $\langle \text{sign} \rangle = 0.30$, the data shown in Fig. 6.11 yield $\tau_{CL} = 55$, $\tau_{SMC} = 16$, and $\tilde{\tau}_{SMC} = 180$. With an average sign this small, the complex Langevin technique is three times more efficient than SMC for evaluating thermodynamic averages. Finally, for $C = 1$, where the average sign is essentially zero, the simulation data provide $\tau_{CL} = 111$ and $\tau_{SMC} = 21$. In spite of maintaining a short correlation time for sampling the real distribution $P(\omega)$, SMC is virtually useless in this case for computing average properties because $\tilde{\tau}_{SMC} \approx \infty$. These examples serve to illustrate the power of the CL simulation technique. While $\tau_{CL}$ grows modestly upon lowering the polymer concentration (and therefore with increasing field fluctuation amplitude), the CL method does not exhibit the catastrophic consequences of the sign problem (divergence of $\tilde{\tau}_{SMC}$) suffered by SMC and other Monte Carlo techniques.

There are a wide variety of alternative numerical algorithms that have been proposed for time integration of stochastic differential equations with the same form as the CL eqns (6.171) (Kloeden and Platen, 1992; Platen, 1999). Unfortunately most higher-order schemes having systematic errors that are $O((\Delta t)^2)$ or smaller require multiple evaluations of the complex force $\mathbf{F}(\mathbf{z})$ at each time step, or worse, higher derivatives of the force – both of which are expensive for large $M$. Since many of these schemes also have stability problems in the high-dimensional context, we favor the use of semi-implicit algorithms such as eqn (6.177). Nevertheless, the application of the CL simulation technique to inhomogeneous polymers is a recent development (Ganesan and Fredrickson, 2001), so improved algorithms will undoubtedly emerge as the subject matures.

There are several extensions of the standard CL equations that could prove important in the polymer field theory context. As described in Appendix D, a *generalized complex Langevin* scheme is

$$\frac{d}{dt}\mathbf{w}(t) = -\boldsymbol{\lambda} \cdot \mathrm{Re}\left[\frac{\partial H(\mathbf{z})}{\partial \mathbf{z}(t)}\right] + \boldsymbol{\eta}_R(t)$$
$$\frac{d}{dt}\mathbf{v}(t) = -\boldsymbol{\lambda} \cdot \mathrm{Im}\left[\frac{\partial H(\mathbf{z})}{\partial \mathbf{z}(t)}\right] + \boldsymbol{\eta}_I(t) \qquad (6.179)$$

where $\boldsymbol{\lambda}$ is an arbitrary symmetric, positive definite $M \times M$ matrix. Gaussian random force terms have also been added to the equations for *both* the real and imaginary components. These random forces are white noise with vanishing means and covariance matrices given by eqn (D.32), i.e.

$$\langle \boldsymbol{\eta}_R(t)\boldsymbol{\eta}_R(t')\rangle = 2(\boldsymbol{\lambda} + \boldsymbol{\epsilon})\,\delta(t - t')$$
$$\langle \boldsymbol{\eta}_I(t)\boldsymbol{\eta}_I(t')\rangle = 2\boldsymbol{\epsilon}\,\delta(t - t')$$
$$\langle \boldsymbol{\eta}_R(t)\boldsymbol{\eta}_I(t')\rangle = 0 \qquad (6.180)$$

The imaginary noise strength $\boldsymbol{\epsilon}$ is a symmetric, positive semi-definite $M \times M$ matrix that can be chosen arbitrarily. As discussed in Appendix D, the inclusion of the random force $\boldsymbol{\eta}_I(t)$ has a *smoothing effect* on the steady state distribution function $P(\mathbf{w}, \mathbf{v})$ of the CL process because it introduces an extra diffusive term

$$\frac{\partial}{\partial \mathbf{v}} \cdot \boldsymbol{\epsilon} \cdot \frac{\partial}{\partial \mathbf{v}} P(\mathbf{w}, \mathbf{v}, t)$$

in the generalized Fokker–Planck equation analogous to eqn (6.174). It can be proven that the generalized CL eqns (6.179) and (6.180) have the same theoretical foundation as the standard CL equations, including the applicability of Theorem 6.1.

The simplest choice of the matrices appearing in the generalized CL equations are $\boldsymbol{\lambda} = \lambda\mathbf{1}$ and $\boldsymbol{\epsilon} = \epsilon\mathbf{1}$, where $\lambda > 0$ and $\epsilon \geq 0$ are real scalars. For $\epsilon = 0$ this reduces to the standard CL scheme. However, the use of small positive values of $\epsilon$ could influence (and possibly improve) the performance of CL algorithms due to the dispersive effect on the steady state distribution.

Another potentially important choice of the matrices $\boldsymbol{\lambda}$ and $\boldsymbol{\epsilon}$ are translationally invariant forms that can be diagonalized by a discrete Fourier transform, thereby preserving the ability to conduct efficient pseudo-spectral simulations (Batrouni et al., 1985). By this technique it is possible to employ noise statistics with a power spectrum weighted towards low spatial wavenumbers $k$. Preliminary results (Lennon et al., 2005) indicate that such a "Fourier acceleration" strategy produces smoother field configurations and better performance of discrete time algorithms.

It is important to emphasize that the complex Langevin dynamics scheme is not designed to be a physically realistic polymer dynamics. Nonetheless, the CL method lends itself more naturally to non-equilibrium extensions than Monte Carlo techniques. Specifically, the diffusive dynamics approaches of Section 5.3.6 can be mated with the CL technique to embed exact equilibrium properties into a time-dependent description (Fredrickson, 2002). A complex Langevin extension of the "$\mu$-explicit" dynamics eqn (5.147), for example, is the CL scheme

$$\frac{\partial}{\partial t}w(\mathbf{r}, t) = \Lambda\nabla^2 \operatorname{Re}\left[\frac{\delta H[z]}{\delta z(\mathbf{r}, t)}\right] + \eta(\mathbf{r}, t)$$
$$\frac{\partial}{\partial t}v(\mathbf{r}, t) = \Lambda\nabla^2 \operatorname{Im}\left[\frac{\delta H[z]}{\delta z(\mathbf{r}, t)}\right] \tag{6.181}$$

with

$$\langle\eta(\mathbf{r}, t)\rangle = 0$$
$$\langle\eta(\mathbf{r}, t)\eta(\mathbf{r}', t')\rangle = -2\Lambda\,\nabla^2\delta(\mathbf{r} - \mathbf{r}')\,\delta(t - t') \tag{6.182}$$

Indeed, upon spatial discretization, these CL equations are seen to be a special case of the generalized CL eqns (6.179)–(6.180) with $\epsilon = 0$ and a translationally invariant matrix $\boldsymbol{\lambda}$ that can be diagonalized by Fourier transformation.

## 6.5 Other numerical techniques

There are a number of other numerical techniques that have proven effective against the sign problem in specific applications. In the context of quantum chemistry, for example, specialized methods have been developed that invoke filtering and sparse sampling methods to suppress phase oscillations in simulations of chemical reaction processes (Filinov, 1986; Doll and Freedman, 1988; Makri and Miller, 1987; Sabo et al., 2002). Here we discuss a few techniques that have potentially broad applicability in the polymer and complex fluids context.

### 6.5.1  *Optimal path sampling*

In Section 6.3 we described Monte Carlo algorithms for simulating polymer field theories that exploited a rather simple strategy for fighting the sign problem; namely, shifting the integration path of the field $w(\mathbf{r})$ from the real axis to a parallel path that passes through through a saddle point of the model. This shift produces a new theory expressed in terms of a field $\omega(\mathbf{r})$, defined according to eqn (6.141), which is also integrated along the real axis. To the extent that the parallel path through the saddle point is locally a constant phase ascent path, the saddle point is the relevant one, and fluctuations about the saddle point remain small, this strategy can be effective.

Unfortunately, the restriction of the shifted path technique to situations where fluctuations are weak, and the requirement of pre-computing saddle point configurations severely limits its utility. It would clearly be desirable to have a Monte Carlo procedure designed so that the shifted path is not fixed in advance by the user, but is *adaptively* identified during the course of a simulation so as to minimize the problematic phase oscillations. With such a scheme, the optimal path need not pass through a saddle point, especially if fluctuations are strong, but rather correspond to a sampling path in the complex plane along which the phase $H_I$ is as constant as possible in some average sense.

An *optimal path sampling (OPS)* method with the above characteristics[110] was recently devised by Moreira *et al.* (2003). The basic idea is to exploit the same *parallel* shift in integration contour that was introduced in Section 6.3.1, but onto a contour $\Gamma_\mathbf{v}$ defined by an arbitrary imaginary field component $\mathbf{v}$:

$$\mathbf{z} = \mathbf{w} + i\mathbf{v}, \quad \mathbf{w} \in \mathbb{R}^M, \quad \mathbf{v} \in \mathbb{R}^M \text{ prescribed} \qquad (6.183)$$

Thus, for a discrete field theory defined on the shifted integration path $\Gamma_\mathbf{v}$, each component of the $M$-vector $\mathbf{w}$ is integrated along the whole real axis while the corresponding component of $\mathbf{v}$ is fixed at a value not necessarily coincident with a saddle point, $i\mathbf{v} = \mathbf{w}^*$. The parallel contour $\Gamma_\mathbf{v}$ can be either homogeneous or inhomogeneous in the components of $\mathbf{v}$, depending on whether $v_k$ varies with $k = 1, 2, ..., M$. Because $\exp[-H(\mathbf{z})]$ is an analytic function of $\mathbf{z}$ in the strip bounding the real axis and the contour $\Gamma_\mathbf{v}$ for all the models considered in Chapter 4, Cauchy's theorem (Ahlfors, 1979) can be used to rewrite the discrete partition function of eqn (6.124) as[111]

$$\mathcal{Z} \propto \int d\mathbf{w} \, \exp[-H(\mathbf{w})] = \int_{\Gamma_0} d\mathbf{z} \, \exp[-H(\mathbf{z})]$$

$$= \int_{\Gamma_\mathbf{v}} d\mathbf{z} \, \exp[-H(\mathbf{z})] \qquad (6.184)$$

---

[110]A method that is similar in spirit to the OPS, but that lacks a variational basis, is the GER technique of Baeurle (2002).

[111]This result also relies on $\exp[-H(\mathbf{z})]$ decaying sufficiently rapidly for $w_k \to \pm\infty$ on the contour $\Gamma_\mathbf{v}$.

where we do not distinguish the canonical and grand canonical cases. Thus, the partition function is independent of the contour shift $\mathbf{v}$.

If one follows a similar argument, the ensemble average of an analytic observable $G(\mathbf{w})$ can be rewritten in a form analogous to eqn (6.129):

$$
\begin{aligned}
\langle G(\mathbf{w}) \rangle &= \mathcal{Z}^{-1} \int_{\Gamma_0} d\mathbf{z}\, G(\mathbf{z}) \exp[-H(\mathbf{z})] \\
&= \frac{\langle G(\mathbf{w} + i\mathbf{v}) \exp[-iH_I(\mathbf{w}, \mathbf{v})] \rangle_{\mathbf{v}}}{\langle \exp[-iH_I(\mathbf{w}, \mathbf{v})] \rangle_{\mathbf{v}}}
\end{aligned}
\tag{6.185}
$$

where the average $\langle ... \rangle_{\mathbf{v}}$ is defined by

$$
\langle h(\mathbf{w}) \rangle_{\mathbf{v}} \equiv \int d\mathbf{w}\, P_{\mathbf{v}}(\mathbf{w}) h(\mathbf{w})
\tag{6.186}
$$

using the normalized real, non-negative probability density

$$
P_{\mathbf{v}}(\mathbf{w}) \equiv \frac{\exp[-H_R(\mathbf{w}, \mathbf{v})]}{\int d\mathbf{w}\, \exp[-H_R(\mathbf{w}, \mathbf{v})]}
\tag{6.187}
$$

The significance of eqn (6.185) is that it holds for *any* $\mathbf{v}$ and it provides a recipe for computing a desired average using a probability weight $P_{\mathbf{v}}(\mathbf{w})$ that can be sampled using Monte Carlo methods.

The key to the OPS method is an *optimal* choice of contour shift $\mathbf{v}$. We have already encountered two naive choices, $\mathbf{v} = 0$ and $i\mathbf{v} = \mathbf{w}^*$. Setting $\mathbf{v} = 0$ recovers eqn (6.129), which we have already noted has a serious sign problem. A better choice is to shift through a relevant saddle point so that $i\mathbf{v} = \mathbf{w}^*$ and eqn (6.185) reduces to eqn (6.139). While this scheme can perform well for weakly fluctuating systems, the constant phase condition breaks down when fluctuations become large in amplitude and the sign problem reemerges. In the OPS method we instead seek an optimal shift $\bar{\mathbf{v}}$ that adjusts to the strength of the fluctuations and attempts to minimize phase oscillations in a *global* way using the *statistically relevant* part of the sampling path. The basis for the method is a variational principle for the following generating function (cf. Appendix B):

$$
\mathcal{G}(\mathbf{v}) = \ln \int d\mathbf{w}\, \exp[-H_R(\mathbf{w}, \mathbf{v})]
\tag{6.188}
$$

The first derivative of this function can be written

$$
\frac{\partial \mathcal{G}(\mathbf{v})}{\partial \mathbf{v}} = - \left\langle \frac{\partial}{\partial \mathbf{v}} H_R(\mathbf{w}, \mathbf{v}) \right\rangle_{\mathbf{v}} = \left\langle \frac{\partial}{\partial \mathbf{w}} H_I(\mathbf{w}, \mathbf{v}) \right\rangle_{\mathbf{v}}
\tag{6.189}
$$

where the Cauchy–Riemann equations have been invoked in the final expression (Ahlfors, 1979), assuming that $H(\mathbf{z})$ is analytic[112] on the contour $\Gamma_{\mathbf{v}}$. Similarly,

---

[112]We remind the reader that the Hamiltonian for any model formulated in the grand canonical ensemble is an analytic function of $\mathbf{z}$. In the canonical ensemble, it may be necessary to adjust the branch cut in the term $\ln Q(i\mathbf{z})$ to render $H(\mathbf{z})$ analytic along the paths $\Gamma_{\mathbf{v}}$ of interest.

by repeated application of the Cauchy–Riemann equations and an integration by parts, the $M \times M$ matrix of second derivatives of $\mathcal{G}$ can be written

$$\frac{\partial^2 \mathcal{G}(\mathbf{v})}{\partial \mathbf{v} \partial \mathbf{v}} = \left\langle \left[ \frac{\partial}{\partial \mathbf{w}} H_I - \left\langle \frac{\partial}{\partial \mathbf{w}} H_I \right\rangle_{\mathbf{v}} \right] \left[ \frac{\partial}{\partial \mathbf{w}} H_I - \left\langle \frac{\partial}{\partial \mathbf{w}} H_I \right\rangle_{\mathbf{v}} \right] \right\rangle_{\mathbf{v}}$$
$$+ \left\langle \frac{\partial}{\partial \mathbf{w}} H_R \frac{\partial}{\partial \mathbf{w}} H_R \right\rangle_{\mathbf{v}} \tag{6.190}$$

This expression is positive definite, because it is the sum of two positive definite forms. Thus, we conclude that the function $\mathcal{G}(\mathbf{v})$ is a *convex* function of the $M$-vector $\mathbf{v}$.

The significance of the above result is that $\mathcal{G}(\mathbf{v})$ has a *unique minimum* for some vector $\bar{\mathbf{v}}$ satisfying

$$\left. \frac{\partial \mathcal{G}(\mathbf{v})}{\partial \mathbf{v}} \right|_{\bar{\mathbf{v}}} = \left\langle \frac{\partial}{\partial \mathbf{w}} H_I(\mathbf{w}, \bar{\mathbf{v}}) \right\rangle_{\bar{\mathbf{v}}} = 0 \tag{6.191}$$

We see from this expression that the extremum path $\Gamma_{\bar{\mathbf{v}}}$ with shift $\mathbf{v} = \bar{\mathbf{v}}$ is an attractive sampling path because $H_I$ has vanishing gradients in an average sense over the path. This is a type of *global stationary phase condition*, as opposed to the *local* stationary phase condition that is familiar in steepest descent analysis (Bender and Orszag, 1978). Indeed, this choice of sampling path is *optimal* because because the sign problem associated with the evaluation of eqn (6.185) is minimized when the global stationary phase condition is met.

There is a close connection between the OPS strategy and the complex Langevin approach discussed in the last section. In particular, if a CL simulation based on eqns (6.171) settles down to a near steady state value of $\mathbf{v}$ satisfying the condition (6.173), then that $\mathbf{v}$ should approximately correspond to $\bar{\mathbf{v}}$. The CL sampling trajectories, however, are *not restricted to parallel paths along the real axis*, so the CL technique is conceptually more powerful than the OPS method. Nevertheless, the OPS method has the advantage of permitting Monte Carlo sampling and thus avoiding the accuracy and stability concerns of CL simulations.

For bulk systems described by models with translationally invariant Hamiltonians $H(\mathbf{w})$, the optimal contour shift $\bar{\mathbf{v}}$ is *homogeneous*, which amounts to a constant shift in field. However, for bound systems, such as Model A confined to a slit, the optimal shift $\bar{\mathbf{v}}$ will be *inhomogeneous* with a non-trivial dependence of $\bar{v}_k$ on field component $k$.

A remaining task is to discuss how to identify the optimal sampling path $\bar{\mathbf{v}}$ during the course of an OPS simulation. This is best explained in the context of an algorithm (Moreira *et al.*, 2003):

### Optimal path sampling Algorithm 6.3

1. Initialize the field configuration $\mathbf{w}$ and make an initial guess for the shift that defines the parallel sampling path, $\mathbf{v} = \mathbf{v}^j$ with $j = 0$.

2. At fixed sampling path $\mathbf{v}^j$, conduct a stochastic simulation to importance sample the distribution $P_{\mathbf{v}^j}(\mathbf{w})$ and generate a Markov chain of $\mathbf{w}$ field configuration states with length $N_T$. The simulation method could be the Metropolis MC Algorithm 6.1, the Smart MC Algorithm 6.2, or a Brownian dynamics scheme, among others.

3. Evaluate the derivative $\partial \mathcal{G}(\mathbf{v}^j)/\partial \mathbf{v}^j$ by approximating the ensemble average in eqn (6.189) with a "time" average over the $N_T$ states accumulated in the simulation.

4. Update $\mathbf{v}$ to approach the optimal shift $\bar{\mathbf{v}}$ by making a steepest descent step

$$\mathbf{v}^{j+1} = \mathbf{v}^j - \lambda_v \frac{\partial \mathcal{G}(\mathbf{v}^j)}{\partial \mathbf{v}^j}$$

where $\lambda_v > 0$ is an adjustable real relaxation parameter.

5. Repeat steps 2–4 for $j = 1, 2, 3, \ldots$ until the sequence of $\mathbf{v}^j$ converges to $\bar{\mathbf{v}}$ to within some prescribed tolerance. The simulation has now equilibrated.

6. Carry out a long stochastic simulation (production run) with $\mathbf{v} = \bar{\mathbf{v}}$ to sample the optimal distribution $P_{\bar{\mathbf{v}}}(\mathbf{w})$ and generate a total of $N_C \gg N_T$ field configuration states.

7. Compute averages of quantities of interest using eqn (6.185) and approximating the ensemble averages in numerator and denominator by time averages over the $N_C$ state Markov chain.

The above algorithm is fully adaptive because the optimal sampling path need not be specified in advance, but is computed during the course of a simulation. The parameters $\lambda_v$ and $N_T$ can be adjusted to accelerate convergence to the optimal path, while maintaining numerical stability. It should be noted that the derivative required in step 3 amounts to an average of the real part of the complex segment density operator $\rho(i\mathbf{z})$ over the sampling path. For example, in the case of the canonical Model A, it follows from eqn (6.175) that

$$\frac{\partial \mathcal{G}(\mathbf{v})}{\partial \mathbf{v}} = \left\langle \frac{\partial}{\partial \mathbf{w}} H_I \right\rangle_{\mathbf{v}} = \left\langle \mathrm{Im} \frac{\partial H(\mathbf{z})}{\partial \mathbf{z}} \right\rangle_{\mathbf{v}}$$
$$= (\Delta x)^3 \left[ \frac{1}{u_0} \mathbf{v} + \langle \boldsymbol{\rho}_R(i\mathbf{z}) \rangle_{\mathbf{v}} \right] \tag{6.192}$$

Thus it is necessary to calculate the complex density operator for the field configurations generated in step 2 of the algorithm. Since the deterministic force along the sampling path involves the imaginary part of the complex density operator, which is a byproduct of this calculation, cf. eqn (6.143), there is a strong incentive to use a force-bias MC scheme during the *equilibration* phase. However, once the optimal path has been identified, a Metropolis algorithm can be used for the production run (step 6) if the extra computational burden of the force-bias scheme is not compensated by proportionally more efficient sampling.

While the OPS method performs quite well for relatively simple models of the Model A type that have only homogeneous saddle points in the bulk, it is not nearly as effective as the complex Langevin technique for models such as Models C, D, E, and J that possess additional inhomogeneous saddle points. One explanation is that the optimal shift $\bar{\mathbf{v}}$ is *homogeneous* in a bulk simulation with periodic boundary conditions, even if the relevant saddle point is inhomogeneous. In contrast, the CL technique generates inhomogeneous $\mathbf{v}$ fields in such situations. Secondly, as is shown in Section 6.5.2 below, parallel paths passing through the saddle points of such models may not be locally constant phase paths, so the restriction to parallel paths may be limiting the performance of the OPS method. In general, the CL method is recommended over the OPS technique for field-theoretic simulations of models that exhibit structured inhomogeneous phases, including mesophases and mixed states of coexisting macrophases and/or mesophases.

### 6.5.2 *Partial saddle point approximation*

Another useful numerical strategy derives from an analytical approximation that we shall refer to as the *partial saddle point approximation*. This approximation is potentially applicable to models in the class of Models B, C, D, E, G, and J that have one or more "pressure-like" fields and one or more "exchange-like" potential fields. In the case of a homopolymer blend, for example, Model C involves a fluctuating pressure field $w_+(\mathbf{r})$ that serves to enforce the local incompressibility constraint and a second exchange potential $w_-(\mathbf{r})$ that determines the instantaneous composition pattern in the melt. We recall that pressure-like fields arise from decoupling purely repulsive interactions and have saddle point values that are *pure imaginary*. Correspondingly, exchange-like fields arise from attractive interactions and have saddle point values that are *purely real*. The partial saddle point approximation amounts to an approximation of a field theory where the functional integrals over the pressure-like fields are evaluated using the saddle point (mean-field) approximation, while the exchange potential integrals are left in original form. For example, in the case of the canonical Model C, the partial saddle point approximation amounts to

$$\mathcal{Z}_C = \mathcal{Z}_0 \int \mathcal{D}w_+ \int \mathcal{D}w_- \ \exp(-H[w_+, w_-])$$

$$\approx \mathcal{Z}_0 \int \mathcal{D}w_- \ \exp(-H_p[w_-]) \tag{6.193}$$

where $H_p[w_-]$ is an effective Hamiltonian for the approximate theory defined by

$$H_p[w_-] \equiv H[w_+^*, w_-] \tag{6.194}$$

In this expression the *partial saddle point* $w_+^*(\mathbf{r}; [w_-])$ is obtained as the solution of

$$\left. \frac{\delta H[w_+, w_-]}{\delta w_+(\mathbf{r})} \right|_{w_+ = w_+^*} = 0 \tag{6.195}$$

for an arbitrary, but prescribed, real value of the field $w_-(\mathbf{r})$. In the case of Model C, this equation amounts to

$$\tilde{\rho}_A(\mathbf{r}; [iw_+^* - w_-]) + \tilde{\rho}_B(\mathbf{r}; [iw_+^* + w_-]) = \rho_0 \tag{6.196}$$

and enforces the local incompressibility condition.

A notable feature of the above approximation is that $w_+^*$ is pure imaginary, which renders the Hamiltonian $H_p[w_-]$ strictly *real*. The *sign problem is thereby eliminated*, and the approximate theory, upon spatial discretization, can be simulated using conventional Monte Carlo or Brownian dynamics techniques. An important question is whether $H_p$ is well defined, i.e. whether a *unique* solution $w_+^*$ of eqn (6.195) exists for any prescribed exchange field $w_-$. Numerical evidence supports the uniqueness of $w_+^*$, although we are not aware of a mathematical proof at the present time. Physically, a unique mean-field pressure is expected for a defined composition pattern $\sim \tilde{\rho}_A(\mathbf{r}) - \tilde{\rho}_B(\mathbf{r})$. The multiplicity of solutions of the *full* saddle point equations

$$\left. \frac{\delta H[w_+, w_-]}{\delta w_+(\mathbf{r})} \right|_{w_+^*, w_-^*} = \left. \frac{\delta H[w_+, w_-]}{\delta w_-(\mathbf{r})} \right|_{w_+^*, w_-^*} = 0 \tag{6.197}$$

arises because of the existence of metastable configurations of the exchange field $w_-$ that dictates the composition pattern. For a prescribed $w_-$ it is therefore reasonable to expect a unique solution of the partial saddle point eqn (6.195).

Of more serious concern is the *validity* of the partial saddle point approximation. The approximation has been applied in the literature in a few instances without justification (Reister *et al.*, 2001) and has been quantitatively examined only in the context of an incompressible ternary blend model (Düchs *et al.*, 2003). In general we expect that the accuracy of the approximation will depend on the nature of the model, as well as on specific model parameters. To gain some insight into the approximation for Model C, it is helpful to expand the theory about the homogeneous (disordered phase) saddle point. In the symmetric case of $b_A = b_B \equiv b$ and $N_A = N_B \equiv N$, the methods of Section 6.1.1 can be used to show that Model C can rewritten as

$$\mathcal{Z}_C \propto \int \mathcal{D}\Omega_+ \int \mathcal{D}\Omega_- \exp(-H_S[\Omega_+, \Omega_-]) \tag{6.198}$$

where $\Omega_\pm(\mathbf{r})$ are shifted versions of the scaled fields $Nw_\pm(\mathbf{r})$ so that the homogeneous saddle point is located at the origin. When expanded to quadratic order about the saddle point, the shifted Hamiltonian can be written

$$\begin{aligned} H_S[\Omega_+, \Omega_-] = {} & \frac{C}{2\widetilde{V}} \sum_{\mathbf{k}} \hat{g}_D(k^2)\, \hat{\Omega}_+(\mathbf{k})\hat{\Omega}_+(-\mathbf{k}) \\ & + \frac{C}{2\widetilde{V}} \sum_{\mathbf{k}} \left[ \frac{2}{\chi_{AB}N} - \hat{g}_D(k^2) \right] \hat{\Omega}_-(\mathbf{k})\hat{\Omega}_-(-\mathbf{k}) \\ & + \frac{iC(\phi_{A0} - \phi_{B0})}{\widetilde{V}} \sum_{\mathbf{k}} \hat{g}_D(k^2)\, \hat{\Omega}_+(\mathbf{k})\hat{\Omega}_-(-\mathbf{k}) \\ & + O(\hat{\Omega}_\pm^3) \end{aligned} \tag{6.199}$$

where the dimensionless variables of Section 6.1.1 have been employed. In this case the dimensionless chain concentration is defined as $C \equiv \rho_0 R_{g0}^3/N$, and $\phi_{K0}$ denotes that average volume fraction of segment species $K = A$ or $B$ in the blend. It is important to note that the integration path for the fields $\Omega_\pm$ is the real axis. The final term in eqn (6.199) shows that *only* in the case of a symmetric blend composition, $\phi_{A0} = \phi_{B0} = 1/2$, is this path through the homogeneous saddle point a constant phase path with $H_{SI} = 0$. Thus, without a further rotation of the fields onto a constant phase ascent path, the sign problem is in evidence for Model C already at quadratic order.

If we assume weak fluctuations near the homogenous saddle point, the partial saddle point associated with Model C is to linear order

$$i\hat{\Omega}_+^*(\mathbf{k}; [\Omega_-]) = (\phi_{A0} - \phi_{B0})\hat{\Omega}_-(\mathbf{k}) + O(\hat{\Omega}_-^2) \tag{6.200}$$

and the partial saddle point approximation amounts to

$$\mathcal{Z}_C \approx \int \mathcal{D}\Omega_- \ \exp(-H_p[\Omega_-]) \tag{6.201}$$

$$H_p[\Omega_-] = \frac{C}{\tilde{V}} \sum_{\mathbf{k}} \left[ \frac{1}{\chi_{AB}N} - 2\phi_{A0}\phi_{B0}\,\hat{g}_D(k^2) \right] \hat{\Omega}_-(\mathbf{k})\hat{\Omega}_-(-\mathbf{k})$$
$$+ O(\hat{\Omega}_-^3) \tag{6.202}$$

It is easily confirmed that to the order shown, averages such as $\langle \hat{\Omega}_-(\mathbf{k})\hat{\Omega}_-(-\mathbf{k}) \rangle$ computed with the statistical weight of the approximate theory (6.201) exactly correspond to averages computed with the full theory (6.198). Beyond a Gaussian treatment of fluctuations, this exact equivalence is no longer expected to hold.

An important feature of eqn (6.199) is that the parameter $C$ multiplies all of the terms in the effective Hamiltonian. This is actually a more general result that holds to all orders in $\hat{\Omega}_\pm$. The magnitude of $C$ thus dictates the strength of the fluctuations in *both* the $\Omega_+$ and $\Omega_-$ fields, and only for $C \to \infty$ are fluctuations in the two fields completely suppressed. At least for Model C, there is apparently no independent "Ginzburg parameter" that can be used to justify making a partial saddle point approximation on the $\Omega_+$ field, while retaining full fluctuations in the $\Omega_-$ field.

In spite of this disappointing analytical observation, numerical studies (Düchs et al., 2003) suggest that the partial saddle point approximation can be quite accurate, at least for copolymer melt and blend models that are composition-ally symmetric and where $C$ is reasonably large ($C \gtrsim 50$). In new applications, it is important that results obtained using the partial saddle point approxima-tion be confirmed by means of independent MC or CL simulations on the full (unapproximated) field theory.

Finally, we discuss how to implement the partial saddle point approximation. A force-bias, or hybrid Monte Carlo, simulation is recommended for importance sampling of the $\mathbf{w}_-$ field configurations in a spatially discrete theory. For each trial configuration $\mathbf{w}_-$, eqn (6.195) can be solved for the real field $\boldsymbol{\mu}_+^* \equiv i\mathbf{w}_+^*$ by applying the relaxational dynamics of eqn (5.98). Specifically, a few iterations of the semi-implicit scheme (5.104) should be sufficient to update the partial saddle point approximation to the pressure $\boldsymbol{\mu}_+^*$ from its value at the previous $\mathbf{w}_-$ configuration. The energy can subsequently be evaluated according to $H_p(\mathbf{w}_-) = H(\boldsymbol{\mu}_+^*, \mathbf{w}_-)$.

From the standpoint of computational cost, the multiple solutions of the diffusion equation and force (density operator) evaluations required to satisfy eqn (6.195) are limiting. If 5–10 iterations of eqn (5.104) are required for each $\mathbf{w}_-$ configuration, then a partial saddle point scheme implemented with smart MC would be 5–10 times more expensive than a corresponding CL simulation of the full field theory (for equal numbers of MC steps and CL time steps). This would be worthwhile only if the smart MC algorithm were able to explore phase space 5–10 times more efficiently than CL sampling. Thus, in spite of having avoided the sign problem in the B–E class of models, it remains to be seen whether the partial saddle point approximation offers any true computational advantage. Moreover, the validity of the approximation must be tested in any case.

## 6.6 Lattice cutoff and finite size effects

Many of the considerations and techniques for preparing, running, and analyzing computer simulations of particle-based models and lattice spin models (Allen and Tildesley, 1987; Frenkel and Smit, 1996; Landau and Binder, 2000) carry over to numerical simulations of polymer field theories. However, field-theoretic simulations require additional attention to *spatial discretization effects*, specifically the role of the short-distance cutoff imposed by the computational grid and the delineation of universal and non-universal average properties. In the case of universal properties, it is also desirable to confirm that results obtained from a series of simulations conducted with increasing spatial resolution converge to a well-defined continuum limit.

### 6.6.1 *Lattice cutoff effects*

We observed in Section 6.1 that ultraviolet divergences are present in continuum polymer field theories, even within a Gaussian treatment of field fluctuations. These divergences arise from a breakdown of the mesoscopic models at a length scale comparable to and below the segment size. Beneath this scale, atomistic details become important and harsh repulsive interactions set up a rich liquid structure (Hansen and McDonald, 1986) that is not accurately described by a coarse-grained model. In the analytical calculations of Section 6.1.1, we removed the ultraviolet divergences, i.e. "regularized" the continuum field theory, by introducing an upper momentum cutoff $\Lambda$ and retaining only the Fourier modes of the fluctuating field with wavenumbers $k = |\mathbf{k}|$ inside the sphere $k = \Lambda$. By means of this hard cutoff in Fourier space, the fields were constrained to fluctuate with a minimum wavelength of $2\pi/\Lambda$. The Wilson-style renormalization group theory described in Section 6.1.5.2 further showed how to systematically distinguish universal properties from non-universal properties in such a regularized theory by studying their $\Lambda$-dependence upon repeated application of the renormalization group.

In a field-theoretic simulation, an ultraviolet cutoff is automatically imposed due to the computational necessity of a finite representation of each field. When one applies finite differences, the cutoff is naturally imposed by the underlying lattice in real space, while the magnitude of the largest reciprocal lattice vector fixes the cutoff in a discrete spectral approximation to a field. In the pseudo-spectral approach that we favor, the distinction between real and reciprocal space is blurred because the lattice constant $\Delta x$ of an underlying uniform collocation grid also sets the maximum wavenumber of field fluctuations. Thus, in the subsequent discussion we shall refer to $\Delta x$ as the ultraviolet or lattice cutoff.

To date, only one systematic study of lattice cutoff effects in field-theoretic polymer simulations has been conducted (Alexander-Katz *et al.*, 2005). These authors used the complex Langevin technique to simulate the canonical Model A in two situations: (i) confined to a cubic simulation cell with periodic boundary conditions and (ii) confined to a three-dimensional slit with Dirichlet boundary conditions applied on two opposing faces of the cell and periodic boundary conditions on the remaining four faces. With the introduction of the scalings and dimensionless variables of Section 6.1.1, average properties for the bulk simulations depend on *four* dimensionless parameters: the chain concentration $C$, the excluded volume parameter $B$, the length of a side of the cell $L$ (in units of $R_{g0}$), and the collocation grid spacing $\Delta x$ (also in units of $R_{g0}$). In the case of the slit simulations, there is an additional fifth parameter associated with the lateral size $L_{\parallel}$ of the cell, while $L$ denotes the separation between the surfaces that bound the slit. The contour step $\Delta s$ used in the integration of the diffusion eqn (6.5) in all cases was refined to the accuracy of the simulation, so it does not appear as an independent parameter.

The bulk simulations conducted by Alexander-Katz *et al.* (2005) were used to investigate a dimensionless "excess" chemical potential $\mu_E$, defined in accordance with eqn (4.34) as

$$\mu_E \equiv -\langle \ln Q[iW] \rangle \qquad (6.203)$$

There is a mean-field contribution to the excess potential, $\mu_{MF} = BC$ [cf. eqn (6.27)], so the contribution to the excess chemical potential from $W$ field *fluctuations* is

$$\mu_{fluc} \equiv \mu_E - \mu_{MF} = -\langle \ln Q[iW] \rangle - BC \qquad (6.204)$$

It is important to emphasize that $\mu_E$ and $\mu_{fluc}$ are referenced to the infinite dilution (ideal gas) state of non-interacting polymers *without* excluded volume. In contrast, the excess chemical potential $\tilde{\mu}_E$ introduced in eqn (6.28) has as its reference the infinite dilution state of swollen polymers *with* excluded volume.

Some insight into the ultraviolet divergences expected in a numerical simulation of Model A in three dimensions can be gained from the Gaussian fluctuation analysis of Section 6.1.1. In particular, eqn (6.27) predicts that for weak fluctuations

$$\mu_{fluc} \approx \frac{B}{2L^3} \sum_{\mathbf{k}} \frac{\hat{g}_D(k^2)}{1 + BC\,\hat{g}_D(k^2)} \tag{6.205}$$

where the $\mathbf{k} = 0$ term in the sum is omitted. When the reciprocal lattice sum is performed on a cubic collocation grid of spacing $\Delta x$, this expression predicts a linear scaling $\mu_{fluc} \sim B/\Delta x$ for $\Delta x \to 0$. Complex Langevin simulations for bulk ($L = 6.4$) and slit ($L = L_{\parallel} = 6.4$) geometries indicate that this scaling relation holds over a wide range of $C$ and $B/\Delta x$, which is surprising given that the Gaussian approximation is valid only in the concentrated regime. Indeed, we see in Fig. 6.12 that this scaling holds over several orders of magnitude of $B/\Delta x$, even at infinite dilution, $C \to 0$. The figure also shows quantitative agreement with the numerical predictions of eqn (6.205) at infinite dilution.

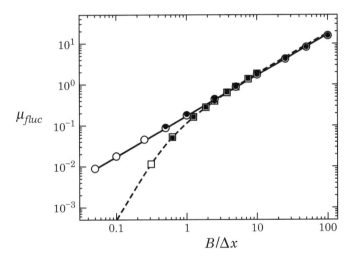

FIG. 6.12. Fluctuation contribution to the chemical potential for $C \to 0$ as a function of the parameter $B/\Delta x$. The filled (slit) and open (bulk) circles correspond to data from CL simulations performed at a fixed lattice constant $\Delta x = 0.2$. The filled (slit) and open (bulk) squares are data from simulations at fixed $B = 1$. The solid and dashed curves were obtained by evaluating the lattice sum in eqn (6.205) for a bulk system numerically. Reproduced from Alexander-Katz *et al.* (2005).

At any concentration $C \geq 0$, CL simulations reveal ultraviolet divergences of the quantities $\mu_E$ and $\mu_{fluc}$ for $\Delta x \to 0$. These divergences can be removed by a simple subtraction analogous to that used in deriving eqn (6.28). In particular, regularized versions of the chemical potentials (in units of $k_B T$) are obtained by referencing the corresponding infinite dilution potentials

$$\tilde{\mu}_E(B, C, L, \Delta x) \equiv \mu_E(B, C, L, \Delta x) - \mu_E(B, 0, L, \Delta x)$$
$$\tilde{\mu}_{fluc}(B, C, L, \Delta x) \equiv \mu_{fluc}(B, C, L, \Delta x) - \mu_{fluc}(B, 0, L, \Delta x) \quad (6.206)$$

These regularized potentials are no longer divergent for $\Delta x \to 0$, as is confirmed in Fig. 6.13 by CL simulation data for the case of $B = C = 1$. The Gaussian approximations for these quantities, e.g.

$$\tilde{\mu}_{fluc} \approx -\frac{B^2 C}{2L^3} \sum_{\mathbf{k}} \frac{[\hat{g}_D(k^2)]^2}{1 + BC\, \hat{g}_D(k^2)}, \quad (6.207)$$

are consistent with the observation that $\tilde{\mu}_E$ and $\tilde{\mu}_{fluc}$ are finite for $\Delta x \to 0$ and are thus *universal*. The inset to the figure, however, indicates a significant departure of the simulation data for $\tilde{\mu}_{fluc}$ from the $\Delta x$ dependence predicted by eqn (6.207).

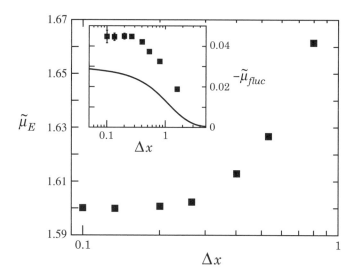

FIG. 6.13. Regularized excess chemical potential $\tilde{\mu}_E$ for $B = C = 1$ and $L = 6.4$ as a function of the lattice cutoff $\Delta x$. Inset: regularized fluctuation contribution to the chemical potential $\tilde{\mu}_{fluc}$. The solid curve corresponds to eqn (6.207). Reproduced from Alexander-Katz *et al.* (2005).

Figure 6.13 not only shows that $\tilde{\mu}_E$ and $\tilde{\mu}_{fluc}$ are universal quantities, but provides physical insight into the lattice spacing $\Delta x$ necessary to obtain results that closely approximate the continuum theory. We note from the figure that both functions plateau at $\Delta x \approx 0.25$. By fitting CL simulation data for the average segment density $\rho(z) = \langle \tilde{\rho}(z; [iW]) \rangle$ of the slit model to an analytical expression based on the ground state dominance approximation, cf. Fig. 6.2, Alexander-Katz *et al.* (2005) obtained an approximation for the bulk correlation length $\xi$ as a function of $\Delta x$. Both $\rho(z)$ and $\xi$ were observed to be universal quantities and for $B = C = 1$, $\xi$ was found to be independent of $\Delta x$ for $\Delta x \lesssim 0.25$. Since the value of $\xi$ was approximately 1.0 in units of $R_{g0}$, we arrive at the following rule of thumb: spatial discretization errors become negligible in Model A when the grid spacing for pseudo-spectral collocation $\Delta x$ is approximately *one quarter* of the bulk correlation length $\xi$. This rule of thumb is fairly insensitive to the precise values of $B$, $C$, and $L$ (provided that $L \gg \xi$) and is consistent with the well known thermodynamic significance of $\xi$ (de Gennes, 1979).[113]

Analogous studies of lattice cutoff effects have not yet been conducted in simulations of polymer models other than Model A. Such investigations will be necessary to fully understand the role of spatial discretization across the full range of models described in Chapter 4.

### 6.6.2 *Finite size effects*

Finite size effects arise from the fact that computer simulations are conducted in a computational cell of finite, rather than infinite, size. This can be an issue if bulk thermodynamic properties are of primary interest and especially if the system has long-ranged interactions or a large correlation length. Phase transitions, both first- and second-order, can also be profoundly influenced by the use of a finite cell size $L$ (Landau and Binder, 2000).

A judicious choice of boundary conditions on the cell can be helpful in minimizing finite size effects. Periodic boundary conditions are usually applied when performing bulk fluid simulations because they tend to produce a faster decay with $L$ of finite size corrections to intensive thermodynamic properties (Allen and Tildesley, 1987; Frenkel and Smit, 1996). Often such corrections decay *exponentially* with $L$ when periodic boundary conditions are imposed. For example, Alexander-Katz *et al.* (2005) observed finite size corrections to the excess chemical potential $\mu_E$ of the canonical Model A consistent with an exponential decay, i.e.

$$\mu_E(B, C, L, \Delta x) \approx \mu_{E\infty}(B, C, \Delta x)$$
$$+ A_E(B, C, \Delta x) \exp\left[-\frac{L}{\xi_D(B, C, \Delta x)}\right] \quad (6.208)$$

where $\mu_{E\infty}$ is the limiting bulk excess chemical potential and $A_E$ is independent of $L$. The finite size decay length $\xi_D$ was found to be comparable to the bulk correlation length $\xi$ at the same values of $B$ and $C$.

*Algebraic* dependence on the system size $L$ of both intensive and extensive quantities is to be expected when conducting computer simulations of fluids near phase transitions, or in other situations where long-ranged interactions or

---

[113]This is discussed in the context of eqn (6.54).

correlations are present. The analysis of finite size effects in these situations is based on the theory of *finite size scaling* (Goldenfeld, 1992). In such analysis, there is nothing that distinguishes simulation data of continuum field theory models from data of lattice spin models or particle-based fluid models. Thus we refer the interested reader to the extensive literature on finite size scaling applied to Monte Carlo simulations (Binder and Heermann, 1988; Landau and Binder, 2000).

## 6.7 Numerical renormalization group

Soon after the advent of renormalization group theory it was recognized that some of the limitations of analytical RG calculations can be overcome by combining RG transformations with Monte Carlo simulations (Ma, 1976; Swendsen, 1979; Swendsen, 1982). In particular, such *Monte Carlo renormalization group (MCRG)* techniques permit RG transformations to be conducted directly in the space dimension of interest, as opposed to relying on approximate epsilon expansions about an upper critical dimension (e.g. $\epsilon = 4 - d$). MCRG techniques also permit a large number of field couplings and interaction parameters to be included in the model Hamiltonian, thus allowing for more accurate calculations of scaling exponents than can be obtained from analytical RG methods.

In the case of polymer field theories we believe that *numerical* RG transformations[114] can be useful in two contexts:

- to evaluate scaling exponents
- to systematically coarse-grain a model, thereby enabling larger-scale simulations than would otherwise be possible

For these purposes it is convenient to express the Hamiltonian of the model in an expansion that is linear in a set of coupling parameters $K_\alpha$, i.e.

$$H[w] = \sum_\alpha K_\alpha S_\alpha[w] \qquad (6.209)$$

where the $S_\alpha[w]$ are functionals of $w(\mathbf{r})$ that describe the types of field interactions in the model. In most MCRG studies to date the $S_\alpha[w]$ reflect local interactions of neighboring spin or discretized field variables. They are thus linear, quadratic, cubic, or quartic forms in the field variables. In contrast, the effective Hamiltonians defining the polymer field theories of Chapter 4, due to contributions from the single-chain partition function $Q[iw]$, have interaction terms of all orders when expanded in powers of $w$. In such cases it is more convenient to attempt to scale the model parameters outside of $Q[iw]$, resulting in a highly non-local contribution to $S_\alpha[w]$. For example, using the scalings of Section 6.1.1 and an $M$-component discrete approximation to the field $w$, $\mathbf{W} \equiv N\mathbf{w} = (W_1, W_2, ..., W_M)^T$, the three-dimensional canonical Model A

---

[114]We shall use the term "numerical RG" rather than "Monte Carlo RG," because complex Langevin simulations can also be applied to implement RG transformations of polymer field theories.

Hamiltonian can be written in the form of eqn (6.209) with two coupling parameters

$$K_1 = (\Delta x)^3/(2B), \quad K_2 = C(\Delta x)^3 \qquad (6.210)$$

and the field interaction terms

$$S_1(\mathbf{W}) = \sum_{j=1}^{M} W_j^2$$
$$S_2(\mathbf{W}) = -M \ln Q(i\mathbf{W}) \qquad (6.211)$$

The partition function $Q(i\mathbf{W})$ can be obtained from the numerical solution of
eqn (6.5) on an $M$-site grid with lattice spacing $\Delta x$.

The numerical RG approach is based on an RG transformation of the form

$$\exp[-H'(\mathbf{W}')] = \int d\mathbf{W}\, T(\mathbf{W}', \mathbf{W}) \exp[-H(\mathbf{W})] \qquad (6.212)$$

where $T(\mathbf{W}', \mathbf{W})$ is a transformation operator that maps the field variables $\mathbf{W}$
of the original field theory onto a reduced set of coarse-grained and rescaled variables $\mathbf{W}'$ in the renormalized theory. The transformation operator is normalized
so that

$$\int d\mathbf{W}'\, T(\mathbf{W}', \mathbf{W}) = 1 \qquad (6.213)$$

which ensures that the partition function is unchanged by the RG transformation. There are several convenient choices for $T$, the most common of which is
the "block spin" approach (Burkhardt and van Leeuwen, 1982)

$$T(\mathbf{W}', \mathbf{W}) = \prod_n \delta \left[ W_n' - \ell^{-3} \sum_{j \in n} W_j \right] \qquad (6.214)$$

where field variables belonging to $\ell^3$ neighboring sites on the lattice are grouped
into "blocks" and $W_n'$ for the $n$th block is obtained by averaging over the $W_j$
values belonging to that block (see Fig. 6.14). A typical choice for a cubic lattice
is $\ell = 2$, so that field values at 8 neighboring grid points are averaged to obtain
each block field variable $W_n'$. It is understood that the original lattice spacing
$\Delta x$ of the computational grid is preserved in this transformation. However, the
overall size of the lattice is reduced by a factor of $\ell^{-1}$ in each direction and
the total number of lattice sites is reduced from $M$ to $M\ell^{-3}$. As an alternative
to eqn (6.214), a momentum space RG transformation analogous to eqn (6.90)
can be implemented by computing the discrete Fourier transform of the field
$\mathbf{W}$, $\hat{\mathbf{W}}$, and then packing $\hat{\mathbf{W}}'$ with the $M\ell^{-3}$ components of $\hat{\mathbf{W}}$ having the
*smallest* wavenumbers $k = |\mathbf{k}|$. Wavevectors are subsequently rescaled according
to $\mathbf{q} = \ell\mathbf{k}$ in order to preserve the original spatial resolution $\Delta x$.

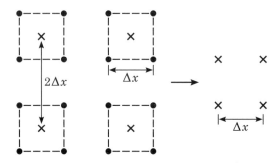

FIG. 6.14. A position space RG transformation for a two-dimensional square lattice with lattice spacing $\Delta x$ and $\ell = 2$. The field variables $W_j$ of the original theory are assigned to the grid points defining the finer lattice (solid dots). The coarse-graining step of the RG transformation is carried out by averaging the field values $W_j$ over the $n$th "block" of four sites according to $W'_n = \ell^{-2} \sum_{j \in n} W_j$. The block field values $W'_n$ are assigned to the staggered lattice denoted by the crosses with lattice constant $2\Delta x$. In the second step of the RG transformation, the lattice spacing of the block variables is reduced by a factor of $\ell = 2$ to recover the original lattice constant.

The renormalized Hamiltonian $H'(\mathbf{W}')$ defined by eqn (6.212) will in general have additional field couplings not present in the starting model $H(\mathbf{W})$. Assuming that it can be expressed in the same type of expansion, we can write

$$H'(\mathbf{W}') = \sum_\alpha K'_\alpha S'_\alpha(\mathbf{W}') \tag{6.215}$$

Truncating both eqn (6.215) and the analogous expression for $H(\mathbf{W})$ after some number of terms $m$ defines an approximate mapping from an $m$-vector of parameters $\mathbf{K}$ in the starting model to an $m$-vector $\mathbf{K}'$ in the renormalized model. For example, a two-term approximation to $H'$ analogous to eqns (6.210)–(6.211) would define a mapping $(B, C) \rightarrow (B', C')$ for the parameters of the canonical Model A.

Ensemble averages can be defined with the original and renormalized theories by applying $\exp(-H)$ and $\exp(-H')$, respectively, as statistical weights. For example, the ensemble average of the operator $S'_\alpha$ in the renormalized theory is defined as

$$\langle S'_\alpha(\mathbf{W}') \rangle = \frac{\int d\mathbf{W}' \, S'_\alpha(\mathbf{W}') \exp[-H'(\mathbf{W}')]}{\int d\mathbf{W}' \, \exp[-H'(\mathbf{W}')]} \tag{6.216}$$

Key to the numerical RG procedure is that such an average can be approximately evaluated by a computer simulation of the *original theory*. By running a Monte Carlo or complex Langevin simulation for the original model at fixed parameters $\mathbf{K}$, a set of $N_C$ field configuration states $\mathbf{W}^j$, $j = 1, 2, ..., N_C$, can be obtained. For each of these states, a corresponding state $(\mathbf{W}')^j$ representative of the renormalized theory is obtained by applying a block spin or momentum space RG transformation. The ensemble average in eqn (6.216) is then approximated as a "time" average over states according to

$$\langle S'_\alpha(\mathbf{W}') \rangle \approx \frac{1}{N_C} \sum_{j=1}^{N_C} S'_\alpha((\mathbf{W}')^j) \tag{6.217}$$

It is understood that to address the sign problem, a complex phase factor may have to be included in eqn (6.217) when applying an MC simulation, or a displacement of $\mathbf{W}'$ off the real axis to $\mathbf{Z}' = \mathbf{W}' + i\mathbf{Y}'$ in the case of a CL simulation.

If the fixed point values of the coupling parameters are known, then a simulation of the critical theory, i.e. with $\mathbf{K} = \mathbf{K}_f$, can be used to compute scaling exponents (Swendsen, 1979). Using the chain rule, we can write

$$\frac{\partial \langle S'_\gamma \rangle}{\partial K_\beta} = \sum_\alpha \frac{\partial \langle S'_\gamma \rangle}{\partial K'_\alpha} \frac{\partial K'_\alpha}{\partial K_\beta} \tag{6.218}$$

However, it follows from eqn (6.212) and explicit differentiation that two of the terms in this expression can be expressed as correlation functions

$$\frac{\partial \langle S'_\gamma \rangle}{\partial K_\beta} = -\langle S'_\gamma(\mathbf{W}') S_\beta(\mathbf{W}) \rangle + \langle S'_\gamma(\mathbf{W}') \rangle \langle S_\beta(\mathbf{W}) \rangle \tag{6.219}$$

$$\frac{\partial \langle S'_\gamma \rangle}{\partial K'_\alpha} = -\langle S'_\gamma(\mathbf{W}') S'_\alpha(\mathbf{W}') \rangle + \langle S'_\gamma(\mathbf{W}') \rangle \langle S'_\alpha(\mathbf{W}') \rangle \tag{6.220}$$

The right-hand sides of these expressions can be evaluated from configurations $\mathbf{W}^j$ and blocked configurations $(\mathbf{W}')^j$ obtained from simulations of the *original* theory at $\mathbf{K} = \mathbf{K}_f$ as described above. It follows that the $m \times m$ RG transformation matrix

$$\mathcal{R}_{\alpha\beta} \equiv \frac{\partial K'_\alpha}{\partial K_\beta} \tag{6.221}$$

can be determined numerically by solving the linear equation set (6.218). If the effective Hamiltonians $H$ and $H'$ are at the fixed point, then the *eigenvalues* of $\mathcal{R}$ provide an approximation to the critical scaling exponents of the model. The quality of this approximation depends on the number $m$ and type of coupling terms that are retained in the expansions (6.209) and (6.215), and the accuracy to which the fixed point coupling parameters are known. Impressive numerical results have been demonstrated for a variety of magnetic spin models (Swendsen, 1982), but to date this procedure has not been applied to polymer field theories.

In the polymer context, a potentially more useful type of numerical RG calculation is concerned with directly computing the renormalized coupling parameters $\mathbf{K}'$, rather than the transformation matrix of eqn (6.221). By monitoring the flow of parameters in RG transformations conducted with states from a "fine scale" field-theoretic simulation, one should be able to systematically coarse-grain a polymer model to a scale $\xi_c$ that is much larger than the grid spacing $\Delta x_1$ used in the simulation. A second simulation of a very large system would then be possible by employing a coarse grid with a spacing $\Delta x_2$ comparable to $\xi_c$ and a Hamiltonian containing renormalized parameters obtained from the first simulation.[115] By such means, it should be possible to tackle simulations of inhomogeneous polymers possessing strong fluctuations over many decades of length scales.

Several methods have been developed for the direct computation of renormalized coupling parameters in MCRG studies (Wilson, 1980; Shenker and Tobochnik, 1980; Hasenfratz and Margaritis, 1983; Hasenfratz *et al.*, 1984; Swendsen, 1984; Callaway and Petronzio, 1984), most of which work only in the close neighborhood of a fixed point. An alternative technique proposed by Falcioni *et al.* (1986) does not have this limitation and appears to be well suited to field-theoretic polymer simulations. The method of Falcioni *et al.* (1986) relies on a discrete version of the familiar Schwinger–Dyson equations, namely the identity

$$\int d\mathbf{W} \frac{\partial}{\partial W_j} (W_{j_1} W_{j_2}...W_{j_n} \exp[-H(\mathbf{W})]) = 0 \qquad (6.222)$$

where we assume that the boundary conditions imposed on the theory, e.g. periodic boundaries, are consistent with the vanishing of surface terms. Equation (6.222) can be rewritten for the case of $j_1 \neq j_2 \neq ... \neq j_n$ as

$$\sum_{k=1}^{n} \delta_{jj_k} \langle W_{j_1}...W_{j_{k-1}} W_{j_{k+1}}...W_{j_n} \rangle - \left\langle W_{j_1}...W_{j_n} \frac{\partial H(\mathbf{W})}{\partial W_j} \right\rangle = 0 \qquad (6.223)$$

For the canonical Model A, the "force" appearing in this expression is

$$F_j(\mathbf{W}) \equiv -\frac{\partial H(\mathbf{W})}{\partial W_j} = -2K_1 W_j - iK_2 \, \phi_j(i\mathbf{W}) \qquad (6.224)$$

where the $K_\alpha$ are related to the parameters $B$ and $C$ by eqn (6.210) and the complex $M$-vector $\boldsymbol{\phi}(i\mathbf{W})$ is obtained by collocating the dimensionless density operator

$$\tilde{\phi}(\mathbf{x}; [iW]) \equiv \frac{1}{Q[iW]} \int_0^1 ds \, q(\mathbf{x}, s; [iW]) q(\mathbf{x}, 1 - s; [iW]) \qquad (6.225)$$

at $M$ spatial grid points using pseudo-spectral solutions of the complex diffusion eqn (6.5).

---

[115]In such a multi-scale simulation strategy, the second rescaling step of the RG transformation would be omitted.

Substitution of such force expressions into eqn (6.223) leads to a (over-determined) set of *linear* equations that can be solved for the coupling parameters. In the specific case of Model A, two linearly independent equations for $n = 0$ and $n = 1$ correspond to

$$2K_1\langle W_j\rangle + iK_2\langle\phi_j(i\mathbf{W})\rangle = 0 \tag{6.226}$$

$$1 - 2K_1\langle W_j^2\rangle - iK_2\langle W_j\phi_j(i\mathbf{W})\rangle = 0 \tag{6.227}$$

for any lattice site $j$. It is easily confirmed that the first of these equations is satisfied in the saddle point approximation where $W_j = -iCB$ and $\phi_j = 1$. Equation (6.227), however, is an exact fluctuation formula that is violated in the mean-field approximation.

Equations (6.226)–(6.227) can be viewed as a system of equations that can be solved for $K_1$ and $K_2$, given analytical or numerical results for the four average quantities appearing in the equations. One way to apply these equations is as a consistency check on field-theoretic simulations. By simulating Model A for some fixed set of parameters $(B, C)$ and hence $(K_1, K_2)$, the four averages can be numerically estimated. Equations (6.226)–(6.227) can then be solved for $K_1$ and $K_2$. The extent to which these derived parameters differ from the original imposed parameters is a measure of the accuracy of the simulation.

A more profound way to use such formulas is to recognize that eqns (6.222) and (6.223) hold also for the renormalized theory (Falcioni *et al.*, 1986). Thus, there is an analogous set of linear equations for Model A involving renormalized couplings and correlation functions of block field variables, i.e.

$$2K_1'\langle W_j'\rangle + iK_2'\langle\phi_j'(i\mathbf{W}')\rangle = 0 \tag{6.228}$$

$$1 - 2K_1'\langle(W_j')^2\rangle - iK_2'\langle W_j'\phi_j'(i\mathbf{W}')\rangle = 0 \tag{6.229}$$

The operator $\phi_j'(i\mathbf{W}')$ is obtained by solving eqn (6.5) on the smaller block variable grid with $\mathbf{W} \to \mathbf{W}'$, but with the same lattice spacing $\Delta x$ as used in the simulations to generate the $\mathbf{W}$ field configurations. We have already discussed how averages and correlation functions of block field variables can be computed from simulations conducted on the original (bare) theory. Thus, by simulating the original field theory with prescribed couplings $(K_1, K_2)$, it is possible to numerically estimate the four averages appearing in eqns (6.228)–(6.229). The solution of these linear equations immediately yields estimates for the renormalized parameters $(K_1', K_2')$. By blocking the original field configurations multiple times, it is thus possible to study the flow of the renormalized coupling parameters upon successive applications of the renormalization group. Numerical RG calculations of this type are currently underway for Model A in both bulk and confined geometries (Katsov and Fredrickson, 2005).

The above numerical renormalization scheme is expected to be particularly useful when simulating *solutions* of polymers that are capable of self-assembly on multiple length scales, such as polyelectrolytes, solution blends, and block and graft copolymers. For example, in the case of a semi-dilute solution of A and B homopolymers in a non-selective solvent S, numerical RG simulations conducted on a fine grid could be used to parameterize a discrete field theory model on a coarse grid with spacing $\Delta x$ comparable to the correlation length $\xi$.[116] Large-scale simulations of this coarse-grained model would then provide access to two-phase morphologies, interfacial properties, and other thermodynamic quantities of interest. Furthermore, it might be possible to invoke the mean-field approximation (SCFT) to lower the computational cost of the coarse-grid simulations, because the "dangerous" excluded volume correlations beneath the $\xi$ scale have already been accounted for in the RG step. Such a procedure amounts to a first-principles strategy for "deriving" the results of Broseta *et al.* (1987) for this ternary system, without invoking the phenomenological renormalization methods of Section 6.1.6.1.

## 6.8 Applications

The field-theoretic computer simulation methods described in the previous sections have only recently been applied to analyze fluctuation effects in polymeric fluids. Thus, the list of applications is not extensive.

The most comprehensive FTS studies to date of any polymer field theory are the investigations by Alexander-Katz and coworkers (Alexander-Katz *et al.*, 2003; Alexander-Katz *et al.*, 2005) of the canonical Model A in bulk, and confined to a three-dimensional slit. Much is known analytically about this model in the homogeneous case, as has been described in Section 6.1, so it is a good testing ground for numerical techniques. In the slit geometry, the mean-field segment density profiles of Model A (Fig. 6.2) depend on the slit width $L$ and the combined parameter $BC$, the latter determining the Edwards correlation length $\xi_E$. By means of complex Langevin simulations, Alexander-Katz *et al.* (2003) showed that when field fluctuations are included, the average segment density is depleted near the walls and enhanced near the center of the slit, in comparison with the mean-field profiles. This observation is consistent with the anticipated excluded volume swelling of the semi-dilute correlation length $\xi$ to values exceeding the Edwards length (de Gennes, 1979). A convenient feature of the slit geometry is that the rise of the average segment density away from a wall provides a straightforward means of measuring the correlation length $\xi$ without the need to analyze spatial correlation functions. Figure 6.15 shows the dependence of the effective correlation length determined in this way on the parameters $B$ and $C$. For $C \gg B$, we observe the expected collapse onto a line that represents the correlation length of the Gaussian theory, i.e. the Edwards length given by eqn (6.20). For $C \ll B$, the simulation data are consistent with the semi-dilute scaling law eqn (6.55).

The first application of the complex Langevin simulation method to a polymer field theory was the study of the order–disorder transition (ODT) of a symmetric

---

[116]Again we note that the rescaling step of the RG transformation would be omitted in such numerical RG calculations.

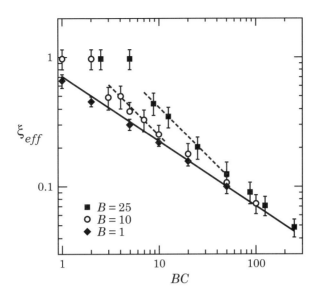

FIG. 6.15. Effective correlation length $\xi_{eff}$ for $B = 1$ (filled diamonds), $B = 10$ (open circles), and $B = 25$ (filled squares) as a function of the parameter $BC$. The solid line denotes the Edwards correlation length $\xi_E$. The dashed lines have a slope of $-0.75$. Reproduced from Alexander-Katz *et al.* (2003).

AB diblock copolymer melt by Ganesan and Fredrickson (2001). These authors used CL simulations to investigate the ODT behavior of the canonical Model E (with $f = 1/2$) in two dimensions. We recall from the discussion of Section 6.1.5.1 that thermal fluctuations render the ODT a fluctuation-induced, weakly first-order phase transition in this symmetric case, unlike the mean-field prediction of a continuous second-order transition (Leibler, 1980). The asymptotic Hartree analysis, which is valid only in the $C \to \infty$ limit, can also be conducted in the two-dimensional case.[117] The result is again an $O(C^{-2/3})$ correction to the mean-field prediction of $(\chi N)_{ODT} = 10.495$, although the numerical prefactor differs from the three-dimensional value given in eqn (6.78). By means of CL simulations, Ganesan and Fredrickson (2001) confirmed that the ODT is first order and examined the dependence of the lamellar phase melting temperature on the Ginzburg parameter $C$. Figure 6.16 shows the fluctuation-induced shift in $(\chi N)_{ODT}$ and confirms the expected $\sim C^{-2/3}$ scaling at large $C$. Unlike the Hartree analysis, which requires tedious analytical computations and is restricted to $C \gg 1$ and $f \approx 1/2$, the CL simulations are straightforward to implement and are not restricted to specific parameter values.

---

[117]In two dimensions, the parameter $C$ is defined according to $C = (n/A)R_{g0}^2$, where $A$ is the area of the system.

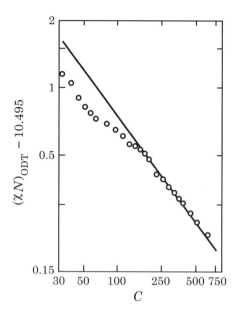

FIG. 6.16. The fluctuation-induced shift in the ODT as a function of the Ginzburg parameter $C$ for the symmetric Model E in two dimensions. The best fit line for $C \geq 128$ yields a slope of $-0.68$. The asymptotic Hartree prediction corresponds to a slope of $-2/3$. Reproduced from Ganesan and Fredrickson (2001).

Another recent application of field-theoretic simulations was the investigation by Düchs *et al.* (2003) of fluctuation phenomena in a ternary melt blend model of A and B homopolymers mixed with AB diblock copolymer. Within the context of mean-field theory, such ternary blends are known (Broseta and Fredrickson, 1990) to possess a special type of multicritical point referred to as an *isotropic Lifshitz point* (Hornreich *et al.*, 1975). At such a point, the competing tendencies for macrophase separation and microphase separation are balanced. Most theoretical and experimental work has focused on compositionally symmetric diblock copolymers ($f = 1/2$) of chain length $N$, mixed with equal proportions ($\phi_{A0} = \phi_{B0}$) of A and B homopolymers having the same chain lengths $\alpha N$. In mean-field theory, such a blend has an isotropic Lifshitz point for $\alpha < 1$, corresponding to homopolymers that are lower in molecular weight than the diblock copolymer. The mean-field (SCFT) phase diagram for an incompressible model of such an idealized blend (closely related to Models C and E) with $\alpha = 0.2$ is shown in Fig. 6.17.

According to mean-field theory, a homogeneous disordered phase and a lamellar mesophase dominate this cut (the "isopleth") through the ternary phase diagram. When the total homopolymer volume fraction $\phi_H = \phi_{A0} + \phi_{B0}$ vanishes,

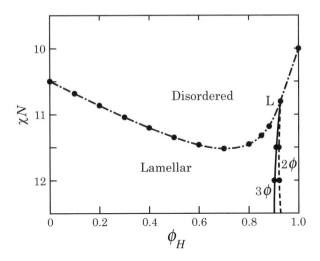

FIG. 6.17. Mean-field theoretical phase diagram for the isopleth of a symmetric $AB + A + B$ blend with $\alpha = 0.2$. Disordered and lamellar phases are indicated explicitly. The Lifshitz point is indicated as L. $2\phi$ denotes a region of two-phase coexistence between A-rich and B-rich homogeneous phases, while the narrow three-phase region ($3\phi$, bounded by the solid and dashed curves) reflects a coexistence between two such homogeneous phases and a third lamellar phase. The $\chi N$ scale has been inverted to correspond more closely to the experimental phase diagram shown in Fig. 6.18, noting that $\chi \propto 1/T$. Reproduced from Düchs et al. (2003).

the ternary alloy reduces to a single-component symmetric diblock copolymer melt that has an ODT at $\chi N = 10.495$. In the opposite composition extreme of $\phi_H = 1$, corresponding to a symmetric binary homopolymer blend, the disordered phase de-mixes into two homogeneous phases at a critical point consistent with the predictions of Flory–Huggins theory (de Gennes, 1979), i.e. at $\chi N = 2/\alpha$. The curve (dot-dashed) in Fig. 6.17 connecting the ODT at one composition extreme with the critical point at the other represents a family of critical points and continuous order–disorder transitions that merge at the Lifshitz point L. The Lifshitz point is located at a total homopolymer fraction of (Broseta and Fredrickson, 1990) $\phi_L = 1/(1 + 2\alpha^2) \approx 0.926$ and an incompatibility of $(\chi N)_L = 2(1 + 2\alpha^2)/\alpha \approx 10.8$, where the approximate values correspond to the $\alpha = 0.2$ case.

The Minnesota group has conducted a beautiful series of experiments on symmetric polymer blends corresponding to the above model with $\alpha \approx 0.2$ (Bates et al., 1995; Bates et al., 1997; Morkved et al., 1999; Hillmyer et al., 1999). As shown in Fig. 6.18, the experimental phase diagram for a PDMS-PEE/PDMS/PEE system is strikingly similar to the mean-field phase diagram, except at homopolymer fractions close to the predicted Lifshitz composition

$\phi_L \approx 0.926$. Indeed, in the experiments there was no evidence of a Lifshitz point, but rather a deep cusp in the phase diagram, centered just to the left of the Lifshitz composition, that allows the disordered phase to be stabilized at low temperatures. Within this cusp the disordered phase is highly structured and possesses large-amplitude, nearly isotropic composition fluctuations. Such a structured, fluctuating disordered phase with equal A-rich and B-rich phase volumes is commonly referred to as a *bicontinuous microemulsion*. Bicontinuous microemulsion phases are of practical interest because they provide an *equilibrium* strategy for creating co-continuous polymer morphologies (Fredrickson and Bates, 1997), rather than the non-equilibrium methods that are commonly applied in technological applications (Cao *et al.*, 1992; Potschke and Paul, 2003). By comparison with the mean-field diagram in Fig. 6.17, it is apparent that thermal fluctuations have destroyed the Lifshitz point in the experimental system, creating instead a region of bicontinuous microemulsion.

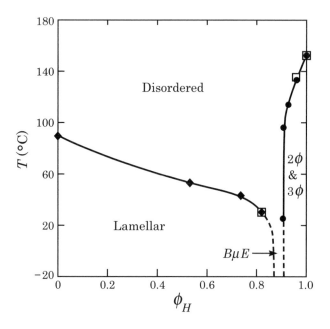

FIG. 6.18. Experimental phase diagram of a PDMS-PEE/PDMS/PEE blend with $\alpha \approx 0.2$. The region denoted by $B\mu E$ corresponds to a bicontinuous microemulsion phase. Reproduced from Morkved *et al.* (1999).

A renormalization group analysis of phenomenological models of isotropic Lifshitz points reveals an upper critical dimension of $d_u = 8$ (Hornreich, 1980), so $\epsilon = 8 - d$ expansions are of little practical use in three-dimensional systems. Although there is some controversy in the literature regarding the lower critical dimension of an isotropic Lifshitz point, it appears that $d_l = 4$ (Erzan and

Stell, 1977), so that thermal fluctuations destroy the ability of a system at the Lifshitz composition to order at a finite temperature for any dimension $d < 4$. Unfortunately, the available analytical and numerical tools for handling such a strong fluctuation scenario in three dimensions are very limited. Monte Carlo simulations have been used to examine a lattice polymer model of a ternary blend (Müller and Schick, 1996), but this approach is restricted to very short polymers. In contrast, the FTS tools of the present chapter have no such restriction and are well suited to theoretical investigations of Lifshitz fluctuation phenomena and microemulsion formation in polymer blends.

Acting on this opportunity, Düchs *et al.* (2003) applied CL and Metropolis Monte Carlo simulations to investigate the same incompressible ternary blend model that was used to obtain the mean-field phase diagram shown in Fig. 6.17. The MC simulations invoked the partial saddle point approximation of Section 6.5.2 and both sets of simulations were conducted in two dimensions to reduce the computational cost. Beyond the parameters of the mean-field theory, $\chi N$, $\alpha$, and $\phi_H$, the full field theory depends on a Ginzburg parameter $C \equiv R_{g0}^2/(Nv_0)$, where $v_0$ is the area occupied per segment and $Nv_0$ is the area per diblock copolymer in the two-dimensional melt. Again, the mean-field approximation becomes asymptotically exact for $C \to \infty$. By monitoring various order parameters in simulations conducted for the case of $C = 50$, Düchs *et al.* (2003) were able to construct a "fluctuation-corrected" phase diagram that is shown in Fig. 6.19. The qualitative features of this diagram in the vicinity of the Lifshitz composition are fully consistent with the experimental observations of microemulsion formation and fluctuation suppression of the Lifshitz point. Indeed, there is a striking similarity between Fig. 6.19 and the experimental diagram shown in Fig. 6.18. At least for $C = 50$, the partial saddle point approximation appears to be reasonably accurate for this incompressible blend model, since the results obtained from the two types of simulations are indistinguishable.

In concluding this section, we note that many of these "early" field-theoretic simulations did not employ the recommended numerical methods described in this book, namely the pseudo-spectral techniques for solving Fokker–Planck equations and evaluating density operators, and semi-implicit algorithms for time-stepping the CL equations. Coupled with the parallelization scheme described in Section 3.6.6 and access to a distributed memory compute cluster, it is now possible to perform high-resolution, three-dimensional field-theoretic simulations of a wide variety of inhomogeneous polymer models. We expect that reports of such simulations will be widespread as more groups gain experience with FTS techniques, as numerical methods continue to develop, and as computational resources become more powerful and less expensive.

## 6.9 Perspectives

It should be clear at this stage that analytical methods for treating mesoscopic models of inhomogeneous polymers have undergone relatively modest development over the past two decades, whereas there have been much more signifi-

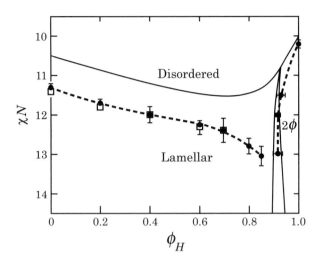

FIG. 6.19. Fluctuation-corrected phase diagram at $C = 50$. The circles show the results from MC simulations with the partial saddle point approximation and the squares correspond to data from CL simulations. The thin solid curves indicate the mean-field phase boundaries shown in Fig. 6.17. Reproduced from Düchs *et al.* (2003).

cant improvements in *numerical techniques* over the same period. The advent of spectral methods for solving the mean-field SCFT equations in 1994 by Matsen and Schick was a major milestone, as was the pioneering work of Fraaije and coworkers demonstrating that large-cell, real-space solutions of the SCFT equations were feasible. Subsequently, our group showed that polymer field theory models could be numerically tackled beyond the confines of the mean-field approximation, opening up the possibility of *field-theoretic polymer simulations*. The present chapter was intended to serve as an introduction to this exciting new subject, while connecting to more established analytical strategies for assessing the role of fluctuation phenomena.

In the coming decade, we believe that the numerical simulation techniques described in Chapter 5 (SCFT) and Chapter 6 (beyond SCFT) will become the predominant theoretical methods for investigating the properties of mesoscopic models of inhomogeneous polymers at equilibrium. Analytical methods, while important for establishing physical insights, are cumbersome when applied to spatially non-uniform systems, and the approximations that they invoke can be unreliable over large regions of parameter space. Moreover, "analytical" approximation schemes, such as the Gaussian fluctuation theory of Section 6.1.4, often lead to expressions that require numerical evaluation. The cost of implementation in some cases can exceed that of a direct field-theoretic simulation. As numerical methods and the computational resources available for conducting

FTS simulations continue to improve, barring an unanticipated theoretical development, it is apparent that direct simulations of polymer field theories will gain in importance.

### 6.9.1 *Particles or fields?*

This entire book has been devoted to field-theoretic descriptions of polymeric fluids, rather than more conventional "particle-based" descriptions based on atomic or coarse-grained segment positions (and possibly momenta). As a result, we have not addressed a large and significant body of work dealing with computer simulations of lattice or continuum particle-based models of inhomogeneous polymers. This omission was intentional; there exist excellent reviews on Monte Carlo and other simulation methods for studying homogeneous and inhomogeneous polymers at equilibrium (Binder, 1995; Binder and Müller, 2000*a*; Binder and Müller, 2000*b*; Attig *et al.*, 2004; Kotelyanskii and Theodorou, 2004), as well as reviews of particle-based simulation methods for broader classes of condensed matter systems (Binder and Ciccotti, 1996; Landau and Binder, 2000). These techniques have been under development for decades, and highly advanced methods and algorithms are in place. Certainly the importance of particle simulation approaches will not be diminished by the advent of FTS techniques!

Given the availability of both particle-based and field-based simulation methods, how does one select the method to apply in a particular circumstance? This is difficult to answer in general, but there are a few guidelines that should be helpful to the reader facing such a decision:

- *Are atomistic details important?* If the phenomena under investigation are controlled by atomic-scale physics, rather than mesoscopic physics, then particle-based methods should be applied. For example, the sequence-specific folding behavior of proteins is not something that can be addressed by a coarse-grained field theory model. Another example of a problem requiring an atomistic approach is the determination of the glass transition temperature $T_g$ of a new polymer structure. While field-theoretic models with embedded atomic details can in principle be constructed for such systems, it would normally require many more degrees of freedom and much more computer time to simulate a field theory at atomic resolution than to conduct an atomistic particle-based simulation.

- *Is it necessary to apply potentials with "hard cores"?* This is closely related to the previous question, since atomic-scale potentials and potentials of mean force are usually harshly repulsive at small separations. However, hard-core potentials can be important to apply even in mesoscopic models; for instance, when incorporating rigid nanoparticles or colloids into a complex fluid. Hard-core potentials create strong density correlations that can be difficult to resolve with a field-based approach. Thus, in circumstances where such potentials need be applied we recommend either a complete particle-based description, e.g. in the case of an atomistic model of a fluid, or a *hybrid description*, wherein certain degrees of freedom are treated by a field-theoretic approach and other degrees of freedom are addressed by retaining explicit particle coordinates. An example of such a hybrid particle–field simulation strategy applied to colloidal and nanoparticle dispersions will be described in the following section.

- *Are long-ranged interactions present?* If long-ranged interactions are expected to dominate the thermodynamic properties and structure of fluid, then a field-based description should be given serious consideration. For example, many polyelectrolyte problems can be addressed with a field-theoretic description in which a fluctuating electrostatic potential serves to decouple Coulomb interactions between charges (Tsonchev *et al.*, 2000). In Model F we observed that this procedure effectively replaces the long-ranged Coulomb potential by a much shorter-ranged inverse operator. Simulations of such field theories can be advantaged over particle-based simulations, which require intricate and computationally expensive methods to avoid finite size effects associated with the Coulomb potential (Frenkel and Smit, 1996).

- *Is the system dense in polymers and are those polymers long?* Particle-based simulations of polymeric fluids generally become more difficult as the density of polymers is increased and as the polymer molecular weight increases. In contrast, field-based simulations actually become easier at high densities and high molecular weight because the mean-field approximation is then more accurate. Under such conditions it is often possible to search for just *one* saddle point field configuration $w^*$, i.e. implement SCFT, rather than sample $\sim 10^6$ or more $w$ field configurations to conduct a full FTS simulation. Furthermore, field-based simulations can be used to investigate a perfectly incompressible fluid model, which would be considerably more difficult (if not impossible) with a particle-based approach.

- *Are time-dependent properties of interest?* If time-dependent phenomena, either near or far from equilibrium, are the primary focus of an investigation, then particle-based simulation methods are significantly advantaged. As described in Section 5.3.6, non-equilibrium extensions of the field-theoretic approach are in an early stage of development and it will be some time until architecture-dependent entanglement effects, mode couplings to energy and momentum densities, and viscoelastic and slip phenomena can be quantitatively captured in a field-theoretic framework. Moreover, even if a physically realistic non-equilibrium framework could be constructed for inhomogeneous polymers, it is not obvious at this stage whether such field theory models would be computationally feasible, or competitive with coarse-grained particle-based approaches.

In spite of the apparent advantage of field-theoretic methods in addressing the *equilibrium* properties of molten block and graft copolymers and polymer alloys, there is a preexisting and extensive literature on Monte Carlo and molecular dynamics simulations applied to coarse-grained models of such systems

(Binder, 1994; Müller, 1999; Binder, 1999; Murat *et al.*, 1999; Binder and Müller, 2000*a*; Vassiliev and Matsen, 2003). A recent trend in this work is to compare the particle-based simulation data to the results of mean-field (SCFT) calculations conducted with the same chain model used in the simulations (Müller and Binder, 1998). It remains to be seen if direct field-theoretic simulations will supplant coarse-grained particle simulations for treating this important class of inhomogeneous polymers in the coming years.

### 6.9.2  *Emerging applications for field-theoretic simulations*

*Colloidal suspensions and dispersions* have long been the basis for a number of important technologies and commercial products such as paints and coatings (Morrison and Ross, 2002). Nevertheless, our understanding of particle–particle interactions (potentials of mean force) in such systems is quite limited because of the complex nature of the surrounding polymeric fluid in a typical formulation. If the suspending fluid is itself capable of structuring, e.g. into mesophases, inter-particle interactions can be even more mysterious. The presence of surface-grafted polymers for "steric stabilization" can further complicate the picture.

Traditional particle-based computer simulation techniques are ill suited for studying colloidal dispersions, because of the large disparity in sizes between the colloidal particles and the molecules comprising the suspending fluid. As a result, it is difficult to simulate a system containing more than a few colloids if the atomic or even coarse-grained coordinates of all the fluid components are retained. A similar difficulty is encountered when attempting to model dispersions of inorganic or organic nanoparticles in complex fluid media, although the size discrepancy is less acute in this case. It would seem that a *hybrid* strategy of retaining explicit colloid or nanoparticle coordinates, but describing the suspending complex fluid in terms of fields, could be a versatile and efficient platform for conducting numerical simulations of such systems.

Some work along the lines of a hybrid theory has been recently reported by Thompson and coworkers (Thompson *et al.*, 2001; Thompson *et al.*, 2002; Lee *et al.*, 2002). These authors coupled a density functional theory for hard sphere nanoparticles to an incompressible SCFT description of diblock copolymers in order to examine the self-assembly behavior of block copolymer–nanoparticle composites. In such an approach the nanoparticle coordinates are not treated explicitly, but rather particle–particle correlations representative of a hard sphere fluid are embedded by means of an approximate density functional. A true hybrid approach that explicitly retains particle coordinates was recently applied by Reister and Fredrickson (2004) to examine the potential of mean force between two nanoparticles placed in a diblock copolymer melt. This formalism was recently extended to many-particle dispersions so that Monte Carlo moves of nanoparticles or colloids can be conducted, while simultaneously performing MC or CL updates of the fields describing the suspending fluid (Sides *et al.*, 2005*b*). With such an approach it is straightforward to treat particles of any shape, size, and surface treatment, as well as retain the flexibility and computational advantage of a field-theoretic description of the polymeric fluid.

In the hybrid particle–field strategy just described (Reister and Fredrickson, 2004; Sides *et al.*, 2005*b*), the fluid fields are defined everywhere in the dispersion, including the particle interiors. The exclusion of fluid from the inside of a particle is accounted for by introducing a "cavity function" $h(\mathbf{r})$, which is a smooth differentiable function that has a value of 1 inside the particle and a value of 0 outside the particle. In the case of a spherical particle, the cavity function has spherical symmetry, i.e. $h(\mathbf{r}) = h(r)$. Given a set of $n$ *non-overlapping* particle coordinates $\mathbf{r}^n \equiv (\mathbf{r}_1, \mathbf{r}_2, ..., \mathbf{r}_n)$, the function

$$\rho(\mathbf{r}; \mathbf{r}^n) = \rho_0 \left[ 1 - \sum_{j=1}^n h(|\mathbf{r} - \mathbf{r}_j|) \right] \tag{6.230}$$

represents the instantaneous fluid density at point $\mathbf{r}$, and $\rho_0$ denotes the bulk fluid density. Notice that $\rho(\mathbf{r}; \mathbf{r}^n)$ vanishes for $\mathbf{r}$ inside any particle and is constant at $\rho_0$ for $\mathbf{r}$ outside a particle. The function $h(r)$ will normally contain two length scales: a large scale $R_p$ setting the radius of the particles and a finer scale $\xi_p$ describing the thickness of the diffuse particle–fluid interfaces.

Adopting a mesoscopic description of the *fluid* in such a dispersion, one finds that it is convenient to impose a local incompressibility constraint analogous to that imposed in Models B, C, and E, namely

$$\prod_{\mathbf{r}} \delta(\hat{\rho}(\mathbf{r}) - \rho(\mathbf{r}; \mathbf{r}^n)) \tag{6.231}$$

where $\hat{\rho}(\mathbf{r})$ is the microscopic density of all fluid components. Equation (6.231) ensures that the fluid density vanishes in the particle interiors for all microscopic configurations of the fluid.[118] Exponentiating this constraint through the introduction of a "pressure" field $w_+$, as in Chapter 4, leads to a contribution to the effective Hamiltonian of the hybrid model reflecting particle–field couplings:

$$H_{pf}(\mathbf{r}^n; [w_+]) = -i \int d\mathbf{r} \, \rho(\mathbf{r}; \mathbf{r}^n) w_+(\mathbf{r}) \tag{6.232}$$

This coupling term is linear in the $w_+$ field; similar linear couplings to the exchange potential field $w_-$ arise by imposing preferential surface interactions between the particles and one of the fluid components, e.g. the A block of an AB diblock melt.

Hybrid simulations can be conducted pseudo-spectrally by sampling the fluid fields, e.g. $w_\pm$, on a collocation grid in real space. Particle centers are restricted to lie on grid points and particle overlaps, i.e. overlaps of the cavity functions, can be avoided by imposing an extra direct "hard-core" interaction $H_{pp}(\mathbf{r}^n)$ among

---

[118]The reader should note that this cavity approach to describing colloids and nanoparticles is closely related to the "masking" technique described in Section 4.9.5.

particles (Sides *et al.*, 2005*b*). Trial moves of particles to new grid points are best accomplished as global sweeps of all particle coordinates, rather than single particle moves, because the cost of each field update between particle moves is the same regardless of whether one or all particles are displaced. Field updates can be accomplished by using the Monte Carlo or complex Langevin procedures discussed in Sections 6.3 and 6.4.

To improve the acceptance ratio for trial Monte Carlo particle moves, it is helpful to have an analytical expression for the force exerted on each particle by the surrounding fluid. The force on the $j$th particle can be computed explicitly from eqn (6.232) according to

$$\mathbf{F}_j \equiv -\frac{\partial H_{pf}}{\partial \mathbf{r}_j} = -i\rho_0 \int d\mathbf{r} \, \mathbf{g}(\mathbf{r}_j - \mathbf{r}) \, w_+(\mathbf{r}) \qquad (6.233)$$

where $\mathbf{g}(\mathbf{r})$ is a "virial" function defined by

$$\mathbf{g}(\mathbf{r}) \equiv \frac{1}{r} \frac{dh(r)}{dr} \mathbf{r} \qquad (6.234)$$

Because eqn (6.233) has the form of a convolution integral, the force on *all* particles can be computed efficiently in a single sweep by applying an FFT, multiplication, and inverse FFT on the field collocation grid, and then extracting the values of the force field at the grid points coinciding with the $n$ particle centers. With such information in hand, a force-bias Monte Carlo technique can be used to conduct trial particle moves in a way that enhances their probability of acceptance.

Fields and particles need not be evolved at the same rate. Indeed, a useful approximation scheme is to slave the fields to their saddle point configuration consistent with the instantaneous particle configuration $\mathbf{r}^n$. This sidesteps the sign problem and amounts to an SCFT computation of the $n$-body potential of mean force for each particle configuration $\mathbf{r}^n$. A snapshot of a nanoparticle/AB diblock copolymer dispersion from a simulation conducted in this way is shown in Fig. 6.20. The particles in this case have a preferential affinity for the A block of the copolymer melt and are thus localized in the A domains of the block copolymer. Such localization is typically observed in experiments on block copolymer nanocomposites, unless special care is taken to neutralize the surface interactions (Chiu *et al.*, 2005).

*Polyelectrolyte solutions* represent another major class of complex fluid systems that is ripe for investigation with field-theoretic simulation techniques. The advent of controlled free radical polymerization techniques has made it possible to synthesize a wide variety of block, graft, and statistical copolymers containing groups that dissociate in water. The self-assembly behavior of aqueous solutions of such polymers can be extremely complex (Bendejacq *et al.*, 2005), especially when the polyelectrolytes have intrinsically hydrophobic backbones and are mixed with other water soluble polymers or surfactants. Field-theoretic models of inhomogeneous polyelectrolytes, similar to Model F, have been known

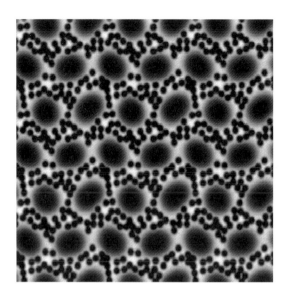

FIG. 6.20. Snapshot of a two-dimensional hybrid particle–field simulation for a block copolymer/nanoparticle composite assuming an incompressible, symmetric ($f = 1/2$) AB diblock copolymer melt with a preferential affinity of the particles for A segments. The particles are indicated as small solid circles, while A and B copolymer domains are shaded light and dark, respectively. The area per particle is $0.212\,R_{g0}^2$ and the area fraction of particles is 0.20. The particles have driven a transition from the lamellar to the hexagonal mesophase; the actual computational cell is replicated 3 times in each direction to emphasize the hexagonal order. Figure courtesy of S. W. Sides.

for some time (Barrat and Joanny, 1996; Netz and Andelman, 2003), but only recently have numerical mean-field solutions and Gaussian fluctuation treatments become available (Shi and Noolandi, 1999; Tsonchev *et al.*, 2000; Wang *et al.*, 2004). The advent of the FTS method provides a powerful new tool by which self-assembly behavior and fluctuation phenomena in mesostructured polyelectrolyte solutions, complexes, and mixtures can be analyzed. We expect this to be an active and fascinating area of investigation in the coming years.

*Mixtures of liquid crystals and polymers, liquid crystalline polymers, conjugated polymers, and block and graft copolymers with liquid crystalline or conjugated segments* also represent active areas of inhomogeneous polymer research driven by the rapid technological developments in organic electronics. If on builds on Model G, it is evidently possible to construct mesoscopic field theory models of a wide variety of semiflexible polymers, including block and graft architectures, and to incorporate anisotropic interactions that produce liquid crystalline order. Only mean-field SCFT solutions have been generated for any such models to date, and these computations have largely restricted the inhomogeneities to one or two dimensions (Morse and Fredrickson, 1994; Matsen, 1996; Düchs and Sullivan, 2002; Hamm *et al.*, 2001; Hamm *et al.*, 2003). Clearly much opportunity exists for developing improved numerical methods to enable higher-dimensional calculations, and for developing simulation strategies that go beyond the mean-field approximation. Hybrid simulations of polymer liquid crystal/nanoparticle mixtures also promise interesting defect structures and couplings between liquid crystalline and compositional order.

# APPENDIX   A

## FOURIER SERIES AND TRANSFORMS

Fourier analysis is used as a tool throughout this book. Here we provide some definitions and useful formulas. Readers interested in more details can consult general applied mathematics texts (Riley *et al.*, 1998; Arfken, 1985) or more specialized books on Fourier analysis (Tolstov, 1976).

Consider a function $f(x)$ defined over a finite interval $-L/2 \leq x \leq L/2$. One way of extending the definition of $f(x)$ outside the interval of interest is to declare that it be a periodic function with period $L$, i.e. that $f(x + L) = f(x)$ on the whole real line. Further, if over $x \in [-L/2, L/2]$ $f(x)$ is absolutely integrable, is single-valued and continuous, except possibly at a finite number of finite discontinuities, and has only a finite number of maxima and minima, then $f(x)$ can be represented by a *Fourier series*. For the purposes of this book, we employ the *complex form* of the Fourier series and write

$$f(x) = \sum_{j=-\infty}^{\infty} a_j \exp(i2\pi jx/L) \tag{A.1}$$

where the $a_j$ are complex Fourier coefficients with index $j = 0, \pm1, \pm2, ..., \pm\infty$ and $i \equiv \sqrt{-1}$. The coefficients can be evaluated by multiplying both sides of eqn (A.1) by $\exp(-i2\pi nx/L)$ and integrating $x$ over $[-L/2, L/2]$. Application of the orthogonality relation

$$\int_{-L/2}^{L/2} dx \, \exp(-i2\pi nx/L) \exp(i2\pi jx/L) = L \, \delta_{nj} \tag{A.2}$$

where the Kronecker delta is defined as $\delta_{nj} = 1$ for $n = j$, $\delta_{nj} = 0$ for $n \neq j$, leads to

$$a_j = \frac{1}{L} \int_{-L/2}^{L/2} dx \, f(x) \exp(-i2\pi jx/L) \tag{A.3}$$

The convergence properties of Fourier series are of particular interest. It is convenient to define a partial sum of the series (A.1) that involves truncation after $M = 2P + 1$ terms:

$$f_M(x) = \sum_{j=-P}^{P} a_j \exp(i2\pi jx/L) \tag{A.4}$$

Provided that $f(x)$ has the properties described above, then the *pointwise convergence theorem* of Fourier series states that $f_M(x) \to f(x)$ as $M \to \infty$ for any

$x \in [-L/2, L/2]$ that is not a point of discontinuity. At a point $x_0$ corresponding to a jump discontinuity, then $f_M(x_0) \to [f(x_{0+}) + f(x_{0-})]/2$ as $M \to \infty$.

The *rate* of convergence of a Fourier series is very sensitive to the smoothness of the function over the interval (including the end points). If the function $f(x)$ is continuous and infinitely smooth (derivatives of all orders exist) over the interval, then the remarkable Theorem 3.1 applies: the absolute error $|f_M(x) - f(x)|$ decays to zero faster than any finite power of $1/M$ as $M \to \infty$. At the other extreme, if $f(x)$ has a point of discontinuity over $x \in [-L/2, L/2]$, then $|f_M(x) - f(x)|$ will decay only as fast as $M^{-1}$. Similarly, if $f(x)$ is continuous, but $f'(x)$ is discontinuous somewhere in the interval, then $|f_M(x) - f(x)|$ will decay as $M^{-2}$ for $M \to \infty$.

Fourier transforms provide a useful representation for functions $f(x)$ defined over the whole real line, $x \in [-\infty, \infty]$. We can approach this situation by considering the Fourier series representation (A.1) in the limit that $L \to \infty$. In this limit, each successive term in the sum increments the argument of the exponential factor by an amount that is $O(1/L)$. Thus, with vanishingly small error for $L \to \infty$, the sum can be replaced by an integral to obtain

$$f(x) = \int_{-\infty}^{\infty} dj \; a_j \exp(i2\pi jx/L)$$
$$= \frac{1}{2\pi} \int_{-\infty}^{\infty} dk \; \hat{f}(k) \exp(ikx) \tag{A.5}$$

In the second line the integration variable has been changed from the index $j$ to a *wavevector* $k = 2\pi j/L$ and we have redefined the Fourier coefficient as $\hat{f}(k) = La_j$. A similar rewriting of eqn (A.3) in the $L \to \infty$ limit leads to

$$\hat{f}(k) = \int_{-\infty}^{\infty} dx \; f(x) \exp(-ikx) \tag{A.6}$$

Equation (A.6) defines the so-called *Fourier transform* $\hat{f}(k)$ of the function $f(x)$. The second line of eqn (A.5) is the so-called *inverse Fourier transform* because given the transform $\hat{f}(k)$, it provides a formula for recovering the function $f(x)$.

Fourier transforms have a number of useful properties. For example, upon integrating by parts, the Fourier transform of the first derivative $f'(x)$ of a function $f(x)$ can be shown to be equal to $ik\hat{f}(k)$, provided that $f(x) \to 0$ for $x \to \pm\infty$. Similarly, the Fourier transform of the second derivative $f''(x)$ is given by $-k^2\hat{f}(k)$, if $f$ and $f'$ both vanish for $x \to \pm\infty$.

Another useful property is the *convolution theorem*. The "convolution" of two functions $g(x)$ and $h(x)$ can be defined by

$$g * h = \int_{-\infty}^{\infty} dx' \; g(x - x')h(x') \tag{A.7}$$

The convolution theorem states that the Fourier transform of the convolution of two functions is the product of their respective Fourier transforms, i.e.

$$\int_{-\infty}^{\infty} dx \, (g * h) \exp(-ikx) = \hat{g}(k)\hat{h}(k) \tag{A.8}$$

The notions of Fourier series and transforms are easily extended to functions of more than one variable. For example, in the case of a function $f(\mathbf{r})$ defined in a cubic domain of side length $L$, i.e. $r_i \in [-L/2, L/2]$ for $i = 1, 2, 3$, we can extend the definition of $f$ to all of $\mathbb{R}^3$ by imposing the periodic boundary conditions

$$f(\mathbf{r} + L\hat{\mathbf{e}}) = f(\mathbf{r}) \tag{A.9}$$

Here, $\hat{\mathbf{e}}$ is a unit vector taken along any edge of the cube. If one adopts the notation of solid state physics (Ashcroft and Mermin, 1976), it is convenient to express the three-dimensional Fourier series in the form

$$f(\mathbf{r}) = \frac{1}{V} \sum_{\mathbf{k}} \hat{f}(\mathbf{k}) \exp(i\mathbf{k} \cdot \mathbf{r}) \tag{A.10}$$

where $V = L^3$ is the volume of the domain and the $\hat{f}(\mathbf{k})$ are the Fourier coefficients. The sum over *reciprocal lattice vectors* $\mathbf{k}$ in this expression denotes summing over the vectors

$$\mathbf{k} = \frac{2\pi}{L} (j_1\hat{\mathbf{x}} + j_2\hat{\mathbf{y}} + j_3\hat{\mathbf{z}}) \tag{A.11}$$

where $j_i = 0, \pm 1, \pm 2, ..., \pm\infty$ for $i = 1, 2, 3$. The vectors $\hat{\mathbf{x}}$, $\hat{\mathbf{y}}$, and $\hat{\mathbf{z}}$ are Cartesian unit vectors defining the edges of the cell. By invoking orthogonality of the plane waves, one obtains the Fourier coefficients by means of the formula

$$\hat{f}(\mathbf{k}) = \int_V d\mathbf{r} \, f(\mathbf{r}) \exp(-i\mathbf{k} \cdot \mathbf{r}) \tag{A.12}$$

where the integral is over the cube.

The *three-dimensional Fourier transform* for an absolutely integrable function defined in $\mathbb{R}^3$ can be deduced by taking the $L \to \infty$ limit of eqn (A.12)

$$\hat{f}(\mathbf{k}) = \int d\mathbf{r} \, f(\mathbf{r}) \exp(-i\mathbf{k} \cdot \mathbf{r}) \tag{A.13}$$

where the integral is now over all of three-dimensional space. The inverse Fourier transform is obtained by again converting the sum over wavevectors in eqn (A.10) into an integral appropriate in the $V \to \infty$ limit:

$$f(\mathbf{r}) = \frac{1}{(2\pi)^3} \int d\mathbf{k} \, \hat{f}(\mathbf{k}) \exp(i\mathbf{k} \cdot \mathbf{r}) \tag{A.14}$$

Finally, we discuss a useful Fourier representation of the *Dirac delta function* (Arfken, 1985) that will be used throughout the book. The Dirac delta function

$\delta(x)$ is defined by the relation $f(x) = \int dx'\ \delta(x - x')f(x')$ for any $f(x)$. Substitution of eqn (A.6) into eqn (A.5), essentially a statement of completeness of the complex exponential Fourier basis, produces the following representation of $\delta(x)$:

$$\delta(x) = \frac{1}{2\pi} \int_{-\infty}^{\infty} dk\ \exp(ikx) \tag{A.15}$$

The analogous definition of the three-dimensional Dirac delta function is $f(\mathbf{r}) = \int d\mathbf{r}'\ \delta(\mathbf{r} - \mathbf{r}')f(\mathbf{r}')$. It follows by substitution of eqn (A.13) into eqn (A.14) that in three dimensions

$$\delta(\mathbf{r}) = \frac{1}{(2\pi)^3} \int d\mathbf{k}\ \exp(i\mathbf{k} \cdot \mathbf{r}) \tag{A.16}$$

# APPENDIX B

## GAUSSIAN INTEGRALS AND PROBABILITY THEORY

Gaussian integrals are fundamental to many fields, including probability theory, statistical mechanics, and field theory. A few basic formulas that will be used throughout the monograph are presented here. More details can be found in a variety of sources (Riley *et al.*, 1998; Zee, 2003).

A basic one-dimensional Gaussian integral is the following:

$$\int_{-\infty}^{\infty} dx \, \exp(-ax^2/2) = \left(\frac{2\pi}{a}\right)^{1/2} \tag{B.1}$$

Two other important integrals are

$$\int_{-\infty}^{\infty} dx \, \exp(-ax^2/2 + Jx) = \left(\frac{2\pi}{a}\right)^{1/2} \exp[J^2/(2a)] \tag{B.2}$$

$$\int_{-\infty}^{\infty} dx \, \exp(-ax^2/2 + iJx) = \left(\frac{2\pi}{a}\right)^{1/2} \exp[-J^2/(2a)] \tag{B.3}$$

where $i = \sqrt{-1}$. It is important to note that the arguments of the exponentials in these two expressions differ only by a sign.

A one-dimensional Gaussian *probability distribution function* or *probability density* $P(x)$ can be defined by

$$P(x) = \left(\frac{a}{2\pi}\right)^{1/2} \exp(-ax^2/2) \tag{B.4}$$

It is normalized in that $\int_{-\infty}^{\infty} dx \, P(x) = 1$. Even *moments* of this distribution are the following:

$$\langle x^{2n} \rangle \equiv \int_{-\infty}^{\infty} dx \, P(x) x^{2n} = a^{-n}(2n-1) \cdot (2n-3) \cdots 5 \cdot 3 \cdot 1$$

$$= a^{-n}(2n-1)!! \tag{B.5}$$

for $n \geq 1$ an integer. Odd moments, $\langle x^{2n+1} \rangle$ $(n \geq 0)$, all vanish. A powerful way to derive such moment formulas is to introduce a *moment generating function*

$$G(J) \equiv \langle \exp(Jx) \rangle = \int_{-\infty}^{\infty} dx \, P(x) \exp(Jx) = \exp[J^2/(2a)] \tag{B.6}$$

where we have used eqn (B.2) to derive the final expression. Moments of the probability distribution $P(x)$ are obtained from $G(J)$ according to

$$\langle x^n \rangle = \frac{d^n G(J)}{dJ^n}\bigg|_{J=0} \tag{B.7}$$

Equation (B.5) and the fact that all odd moments vanish then follow immediately by application of this formula to the explicit form $G(J) = \exp[J^2/(2a)]$.

A closely related generating function is the *cumulant generating function* defined by

$$H(J) \equiv \ln\langle\exp(Jx)\rangle = \ln \int_{-\infty}^{\infty} dx\, P(x)\exp(Jx) \tag{B.8}$$

The formula

$$\langle x^n \rangle_c = \frac{d^n H(J)}{dJ^n}\bigg|_{J=0} \tag{B.9}$$

defines the so-called *cumulant* or *connected* moments of the probability distribution $P(x)$. These are related to the ordinary moments by expressions such as

$$
\begin{aligned}
\langle x \rangle_c &= \langle x \rangle \\
\langle x^2 \rangle_c &= \langle x^2 \rangle - \langle x \rangle^2 \\
\langle x^3 \rangle_c &= \langle x^3 \rangle - 3\langle x^2 \rangle\langle x \rangle + 2\langle x \rangle^3 \\
\langle x^4 \rangle_c &= \langle x^4 \rangle - 4\langle x^3 \rangle\langle x \rangle + 12\langle x^2 \rangle\langle x \rangle^2 - 3\langle x^2 \rangle^2 - 6\langle x \rangle^4
\end{aligned}
\tag{B.10}
$$

For the Gaussian distribution (B.4), $H(J) = J^2/(2a)$, so it follows from eqn (B.9) that only the second cumulant moment is non-vanishing and $\langle x^2 \rangle_c = \langle x^2 \rangle = 1/a$. All higher even moments of the Gaussian distribution can be related to the second cumulant $\langle x^2 \rangle_c$ by a relationship known as *Wick's theorem* (Zee, 2003). The theorem states that $\langle x^{2n} \rangle$ is equal to $\langle x^2 \rangle_c^n$ multiplied by a combinatorial factor $(2n-1)!!$ that is equal to the number of distinct ways of grouping the $2n$ factors of $x$ into pairs. Wick's theorem is evidently consistent with eqn (B.5).

These results can be easily generalized to higher-dimensional Gaussian integrals. Consider extending $x$ and $J$ to column vectors, i.e. $\mathbf{x} = (x_1, x_2, ..., x_N)^T$, and $a$ to an $N \times N$ real symmetric matrix $\mathbf{A}$ with positive eigenvalues (positive definite matrix). The generalizations of eqns (B.1)–(B.3) are

$$\int d\mathbf{x}\, \exp(-\mathbf{x}^T\mathbf{A}\mathbf{x}/2) \equiv \int_{-\infty}^{\infty} dx_1 \,...\, \int_{-\infty}^{\infty} dx_N\, \exp(-\mathbf{x}^T\mathbf{A}\mathbf{x}/2)$$

$$= \frac{(2\pi)^{N/2}}{(\det \mathbf{A})^{1/2}} \tag{B.11}$$

$$\int d\mathbf{x}\, \exp(-\mathbf{x}^T\mathbf{A}\mathbf{x}/2 + \mathbf{J}^T\mathbf{x}) = \frac{(2\pi)^{N/2}}{(\det \mathbf{A})^{1/2}}\exp(\mathbf{J}^T\mathbf{A}^{-1}\mathbf{J}/2) \tag{B.12}$$

$$\int d\mathbf{x}\, \exp(-\mathbf{x}^T\mathbf{A}\mathbf{x}/2 + i\mathbf{J}^T\mathbf{x}) = \frac{(2\pi)^{N/2}}{(\det \mathbf{A})^{1/2}}\exp(-\mathbf{J}^T\mathbf{A}^{-1}\mathbf{J}/2) \tag{B.13}$$

Probability distribution functions and generating functions can also be readily extended to the multivariate case. A joint Gaussian probability distribution function for a set of random variables $\mathbf{x} = (x_1, x_2, ..., x_N)^T$ with zero mean can be written

$$P(\mathbf{x}) = \frac{(\det \mathbf{A})^{1/2}}{(2\pi)^{N/2}} \exp(-\mathbf{x}^T \mathbf{A} \mathbf{x}/2) \tag{B.14}$$

and is normalized so that $\int d\mathbf{x}\ P(\mathbf{x}) = 1$. The only non-vanishing cumulant moment of this distribution is the second cumulant

$$\langle x_i x_j \rangle_c = \langle x_i x_j \rangle \equiv \int d\mathbf{x}\ x_i x_j P(\mathbf{x}) = (\mathbf{A}^{-1})_{ij} \tag{B.15}$$

and all odd moments vanish. Higher even moments are related to the second cumulant by a version of Wick's theorem. For example, in the case of the fourth moment

$$\begin{aligned}
\langle x_i x_j x_k x_l \rangle &= (\mathbf{A}^{-1})_{ij}(\mathbf{A}^{-1})_{kl} + (\mathbf{A}^{-1})_{ik}(\mathbf{A}^{-1})_{jl} + (\mathbf{A}^{-1})_{il}(\mathbf{A}^{-1})_{jk} \\
&= \langle x_i x_j \rangle_c \langle x_k x_l \rangle_c + \langle x_i x_k \rangle_c \langle x_j x_l \rangle_c + \langle x_i x_l \rangle_c \langle x_j x_k \rangle_c
\end{aligned} \tag{B.16}$$

# APPENDIX C

## CALCULUS OF FUNCTIONALS

In order to understand the field-based approach to modelling inhomogeneous fluids, it is necessary to have a basic familiarity with the calculus of functionals (Volterra, 1959). This broad subject includes topics such as functional differentiation, functional integration, and min–max problems. These are typically discussed in texts on functional analysis, calculus of variations, optimization theory, and field theory. Here, we provide a brief tutorial. Physically oriented references where more details can be found include (Fetter and Walecka, 1980; Hansen and McDonald, 1986; Parr and Yang, 1989; Zee, 2003).

### C.1 Functionals

In the simplest case, a *functional* is a mapping between a function $f(x)$ defined over some interval $x \in [a, b]$ and a number $F$ that generally depends on the values of the function over *all* points of the interval. For example, a simple linear functional is just the integral

$$F_1[f] = \int_a^b dx\ f(x) \tag{C.1}$$

This formula associates a number $F_1$ with the integral of $f(x)$ over $a \leq x \leq b$. We adopt the functional "square bracket" notation $F_1[f]$ to indicate that $F_1$ depends on $f(x)$ at *all* points over the interval. An example of a *nonlinear* functional is

$$F_2[f] = \int_a^b dx\ [f(x)]^2 \tag{C.2}$$

Both $F_1[f]$ and $F_2[f]$ are referred to as *local* functionals because values of $f(x)$ for different $x$ contribute independently (additively) to the value of the functional.

More generally, a functional can depend on the function and its derivatives over the interval. Such functionals are referred to as *non-local*. For example,

$$F_3[f] = \int_a^b dx\ \left( -[f(x)]^2 + [f(x)]^4 + [f'(x)]^2 \right) \tag{C.3}$$

is a familiar functional that appears in the Landau–Ginzburg theory of phase transitions (Chaikin and Lubensky, 1995; Goldenfeld, 1992). A second example of a non-local functional is the quadratic expression

$$F_4[f] = \int_a^b dx \int_a^b dx'\ f(x)K(x, x')f(x') \tag{C.4}$$

where the kernel $K(x, x')$ is an arbitrary function of $x$ and $x'$.

Functionals can also be defined for multi-variable functions $f(\mathbf{r})$ that, e.g., could represent the chemical potential fields $w(\mathbf{r})$ that are central to the subject of this monograph. For example, the extension of eqn (C.3) to functions defined in three dimensions is the Landau–Ginzburg "square gradient" functional

$$F_5[f] = \int d\mathbf{r} \left(-[f(\mathbf{r})]^2 + [f(\mathbf{r})]^4 + |\nabla f|^2\right) \tag{C.5}$$

## C.2   Functional differentiation

The concept of differentiation of functionals is a straightforward extension of the notion of partial differentiation for multi-variable functions. Consider subjecting a function $f(x)$ defined over $x \in [a, b]$ to an arbitrary small perturbation $\delta f(x)$. The perturbation $\delta f(x)$ is itself a function defined over the same interval of $x$. For some functional $F[f]$, we then consider its value $F[f + \delta f]$ when $f(x) \to f(x) + \delta f(x)$. This quantity can be Taylor-expanded in powers of the perturbation $\delta f(x)$ to yield the general form

$$F[f + \delta f] = F[f] + \int_a^b dx\, \Gamma_1(x)\delta f(x)$$
$$+ \frac{1}{2!} \int_a^b dx \int_a^b dx'\, \Gamma_2(x, x')\delta f(x)\delta f(x') + \ldots \tag{C.6}$$

where the functions $\Gamma_i$ represent Taylor expansion coefficients. For example, in the case of the functional $F_2[f]$, it is straightforward to show that $\Gamma_1(x) = 2f(x)$ and $\Gamma_2(x, x') = 2\delta(x - x')$, where $\delta(x - x')$ is the Dirac delta function defined by $\int dx'\, f(x')\delta(x - x') = f(x)$.

The coefficient functions $\Gamma_i$ are typically written differently so that eqn (C.6) is more suggestive of a *functional Taylor series*. Namely, the first and second *functional derivatives* of $F$ with respect to $f(x)$ are defined by

$$\frac{\delta F[f]}{\delta f(x)} \equiv \Gamma_1(x) \tag{C.7}$$

$$\frac{\delta^2 F[f]}{\delta f(x)\delta f(x')} \equiv \Gamma_2(x, x') \tag{C.8}$$

The first functional derivative $\delta F[f]/\delta f(x)$ is a function of $x$ that dictates the rate of change of the functional $F[f]$ when $f(x)$ is perturbed at the point $x$. Similarly, the second derivative $\delta^2 F[f]/\delta f(x)\delta f(x')$ is a function of $x$ and $x'$ that expresses the rate of change of $F[f]$ when $f(x)$ is simultaneously perturbed at points $x$ and $x'$.

By explicitly working out the Taylor expansion of eqn (C.6) for a prescribed functional $F[f]$, it is possible to derive functional differentiation formulas that closely resemble formulas from ordinary differential calculus. For example, the functional

$$F_6[f] = \int_a^b dx\, [f(x)]^n \tag{C.9}$$

has derivatives

$$\frac{\delta F_6[f]}{\delta f(x)} = n[f(x)]^{n-1}, \qquad \frac{\delta^2 F_6[f]}{\delta f(x)\delta f(x')} = n(n-1)[f(x)]^{n-2}\delta(x-x') \tag{C.10}$$

Similarly, the functional $F_4[f]$ of eqn (C.4) for a symmetric kernel $K(x,x') = K(x',x)$ has derivatives

$$\frac{\delta F_4[f]}{\delta f(x)} = 2\int_a^b dx'\, K(x,x')f(x'), \qquad \frac{\delta^2 F_4[f]}{\delta f(x)\delta f(x')} = 2K(x,x') \tag{C.11}$$

The computation of functional derivatives according to eqns (C.6)–(C.8) for functionals such as $F_3[f]$ that involve derivatives of $f(x)$ proceeds by one or more integrations by parts. These in turn require *boundary conditions* to be imposed on the arbitrary perturbation $\delta f(x)$. For example, we might want to restrict attention to functions $f(x)$ that satisfy the fixed end (Dirichlet) conditions $f(a) = f_a$, $f(b) = f_b$. The variations of a functional $F[f]$ subject to these fixed end conditions can thus be examined by expanding $F[f + \delta f]$ according to eqn (C.6) with the arbitrary perturbation satisfying the *homogeneous* end conditions $\delta f(a) = \delta f(b) = 0$. As an explicit example, variation of the functional

$$F_7[f] = \int_a^b dx\, [f'(x)]^2 \tag{C.12}$$

subject to fixed end conditions produces the functional derivatives

$$\frac{\delta F_7[f]}{\delta f(x)} = -2f''(x), \qquad \frac{\delta^2 F_7[f]}{\delta f(x)\delta f(x')} = \frac{d}{dx}\frac{d}{dx'}\delta(x-x') \tag{C.13}$$

A similar approach can be used to define and compute functional derivatives for functionals of multivariate functions. For example, variation of $F_5[f]$ subject to fixed conditions at the boundary of the **r** domain leads to the functional derivative

$$\frac{\delta F_5[f]}{\delta f(\mathbf{r})} = -2f(\mathbf{r}) + 4[f(\mathbf{r})]^3 - 2\nabla^2 f(\mathbf{r}) \tag{C.14}$$

A variety of other useful functional differentiation formulas can be derived. A particularly important relation is the *chain rule*

$$\frac{\delta F[g]}{\delta f(x)} = \int dx'\, \frac{\delta F[g]}{\delta g(x')}\frac{\delta g(x')}{\delta f(x)} \tag{C.15}$$

Another important expression is

$$\frac{\delta f(x')}{\delta f(x)} = \delta(x' - x) \tag{C.16}$$

Finally, this expression, combined with the the choice of $F = f(x'')$ in eqn (C.15), leads to

$$\int dx' \, \frac{\delta f(x'')}{\delta g(x')} \frac{\delta g(x')}{\delta f(x)} = \delta(x'' - x) \tag{C.17}$$

which shows that $\delta f/\delta g$ and $\delta g/\delta f$ are functional inverses.

## C.3   Min–max problems

An important application of the calculus of functionals is to optimization problems. A typical problem involves finding the function $f(x)$ belonging to some function space that minimizes or maximizes a prescribed functional $F[f]$. For example, in the classical density functional theory of inhomogeneous fluids (Rowlinson and Widom, 1989), $f$ corresponds to a density field and $F$ to a free energy functional. The free energy is minimized for the equilibrium configuration of the density. In Chapter 5, a similar variational principle is used to derive the self-consistent field theory (SCFT) of inhomogeneous polymeric fluids.

The theoretical basis for solving functional min–max problems is the Taylor expansion of eqn (C.6). The *first variation* $\delta F$ of a functional $F[f]$ that is subjected to an arbitrary infinitesimal perturbation $\delta f(x)$ over $x \in [a, b]$ is defined by

$$\delta F \equiv F[f + \delta f] - F[f] = \int_a^b dx \, \frac{\delta F[f]}{\delta f(x)} \delta f(x) \tag{C.18}$$

The functional $F[f]$ attains an extremum value, i.e. a maximum, minimum, or saddle point, when $f(x)$ is adjusted to a function $f^*(x)$ such that the first variation vanishes. Because the perturbation $\delta f(x)$ is arbitrary, this condition implies that the extremum function $f^*(x)$ is determined by the vanishing of the first functional derivative

$$\left. \frac{\delta F[f]}{\delta f(x)} \right|_{f=f^*} = 0 \tag{C.19}$$

Thus, just as we locate minima or maxima of an ordinary function $f(x)$ by setting the first derivative $f'(x)$ to zero, the extremum of a functional corresponds to the function $f^*(x)$ that causes the first functional derivative to vanish. Equation (C.19) is commonly referred to as an *Euler–Lagrange equation* and may be an ordinary differential, a partial differential, or an integral equation to solve for $f^*(x)$ depending on the form of the functional. For example, the Euler–Lagrange equation that arises from variation of the functional $F_3[f]$ subject to the fixed end conditions $f(a) = f_a$, $f(b) = f_b$ is the *ordinary* differential equation

$$\frac{d^2}{dx^2} f^*(x) + f^*(x) - 2[f^*(x)]^3 = 0 \tag{C.20}$$

This equation has a unique solution $f^*(x)$ that depends on the prescribed boundary conditions. Correspondingly, the extremum of the functional $F_5[f]$ subject to

Dirichlet or periodic boundary conditions on $f(\mathbf{r})$ satisfies the *partial* differential equation

$$\nabla^2 f^*(\mathbf{r}) + f^*(\mathbf{r}) - 2[f^*(\mathbf{r})]^3 = 0 \tag{C.21}$$

The Euler–Lagrange equation (C.19) provides a condition for determining a function $f^*(x)$ that corresponds to an extremum of a prescribed functional $F[f]$. In order to establish whether that extremum is a maximum, minimum, or saddle point, the second functional derivative must be analyzed. This analysis involves the solution of the eigenvalue problem

$$\int_a^b dx' \ \left. \frac{\delta^2 F[f]}{\delta f(x)\delta f(x')}\right|_{f=f^*} \phi_i(x') = \Lambda_i\phi_i(x) \tag{C.22}$$

If the eigenvalues $\Lambda_i$ are all positive, then $f^*$ represents a *local minimum* of $F[f]$. Correspondingly, if $\Lambda_i < 0$ for all $i$, $f^*$ is a *local maximum* of $F[f]$. In the intermediate case of eigenvalues of mixed sign, we conclude that $f^*$ corresponds to a *saddle point* of the functional. Establishing whether a particular extremum is a *global*, rather than local, minimum or maximum is a more difficult problem in optimization theory (Nocedal and Wright, 1999) that remains unsolved for arbitrary $F[f]$, although physical intuition is often helpful in specific contexts.

As a final note, the above optimization scheme can be extended to include linear and nonlinear *constraints* through the introduction of Lagrange multipliers. Interested readers should consult the literature (Riley *et al.*, 1998; Weinstock, 1974).

## C.4 Functional integration

In addition to taking the derivative of a functional with respect to a function, it is also possible to define the integral of a functional over all functions belonging to some function space (Feynman and Hibbs, 1965; Simon, 1979; Zee, 2003). Such an integral is referred to as a *functional integral*, or more specifically a *path integral*, if the function $f(x)$ corresponds to the trajectory $q(t)$ of a particle at various times $t$ or the configuration $\mathbf{r}(s)$ of a polymer molecule at various contour locations $s$.

A generic functional integral will be written in the form

$$I = \int \mathcal{D}f \ F[f] \tag{C.23}$$

where the notation $\int \mathcal{D}f$ is understood to represent an integral over all functions $f(x)$ defined over $x \in [a,b]$ belonging to some function space. The relevant function space is determined by smoothness and boundary conditions on $f$. For example, if we were interested in summing over all possible shapes of a polymer that is clamped at both ends, eqn (C.23) could be interpreted as an integral over all continuous and infinitely differentiable functions $f(x)$ that satisfy $f(a) = f(b) = 0$.

How does one define such a functional integral? One approach is to *discretize* the function over the interval. In the clamped polymer example, a sensible strategy would be to sample $f(x)$ at a set of $N$ equally spaced interior points, $x_i = a + i(b-a)/(N+1)$, $i = 1, 2, ..., N$. The function can thus be approximated by an $N$-vector $\mathbf{f} = (f_1, f_2, ..., f_N)$ with components $f_i \equiv f(x_i)$. For a prescribed $N$, the $N$-dimensional integral

$$I_N = \int_{-\infty}^{\infty} df_1 \cdots \int_{-\infty}^{\infty} df_N \ F(\mathbf{f}) \tag{C.24}$$

can thus be viewed as an approximation to the functional integral $I$. In this equation we use the conventional notation $F(\mathbf{f})$ of a *multivariate function* to indicate the discrete approximation to a functional $F[f]$. The formal transition from an ordinary multi-dimensional integral to an infinite-dimensional functional integral is through the limit $\lim_{N \to \infty} I_N = I$. Depending on the form of the functional $F[f]$, this limit may or may not exist. However, in the statistical mechanics of classical fields, we are normally interested in *average* quantities that can be expressed as the *ratio* of two functional integrals. In such cases the limiting procedure usually converges to a finite result for the ratio, even if the limits of the individual integrals do not exist.

A second way to interpret a functional integral such as eqn (C.23) is through a spectral or normal-mode representation of the function. For example, in the tethered polymer situation with $f(a) = f(b) = 0$, a Fourier sine series representation would be appropriate:

$$f(x) = \sum_{n=1}^{\infty} a_n \sin\left(\frac{n\pi(x-a)}{b-a}\right) \tag{C.25}$$

The functional integral would then be interpreted as an integral over all the Fourier coefficients $\mathbf{a} = (a_1, a_2, a_3, ...)$ according to

$$I = \left[\prod_{n=1}^{\infty} \int_{-\infty}^{\infty} da_n\right] F(\mathbf{a}) \tag{C.26}$$

Again, the expression on the right may not exist, but the ratio of two such formulas, corresponding to a thermodynamic average, will usually be well defined.

With the exception of Gaussian integrals, very few functional integrals can be evaluated analytically. Two important Gaussian integral formulas that can be viewed as infinite-dimensional versions of eqns (B.12) and (B.13) are

$$\frac{\int \mathcal{D}f \ \exp\left[-(1/2)\int dx \int dx' \ f(x)A(x,x')f(x') + \int dx \ J(x)f(x)\right]}{\int \mathcal{D}f \ \exp\left[-(1/2)\int dx \int dx' \ f(x)A(x,x')f(x')\right]}$$

$$= \exp\left(\frac{1}{2}\int dx \int dx' \ J(x)A^{-1}(x,x')J(x')\right) \tag{C.27}$$

$$\frac{\int \mathcal{D}f \, \exp\left[-(1/2)\int dx \int dx' \, f(x)A(x,x')f(x') + i \int dx \, J(x)f(x)\right]}{\int \mathcal{D}f \, \exp\left[-(1/2)\int dx \int dx' \, f(x)A(x,x')f(x')\right]}$$

$$= \exp\left(-\frac{1}{2}\int dx \int dx' \, J(x)A^{-1}(x,x')J(x')\right) \tag{C.28}$$

where $A(x,x')$ is assumed to be real, symmetric, and positive definite. The functional inverse of $A$, $A^{-1}$, is defined in accordance with eqn (C.17) by

$$\int dx' \, A(x,x')A^{-1}(x',x'') = \delta(x - x'') \tag{C.29}$$

When these formulas are applied to interacting particle models in classical statistical physics, such as those described in Chapter 4, $J$ represents a microscopic density operator, and $A^{-1}$ is a pair potential function. The function $f$ is an auxiliary potential that serves to decouple particle–particle interactions. In this context, eqns (C.27) and (C.28) are generically referred to as *Hubbard–Stratonovich transformations* (Chaikin and Lubensky, 1995).

# APPENDIX D

## COMPLEX LANGEVIN THEORY

The complex Langevin (CL) simulation method described in Section 6.4 is a versatile tool for bypassing the sign problem that arises in sampling field theories with non-positive definite weights, i.e. theories with a complex Hamiltonian $H[w]$. In this appendix we discuss the theoretical basis for the method.

The complex Langevin technique was devised independently by Parisi (1983) and Klauder (1983) for evaluating averages such as

$$\langle G(x) \rangle = \frac{\int_{-\infty}^{\infty} dx\ G(x) \exp[-H(x)]}{\int_{-\infty}^{\infty} dx\ \exp[-H(x)]} \tag{D.1}$$

where the integration path is along the real axis for the variable $x$, but the Hamiltonian $H(x)$ is a complex (not strictly real) function of $x$. We shall begin by discussing the case where $x$ is a scalar, but then generalize to the more important situation where $x$ is replaced by an $M$-vector so that the integrals in eqn (D.1) are $M$-dimensional integrals taken along the real axis.

A convenient way to rewrite eqn (D.1) is in the form

$$\langle G(x) \rangle = \int dx\ G(x) P_c(x) \tag{D.2}$$

where $P_c(x)$ is a so-called "complex probability weight" defined by

$$P_c(x) = \frac{\exp[-H(x)]}{\int dx\ \exp[-H(x)]} \tag{D.3}$$

and it is understood that the path of integration is the real axis. In spite of its name, $P_c(x)$ is not a true probability density because it is not positive semi-definite for $H(x)$ complex. This also implies that eqn (D.2) cannot be directly tackled by Monte Carlo importance sampling of $P_c(x)$ (Landau and Binder, 2000).[119] The basic idea behind the CL technique is to assume that one can find a *real, non-negative* probability density $P(x, y)$ so that eqn (D.2) can be reexpressed as

$$\langle G(x) \rangle = \int dx \int dy\ G(x + iy) P(x, y) \tag{D.4}$$

Equation (D.4) amounts to the assumption that the *line integral* in eqn (D.2) along the real axis can be exactly rewritten as an *area integral* over the entire

---

[119] Only *stochastic* methods of evaluating integrals such as eqn (D.2) are considered, because our main interest is in the discrete field theory case where the integral is $M$-dimensional with $M \gg 1$.

complex plane of $z = x + iy$. If such a probability density $P(x,y)$ exists, so that eqns (D.2) and (D.4) are equivalent, then eqn (D.4) can be approximately evaluated with the importance sampling formula

$$\langle G(x) \rangle \approx \frac{1}{N_C} \sum_{j=1}^{N_C} G(z^j) \tag{D.5}$$

where $z^j = x^j + iy^j$ for $j = 1, 2, 3, ..., N_C$ are a set of random points in the complex plane selected from the distribution $P(x,y)$. The sign problem discussed in the context of eqn (6.129) in Section 6.3 would thereby be avoided, because no complex phase factor appears in eqn (D.5) multiplying the observable $G$ to be averaged.

For such a strategy to be realized, we require two things:

- proof that eqn (D.2) can be rewritten in the form of eqn (D.4) and that $P(x,y)$ exists for any physically reasonable $H(x)$
- a numerical scheme for importance sampling of the function $P(x,y)$

With regard to the first point, direct comparison of the right-hand sides of the two equations indicates that they are equivalent if a $P(x,y)$ can be found such that

$$
\begin{aligned}
P_c(x) &= \int dy\, P(x - iy, y) \\
&= \int dx' \int dy'\, \delta(x - x' - iy')\, P(x', y')
\end{aligned}
\tag{D.6}
$$

where we have assumed that $G(x)$, $P_c(x)$, and $P(x,y)$ are analytic functions of their $x$ arguments. The second line of this expression will prove especially useful in the following. Necessary and sufficient conditions for the existence of $P(x,y)$ have recently been identified (Salcedo, 1997; Weingarten, 2002). These conditions lead us to expect that most, if not all, physically realistic polymer field theory models will possess a real, non-negative distribution $P$ satisfying eqn (D.6).

The complex Langevin (CL) scheme is a stochastic dynamics that, *if convergent to a steady state*, provides a method for sampling the distribution $P(x,y)$ and verifying that it exists. The method amounts to writing a Langevin equation analogous to eqn (6.144), but generalizing it to trajectories $z(t) = x(t) + iy(t)$ in the complex plane according to (Parisi, 1983; Klauder, 1983)

$$
\begin{aligned}
\frac{d}{dt}x(t) &= -\lambda\, \mathrm{Re}\left[\frac{dH}{dz(t)}\right] + \eta(t) \\
\frac{d}{dt}y(t) &= -\lambda\, \mathrm{Im}\left[\frac{dH}{dz(t)}\right]
\end{aligned}
\tag{D.7}
$$

In these equations, Re and Im denote the operations of taking the real and imaginary parts of a complex function and $dH(z)/dz$ is the complex derivative

for an analytic Hamiltonian $H(z)$ (Ahlfors, 1979).[120] The random force $\eta(t)$ is a
*real, Gaussian white noise* defined by (van Kampen, 1981; Kloeden and Platen,
1992)

$$\langle \eta(t) \rangle = 0$$
$$\langle \eta(t) \eta(t') \rangle = 2\lambda \, \delta(t - t') \tag{D.8}$$

The "kinetic coefficient" $\lambda$ appearing in eqns (D.7) and (D.8) must be real and
positive, although its value is arbitrary and can be absorbed into the time variable. Here we keep it explicit for reasons that will become apparent.

There are several notable features of the CL eqns (D.7). The first is the
asymmetry with respect to the addition of the random force – the force is added
*only* to the equation for the *real* component $x(t)$ of the complex trajectory $z(t)$.
This asymmetry is necessary to preserve the broken symmetry of the original
model in the complex $x$–$y$ plane; namely, the fact that the integral in eqn (D.2)
is taken along the real axis. The noise covariance in eqn (D.8) is consistent with
the usual *fluctuation–dissipation theorem* for Brownian dynamics (van Kampen,
1981; McQuarrie, 1976), which states that the noise strength should be twice
the dissipative coefficient $\lambda$ appearing in front of the force terms in a Langevin
equation. Another important feature of eqns (D.7) is that with the random force
$\eta(t)$ removed, the equations constitute a relaxational dynamics towards a saddle
point $z^*$ of the model satisfying

$$\left. \frac{dH(z)}{dz} \right|_{z=z^*} = 0 \tag{D.9}$$

Indeed, without the random force, the CL eqns (D.7) reduce to the relaxation
scheme eqn (5.106) presented in Chapter 5 for the numerical computation of
saddle points.

The physical content of eqns (D.7) should now be clear. In the absence of the
noise, the CL equations evolve deterministically towards a nearby saddle point.
However, with the random force present, the second of the two equations drives
the stochastic sampling path to a value of $y$ that is approximately consistent
with the *local constant phase condition*[121]

$$\mathrm{Im} \frac{dH}{dz} = \frac{\partial}{\partial x} H_I(x, y) = 0 \tag{D.10}$$

The second equation in (D.7) thus attempts to maintain the dynamic trajectory
$z(t)$ on a locally constant phase path by adjusting the imaginary component
$y(t)$. In contrast, the first equation stochastically drives the trajectory along the
path through the action of the random force on the real component $x(t)$. As a
result, if the Langevin dynamics converge to a stationary distribution $P(x, y)$

---

[120]We shall assume throughout this appendix that $H(z)$ is an analytic function of $z$.

[121]Subscripts $R$ and $I$ denote the real and imaginary parts, respectively, of a complex function.

in the complex plane, we expect $P(x, y)$ to have *maximum intensity centered around a constant phase ascent path passing through one or more saddle points of the model*. The beauty of the technique is that it is *fully adaptive* – namely, the dominant saddle point and constant phase path need not be determined in advance of running a CL simulation!

Our next task is to use eqns (D.7) to derive a Fokker–Planck equation (van Kampen, 1981) for the time-dependent probability distribution $P(x, y, t)$ implied by the CL stochastic dynamics. The steady state solution of this equation, if it exists, is the real probability density $P(x, y)$. Integrating both sides of eqns (D.7) from $t$ to $t + \Delta t$ leads to

$$\Delta x \equiv x(t + \Delta t) - x(t) = \lambda \int_t^{t+\Delta t} ds \, F_R(x(s), y(s)) + \mu$$

$$\Delta y \equiv y(t + \Delta t) - y(t) = \lambda \int_t^{t+\Delta t} ds \, F_I(x(s), y(s)) \qquad \text{(D.11)}$$

where $F(z) \equiv -dH(z)/dz$ is the complex force. The quantity $\mu \equiv \int_t^{t+\Delta t} ds \, \eta(s)$ is a new Gaussian random force acting over the timestep with mean and variance that follow immediately from eqn (D.8):

$$\langle \mu \rangle = 0$$

$$\langle \mu^2 \rangle = \int_t^{t+\Delta t} ds \int_t^{t+\Delta t} ds' \, \langle \eta(s)\eta(s') \rangle = 2\lambda \Delta t \qquad \text{(D.12)}$$

It is important to note that $\mu$ is characteristically $O((\Delta t)^{1/2})$. Assuming continuity of $dH/dz$, eqns (D.11) can be approximated by

$$\Delta x = \lambda \Delta t \, F_R + \mu$$

$$\Delta y = \lambda \Delta t \, F_I \qquad \text{(D.13)}$$

with errors that are $O((\Delta t)^2)$. Using these equations, it is straightforward to show that the first two moments of the random variables $\Delta x$ and $\Delta y$, averaged over all realizations of the Gaussian force $\mu$, are given to $O(\Delta t)$ by

$$\langle \Delta x \rangle = \lambda \Delta t \, F_R, \quad \langle \Delta y \rangle = \lambda \Delta t \, F_I$$

$$\langle (\Delta x)^2 \rangle = 2\lambda \Delta t, \quad \langle (\Delta y)^2 \rangle = \langle \Delta x \Delta y \rangle = 0 \qquad \text{(D.14)}$$

These results can now be used to derive a Fokker–Planck equation for the probability density $P(x, y, t)$. The starting point is a Chapman–Kolmogorov (CK) equation strictly analogous to eqn (2.58) for the continuous Gaussian chain. Defining a two-component state vector according to $\mathbf{x} = (x, y)^T$, the CK equation can be written

$$P(\mathbf{x}, t + \Delta t) = \int d(\Delta \mathbf{x}) \, \Phi(\Delta \mathbf{x}; \mathbf{x} - \Delta \mathbf{x}) \, P(\mathbf{x} - \Delta \mathbf{x}, t) \qquad \text{(D.15)}$$

where $\Phi(\Delta \mathbf{x}; \mathbf{x})$ is the transition probability density for a displacement $\Delta \mathbf{x}$ in the complex plane, starting at the point $\mathbf{x}$, over a time interval of $\Delta t$. This function is normalized so that $\int d(\Delta\mathbf{x})\,\Phi = 1$ and its first two moments are summarized by eqn (D.14). Following the procedure outlined in Section 2.4, eqn (D.15) can be converted to a Fokker–Planck equation by expanding the left-hand side in powers of $\Delta t$ to $O(\Delta t)$ and expanding the right-hand side in powers of $\Delta\mathbf{x}$ to $O((\Delta\mathbf{x})^2) = O(\Delta t)$. This leads to

$$\Delta t \frac{\partial}{\partial t} P(\mathbf{x}, t) = -\nabla_\mathbf{x} \cdot [\langle\Delta\mathbf{x}\rangle P(\mathbf{x}, t)]$$

$$+ \frac{1}{2!} \nabla_\mathbf{x} \nabla_\mathbf{x} : [\langle\Delta\mathbf{x}\Delta\mathbf{x}\rangle P(\mathbf{x}, t)] + O((\Delta t)^2) \qquad \text{(D.16)}$$

Finally, substituting eqn (D.14) for the moments of $\Phi$ produces the desired Fokker–Planck equation for the CL process

$$\frac{\partial}{\partial t} P(x, y, t) = -\lambda \frac{\partial}{\partial x} [F_R(x, y) P(x, y, t)] - \lambda \frac{\partial}{\partial y} [F_I(x, y) P(x, y, t)]$$

$$+ \lambda \frac{\partial^2}{\partial x^2} P(x, y, t) \qquad \text{(D.17)}$$

In spite of its linearity, this Fokker–Planck equation apparently has no closed form solution for an arbitrary force $F(z)$, even in the steady state limit where $P(x, y, t) \to P(x, y)$.

Our final task is to prove that if a steady state solution $P(x, y)$ of the above equation exists, then averages computed with this solution using eqn (D.4) are equivalent to averages computed with eqn (D.2) using the complex weight $P_c(x)$ (Schoenmaker, 1987; Lee, 1994). This can be shown by combining eqns (D.6) and (D.17). Specifically, applying the operation $\int dx' \int dy'\, \delta(x - x' - iy')$ to both sides of the Fokker–Planck equation written for $P(x', y', t)$ leads to

$$\frac{\partial}{\partial t} P_c(x, t) = T_1(x, t) + T_2(x, t) + T_3(x, t) \qquad \text{(D.18)}$$

where $P_c(x, t) \equiv \int dy\, P(x - iy, y, t)$ and $T_j(x, t)$ is the function obtained by applying the indicated operation to the $j$th term on the right-hand side of eqn (D.17). $T_1$ can be manipulated as follows:

$$T_1(x, t) = \lambda \int dx' \int dy'\, \delta(x - x' - iy') \frac{\partial}{\partial x'} \left[ \text{Re}\left( \frac{dH(x' + iy')}{d(x' + iy')} \right) P(x', y', t) \right]$$

$$= \lambda \int dy' \frac{\partial}{\partial x} \left[ \text{Re}\left( \frac{dH(x)}{dx} \right) P(x - iy', y', t) \right]$$

$$= \lambda \frac{\partial}{\partial x} \left[ \text{Re}\left( \frac{dH(x)}{dx} \right) P_c(x, t) \right] \qquad \text{(D.19)}$$

Similarly, the second term can be written

$$T_2(x,t) = \lambda \int dx' \int dy' \; \delta(x - x' - iy') \frac{\partial}{\partial y'} \left[ \mathrm{Im} \left( \frac{dH(x' + iy')}{d(x' + iy')} \right) P(x', y', t) \right]$$

$$= i\lambda \int dx' \int dy' \; \delta(x - x' - iy') \frac{\partial}{\partial x} \left[ \mathrm{Im} \left( \frac{dH(x)}{dx} \right) P(x', y', t) \right]$$

$$= i\lambda \frac{\partial}{\partial x} \left[ \mathrm{Im} \left( \frac{dH(x)}{dx} \right) P_c(x, t) \right] \tag{D.20}$$

Finally, the last term is

$$T_3(x,t) = \lambda \int dx' \int dy' \; \delta(x - x' - iy') \frac{\partial^2}{(\partial x')^2} P(x', y', t)$$

$$= \lambda \int dy' \; \frac{\partial^2}{\partial x^2} P(x - iy', y', t)$$

$$= \lambda \frac{\partial^2}{\partial x^2} P_c(x, t) \tag{D.21}$$

Combining these results, we see that the function $P_c(x, t)$ satisfies the following *complex* Fokker–Planck (FP) equation:

$$\frac{\partial}{\partial t} P_c(x, t) = \frac{\partial}{\partial x} \lambda \left[ \frac{\partial}{\partial x} + \frac{dH(x)}{dx} \right] P_c(x, t) \tag{D.22}$$

This equation has a complex steady state solution $P_c(x) \propto \exp[-H(x)]$ corresponding to eqn (D.3), but also a second "spurious" steady state (Lee, 1994)

$$P_{spur}(x) \propto \exp[-H(x)] \int^x dy \; \exp[H(y)] \tag{D.23}$$

This spurious solution is usually not relevant, because it leads to expectation values that are incompatible with the most common choices of boundary conditions, e.g. periodic. Thus, if the real FP eqn (D.17) *converges to a steady state* $P(x, y)$, then the associated function $P_c(x) = \int dy \, P(x - iy, y)$ corresponds to the desired complex probability given by eqn (D.3). When this condition is met, a "time average" computed with eqn (D.5) along a complex Langevin trajectory will converge to the ensemble average (D.2) in the limit of $N_C \to \infty$.

Unfortunately, an analytical proof that eqn (D.17) has a steady state solution is not in hand. Nevertheless, there is a simple test to ensure that a numerical CL simulation is working properly (Gausterer and Lee, 1993; Lee, 1994). *If the expectation values for analytic observables $G(x)$ become time independent over the course of a CL simulation, then it can be proven that these values are correct and in agreement with eqn (D.2).* In practice, we have not encountered convergence problems in CL simulations for any of the models considered in Chapter 4.

The complex Langevin equations given in (D.7) constitute the "standard" CL approach. However, there are several extensions of the formalism that are potentially useful in conducting field-theoretic simulations. The first of these is a generalization to include a noise source acting on both the real *and imaginary* parts of the field:[122]

$$\frac{d}{dt}x(t) = -\lambda \, \mathrm{Re} \left[ \frac{dH}{dz(t)} \right] + \eta_R(t)$$

$$\frac{d}{dt}y(t) = -\lambda \, \mathrm{Im} \left[ \frac{dH}{dz(t)} \right] + \eta_I(t) \qquad (\mathrm{D}.24)$$

The random forces $\eta_R(t)$ and $\eta_I(t)$ can be taken to be real, Gaussian white noise processes with vanishing mean values, $\langle \eta_R(t) \rangle = \langle \eta_I(t) \rangle = 0$. The covariance of the noise can be further generalized from eqn (D.8) to

$$\langle \eta_R(t)\eta_R(t') \rangle = 2(\lambda + \epsilon) \, \delta(t - t')$$
$$\langle \eta_I(t)\eta_I(t') \rangle = 2\epsilon \, \delta(t - t')$$
$$\langle \eta_R(t)\eta_I(t') \rangle = 0 \qquad (\mathrm{D}.25)$$

where $\lambda > 0$ and $\epsilon \geq 0$ are real parameters that determine the relative strengths of the two noise components. Equations (D.24) and (D.25) evidently reduce to the "standard" CL eqns (D.7)–(D.8) in the special case of $\epsilon = 0$. Since the parameter $\lambda$ can be absorbed into the time scale of the stochastic process, $\epsilon$ is effectively a free parameter that can be adjusted to optimize the performance of a CL simulation.

The role of $\epsilon$ can be established by deriving the Fokker–Planck equation corresponding to these *generalized CL equations*. By repeating the steps leading to eqn (D.17) it is straightforward to show that eqns (D.24)–(D.25) are consistent with the Fokker–Planck equation

$$\frac{\partial}{\partial t}P(x,y,t) = -\lambda \frac{\partial}{\partial x}[F_R(x,y)P(x,y,t)] - \lambda \frac{\partial}{\partial y}[F_I(x,y)P(x,y,t)]$$
$$+ (\lambda + \epsilon)\frac{\partial^2}{\partial x^2}P(x,y,t) + \epsilon\frac{\partial^2}{\partial y^2}P(x,y,t) \qquad (\mathrm{D}.26)$$

Moreover, it can be shown that if this generalized Fokker–Planck equation has a steady state $P(x,y)$, then the associated function $P_c(x) = \int dy \, P(x - iy, y)$ reduces to eqn (D.3). Thus, the generalized CL equations are a suitable alternative to the standard CL approach. For non-zero $\epsilon$, however, eqn (D.26) shows that the generalized scheme contains an extra dispersive term $\epsilon\partial_y^2 P$ that will tend to *smooth* the steady state distribution in the variable $y$.

To illustrate the role of finite $\epsilon$, we return to the "toy" model of eqn (5.7) for the case of $p = 1$, i.e. $H(x) = ix + x^2/2$. The steady state solution of eqn (D.26) for this simple quadratic Hamiltonian is

$$P(x,y) \sim \exp \left( -\frac{\lambda x^2}{2(\lambda + \epsilon)} - \frac{\lambda (y+1)^2}{2\epsilon} \right) \qquad (\mathrm{D}.27)$$

---

[122]We are not aware of this generalization having been reported previously in the literature.

apart from a normalization constant. The distribution function has a Gaussian ridge of maximum probability centered on the line $y = -1$ in the complex plane that passes through the saddle point, $z^* = -i$. This line is, of course, the constant-phase ascent path for the model. The width of the ridge normal to this line is $O(\epsilon^{1/2})$, while the decay in probability density away from the saddle point along the line $y = -1$ is much slower for $\epsilon \ll \lambda$. For $\epsilon \to 0+$, the conventional CL theory is recovered and eqn (D.27) reduces to the singular distribution

$$P(x, y) \sim \delta(y + 1) \exp\left(-x^2/2\right) \tag{D.28}$$

At least for this simple model, it is clear that the generalized CL theory with $\epsilon > 0$ produces a smoother steady state distribution function that is centered on the constant-phase ascent path.

The CL theory can be immediately extended to the *multi-dimensional case* of model Hamiltonians $H(\mathbf{x})$, where $\mathbf{x} = (x_1, x_2, ..., x_M)^T$ is a real $M$-vector and $H$ is complex valued. In particular, the generalized CL eqns (D.24)–(D.25) can be written in the multi-variate case as

$$\frac{d}{dt}\mathbf{x}(t) = -\lambda \operatorname{Re}\left[\frac{\partial H}{\partial \mathbf{z}(t)}\right] + \boldsymbol{\eta}_R(t)$$

$$\frac{d}{dt}\mathbf{y}(t) = -\lambda \operatorname{Im}\left[\frac{\partial H}{\partial \mathbf{z}(t)}\right] + \boldsymbol{\eta}_I(t) \tag{D.29}$$

where $\mathbf{z} = \mathbf{x} + i\mathbf{y}$ is a complex $M$-vector. The Gaussian random forces $\boldsymbol{\eta}_R(t)$ and $\boldsymbol{\eta}_I(t)$ are $M$-vectors with vanishing mean and covariance matrices given by

$$\langle \boldsymbol{\eta}_R(t)\boldsymbol{\eta}_R(t')\rangle = 2(\lambda + \epsilon)\,\mathbf{1}\,\delta(t - t')$$
$$\langle \boldsymbol{\eta}_I(t)\boldsymbol{\eta}_I(t')\rangle = 2\epsilon\,\mathbf{1}\,\delta(t - t')$$
$$\langle \boldsymbol{\eta}_R(t)\boldsymbol{\eta}_I(t')\rangle = 0 \tag{D.30}$$

where $\mathbf{1}$ is the $M \times M$ unit tensor. By setting $\epsilon = 0$, the above scheme reduces to the standard multi-variate CL theory.

Finally, in the multi-dimensional case, the parameters $\lambda$ and $\epsilon$ can be replaced by a positive definite, $M \times M$ "kinetic coefficient" matrix $\boldsymbol{\lambda}$ and a positive semi-definite, $M \times M$ "noise" matrix $\boldsymbol{\epsilon}$. This generalization amounts to the scheme

$$\frac{d}{dt}\mathbf{x}(t) = -\boldsymbol{\lambda} \cdot \operatorname{Re}\left[\frac{\partial H}{\partial \mathbf{z}(t)}\right] + \boldsymbol{\eta}_R(t)$$

$$\frac{d}{dt}\mathbf{y}(t) = -\boldsymbol{\lambda} \cdot \operatorname{Im}\left[\frac{\partial H}{\partial \mathbf{z}(t)}\right] + \boldsymbol{\eta}_I(t) \tag{D.31}$$

with

$$\langle \boldsymbol{\eta}_R(t)\boldsymbol{\eta}_R(t')\rangle = 2(\boldsymbol{\lambda} + \boldsymbol{\epsilon})\,\delta(t - t')$$
$$\langle \boldsymbol{\eta}_I(t)\boldsymbol{\eta}_I(t')\rangle = 2\boldsymbol{\epsilon}\,\delta(t - t')$$
$$\langle \boldsymbol{\eta}_R(t)\boldsymbol{\eta}_I(t')\rangle = 0 \tag{D.32}$$

Again it can be proven that if the above CL equations converge to a steady state $P(\mathbf{x}, \mathbf{y})$, the steady state solution is consistent with the proper complex probability weight $P_c(\mathbf{x}) \sim \exp[-H(\mathbf{x})]$. By adjusting the form of the matrices $\boldsymbol{\lambda}$ and $\boldsymbol{\epsilon}$ it may be possible to achieve better performance in numerical simulations than the simplest choice of $\boldsymbol{\lambda} = \lambda\mathbf{1}$ and $\boldsymbol{\epsilon} = \epsilon\mathbf{1}$. A potentially important class of rank-$M$ matrices is those that are translationally invariant on a $d$-dimensional collocation grid and can be diagonalized by a discrete Fourier transform. By employing such matrices, efficient pseudo-spectral numerical solutions of the CL equations can be achieved.

# REFERENCES

Abetz, V. and Goldacker, T. (2000). *Macromol. Rapid Comm.*, **21**, 16.

Adams, J. C. and Swarztrauber, P. N. (2003). Spherepack 3.1: A model development facility, http://www.scd.ucar.edu/css/software/spherepack.

Adams, J. L., Quiram, D. J., Graessley, W. W., Register, R. A., and Marchand, G. R. (1996). *Macromolecules*, **29**, 2929.

Ahlfors, L. V. (1979). *Complex Analysis* (3rd edn). McGraw-Hill, New York.

Alexander, S. (1977). *J. Phys. (Paris)*, **38**, 983.

Alexander-Katz, A., Moreira, A. G., and Fredrickson, G. H. (2003). *J. Chem. Phys.*, **118**, 9030.

Alexander-Katz, A., Moreira, A. G., Sides, S. W., and Fredrickson, G. H. (2005). *J. Chem. Phys.*, **122**, 014904.

Allen, M. P. and Tildesley, D. J. (1987). *Computer Simulation of Liquids*. Oxford University Press, Oxford.

Almdal, K., Mortensen, K., Ryan, A. J., and Bates, F. S. (1996). *Macromolecules*, **29**, 5940.

Almdal, K., Rosedale, J. H., Bates, F. S., Wignall, G. D., and Fredrickson, G. H. (1990). *Phys. Rev. Lett.*, **65**, 1112.

Ames, W. F. (1977). *Numerical Methods for Partial Differential Equations* (2nd edn). Academic, New York.

Amit, D. J. (1984). *Field Theory, The Renormalization Group, and Critical Phenomena* (2nd edn). World Scientific, Singapore.

Amit, D. J. and Zannetti, M. (1973). *J. Stat. Phys.*, **9**, 1.

Anisimov, M. A., Kiselev, S. B., Sengers, J. V., and Tang, S. (1992). *Physica A*, **188**, 487.

Araki, T. and Tanaka, H. (2001). *Macromolecules*, **34**, 1953.

Arfken, G. (1985). *Mathematical Methods for Physicists* (3rd edn). Academic, New York.

Ascher, U. M., Ruuth, S. J., and Wetton, B. T. R. (1995). *SIAM J. Num. Anal.*, **32**, 797.

Ashcroft, N. W. and Mermin, N. D. (1976). *Solid State Physics*. Saunders College, Philadelphia.

Attig, N., Binder, K., Grubmüller, H., and Kremer, K. (ed.) (2004). *Computational Soft Matter: From Synthetic Polymers to Proteins*. NIC, Jülich.

Auschra, C. and Stadler, R. (1993). *Macromolecules*, **26**, 2171.

Baer, R., Head-Gordon, M., and Neuhauser, D. (1998). *J. Chem. Phys.*, **109**, 6219.

Baeurle, S. A. (2002). *Phys. Rev. Lett.*, **89**, 080602.

Baigl, D., Seery, A. P., and Williams, C. E. (2002). *Macromolecules*, **35**, 2318–2326.

Baker, W., Scott, C., and Hu, G.-H. (ed.) (2001). *Reactive Polymer Blending*. Hanser, Munich.

Barrat, J. L. and Fredrickson, G. H. (1991). *J. Chem. Phys.*, **95**, 1281.

Barrat, J.-L., Fredrickson, G. H., and Sides, S. W. (2005). *J. Phys. Chem. B*, **109**, 6694.

Barrat, J.-L. and Joanny, J.-F. (1996). *Adv. Chem. Phys.*, **94**, 1.

Baschnagel, J., Binder, K., Doruker, P., Gusev, A. A., Hahn, O., Kremer, K., Mattice, W. L., Muller-Plathe, F., Murat, M., Paul, W., Santos, S., Suter, U. W., Tries, V., and Abe, A. (2000). *Adv. in Polym. Sci.*, **152**, 41.

Bates, F. S. and Fredrickson, G. H. (1990). *Annu. Rev. Phys. Chem.*, **41**, 525.

Bates, F. S. and Fredrickson, G. H. (1999). *Phys. Today*, **52**(2), 32.

Bates, F. S. and Hartney, M. A. (1985). *Macromolecules*, **18**, 2478.

Bates, F. S., Maurer, W., Lodge, T. P., Schulz, M. F., Matsen, M. W., Almdal, K., and Mortensen, K. (1995). *Phys. Rev. Lett.*, **75**, 4429.

Bates, F. S., Maurer, W. W., Lipic, P. M., Hillmyer, M. A., Almdal, K., Mortensen, K., Fredrickson, G. H., and Lodge, T. P. (1997). *Phys. Rev. Lett.*, **79**, 849.

Bates, F. S., Rosedale, J. H., and Fredrickson, G. H. (1990). *J. Chem. Phys.*, **92**, 6255.

Bates, F. S., Rosedale, J. H., Fredrickson, G. H., and Glinka, C. J. (1988). *Phys. Rev. Lett.*, **61**, 2229.

Bates, F. S., Rosedale, J. H., Stepanek, P., Lodge, T. P., Wiltzius, P., Fredrickson, G. H., and Hjelm, R. P. (1990). *Phys. Rev. Lett.*, **65**, 1893.

Bates, F. S., Schulz, M. F., Khandpur, A. K., Forster, S., Rosedale, J. H., Almdal, K., and Mortensen, K. (1994). *Faraday Discuss.* (98), 7.

Bates, F. S. and Wignall, G. D. (1986). *Phys. Rev. Lett.*, **57**, 1429.

Batrouni, G. G., Katz, G. R., Kronfeld, A. S., Lepage, G. P., Svetitsky, B., and Wilson, K. G. (1985). *Phys. Rev. D*, **32**, 2736.

Belyakov, M. Y. and Kiselev, S. B. (1992). *Physica A*, **190**, 75.

Bendejacq, D. D., Ponsinet, V., and Joanicot, M. (2005). *Langmuir*, **21**, 1712.

Bender, C. M. and Orszag, S. A. (1978). *Advanced Mathematical Methods for Scientists and Engineers*. McGraw-Hill, New York.

Benoit, H. and Hadziioannou, G. (1988). *Macromolecules*, **21**, 1449.

Berne, B. J. and Pecora, R. (1976). *Dynamic Light Scattering*. Wiley-Interscience, New York.

Binder, K. (1984). *Phys. Rev. A*, **29**, 341.

Binder, K. (1994). *Adv. Polym. Sci.*, **112**, 181.

Binder, K. (ed.) (1995). *Monte Carlo and Molecular Dynamics Simulations in Polymer Science*. Oxford University Press, New York.

Binder, K. (1999). *Adv. Polym. Sci.*, **138**, 1.

Binder, K. and Ciccotti, G. (ed.) (1996). *Monte Carlo and Molecular Dynamics of Condensed Matter Systems*. Italian Physical Society, Bologna.

Binder, K. and Heermann, D. W. (1988). *Monte Carlo Simulations in Statistical Physics*. Springer-Verlag, Berlin.

Binder, K. and Müller, M. (2000*a*). *Curr. Opin. Coll. In.*, **5**, 315.

Binder, K. and Müller, M. (2000*b*). *Macromol. Symp.*, **149**, 1.

Binder, K., Müller, M., Schmid, F., and Werner, A. (2001). *Adv. Coll. In.*, **94**, 237.

Bird, R. B., Hassager, O., Armstrong, R. C., and Curtiss, C. F. (1977). *Dynamics of Polymeric Liquids*, Volume 2. John Wiley & Sons, New York.

Blumstein, A. (ed.) (1985). *Polymer Liquid Crystals*. Plenum, New York.

Boffa, L. S. and Novak, B. M. (2000). *Chem. Rev.*, **100**, 1479.

Bohbot-Raviv, Y. and Wang, Z. G. (2000). *Phys. Rev. Lett.*, **85**, 3428.

Borukhov, I., Andelman, D., and Orland, H. (1998). *Eur. Phys. J. B*, **5**, 869.

Bosman, A. W., Sijbesma, R. P., and Meijer, E. W. (2004). *Mat. Today* (4), 34.

Boyd, J. P. (2001). *Chebyshev and Fourier Spectral Methods* (2nd edn). Dover, New York.

Brandt, A. (1977). *Math. Comp.*, **31**, 333.

Brazovskii, S. A. (1975). *Sov. Phys. JETP (USSR)*, **41**, 85.

Brezin, E., Le Guillou, J. C., and Zinn-Justin, J. (1976). In *Phase Transitions and Critical Phenomena* (ed. C. Domb and M. S. Green), Volume 6. Academic, New York.

Briggs, W. L. (1987). *A Multigrid Tutorial*. S.I.A.M., Philadelphia.

Brinkmann-Rengel, S., Abetz, V., Stadler, R., and Thomas, E. L. (1999). *Kaut. Gummi Kunst.*, **52**, 806.

Broseta, D. and Fredrickson, G. H. (1990). *J. Chem. Phys.*, **93**, 2927.

Broseta, D., Leibler, L., Kaddour, L. O., and Strazielle, C. (1987). *J. Chem. Phys.*, **87**, 7248.

Brydson, J. A. (1995). *Plastic Materials* (6th edn). Butterworth-Heinemann, Oxford.

Burkhardt, T. W. and van Leeuwen, J. M. J. (ed.) (1982). *Real-Space Renormalization*, Volume 30 of *Topics in Current Physics*. Springer-Verlag, Berlin.

Caillol, J.-M. (2003). *Mol. Phys.*, **101**, 1617.

Callaway, D. J. E. and Petronzio, R. (1984). *Phys. Lett. B*, **139**, 189.

Cao, Y., Smith, P., and Heeger, A. J. (1992). *Synthetic Met.*, **48**, 91.

Cates, M. E. (1988). *J. Phys. (Paris)*, **49**, 1593.

Catterall, S. and Karamov, S. (2002). *Phys. Lett. B*, **528**, 301.

Ceniceros, H. D. and Fredrickson, G. H. (2004). *Mult. Mod. Simulat.*, **2**, 452.

Ceperley, D., Chester, C. V., and Kalos, M. H. (1977). *Phys. Rev. B*, **16**, 3081.

Chaikin, P. M. and Lubensky, T. C. (1995). *Principles of Condensed Matter Physics*. Cambridge University Press, Cambridge.

Chandler, D. (1982). In *The Liquid State of Matter* (ed. E. W. Montroll and J. L. Lebowitz), pp. 275–340. North-Holland, Amsterdam.

Chandler, D. (1987). *Introduction to Modern Statistical Mechanics*. Oxford University Press, New York.

Chandler, D. (1993). *Phys. Rev. E*, **48**, 2898.

Chandrasekhar, S. (1943). *Rev. Mod. Phys.*, **15**, 1.

Cheng, J. Y., Ross, C. A., Chan, V. Z. H., Thomas, E. L., Lammertink, R. G. H., and Vancso, G. J. (2001). *Adv. Mater.*, **13**, 1174.

Cherayil, B. J., Douglas, J. F., and Freed, K. (1985). *J. Chem. Phys.*, **83**, 5293.

Chiu, J. J., Kim, B. J., Kramer, E. J., and Pine, D. J. (2005). *J. Am. Chem. Soc.*, **127**, 5036.

Chong, B. Y. K., Le, T. P. T., Moad, G., Rizzardo, E., and Thang, S. H. (1999). *Macromolecules*, **32**, 2071.

Chu, B., Ying, Q., Linliu, K., Xie, P., Gao, T., Li, Y., Nose, T., and Okada, M. (1992). *Macromolecules*, **25**, 7382.

Chung, T. J. (2002). *Computational Fluid Mechanics*. Cambridge University Press, Cambridge.

Ciferri, A., Krigbaum, W. R., and Meyer, R. B. (ed.) (1982). *Polymer Liquid Crystals*. Academic, New York.

Coates, G. W., Hustad, P. D., and Reinartz, S. (2002). *Angew. Chem. Int. Edit.*, **41**, 2237.

Cochran, E. W., Garcia-Cervera, C., and Fredrickson, G. H. (2005). Unpublished work.

Cohen, Y., Albalak, R. J., Dair, B. J., Capel, M. S., and Thomas, E. L. (2000). *Macromolecules*, **33**, 6502.

Danielson, E., Golden, J. H., McFarland, E. W., Reaves, C. M., Weinberg, W. H., and Wu, X. D. (1997). *Nature*, **389**, 944.

Daoud, M., Cotton, J. P., Farnoux, B., Jannink, G., Sarma, G., Benoit, H., Duplessix, R., Picot, C., and de Gennes, P. G. (1975). *Macromolecules*, **8**, 804.

Dautzenberg, H., Jaeger, W., Kotz, J., Philipp, B., Seidel, Ch., and Stscherbina, D. (1994). *Polyelectrolytes: Formation, Characterization, and Application*. Hanser Gardner, Munich.

de Gennes, P. G. (1969). *Rep. Prog. Phys.*, **32**, 187.

de Gennes, P. G. (1977). *J. Phys. Lett. (Paris)*, **38**, L–441.

de Gennes, P. G. (1979). *Scaling Concepts in Polymer Physics*. Cornell University Press, Ithaca, NY.

de Gennes, P. G. (1980*a*). *Macromolecules*, **13**, 1069.

de Gennes, P. G. (1980*b*). *C. R. Acad. Sci. Paris*, **290 B**, 509.

de Gennes, P. G. and Prost, J. (1993). *The Physics of Liquid Crystals*. Oxford University Press, New York.

de la Cruz, M. O. (1989). *J. Chem. Phys.*, **90**, 1995.

de la Cruz, M. O. (1991). *Phys. Rev. Lett.*, **67**, 85.

de la Cruz, M. O., Edwards, S. F., and Sanchez, I. C. (1988). *J. Chem. Phys.*, **89**, 1704.

de Raedt, H. and Lagendijk, A. (1981). *Phys. Rev. Lett.*, **46**, 77.

Dennis, J. E. Jr. and Schnabel, R. B. (1996). *Numerical Methods for Unconstrained Optimization and Nonlinear Equations*. Prentice Hall, New York.

des Cloizeaux, J. and Jannink, G. (1990). *Polymers in Solution. Their Modelling and Structure*. Clarendon Press, Oxford.

Desrosiers, P., Guan, S. H., Hagemeyer, A., Lowe, D. M., Lugmair, C., Poojary, D. M., Turner, H., Weinberg, H., Zhou, X. P., Armbrust, R., Fengler, G., and Notheis, U. (2003). *Catal. Today*, **81**, 319.

Diamanti, S. J., Ghosh, P., Shimizu, F., and Bazan, G. C. (2003). *Macromolecules*, **36**, 9731.

Diana, F. S., Lee, S. H., Petroff, P. M., and Kramer, E. J. (2003). *Nano Letters*, **3**, 891.

Dobrynin, A. V. and Leibler, L. (1997). *Macromolecules*, **30**(16), 4756.

Dobrynin, A. V. and Rubinstein, M. (2001). *Macromolecules*, **34**, 1964.

Doi, M. (2003). The octa project, http://octa.jp/.

Doi, M. and Edwards, S. F. (1986). *The Theory of Polymer Dynamics*. Oxford University Press, New York.

Doi, M. and Onuki, A. (1992). *J. Phys. II*, **2**, 1631.

Dolan, A. K. and Edwards, S. F. (1975). *Proc. R. Soc. Lond. Ser. A*, **343**, 427.

Doll, J. D. and Freedman, D. L. (1988). *Adv. Chem. Phys.*, **73**, 289.

Donley, J. P., Rajasekaran, J. J., McCoy, J. D., and Curro, J. G. (1995). *J. Chem. Phys.*, **103**, 5061.

Dormidontova, E. E. and Lodge, T. P. (2001). *Macromolecules*, **34**, 9143.

Driscoll, J., Healy, D. M., and Rockmore, D. (1997). *SIAM J. Comput.*, **26**, 1066.

Drolet, F. and Fredrickson, G. H. (1999). *Phys. Rev. Lett.*, **83**, 4317.

Drolet, F. and Fredrickson, G. H. (2001). *Macromolecules*, **34**, 5317.

Duane, S., Kennedy, A. D., Pendleton, B. J., and Roweth, D. (1987). *Phys. Lett. B*, **195**, 216.

Düchs, D., Ganesan, V., Fredrickson, G. H., and Schmid, F. (2003). *Macromolecules*, **36**, 9237.

Düchs, D. and Sullivan, D. E. (2002). *J. Phys. Cond. Mat.*, **14**, 12189.

Duplantier, B. (1982). *J. Phys. (Paris)*, **43**, 991.

Duque, D., Katsov, K., and Schick, M. (2002). *J. Chem. Phys.*, **117**, 10315.

Duque, D. and Schick, M. (2000). *J. Chem. Phys.*, **113**, 5525.

Edmonds, A. R. (1974). *Angular Momentum in Quantum Mechanics*. Princeton University Press, Princeton, NJ.

Edwards, S. F. (1959). *Philos. Mag.*, **4**, 1171.

Edwards, S. F. (1965). *Proc. Phys. Soc.*, **85**, 613.

Edwards, S. F. (1966). *Proc. Phys. Soc.*, **88**, 265.

Edwards, S. F. (1975). *J. Phys. A.–Math. Gen.*, **8**, 1670.

Epps, T. H., Cochran, E. W., Hardy, C. M., Bailey, T. S., Waletzko, R. S., and Bates, F. S. (2004). *Macromolecules*, **37**, 7085.

Ermak, D. L. (1975). *J. Chem. Phys.*, **62**, 4189.

Erzan, A. and Stell, G. (1977). *Phys. Rev. B*, **16**, 4146.

Eyges, L. (1972). *The Classical Electromagnetic Field*. Dover, New York.

Falcioni, M., Martinelli, G., Paciello, M. L., Parisi, G., and Taglienti, B. (1986). *Nucl. Phys. B*, **265**, 187.

Feit, M. D., Fleck, J. A. J., and Steiger, A. (1982). *J. Comput. Phys.*, **47**, 412.

Feng, E. and Fredrickson, G. H. (2005). Unpublished work.

Fetter, A. L. and Walecka, J. D. (1980). *Theoretical Mechanics of Particles and Continua*. McGraw-Hill, New York.

Feynman, R. P. and Hibbs, A. R. (1965). *Quantum Mechanics and Path Integrals*. McGraw-Hill, New York.

Filinov, V. S. (1986). *Nucl. Phys. B*, **271**, 717.

Fleck, J. A. J., Morris, J. R., and Feit, M. D. (1976). *Appl. Phys.*, **10**, 129.

Fleer, G. J., Cohen-Stuart, M. A., Scheutjens, J. M. H. M., Cosgrove, T., and Vincent, B. (1993). *Polymers at Interfaces*. Chapman & Hall, New York.

Flory, P. J. (1953). *Principles of Polymer Chemistry*. Cornell University Press, Ithaca, NY.

Flory, P. J. (1969). *Statistics of Chain Molecules*. Interscience Publishers, New York.

Förster, S. and Schmidt, M. (1995). *Adv. Polym. Sci.*, **120**, 50.

Foster, D. P., Jasnow, D., and Balazs, A. C. (1995). *Macromolecules*, **28**, 3450.

Fraaije, J. G. E. M. (1993). *J. Chem. Phys.*, **99**, 9202.

Fraaije, J. G. E. M. and Sevink, G. J. A. (2003). *Macromolecules*, **36**, 7801.

Fraaije, J. G. E. M., van Vlimmeren, B. A. C., Maurits, N. M., Postma, M., Evers, O. A., Hoffmann, C., Altevogt, P., and GoldbeckWood, G. (1997). *J. Chem. Phys.*, **106**, 4260.

Fraaije, J. G. E. M., Zvelindovsky, A. V., and Sevink, G. J. A. (2004). *Mol. Simulat.*, **30**, 225.

Fredrickson, G. H. (2002). *J. Chem. Phys.*, **117**, 6810.

Fredrickson, G. H., Ajdari, A., Leibler, L., and Carton, J. P. (1992). *Macromolecules*, **25**, 2882.

Fredrickson, G. H. and Bates, F. S. (1997). *J. Polym. Sci. Polym. Phys.*, **35**, 2775.

Fredrickson, G. H. and Binder, K. (1989). *J. Chem. Phys.*, **91**, 7265.

Fredrickson, G. H., Ganesan, V., and Drolet, F. (2002). *Macromolecules*, **35**, 16.

Fredrickson, G. H. and Helfand, E. (1987). *J. Chem. Phys.*, **87**, 697.

Fredrickson, G. H. and Leibler, L. (1989). *Macromolecules*, **22**, 1238.

Fredrickson, G. H. and Milner, S. T. (1991). *Phys. Rev. Lett.*, **67**, 835.

Fredrickson, G. H., Milner, S. T., and Leibler, L. (1992). *Macromolecules*, **25**, 6341.

Fredrickson, G. H. and Sides, S. W. (2003). *Macromolecules*, **36**, 5415.

Freed, K. (1972). *Adv. Chem. Phys.*, **22**, 1.

Freed, K. (1987). *Renormalization Group Theory of Macromolecules*. Wiley, New York.

Frenkel, D. and Smit, B. (1996). *Understanding Molecular Simulation*. Academic, New York.

Frigo, M. and Johnson, S. G. (1998). *Proc. ICASSP*, **1998**, 1381.

Frischknecht, A. L., Weinhold, J. D., Salinger, A. G., Curro, J. G., Frink, L. J. D., and McCoy, J. D. (2002). *J. Chem. Phys.*, **117**, 10385.

Fuchs, M. and Schweizer, K. S. (2002). *J. Phys.–Cond. Matter*, **14**, R239.

Ganesan, V. and Fredrickson, G. H. (2001). *Europhys. Lett.*, **55**, 814.

Garcia-Cervera, C. (2005). Personal communication.

Gausterer, H. and Lee, S. (1993). *J. Stat. Phys.*, **73**, 147.

Gido, S. P. and Thomas, E. L. (1994). *Macromolecules*, **27**, 6137.

Ginzburg, V. L. (1960). *Sov. Phys. Solid State*, **2**, 1824.

Goldacker, T., Abetz, V., Stadler, R., Erukhimovich, I., and Leibler, L. (1999). *Nature*, **398**, 137.

Goldenfeld, N. (1992). *Lectures on Phase Transitions and the Renormalization Group*. Addison-Wesley, New York.

Gottfried, A. C. and Brookhart, M. (2003). *Macromolecules*, **36**, 3085.

Gottlieb, D. and Orszag, S. A. (1977). *Numerical Analysis of Spectral Methods: Theory and Applications*. Society for Industrial and Applied Mathematics (SIAM), Philadelphia.

Gottschalk, M., Linse, P., and Piculell, L. (1998). *Macromolecules*, **31**, 8407.

Grason, G. M. and Kamien, R. D. (2004). *Macromolecules*, **37**, 7371.

Greer, S. C. (1996). *Adv. Chem. Phys.*, **96**, 261.

Grest, G. S. and Murat, M. (1995). In *Monte Carlo and Molecular Dynamics Simulations in Polymer Science* (ed. K. Binder), pp. 476–578. Oxford University Press, New York.

Grinstein, G. (1985). In *Fundamental Problems in Statistical Mechanics IV* (ed. E. G. D. Cohen). North-Holland, Amsterdam.

Gropp, W., Lusk, W., and Skjellum, A. (1999). *Using MPI*. MIT Press, Cambridge, MA.

Guenza, M. and Schweizer, K. S. (1997). *J. Chem. Phys.*, **106**, 7391.

Gunton, J. D. and Droz, M. (1983). *Introduction to the Theory of Metastable and Unstable States*, Volume 183 of *Lecture Notes in Physics*. Springer-Verlag, Berlin.

Gupta, A. M. and Edwards, S. F. (1993). *J. Chem. Phys.*, **98**, 1588.

Hair, D. W., Hobbie, E. K., Nakatani, A. I., and Hahn, C. C. (1992). *J. Chem. Phys.*, **96**, 9133.

Hajduk, D. A., Harper, P. E., Gruner, S. M., Honeker, C. C., Kim, G., and Thomas, E. L. (1994). *Macromolecules*, **27**, 4063.

Hajduk, D. A., Harper, P. E., Gruner, S. M., Honeker, C. C., Kim, G., Thomas, E. L., and Fetters, L. J. (1995). *Macromolecules*, **28**, 2570.

Hajduk, D. A., Takenouchi, H., Hillmyer, M. A., Bates, F. S., Vigild, M. E., and Almdal, K. (1997). *Macromolecules*, **30**, 3788.

Halperin, A., Tirrell, M., and Lodge, T. P. (1992). *Adv. Polym. Sci.*, **100**, 31.

Hamley, I. W. (1998). *The Physics of Block Copolymers*. Oxford University Press, Oxford.

Hamley, I. W. and Podneks, V. E. (1997). *Macromolecules*, **30**, 3701.

Hamm, M., Goldbeck-Wood, G., Zvelindovsky, A. V., and Fraaije, J. G. E. M. (2003). *J. Chem. Phys.*, **118**, 9401.

Hamm, M., Goldbeck-Wood, G., Zvelindovsky, A. V., Sevink, G. J. A., and

Fraaije, J. G. E. M. (2001). *Macromolecules*, **34**, 8378.

Hammond, M. R., Sides, S. W., Fredrickson, G. H., Kramer, E. J., Ruokolainen, J., and Hahn, S. F. (2003). *Macromolecules*, **36**, 8712.

Hansen, J.-P. and McDonald, I. R. (1986). *Theory of Simple Liquids*. Academic Press, New York.

Hara, M. (1993). *Polyelectrolytes: Science and Technology*. Marcel Dekker, New York.

Harris, R. A. and Hearst, J. E. (1966). *J. Chem. Phys.*, **44**, 2595.

Hasegawa, H. and Doi, M. (1997). *Macromolecules*, **30**, 3086.

Hasenfratz, A., Hasenfratz, P., Heller, U., and Karsch, F. (1984). *Phys. Lett. B*, **140**, 76.

Hasenfratz, A. and Margaritis, A. (1983). *Phys. Lett. B*, **133**, 211.

Hawker, C. J., Bosman, A. W., and Harth, E. (2001). *Chem. Rev.*, **101**, 3661.

Helfand, E. (1975). *J. Chem. Phys.*, **62**, 999.

Helfand, E. and Fredrickson, G. H. (1989). *Phys. Rev. Lett.*, **62**, 2468.

Helfand, E. and Tagami, Y. (1971). *J. Polym. Sci., Polym. Lett.*, **9**, 741.

Helfand, E. and Wasserman, Z. R. (1976). *Macromolecules*, **9**, 879.

Hermans, J. J. and Ullman, R. (1952). *Physica*, **18**, 951.

Higham, D. J. (2001). *SIAM Rev.*, **43**, 525.

Hill, T. L. (1960). *Statistical Thermodynamics*. Addison-Wesley, Reading, MA.

Hillmyer, M. A., Bates, F. S., Almdal, K., Mortensen, K., Ryan, A. J., and Fairclough, J. P. A. (1996). *Science*, **271**, 976.

Hillmyer, M. A., Lipic, P. M., Hajduk, D. A., Almdal, K., and Bates, F. S. (1997). *J. Am. Chem. Soc.*, **119**, 2749.

Hillmyer, M. A., Maurer, W. W., Lodge, T. P., Bates, F. S., and Almdal, K. (1999). *J. Phys. Chem. B*, **103**, 4814.

Hohenberg, P. C. and Halperin, B. I. (1977). *Rev. Mod. Phys.*, **49**, 435.

Hohenberg, P. C. and Kohn, W. (1964). *Phys. Rev. B*, **136**, 864.

Hohenberg, P. C. and Swift, J. B. (1995). *Phys. Rev. E*, **52**, 1828.

Holden, G., Legge, N. R., Quirk, R. P., and Schroeder, H. E. (ed.) (1996). *Thermoplastic Elastomers* (2nd edn). Hanser/Gardner Publications, Cincinnati, OH.

Hong, K. M. and Noolandi, J. (1981). *Macromolecules*, **14**, 727.

Hooper, J. B., Schweizer, K. S., Desai, T. G., Koshy, R., and Keblinski, P. (2004). *J. Chem. Phys.*, **121**, 6986.

Hornreich, R. M. (1980). *J. Magn. Magn. Mater.*, **15**, 387.

Hornreich, R. M., Luban, M., and Shtrikman, S. (1975). *Phys. Rev. Lett.*, **35**, 1678.

Hsieh, H. L. and Quirk, R. P. (1996). *Anionic Polymerization*. Marcel Dekker, New York.

Hubbard, J. (1954). *Phys. Rev. Lett.*, **3**, 77.

Huckstadt, H., Gopfert, A., and Abetz, V. (2000). *Macromol. Chem. Phys.*, **201**, 296.

Huse, D. A., Van Saarloos, W., and Weeks, J. D. (1985). *Phys. Rev. B*, **32**,

233.

Ittel, S. D., Johnson, L. K., and Brookhart, M. (2000). *Chem. Rev.*, **100**, 1169.

Janert, P. K. and Schick, M. (1997). *Macromolecules*, **30**, 3916.

Joanny, J.-F. (1978). *J. Phys. A.*, **11**, 117.

Joanny, J.-F. and Leibler, L. (1990). *J. Phys. (Paris)*, **51**, 545.

Joanny, J.-F., Leibler, L., and Ball, R. (1984). *J. Chem. Phys.*, **81**, 4640.

Kac, M. (1959). *Phys. Fluids*, **2**, 8.

Kamigaito, M., Ando, T., and Sawamoto, M. (2001). *Chem. Rev.*, **101**, 3689.

Katsov, K. and Fredrickson, G. H. (2005). Unpublished work.

Katz, G. R., Batrouni, G. G., Davies, C., Kronfeld, A. S., Lepage, P., Rossi, P., Svetitsky, B., and Wilson, K. G. (1988). *Phys. Rev. D*, **37**, 1589.

Kawasaki, K. and Sekimoto, K. (1987). *Physica A*, **143**, 349.

Kennedy, A. D. (1999). *Parallel Comput.*, **25**, 1311.

Khandpur, A. K., Forster, S., Bates, F. S., Hamley, I. W., Ryan, A. J., Bras, W., Almdal, K., and Mortensen, K. (1995). *Macromolecules*, **28**, 8796.

Kieu, T. D. and Griffin, C. J. (1994). *Phys. Rev. E*, **49**, 3855.

Kim, J. K., Lee, H. H., Sakurai, S., Aida, S., Masamoto, J., Nomura, S., Kitagawa, Y., and Suda, Y. (1999). *Macromolecules*, **32**, 6707.

Kirkpatrick, S. (1984). *J. Stat. Phys.*, **34**, 975.

Klauder, J. R. (1983). *J. Phys. A.*, **16**, L317.

Klauder, J. R. and Petersen, W. P. (1985). *J. Stat. Phys.*, **39**, 53.

Kloeden, P. E. and Platen, E. (1992). *Numerical Solution of Stochastic Differential Equations*, Volume 23 of *Applications of Mathematics*. Springer-Verlag, Berlin.

Knoll, A., Lyakhova, S., Horvat, A., Krausch, G., Sevink, G. J. A., Zvelindovsky, A. V., and Magerle, R. (2004). *Nature Mat.*, **3**, 886.

Koizumi, S., Hasegawa, H., and Hashimoto, T. (1994). *Macromolecules*, **27**, 6532.

Komori, Y., Saito, Y., and Mitsui, T. (1994). *Comput. Math. Appl.*, **28**, 269.

Kosmas, K. (1984). *J. Phys. Lett. (Paris)*, **45**, L889.

Kotelyanskii, M. J. and Theodorou, D. N. (ed.) (2004). *Simulation Methods for Polymers*. Marcel Dekker, New York.

Kratky, O. and Porod, G. (1949). *Rec. Trav. Chim. Pay. B.*, **68**, 1106.

Kreer, T., Metzger, S., Müller, M., Binder, K., and Baschnagel, J. (2004). *J. Chem. Phys.*, **120**, 4012.

Kudlay, A. and Stepanow, S. (2003). *J. Chem. Phys.*, **118**, 4272.

Kyrylyuk, A. V. and Fraaije, J. G. E. M. (2004). *J. Chem. Phys.*, **121**, 2806.

Lanczos, C. (1956). *Applied Analysis*. Prentice-Hall, Englewood Cliffs, NJ.

Landau, D. P. and Binder, K. (2000). *A Guide to Monte Carlo Simulation in Statistical Physics*. Cambridge University Press, Cambridge.

Landau, L. D. and Lifshitz, E. M. (1986). *Theory of Elasticity* (3rd edn), Volume 7. Pergamon, New York.

Langer, J. S. (1980). In *Systems Far from Equilibrium* (ed. L. Garrido), Volume 132 of *Lecture Notes in Physics*. Springer-Verlag, Berlin.

Lansac, Y. and Maissa, P. (1992). *Physica A*, **180**, 53.

Laradji, M., Shi, A. C., Desai, R. C., and Noolandi, J. (1997*a*). *Phys. Rev. Lett.*, **78**, 2577.

Laradji, M., Shi, A. C., Noolandi, J., and Desai, R. C. (1997*b*). *Macromolecules*, **30**, 3242.

Larson, R. G. (1988). *Constitutive Equations for Polymer Melts and Solutions*. Butterworths Series in Chemical Engineering. Butterworth, Stoneham, MA.

Larson, R. G. (1999). *The Structure and Rheology of Complex Fluids*. Oxford University Press, Oxford.

Le Guillou, J. C. and Zinn-Justin, J. (1977). *Phys. Rev. Lett.*, **39**, 95.

Lee, J. Y., Thompson, R. B., Jasnow, D., and Balazs, A. C. (2002). *Macromolecules*, **35**, 4855.

Lee, S. (1994). *Nucl. Phys. B*, **413**, 827.

Leibler, L. (1980). *Macromolecules*, **13**, 1602.

Lennon, E., Katsov, K., Ceniceros, H. D., Garcia-Cervera, C., and Fredrickson, G. H. (2005). Unpublished work.

Levine, I. N. (2000). *Quantum Chemistry* (5th edn). Prentice-Hall, Upper Saddle River, NJ.

Li, X. J. and Schick, M. (2000). *Biophys. J.*, **78**, 34.

Lifshitz, I. M., Grosberg, A. Y., and Khokhlov, A. R. (1978). *Rev. Mod. Phys.*, **50**, 683.

Lin, H. Q. and Hirsch, J. E. (1986). *Phys. Rev. B*, **34**, 1964.

Linse, P. (1993). *J. Phys. Chem.*, **97**, 13896.

Lipic, P. M., Bates, F. S., and Hillmyer, M. A. (1998). *J. Am. Chem. Soc.*, **120**(35), 8963.

Lodge, T. P., Pudil, B., and Hanley, K. J. (2002). *Macromolecules*, **35**(12), 4707.

Loh Jr., E. Y., Gubernatis, J. E., Scalettar, R. T., White, S. R., Scalapino, D. J., and Sugar, R. L. (1990). *Phys. Rev. B*, **41**, 9301.

Löwen, H. (2002). *J. Phys. Cond. Matter*, **14**, 11897.

Ma, S.-k. (1976). *Phys. Rev. Lett.*, **37**, 461.

Maier, W. and Saupe, A. (1958). *Z. Naturforsch*, **A13**, 564.

Makri, N. and Miller, W. H. (1987). *Chem. Phys. Lett.*, **139**, 10.

Matsen, M. W. (1995*a*). *Macromolecules*, **28**, 5765.

Matsen, M. W. (1995*b*). *J. Chem. Phys.*, **103**, 3268.

Matsen, M. W. (1996). *J. Chem. Phys.*, **104**, 7758.

Matsen, M. W. (1997*a*). *J. Chem. Phys.*, **106**, 7781.

Matsen, M. W. (1997*b*). *J. Chem. Phys.*, **107**, 8110.

Matsen, M. W. (1998*a*). *Curr. Opin. Coll. In.*, **3**, 40.

Matsen, M. W. (1998*b*). *J. Chem. Phys.*, **108**, 785.

Matsen, M. W. (2000). *J. Chem. Phys.*, **113**, 5539.

Matsen, M. W. (2002). *J. Phys.–Cond. Matter*, **14**, R21.

Matsen, M. W. (2004). *J. Chem. Phys.*, **121**, 1938.

Matsen, M. W. (2005*a*). *J. Chem. Phys.*, **122**, 144904.

Matsen, M. W. (2005*b*). In *Soft Condensed Matter* (ed. G. Gompper and M. Schick), Volume 1. Wiley-VCH, Berlin.

Matsen, M. W. and Barrett, C. (1998). *J. Chem. Phys.*, **109**, 4108.

Matsen, M. W. and Bates, F. S. (1995). *Macromolecules*, **28**, 7298.

Matsen, M. W. and Bates, F. S. (1996*a*). *Macromolecules*, **29**, 7641.

Matsen, M. W. and Bates, F. S. (1996*b*). *Macromolecules*, **29**, 1091.

Matsen, M. W. and Bates, F. S. (1997). *J. Polym. Sci. Polym. Phys.*, **35**, 945.

Matsen, M. W. and Gardiner, J. M. (2000). *J. Chem. Phys.*, **113**, 1673.

Matsen, M. W. and Schick, M. (1994*a*). *Phys. Rev. Lett.*, **72**, 2660.

Matsen, M. W. and Schick, M. (1994*b*). *Macromolecules*, **27**, 7157.

Matsen, M. W. and Schick, M. (1994*c*). *Macromolecules*, **27**, 6761.

Matsen, M. W. and Schick, M. (1996). *Curr. Opin. Coll. In.*, **1**, 329.

Matsen, M. W. and Thompson, R. B. (1999). *J. Chem. Phys.*, **111**, 7139.

Matyjaszewski, K. and Xia, J. H. (2001). *Chem. Rev.*, **101**, 2921.

Maurer, W. W., Bates, F. S., Lodge, T. P., Almdal, K., Mortensen, K., and Fredrickson, G. H. (1998). *J. Chem. Phys.*, **108**, 2989.

Maurits, N. M. and Fraaije, J. G. E. M. (1997). *J. Chem. Phys.*, **107**, 5879.

Maurits, N. M., Sevink, G. J. A., Zvelindovsky, A. V., and Fraaije, J. G. E. M. (1999). *Macromolecules*, **32**, 7674.

Mayes, A. M. and de la Cruz, M. O. (1991). *J. Chem. Phys.*, **95**, 4670.

McQuarrie, D. A. (1976). *Statistical Mechanics*. Harper & Row, New York.

Metropolis, N., Rosenbluth, A. W., Rosenbluth, M. N., Teller, A. M., and Teller, E. (1953). *J. Chem. Phys.*, **21**, 1087.

Mezard, M., Parisi, G., and Virasoro, M. A. (1987). *Spin Glass Theory and Beyond*. World Scientific, Teaneck, NJ.

Mezzenga, R., Fredrickson, G. H., and Kramer, E. J. (2003). *Macromolecules*, **36**, 4457.

Milner, S. T. (1990). *J. Chem. Soc. Faraday T.*, **86**, 1349.

Milner, S. T. (1991). *Science*, **251**, 905.

Milner, S. T. (1993). *Phys. Rev. E*, **48**, 3674.

Milner, S. T. and Morse, D. C. (1996). *Phys. Rev. E*, **54**, 3793.

Milner, S. T., Witten, T. A., and Cates, M. E. (1988). *Macromolecules*, **21**, 2610.

Mogi, Y., Kotsuji, H., Kaneko, Y., Mori, K., Y., Matsushita, and Noda, I. (1992). *Macromolecules*, **25**, 5408.

Monasson, R. (1995). *Phys. Rev. Lett.*, **75**, 2847.

Montvay, I. and Münster, G. (1994). *Quantum Fields on the Lattice*. Cambridge University Press, Cambridge.

Moore, M. A. (1977). *J. Phys. A – Math. Gen.*, **10**, 305.

Moreira, A. G., Baeurle, S. A., and Fredrickson, G. H. (2003). *Phys. Rev. Lett.*, **91**, 150201.

Morkved, T. L., Chapman, B. R., Bates, F. S., Lodge, T. P., Stepanek, P., and Almdal, K. (1999). *Faraday Discuss.* (112), 335.

Morrison, I. D. and Ross, S. (2002). *Colloidal Dispersions: Suspensions, Emulsions, and Foams*. Wiley-Interscience, New York.

Morse, D. C. (2005). Personal communication.

Morse, D. C. and Fredrickson, G. H. (1994). *Phys. Rev. Lett.*, **73**, 3235.

Müller, M. (1999). *Macromol. Theor. Simul.*, **8**, 343.

Müller, M. (2002). *Phys. Rev. E*, **65**, 030802(R).

Müller, M. and Binder, K. (1998). *Macromolecules*, **31**, 8323.

Müller, M. and Schick, M. (1996). *J. Chem. Phys.*, **105**, 8885.

Müller, M. and Schmid, F. (2005). *Adv. Polym. Sci.*, in press.

Murat, M., Grest, G. S., and Kremer, K. (1999). *Macromolecules*, **32**, 595.

Muthukumar, M. and Edwards, S. F. (1982). *J. Chem. Phys.*, **76**, 2720.

Nahfeh, A. H. (1973). *Perturbation Methods*. Wiley, New York.

Nakanishi, A. and Ohta, T. (1985). *J. Phys. A – Math. Gen.*, **18**, 127.

Naughton, J. R. and Matsen, M. W. (2002). *Macromolecules*, **35**, 5688.

Nelson, D. R. (1983). *Phys. Rev. B*, **28**, 5515.

Nelson, D. R. (2002). *Defects and Geometry in Condensed Matter Physics*. Cambridge University Press, Cambridge.

Netz, R. R. and Andelman, D. (2003). *Phys. Rep.*, **380**, 1.

Netz, R. R., Andelman, D., and Schick, M. (1997). *Phys. Rev. Lett.*, **79**, 1058.

Netz, R. R. and Orland, H. (1999). *Eur. Phys. J. B*, **8**, 81.

Netz, R. R. and Orland, H. (2000). *Eur. Phys. J. E*, **1**, 203.

Netz, R. R. and Schick, M. (1998). *Macromolecules*, **31**, 5105.

Nocedal, J. and Wright, S. J. (1999). *Numerical Optimization*. Springer, New York.

Noda, I., Imai, M., Kitano, T., and Nagasawa, M. (1983). *Macromolecules*, **16**, 425.

Noda, I., Kato, N., Kitano, T., and Nagasawa, M. (1981). *Macromolecules*, **14**, 668.

Noolandi, J., Shi, A. C., and Linse, P. (1996). *Macromolecules*, **29**, 5907.

Odian, G. (1981). *Principles of Polymerization* (2nd edn). John Wiley & Sons, New York.

Odijk, T. (1986). *Macromolecules*, **19**, 2313.

Ohta, T. and Nakanishi, A. (1983). *J. Phys. A – Math. Gen.*, **16**, 4155.

Ohta, T. and Oono, Y. (1982). *Phys. Lett. A*, **89**, 460.

Olmsted, P. D. and Milner, S. T. (1994). *Macromolecules*, **27**, 1964.

Onsager, L. (1949). *Ann. NY Acad. Sci.*, **51**, 627.

Onuki, A. (1990). *J. Phys. Soc. Jpn.*, **59**, 3427.

Oono, Y. (1985). *Adv. Chem. Phys.*, **61**, 301.

Oozawa, F. and Asakura, S. (1975). *Thermodynamics in the Polymerization of Proteins*. Academic, New York.

Orszag, S. A. (1986). Fast eigenfunction transforms. In *Science and Computers* (ed. G. C. Rota), pp. 23–30. Academic, New York.

Otten, R. H. J. M. and van Ginneken, L. P. P. P. (1989). *The Annealing Algorithm*. Kluwer, Boston.

Ottinger, H. C. (1998). *Phys. Rev. E*, **57**, 1416.

Ottinger, H. C. and Grmela, M. (1997). *Phys. Rev. E*, **56**, 6633.

Oxtoby, D. W. (2002). *Annu. Rev. Mater. Res.*, **32**, 39.

Oyerokun, F. T. and Schweizer, K. S. (2004). *J. Chem. Phys.*, **120**, 475.

Pangali, C., Rao, M., and Berne, B. J. (1978). *Chem. Phys. Lett.*, **55**, 413.

Parisi, G. (1983). *Phys. Lett. B*, **131**, 393.

Parisi, G. (1988). *Statistical Field Theory*. Addison-Wesley, New York.

Park, M., Harrison, C., Chaikin, P. M., Register, R. A., and Adamson, D. H. (1997). *Science*, **276**, 1401.

Parr, R. G. and Yang, W. (1989). *Density-Functional Theory of Atoms and Molecules*. Oxford University Press, Oxford.

Parrinello, M. and Rahman, A. (1981). *J. Appl. Phys.*, **52**, 7182.

Patel, D. M. and Fredrickson, G. H. (2003). *Phys. Rev. E*, **68**, 051802.

Pauling, L. (1988). *General Chemistry*. Dover, New York.

Pickett, G. T. and Balazs, A. C. (1997). *Macromolecules*, **30**, 3097.

Pickett, G. T. and Schweizer, K. S. (2000). *J. Chem. Phys.*, **112**, 4881.

Platen, E. (1999). *Acta Numer.*, **8**, 197.

Platonova, O. A., Bronstein, L. M., Solodovnikov, S. P., Yanovskaya, I. M., Obolonkova, E. S., Valetsky, P. M., Wenz, E., and Antonietti, M. (1997). *Colloid Polym. Sci.*, **275**, 426.

Potemkin, I. I. and Panyukov, S. V. (1998). *Phys. Rev. E*, **57**, 6902.

Potschke, P. and Paul, D. R. (2003). *J. Macromol. Sci.*, **C43**, 87.

Press, W. H., Teukolsky, S. A., Vetterling, W. T., and Flannery, B. P. (1992). *Numerical Recipes in Fortran*. Cambridge University Press, New York.

Qi, S. Y., Chakraborty, A. K., Wang, H., Lefebvre, A. A., Balsara, N. P., Shakhnovich, E. I., Xenidou, M., and Hadjichristidis, N. (1999). *Phys. Rev. Lett.*, **82**, 2896.

Qi, S. Y. and Wang, Z.-G. (1997). *Macromolecules*, **30**, 4491.

Rapaport, D. C. (1995). *The Art of Molecular Dynamics Simulation*. Cambridge University Press, Cambridge.

Rasmussen, K. O. and Kalosakas, G. (2002). *J. Polym. Sci. Polym. Phys.*, **40**, 1777.

Ray, J. R. and Rahman, A. (1984). *J. Chem. Phys.*, **80**, 4423.

Reister, E. and Fredrickson, G. H. (2004). *Macromolecules*, **37**, 4718.

Reister, E., Müller, M., and Binder, K. (2001). *Phys. Rev. E*, **64**, 041804.

Richardson, T. L. and Lokensgard, E. (1989). *Industrial Plastics* (3rd edn). Delmar Publishers, New York.

Riley, K. F., Hobson, M. P., and Bence, S. J. (1998). *Mathematical Methods for Physics and Engineering*. Cambridge University Press, Cambridge.

Rossky, P. J., Doll, J. D., and Friedman, H. L. (1978). *J. Chem. Phys.*, **69**, 4628.

Rottler, J., Barsky, S., and Robbins, M. O. (2002). *Phys. Rev. Lett.*, **89**, 8304.

Rowlinson, J. S. and Widom, B. (1989). *Molecular Theory of Capillarity*. Oxford University Press, New York.

Rubinstein, M. and Colby, R. H. (2003). *Polymer Physics*. Oxford University Press, Oxford.

Sabo, D., Doll, J. D., and Freedman, D. L. (2002). *J. Chem. Phys.*, **116**, 3509.

Saito, N. and Namiki, M. (1956). *Prog. Theor. Phys. (Kyoto)*, **16**, 71.

Saito, N., Takahashi, K., and Yunoki, Y. (1967). *J. Phys. Soc. Jpn.*, **22**, 219.

Sakamoto, N., Hashimoto, T., Han, C. D., Kim, D., and Vaidya, N. Y. (1997). *Macromolecules*, **30**, 5321.

Salcedo, L. L. (1997). *J. Math. Phys.*, **38**, 1710.

Schäfer, L. (1982). *Macromolecules*, **15**, 652.

Schäfer, L. and Kappeler, Ch. (1985). *J. Phys.*, **46**, 1853.

Scheutjens, J. M. H. M. and Fleer, G. J. (1979). *J. Chem. Phys.*, **83**, 1619.

Schick, M. and Li, X. J. (2000). *Biophys. J.*, **78**, 487a.

Schiff, L. I. (1968). *Quantum Mechanics*. McGraw-Hill, New York.

Schmid, F. (1998). *J. Phys.: Cond. Matt.*, **10**, 8105.

Schoenmaker, W. J. (1987). *Phys. Rev. D*, **36**, 1859.

Schulz, M. F., Bates, F. S., Almdal, K., and Mortensen, K. (1994). *Phys. Rev. Lett.*, **73**(1), 86.

Schulz, M. F., Khandpur, A. K., Bates, F. S., Almdal, K., Mortensen, K., Hajduk, D. A., and Gruner, S. M. (1996). *Macromolecules*, **29**, 2857.

Schwab, M. and Stühn, B. (1996). *Phys. Rev. Lett.*, **76**, 924.

Schwahn, D., Schmackers, T., and Mortensen, K. (1995). *Phys. Rev. E*, **52**, R1288.

Schweizer, K. S. and Curro, J. G. (1994). *Adv. Polym. Sci.*, **116**, 319.

Schweizer, K. S. and Curro, J. G. (1997). *Adv. Chem. Phys.*, **98**, 1.

Schweizer, K. S. and Singh, C. (1995). *Macromolecules*, **28**, 2063.

Schweizer, K. S. and Yethiraj, A. (1993). *J. Chem. Phys.*, **98**, 9053.

Scott, R. L. (1965). *J. Phys. Chem.*, **69**, 261.

Semenov, A. N. (1985). *Sov. Phys. JETP (USSR)*, **61**, 733.

Semenov, A. N. (1989). *Macromolecules*, **22**, 2849.

Shakhnovich, E. I. and Gutin, A. M. (1989). *J. Phys. (Paris)*, **50**, 1843.

Shankar, R. (1994). *Rev. Mod. Phys.*, **66**, 129.

Shefelbine, T. A., Vigild, M. E., Matsen, M. W., Hajduk, D. A., Hillmyer, M. A., Cussler, E. L., and Bates, F. S. (1999). *J. Am. Chem. Soc.*, **121**, 8457.

Shenker, S. H. and Tobochnik, J. (1980). *Phys. Rev. B*, **22**, 4462.

Shi, A. C. and Noolandi, J. (1994). *Macromolecules*, **27**, 2936.

Shi, A. C. and Noolandi, J. (1995). *Macromolecules*, **28**, 3103.

Shi, A. C. and Noolandi, J. (1999). *Macromol. Theor. Simulat.*, **8**, 214.

Shi, A. C., Noolandi, J., and Desai, R. C. (1996). *Macromolecules*, **29**, 6487.

Shull, K. R. (1993). *Macromolecules*, **26**, 2346.

Shusharina, N. and Linse, P. (2001). *Eur. Phys. J. E*, **4**, 399.

Sides, S. W. and Fredrickson, G. H. (2003). *Polymer*, **44**, 5859.

Sides, S. W. and Fredrickson, G. H. (2004). *J. Chem. Phys.*, **121**, 4974.

Sides, S. W., Katsov, K., and Fredrickson, G. H. (2005*a*). Unpublished work.

Sides, S. W., Kim, B., Kramer, E. J., and Fredrickson, G. H. (2005$b$). Unpublished work.

Simon, B. (1979). *Functional Integration and Quantum Physics*. Academic, New York.

Singh, C., Schweizer, K. S., and Yethiraj, A. (1995). *J. Chem. Phys.*, **102**, 2187.

Sioula, S., Hadjichristidis, N., and Thomas, E. L. (1998). *Macromolecules*, **31**, 5272.

Skvortsov, A. M., Gorbunov, A. A., Pavlushkov, V. A., Zhulina, E. B., Borisov, O. V., and Priamitsyn, V. A. (1988). *Polym. Sci. USSR*, **30**, 1706.

Spakowitz, A. J. and Wang, Z.-G. (2004). *Macromolecules*, **37**, 5814.

SPI (2003). http://www.plasticsdatasource.org.

Stepanow, S. (1995). *Macromolecules*, **28**, 8233.

Stockmayer, W. H. (1949). *J. Chem. Phys.*, **17**, 588.

Stratonovich, R. L. (1958). *Sov. Phys. Solid State*, **2**, 1824.

Subbotin, A. V. and Semenov, A. N. (2002). *Eur. Phys. J. A*, **7**, 49.

Svensson, M., Alexandridis, P., and Linse, P. (1999). *Macromolecules*, **32**, 5435.

Swarztrauber, P. N. (1979). *SIAM J. Numer. Anal.*, **16**, 934.

Swendsen, R. H. (1979). *Phys. Rev. Lett.*, **42**, 859.

Swendsen, R. H. (1982). Monte carlo renormalization. In *Real-Space Renormalization* (ed. T. W. Burkhardt and J. M. J. van Leeuwen). Springer-Verlag, Berlin.

Swendsen, R. H. (1984). *Phys. Rev. Lett.*, **52**, 2321.

't Hooft, G. and Veltman, M. (1972). *Nucl. Phys. B*, **44**, 189.

Tagami, Y. (1969). *Macromolecules*, **2**, 8.

Tanaka, H. (1993). *Phys. Rev. Lett.*, **71**, 3158.

Tang, H. and Freed, K. (1991). *J. Chem. Phys.*, **94**, 1572.

Thomas, E. L., Alward, D. B., Kinning, D. J., Martin, D. C., Handlin, D. L., and Fetters, L. J. (1986). *Macromolecules*, **19**, 2197.

Thomas, E. L., Anderson, D. M., Henkee, C. S., and Hoffman, D. (1988). *Nature*, **334**, 598.

Thompson, R. B., Ginzburg, V. L., Matsen, M. W., and Balazs, A. C. (2001). *Science*, **292**, 2469.

Thompson, R. B., Ginzburg, V. L., Matsen, M. W., and Balazs, A. C. (2002). *Macromolecules*, **35**, 1060.

Thompson, R. B., Rasmussen, K. O., and Lookman, T. (2004). *J. Chem. Phys.*, **120**, 3990.

Thurn-Albrecht, T., Schotter, J., Kastle, C. A., Emley, N., Shibauchi, T., Krusin-Elbaum, L., Guarini, K., Black, C. T., Tuominen, M. T., and Russell, T. P. (2000). *Science*, **290**, 2126.

Tolstov, G. P. (1976). *Fourier Series*. Dover, New York.

Trefethen, L. N. (2000). *Spectral Methods in MATLAB*. SIAM, Philadelphia.

Tsonchev, S., Coalson, R. D., and Duncan, A. (2000). *Phys. Rev. E*, **62**, 799.

Tyler, C. A. and Morse, D. C. (2003$a$). *Macromolecules*, **36**, 8184.

Tyler, C. A. and Morse, D. C. (2003$b$). *Macromolecules*, **36**, 3764.

Tyler, C. A. and Morse, D. C. (2005). *Phys. Rev. Lett.*, **94**, 208302.

Vaia, R. A. and Giannelis, E. P. (2001). *MRS Bulletin*, **26**, 394.

van Kampen, N. G. (1981). *Stochastic Processes in Physics and Chemistry.* North-Holland, New York.

van Vlimmeren, B. A. C., Maurits, N. M., Zvelindovsky, A. V., Sevink, G. J. A., and Fraaije, J. G. E. M. (1999). *Macromolecules*, **32**, 646.

Vassiliev, O. N. and Matsen, M. W. (2003). *J. Chem. Phys.*, **118**, 10293.

Volkenstein, M. V. (1963). *Configurational Statistics of Polymer Chains* (English edn), Volume XVII of *High Polymers*. Interscience, New York.

Volterra, V. (1959). *Theory of Functionals.* Dover, New York.

VSPA (2003). http://www.vnplas.com/explore/worldwide/.

Wallace, D. (1973). *Thermodynamics of Crystals.* Wiley, New York.

Wang, J. F., Wang, Z.-G., and Yang, Y. L. (2005). *Macromolecules*, **38**, 1979.

Wang, Q., Nealey, P. F., and de Pablo, J. J. (2003). *Macromolecules*, **36**, 1731.

Wang, Q., Taniguchi, T., and Fredrickson, G. H. (2004). *J. Phys. Chem. B*, **108**, 6733.

Wang, X. H., Dormidontova, E. E., and Lodge, T. P. (2002). *Macromolecules*, **35**, 9687.

Wang, X. J. and Warner, M. (1986). *J. Phys. A – Math. Gen.*, **19**, 2215.

Wang, Z.-G. (2002). *J. Chem. Phys.*, **117**, 481.

Wax, N. (ed.) (1954). *Selected Papers on Noise and Stochastic Processes.* Dover, New York.

Weber, M. (1999). In *Polymer Blends and Alloys* (ed. G. O. Shonaike and G. P. Simon), pp. 263–264. Marcel Dekker, New York.

Weeks, J. D. (1977). *J. Chem. Phys.*, **67**, 3106.

Weingarten, D. (2002). *Phys. Rev. Lett.*, **89**, 240201.

Weinstock, R. (1974). *Calculus of Variations.* Dover, New York.

Wheeler, J. C., Kennedy, S. J., and Pfeuty, P. (1980). *Phys. Rev. Lett.*, **45**, 1748.

Whitmore, M. D. and Vavasour, J. D. (1992). *Macromolecules*, **25**, 5477.

Wilson, K. G. (1980). In *Recent Developments of Gauge Theories* (ed. G. 't Hooft). Plenum, New York.

Wilson, K. G. and Fisher, M. E. (1972). *Phys. Rev. Lett.*, **28**, 248.

Wilson, K. G. and Kogut, J. (1974). *Phys. Rep. C*, **12**, 75.

Wiltzius, P., Haller, H. R., Cannell, D. S., and Schäfer, D. W. (1983). *Phys. Rev. Lett.*, **51**, 1183.

Winey, K. I., Thomas, E. L., and Fetters, L. J. (1992). *Macromolecules*, **25**, 2645.

Wittmer, J. P., Milchev, A., and Cates, M. E. (1998). *J. Chem. Phys.*, **109**, 834.

Wu, D. T., Fredrickson, G. H., and Carton, J. P. (1996). *J. Chem. Phys.*, **104**, 6387.

Wu, D. T., Fredrickson, G. H., Carton, J. P., Ajdari, A., and Leibler, L. (1995). *J. Polym. Sci. Polym. Phys.*, **33**, 2373.

Wu, Y., Cheng, G., Katsov, K., Sides, S. W., Wang, J., Tang, J., Fredrickson, G. H., Moskovits, M., and Stucky, G. D. (2004). *Nature Mater.*, **3**, 816.

Xiang, H., Shin, K., Kim, T., Moon, S. I., McCarthy, T. J., and Russell, T. P. (2005). *Macromolecules*, **38**, 1055.

Yamakawa, H. (1972). *Modern Theory of Polymer Solutions*. Harper & Row, New York.

Yamakawa, H. (1997). *Helical Wormlike Chains in Polymer Solutions*. Springer-Verlag, Berlin.

Yethiraj, A. and Schweizer, K. S. (1993). *J. Chem. Phys.*, **98**, 9080.

Yeung, C. and Shi, A. C. (1999). *Macromolecules*, **32**, 3637.

Yeung, C., Shi, A. C., Noolandi, J., and Desai, R. C. (1996). *Macromol. Theor. Simulat.*, **5**, 291.

Zee, A. (2003). *Quantum Field Theory in a Nutshell*. Princeton University Press, Princeton.

Zhang, C.-Z. and Wang, Z.-G. (2005). Preprint.

Zhao, J., Majumdar, B., Schulz, M. F., Bates, F. S., Almdal, K., Mortensen, K., Hajduk, D. A., and Gruner, S. M. (1996). *Macromolecules*, **29**, 1204.

Zhulina, E. B., Lyatskaya, Y. V., and Birshtein, T. M. (1992). *Polymer*, **33**, 332.

Zhulina, E. B., Singh, C., and Balazs, A. C. (1996). *Macromolecules*, **29**, 6338.

Zienkiewicz, O. C. and Taylor, R. L. (2000). *The Finite Element Method* (5th edn), Volume 1. Butterworth-Heinemann, Boston.

Zimm, B. H., Stockmayer, W. H., and Fixman, M. (1953). *J. Chem. Phys.*, **21**, 1716.

Zvelindovsky, A. V., Sevink, G. J. A., van Vlimmeren, B. A. C., Maurits, N. M., and Fraaije, J. G. E. M. (1998). *Phys. Rev. E*, **57**, R4879.

Zwanzig, R. (1961). *Phys. Rev.*, **124**, 983.

# INDEX